ANIMALS IN HUMAN HISTORIES

Studies in Comparative History

Essays from the

Shelby Cullom Davis Center for Historical Studies

Animals in Human Histories: The Mirror of Nature and Culture

Edited by Mary J. Henninger-Voss

A Publication of the
Shelby Cullom Davis Center for Historical Studies
Princeton University

ANIMALS IN HUMAN HISTORIES

The Mirror of Nature and Culture

EDITED BY

MARY J. HENNINGER-VOSS

University of Rochester Press

University of Rochester Press
668 Mt. Hope Avenue
Rochester, New York, 14620, USA
Boydell & Brewer, Ltd.
P.O. Box 9, Woodbridge, Suffolk IIP12 3DF, UK
www.urpress.com
ISSN 1539-4905

Library of Congress Cataloging-in-Publication Data

Animals in Human Histories : The Mirror of Nature and Culture / edited by Mary J. Henninger-Voss.
p. cm. (Studies in comparative history, ISSN 1539-4905 ; 1)
Includes bibliographical references and index.
ISBN 1-58046-121-2 (alk. paper)
1. Human-animal relationships—History. 2. Food animals. 3. Pets—Social aspects. 4. Laboratory animals. I. Henninger-Voss, Mary J., 1959- II. Series.

QL85.A557 2002
306—dc21 2002072457

British Library Cataloguing-in-Publication Data

A catalogue record for this book is available from the British Library

This publication is printed on acid-free paper
Printed in the United States of America
Designed and typeset by Straight Creek Bookmakers

Contents

Acknowledgments

This volume came out of a year-long discussion on animals and humans at the Shelby Cullom Davis Center; it owes its scope to the intellectual vision and leadership of William Chester Jordan, who directed the Center from 1992 to 1999. Thanks also go to Kari Hoover for her technical assistance and knowledge of the Center's workings. The clarity and stylistic nicety of these essays were greatly enhanced by the careful editing of Gavin Lewis. Enormous gratitude is owed Angela Creager, who, assisted by Eric Ash, guided this volume through the press. Finally, I thank all the participants in discussions that made the topic of "animal" come alive as an object of scholarly concern, but especially the contributors to this volume for both their thoughtful attention to the pieces collected here, and their patience.

Chapter 6 also appears in Nigel Rothfels, *Savages and Beasts: The Birth of the Modern Zoo* (Baltimore: Johns Hopkins University Press, 2002). Chapter 7 is a revised version of Richard W. Burkhardt Jr., "La ménagerie et la vie du Muséum," in *Le Muséum au premier siècle de son histoire,* ed. C. Blanckaert et al. (Paris: Editions du Muséum national d'histoire naturelle, 1997), pp. 481–508.

Mary J. Henninger-Voss
Princeton, N. J.
April 2000

Introduction

Mary J. Henninger-Voss

Two instances from the recent past serve as a frontispiece for this volume. In 1966, the artist Andy Warhol covered the walls of the Leo Castelli Gallery with "cow wallpaper." The image of the cow was silk-screened repeatedly in a variety of psychedelic colors. Warhol's friend, Ivan Karp, had cut the original photo out of an agriculture magazine, and suggested the cow as a "wonderfully pastoral" and "durable" image in the history of art. The artist himself, and the public at large, saw the large-scale reproduction of the heifer's portrait as an expression of "the end of art."[1] The exhibition did not mark the end of Warhol's art, or anyone else's, but it may have memorialized a disappearing way of life. The show was mounted in the context of an American culture whose recent rural past was giving way to an urbanized world of manufactured goods, and where farm life had begun to look wonderfully pastoral. Warhol's "factory" of art had its counterpart in the sphere of agribusiness where record beef and dairy consumption could be served through mass production of real cows in a factory-like process. Some thirty years later, a Scottish scientist named Ian Wilmut perfected a method by which an adult sheep (the Scottish cultural equivalent of the American cow) could be cloned, literally reproduced, and possibly manufactured on a large scale like so many silk-screened images. *Time* magazine dubbed Wilmut's achievement a "breakthrough of biblical proportions" as the scientific, economic, and moral ramifications of cloning began to reverberate worldwide.[2] The unassuming Dolly, the first mammal cloned from the cell of an adult animal, has taken center stage in the reassessment of age-old questions concerning the nature of life, the possibilities of technology, and the line between animals and humans. The Finn Dorset ewe represents simultaneously the traditional staple of Scottish husbandry and the potential of humankind.

This volume delves into the realm between the representative images and real animals that both these contemporary instances capture. It is a historical inquiry into human interaction with the animals we eat, pamper, experiment on, and imagine, as they have been variously domesticated, slaughtered, loved, studied, and made into icons of human invention. As we stand in the doorway of a new era of human-animal relationships, it is ever clearer how common assumptions and experiences with animals have entered into the functioning and conceptualizing of life, and the extent to which those assumptions and experiences are historically and culturally contingent.

In the year 1996–97, the Shelby Cullom Davis Center for Historical Studies brought scholars together to discuss the subject of humans and animals. The role of animals has changed drastically in the last century, and the fact that most of the papers presented at our meetings focused on the nineteenth century, while scholars of earlier periods and of the twentieth century documented old and new roles that animals have played in our culture, reflects this recent era as a moment of transition in human-animal relations. For millennia animals were visible components of human society. They were omnipresent as resources for meat, hides, and dairy products; providers of muscle power in farming, milling, mining, and transportation; and aids to hunters and herders. It is hard to imagine that their bleats and squawks, their smells, or their necessity were ever distant from the average person. Now largely invisible in the quotidian life of most Western people, animals continue to render resources for processing and consumption, but inhabit our imaginations and affections in new ways. While the natural history of animals, as well as the symbology of animals as emblems of human characteristics, are ancient studies, the scientific, emblematic, and sentimental meanings of animals have been extended and rewritten throughout the age of industrialization.

The authors of the essays chosen for this volume unveil the ways in which human-animal relationships reveal the interhuman structures of the cultures in which they are formed. By using the animal as a lens, they refocus our awareness of the ways in which humans have allotted resources, gathered knowledge, and structured families. If animals are a "mirror of nature," as Epicurus suggested,[3] their image in human society must also be a mirror of the culture which defines itself with respect to nature. The treatment of animals is often a road to a story about the treatment of people within a society, while the perceived "stewardship" of humans over animals has helped shape the broader environment which both human and nonhuman animals share. Indeed the number of ways in which animals enter into our thinking about the human condition are so multifarious that the roles of animals in human society can only be categorized in a very artificial manner.

The approaches taken to animals here can roughly be identified as popular, scientific, or economic. That is, the authors have investigated how animals have figured as the objects of popular conception, objects of scientific understanding, and objects of exchange. As objects in these various spheres, animals have served as a screen on which public morals and social differentiation can be directed; they have been the subject of analyses that may convey an "objective" notion of nature on the one hand, or on the other hand may stand in for human physiological and cognitive functioning. As creatures of value for meat, curiosity, or companionship, animals are marketed as objects for human consumption. No paper in this volume deals exclusively with a single facet of the animal. Most authors have tackled their subject with full recognition of all these levels—popular, scientific, and economic—on which animals are staged. This volume has been arranged neither by approach (which is polysemic) nor by chronology (which would have presented a blotchy picture at best), but by the loci in which human-animal interactions occur. Each place, or set of conditions, offers a view of how popular, scientific, and commercial aspects of animals interconnect.

The essays are arranged approximately in order of increasing subjectification: from Part I, which focuses on the most objective and enduring of animal roles—as foodstuffs—to Part V, which investigates the most recent stage for animals—as surrogate human subjects in the modern laboratory. We journey from the point at which animals are little more than commodities to a terminus in which they are the sacrificial stand-ins for human beings themselves. Of course these two "ends" may describe something closer to a circle than a straight line. Both the meat producer and the laboratory scientist must bring animals under their control in a specialized, unsentimentalized sphere where they may extract from animal bodies food or knowledge for human use. Between these, however, the essays of this volume describe a continuum of relations in which the animal increasingly approaches the line that separates human from nonhuman animal, often with greater degrees of human emotional involvement with the beast. Part II portrays various episodes in the history of hunting as elements within complex cultural arrangements, where the hunted animals can be seen variously as commodities, bestial analogs of human enemies, or increasingly sentimentalized objects of fancy. Part III examines conditions demanded by the display of wild animals in zoos, and relates these to the many ways in which public morality and scientific observation are interwoven in zoos' animal spectacle. And Part IV deals with animals in the domestic sphere where they are tamed, bred, and cared for by human owners as livestock or as pets. Here the line between popular and scientific knowledge of animals can fade, as well as the line between animal and member of the family.

The essays in Part I examine several broadly separated instances of animal management for large-scale consumption. The structures of organization that societies adopt in order to feed themselves are among the most fundamental, and yet among the most easily forgotten, in the historical record.

The stories here trace the effects of the manipulation of huge numbers of animals for food production. In each case, the interaction of the technologies of production with the natural ecology inform a culture of food producers which is itself in dynamic relation to a larger world of consumers. One only has to think of the dramatic consequences of the twentieth-century drive to turn rain forests into cattle grounds in order to understand the cultural, economic, and ecological impact that commercial animal husbandry can have. Each paper in this section analyzes human power relations and cultural understandings of animal food production, as well as the physical ways in which intensive fish/meat production molds natural and human environments.

The interpenetration of technology, social power, and natural ecology in animal food production is forcefully demonstrated in Richard Hoffmann's essay on the development of private and commercial fishing in late medieval Europe. Drawing on a vast array of sources, Hoffmann explicates the structures of demand for fish among medieval Christians, reconstructs the means by which that demand could be met, and analyzes the complex economic and environmental changes wrought by new forms of fish production. Hoffmann details both the creation of artificial fisheries, as well as enormous offshore expansion. Inland fisheries developed to serve the demand for fish among the elite as drainage and engineering allowed the manipulation of aquatic ecosystems—often by flooding or destroying common lands. Meanwhile, organization of technologies in salting and transportation allowed the growth of a capitalist fishing industry that could meet a mass demand. The transformation, in many parts of Europe, of the artisanal craft of fishing into large-scale enterprise not only affected the human diet of the late Middle Ages, but adapted natural environments and favored particular species of fish.

Hoffmann's essay leaves us in the midst of a fishing industry that was one of the first harbingers of European capitalist expansion, in a culture that still prohibited the eating of slaughtered quadrupeds during approximately 40 percent of the days of the year. With the next essay, by John Otto, we are placed in the more familiar territory of beef-eating America—but with an interesting twist. While the cattle culture of the West has become both a key topic in the narrative of nineteenth-and twentieth-century American development, and the stuff of American myth, Otto focuses on the forgotten episode of flourishing commercial cattle trade in the southeastern United States. Otto's work suggests that the framework in which to view the centuries of flourishing cattle trade in areas now known for citrus production and vacation spots is within a Caribbean, not solely American, context. The author recovers the lost rhythms of cattle raising among the "cracker cowboys" of Florida and the Carolinas, and shows how they enlisted practices contributed by both British and Spanish colonists, as well as their West African slaves. Ways of raising cattle in all these societies were combined and adapted to the land of savannah grasses and pine scrub. The cattle culture of the Southeast was both socially and economically a deeply Caribbean one, formed of the animal husbandry traditions of all the peoples transplanted there, and thoroughly reliant on the Caribbean as a primary trading sphere. The cattle industry served major markets by the mid-nineteenth century, but gave way bit by bit, as land competition from timber companies, the development of a stronger Latin American beef trade, and finally sprawling suburbs and a growing national agribusiness drove Florida cowmen out of business.

While the accounts of medieval fisheries and early American cattle raising offer windows into the organization of cultures now past, anthropologist Ken Erickson examines present-day procedures of meat production to expose synthesis and paradox in late twentieth-century America. Erickson describes a contemporary mix of cultures in a southwestern meat-packing plant, exploring the modern process of cattle slaughter and beef packaging through the experience of the workers who transform living steers into boxed food products. He shows us who does the work—predominantly Hispanic, Anglo, and Southeast Asian groups—how the work is divided, and what attitudes prevail. With great attention to ethnicity profiles, the author examines workers' relationships to their essentially gruesome work. How do people engaged in industrial slaughter "get used to it?" The separation of tasks and the use of new technologies have rendered old craft skills of butchers useless. Have these developments also facilitated the "numbing" required to butcher so many animals so quickly, and are such machines also allowing the distinction between "animal" and "human" to blur in the culture at large?

Part II of this volume concerns a more immediate means of animal slaughter—hunting. Hunting has often been a way to procure meat outside the market of commercial exchange. However this group of essays looks at hunting in a variety of nineteenth-century contexts in which the hunt was entirely folded into commercial or imperial directives. Because each narrative concerns hunting by Western people in foreign lands marked for appropriation, and each takes place in approximately the same time frame (the last quarter of the nineteenth century), the three essays taken together offer a broad view of the crossroads where expansionist attitudes toward power over animals and attitudes toward native peoples met. These hunts themselves took place on the borderlands where Europeans (or Euroamericans) penetrated the native cultures they sought to dominate; the indiscriminate killing of animals there achieved both real and symbolic ends for the expansion of power. Annihilation, control, stewardship: hunting contributed to each of these modes of dominating animal and human populations.

Andrew Isenberg's essay is a complex investigation into the destruction of the American bison by enterprising Euroamerican hunters' wholesale killing of the animal. The struggle between the expanding commercial ventures of the United States and the traditional hunting society of the Plains Indians can be summed up in the 1873 comment of Colonel Richard Irving Dodge: "every bison dead is an Indian gone." Indeed ten years later, the bison that had ranged in tremendous numbers at mid-century were almost extinct; Native Americans whose treaty guaranteed hunting rights "as long as [the buffaloes] last" were forced to submit to the reservation system. This speedy extinction of the bison was due to the combined effect of droves of Euroamerican hunters eager to make easy money, the encouragement of capitalists, the imperatives of politicians, the beliefs of naturalists, and the ongoing dialectic between the environment and the life it supports. Isenberg shows how the commercial trade of bison hides and meat intertwined with the extension of railroads across the plains, and dovetailed with political ideologies of manifest destiny and scientific theories of the "natural" extinction of "unfit" creatures. Both the American Indians and their major food source, the bison, were seen as the unfit obstacles to a vigorous and expanding capitalist society.

The actual hunters in Isenberg's story appear as simple agents of a much larger agenda, but as we move across the globe from freewheeling Gilded Age America to late imperial British Africa, we find a scene both analogous and strikingly different. Edward Steinhart also looks at the multilayered contexts of European hunting amidst a traditional society in his essay on turn-of-the-century Kenya. Here, however, the experience and bloodlust of the hunt is foregrounded, while ideology—that of aristocratic English hunters—is articulated primarily in terms of "sportsmanship." Once again the hunting of the animals becomes a surrogate for the hunting of the native peoples, but here the relatively easy target of the commercially hunted American bison is replaced by the "big game hunting" of elephants and other animals. Ivory trade notwithstanding, Steinhart shows how the tales of these hunters were framed in terms of adventure and the "sporting feeling," and how they portrayed Africa as a last preserve for the unhindered exercise of prowess. The author establishes the big game hunter in context of traditional aristocratic monopolies on hunting grounds, and links the hunt to the police and military actions through which England sought to control this far-off domain. Through the reading of hunters' memoirs and diaries, he figures hunting as a "theater of terror" for imperial power.

Concurrent with the exploits of big game hunters outlined by Steinhart were the efforts of men like Carl Hagenbeck to bring exotic animals alive for exhibition in European and American zoos, circuses, or private menageries. For the Hagenbeck company, commerce was in the living animals themselves, and the company was a major supplier of animals worldwide throughout the late nineteenth century. In the final essay of Part II, Nigel Rothfels examines exactly how this major animal purveyor procured his animals, and analyzes the cultural import of Hagenbeck's ventures. Rothfels reconstructs the popular excitement about viewing curious animals on the one end, and the methods used for their capture on the other. Once animal catching was systematized and directed at particular species, many adult animals were killed in order to isolate young apes, elephants, or rhinoceroses, who could then be taken from the side of their dead mothers. In discussing the course of Hagenbeck's career from the mid-nineteenth to the early twentieth century, the author outlines the transformation of Western sentiment toward both animals and the colonial lands from whence they were fetched. Increasingly the sort of bloody adventure that characterized the narratives of those hunters described by Steinhart had to be played down. A sense of stewardship was constructed around "animal preservation," and by the turn of the century, methods of capture became a part of the process to be suppressed. Rothfels explores the paradox of the Hagenbeck company's image as a protector of animals and lover of nature even as Hagenbeck's business depended on the violent capture and sale of animals. Rothfels suggests that the Hagenbeck paradox is shaped by an essential ambivalence toward the encroachment of Western civilization on foreign cultures and natural worlds.

The finale of Hagenbeck's story brings us back to the artificial environments in which wild and exotic animals are displayed to the general public. While hunters may experience the most visceral confrontation between the tools of Western civilization and the wild habits of indigenous animals, the captivity, maintenance, and display of animals presents a whole roster of complex and mediated relationships between culture and nature. The essays in Part III address the zoo as a site of this mediation. At the zoo, public entertainment and scientific research operate together to create a spectacle of constrained nature. On the one hand, zoos have served as a staging area for the observation of animals, a place to extract "objective" knowledge of the natural world. On the other hand, the conditions of exhibition and containment point up the web of cultural values that encase captive animals, the cross-play of projection and observation which make up the understanding of the animal, and the indeterminacy of the culture/nature divide. Zoos display nature's actors, but upon analysis these exhibits—

constructed and viewed by men and women of specific times and places—can easily erupt into what Jonathan Burt calls "a chaos of symbols."

Richard W. Burkhardt examines the construction of one of the first modern zoos, the Paris Menagerie. Constituted in 1794 during the French Revolution, the menagerie was a collecting place for the few remaining animals of the Versailles zoological gardens and the menageries of nobles, the wild beasts that Paris vendors had previously exhibited in the streets to curious onlookers with a few sous to spare, and the exotic animals that were either exchanged as gifts with other state governments or were captured with other military booty in the wars. While the menagerie served as a symbol of state power, it was institutionalized as an arm of scientific research and public education. By the first years of the nineteenth century, the provisional menagerie became established at the nascent Muséum d'Histoire Naturelle. Burkhardt explores the wide panoply of factors that conditioned the development of the zoo, and the multiple functions that the maintenance of these animals filled for a variety of political, popular, and scientific agendas. With special care, however, he reveals the importance of the Paris menagerie for a kind of "comparative psychology" that was played out in both public spectacle and private scientific observation. While the key scientists in what was a golden age of French zoology sought scientific comprehension of all life forms through the comparative anatomy of dead specimens, the experiments in the menagerie displayed animal behavior of live beasts. The Paris menagerie was a gallery for "natural" tastes, passions, and drives. The search for the key to animal nature (and thus the key to human difference) was sought in flashy experiments and quotidian observation. A concert for elephants featured performances of Glück, popular revolutionary songs, and opera choruses in order to induce the animals to exhibit various moods (including, it was hoped, romance); meanwhile, the menagerie's director carefully catalogued the behavior of beavers. While the zoo represents the concurrent popular and scientific construction of the understanding of animals, in the early nineteenth century it also formed a ground against which understanding of human behavior and sociality could be deduced.

It is exactly this set of conditions provided by zoos for both the projection and demarcation of animal/human behavior that comes under analysis in Jonathan Burt's essay on the London Zoo. The zoo dramatizes a boundary between nature and culture, and during the early twentieth-century period with which Burt is most concerned, the London Zoo staged relationships between colonial power and "conserved" nature, between a capitalist industrialized world and a fanciful antithesis of that world. Burt finds a vehicle for this theater of contrasts in his rich analysis of Monkey Hill, an exhibit of baboons on an artificial rockscape, during the 1920s and 1930s. To the secretary of the Zoo, Monkey Hill was a model of contemporary wisdom for public health as applied to zoo inhabitants—the baboons could breathe fresh air and "sun" themselves under ultraviolet lighting. The constant violence of the baboons, however, seemed to belie the social benefits of this healthful atmosphere. For the London press, the hill provided a number of sensationalist stories on sexual aggression and infanticide, although also uncanny displays of respect for life and death. For the scientist S. Zuckerman, Monkey Hill was the site of observations that would become central to primatology. Zuckerman's *The Social Life of Monkeys and Apes* was largely based on the behavior of the baboons at Monkey Hill, although the author himself both argued for the wildness of this animal colony, and suggested ways in which correctives had to be applied for the artificiality of confinement and food provision. The multivalent meanings projected by Zoo displays of animals are captured by Burt's final vignette: while violent baboons exhibited the hierarchies and rituals of a seemingly unfettered animal colony at Monkey Hill, a concurrently running exhibition featured the domestic high-jinks of a "Chimpanzee Tea Party."

Zoos embody an ambivalence between the desire to look at wild, untamed animal behavior and the practical necessity to contain and render harmless wild beasts—a necessity that becomes a virtue, or at least a new sort of entertainment, in idealized and sentimentalized displays of animals acting out human domesticity. The exotic experience of "wild nature" is balanced by the more familiar pleasure of "tamed nature." It is to encounters of the latter sort that Part IV of this volume turns. The scene of domestication is obverse to that of the zoo. In sharp relief to the theatrical delineation of animal/human space, domestic animals are kept and raised within the borders of private homes and estates; they are not valued for their "natural" or "wild" qualities, except insofar as these can be adapted to the convenience of human masters. Ironically, however, it is exactly the raising of animals within the human sphere that makes the learned behaviors and artful practices of humans toward animals appear "natural." The "domestic sphere" signifies two modes taken by our authors: first as a venue in which the line is drawn between "wild" and "tame," and secondly as a human home habitat that includes animals. The three essays that make up this section vary widely in emphasis, but offer a three-dimensional picture of the ways in which "domestication" was constructed. Each author has specifically linked the hierarchies established through manipulation of or caring for animal bodies to larger intellectual or social hierarchies.

Harriet Ritvo emphasizes the biological manipulation to which domestic animals are subject, and her contribution here relates to the use of nomenclature among nineteenth-century breeders and naturalists. Livestock breeders and dog fanciers appropriated the Latin binomial classification of animals used by naturalists and scientists. The use of this scientific naming system gave authority to the sort of knowledge they traded in, and while breeders and naturalists seemingly engaged in very different activities, the shared language allowed passage ways between hands-on breeder knowledge of animals and naturalists' theoretical concerns. Breeders specified "families" and "tribes" to define distinct pedigreed breeds of cattle or dogs rather than to solve the raging question of natural speciation. This suited the ends of the myriad breeder societies and dog clubs that flourished in mid-nineteenth-century Britain, and legitimized their claims to certify the "purity" and even fix the existence of particular breeds. Ritvo explores the different application, reasoning, and agendas involved in the breeders' use of naturalists' language. Breeders were less concerned with nomenclature as a means of analysis and theoretical understanding of animals than with its use as a formal hierarchy of actual breeds and subbreeds that signified the breeders' direct reshaping of animal bodies. Because this shared form of language embodied studied ambiguity between domesticated animals as products of human art and as products of nature, it problematized differentiations between wild and domestic, pure and impure, and complicated developing notions of "racial science" applied to humans. Finally, Ritvo suggests that the difficulty of establishing the stability of breeds except through nomenclature relates to the naturalists' difficulty in establishing the stability of whole animal species—that is, the problem that was finally taken up in earnest by the naturalist and avid follower of breeder journals and correspondence, Charles Darwin.

While Ritvo's study of breeders focuses on the biological manipulation of animals under domestication, the following essay focuses on the social manipulation of domesticated animals. Katherine Grier analyzes the affective side of animal relations during this same period, as pets were increasingly found as members of the family among nineteenth-century middle-class American households. Through a rich analysis of midcentury natural histories, advice literature, religious tracts, children's journals, and personal anecdotes, Grier charts the sentimental shifts toward animals that can also serve as keys to the emotional structure of the middle-class home. Here "domestication"refers not only to the raising of animals as pets, but to children whose natural passions required taming, and to the activities of the Victorian mother who was expected to coordinate the "softening" influences in her home. The discipline of "kindness to animals" curbed the penchant to perform cruel pranks, and taught children a kind of Christian stewardship over animals that could in the future be extended to all creatures (human and animal) weaker than oneself. Grier shows how the co-construction of Protestant liberalism, domesticity, and gentility converged in the formation of values and expectations that have lain behind pet-keeping practices, and broader bourgeois attitudes toward animals. She explores key metaphors used to describe human-animal relations, and reveals how each helps us read the configuration of a larger realm of affection and feeling taking root in the mid-nineteenth century, and how this realm may relate to the more familiar precincts of social history.

The essay by Jacqueline Milliet in effect combines the meanings of domestication discussed in the first two essays of the section—the biological domestication of animals and "domestication" as that process effected by mothers in the raising of young. While Grier's essay might be read (on one axis) as a deconstruction of women's "natural" domesticity in the emotional sphere, Milliet challenges assumptions among traditional anthropologists that women are natural animal domesticators because they have the biological capacity to suckle young wild animals brought within their home domain. The key image associated with this traditional view is that of the primitive tribeswoman suckling a wild pig—an image that figured prominently in early twentieth-century anthropologists' work. Milliet seeks to explode the myth that women may have been largely responsible for the domestication phase of development due to the natural predisposition of their bodies; she shows that in fact women in traditional societies use their bodies in this manner because it is simply the only tool left to them. By exploring a broad range of interactions between humans and animals across a variety of cultures, Milliet discovers that male interactions almost always involve tools—plows, knives, brands—while women, by and large deprived of the use of these tools, use their bodies as their only tools, and sometimes the bodies of domesticated animals themselves.

The essays in Part V look at animals not as domestic equipment, but as laboratory equipment—sentient beings who can be biologically standardized, or incorporated into laboratory machines. While this description may appear to completely objectify these animals, the use of animals in the laboratory paradoxically marks the height of their subjectivity in the sense that they are subjects who stand in for the human body or even the human psyche. Subjected to scientific experimentation, their animal bodies are made to relinquish secrets valuable to human science and medicine. As in the zoo, cultural and scientific meanings also merge in the laboratory; however, here there is no public spectacle, but rather a highly structured professional practice. Through these investigations of the cultural and scientific investments laboratory animals take on within a network of scientific practices and politics, we are faced with the extreme reflexivity of our understanding of the natural laws we share with animals.

Karen Rader's essay examines the factors that contributed to the adoption of mice as a standard laboratory animal in American research. How is it that the mouse became the experimental stand-in for human subjects? Rader analyzes in particular the career of C. C. Little, the scientist responsible for the factory-like production of standardized mice that were to be employed in American biomedical research. Because these animals came of one pure strain and were genetically similar, they could serve the needs of other scientists who needed a standardized biomaterial in order to coordinate and compare results. In the mid-1920s, Little seized on the mouse as a mammal whose genetic make-up could be standardized through inbreeding, and who could serve as a link between the well-known (but essentially useless) genetics of the fruit fly and the little-known (but potentially momentous) genetics of humans. He harnessed the interest of eugenicists to the growing concern over cancer, a concern that began to manifest itself in the outpouring of philanthropic, and later government, funding. The future mogul of mouse research was able to reorient scientific and popular attitudes toward this annoying vermin by deploying culturally loaded rhetoric; he argued for a "New Deal" for mice, and figured them as soldiers on the "battlefield of cancer." By skillfully manipulating not only the breeding of the mice themselves, but also organizational demands and opportunities, Little became a central figure in the structure of American biomedical research by the late 1930s. While sensitive to the cultural specificity of this project, Rader elicits, as did Ritvo, questions concerning the ways in which humans seek to understand nature through the artificial manipulation of its plasticity.

While the little white mouse received the kind of "New Deal" Little had hoped for, the big white rat has had fewer defenders. Kenneth Shapiro takes up the task in his analysis of the laboratory methods and assessment of the practical payoffs of psychological experiments performed on rats. In sharp contrast to the scientists' New Deal, however, Shapiro is concerned with the ethical viability of the experimental practices enacted on laboratory animals. The author uses the analytic tools of the social construction of knowledge to frame animal rights issues. The value of the rat as a "model" for psychological research is not that it is genetically standardized, but that it is a *generic* animal. That status, however, is socially constructed through the assumptions and practices of investigators, Shapiro argues. Laboratory scientists control and manipulate the conditions of the lives of their experimental animals in such a way that the animals can be considered to exhibit neither individual, nor even species-specific behavior. Finally, the animal becomes merely one element in a larger apparatus, thereby stripped even of its animality. Shapiro examines the variety of claims made by experimental psychologists, and assesses the impact of the research through citation analysis and surveys. Shapiro concludes that, rather than revealing useful information, animal experimentation serves primarily rhetorical uses in the modelling of psychological processes.

While Shapiro's essay is frankly motivated by animal rights, it is the more basic perception that dealings with animals invariably reveal important aspects of the human estate that courses through this volume. The rhythms of living among animals, the thrill of exotic beasts, the love of pets, and the knowledge we derive from this "mirror of nature" are all encounters as structured by time and place as human interactions with each other. The very inability of animals to participate in their own representation makes their activities and uses a matter of human interpretation and will. The human imagination, equipped with a lexicon of words, pictures, and practices provided by individual cultures, meets the brute animal in ways almost as multitudinous as the varieties of animal life itself. Through an investigation of these encounters, the essays here have begun to explore that vast borderland between human fancy and physical nature.

NOTES

1. David Bourdon, *Warhol* (New York, 1989), pp. 231–32.
2. Charles Krauthammer, *Time Magazine,* March 10, 1997, p. 60.
3. Quoted in Cicero, *De Finibus,* 2.32.

PART ONE

ANIMAL ECONOMIES/ HUMAN SOCIETIES: FOOD SYSTEMS

I

Carp, Cods, Connections

New Fisheries in the Medieval European Economy and Environment

Richard C. Hoffmann

New Foods, New Animals, New Actions

Reasonably well-off late medieval Parisians who lived on the rue Fromenteau near the king's Louvre palace brushed into their garbage heaps and cesspits food wastes with an unremarkable (high) proportion of fish bones. Analysis of archeologically recovered fish remains from this site finds 62 percent fishes of marine origin—after about 1450 often headless cod—and 30 percent from fresh water—after 1450 nearly all carp of a uniform size. Preliminary counts from another Parisian site of the early sixteenth century show comparably numerous headless cods, herring, carp, and pike.[1] Further south at Orleans, 15,004 fish bones from a latrine structure dated 1490–1568 in the yard of a collegiate church came from thirty species, but 52 percent were carp and 23 percent cod.[2] Monks at contemporary La Charité-sur-Loire also ate more carp than all other fishes combined, but none from the sea.[3] Up in the bustling Low Countries fish bones which sifted through the wooden floor of the late fifteenth-century kitchen at Ename priory, like La Charité a Benedictine house, came mostly from herring that had been opened and gutted before delivery, from various codfishes (*Gadidae*), and from carp.[4] Not far away, the household at Eindhoven castle, where the Countess van Buren then lived, consumed 60 percent fishes of marine origin, half of them cods, and one-third fishes from freshwater, of which one in ten were carp.[5] The same dietary habits are visible in remains of the fifteenth and sixteenth centuries from castles at Bodenteich and Plesse in inland Lower Saxony.[6]

Further south the best records are written, not archeological, and document the massive appearance of preserved fishes from the Atlantic on tables of interior and Mediterranean Spain. Among the hardest evidence are accounts of toll collectors at Seville—where the annual value of hake, conger, and sardines coming from Galicia had reached about five million maravedis by the 1450s and then doubled to more than nine million in the 1490s—

and at Valencia—where they kept track of actual volume of goods. Between seven and eleven million Galician sardines and some two hundred thousand dried hake (a cod relative perhaps twenty times the size of a sardine) came into Valencia each year during the 1490s, twice the quantity of a half-century before; conger shipments stabilized at about twenty-two tonnes.[7] Fishmongers at inland towns conveyed these goods to willing consumers.[8] Just as telling is the first recorded shipment to Italy in 1426 of dried hake and pickled sardines from the Atlantic for the Aragonese garrison in Naples.[9]

In the southern and northwestern European consumption areas just mentioned those types of fishes had been absent or barely known back in 1200. New foods and, in the case of the carp, new animals, were being introduced to meet European demand for fish to eat. This paper contextualizes two central innovations in the fishing industry of medieval Europe, namely the invention and spread of artificial fish culture inland and the large-scale commercial exploitation of offshore and distant-water fisheries on Europe's maritime frontiers. In both inland and coastal Europe the new fisheries established in the high Middle Ages and further developed during the fourteenth and fifteenth centuries formed new and seminal relations between people and environments, extending human use of aquatic life beyond the historic bounds of natural local ecosystems. What one person might think of as economic development, another could label human manipulation, environmental destruction, or ecological revolution. There is always a catch.

The new economic activities and impacts here explored must be seen against older practices and consistent features in medieval European use of aquatic resources. What has been established elsewhere can now receive only brief background summary.[10]

Before the end of the first millennium and with some subtle exceptions, there is little sign of direct and transforming human impact on European waters and the life in them. In present-day terms, human use was sustainable, though by no means absent. Driven by religious taboos against eating terrestrial quadrupeds almost 40 percent of the year, and with tastes shaped by other cultural norms and preferences, medieval Christians actively consumed fish as food. What they ate came overwhelmingly from stocks native to local inland and coastal waters, which were exploited by part-time subsistence fishers and/or dependents working for their lords. It is important to distinguish between the direct subsistence fishing of peasants for their own families and the indirect subsistence fisheries run for elites who commanded the most favored part of what their subordinates caught, but apart from criteria of taste and luxury display—lords liked to show off with big sturgeon, salmon, and pike—high and low used the same nearby ecosystems.

The wave of economic expansion and the development of an exchange sector, which got well under way in eleventh-century Europe, greatly enlarged demand for fish and especially that expressed on the market. Like other manorial artisans, some part-time servile fishers were able to turn themselves into small-scale commercial fishers feeding elites and townspeople who could pay for their dinner. Even as artisanal fishers formed some of the first craft guilds in their regions, all over Europe they kept using the same simple trapping, netting, and even angling techniques which their peasant forebears and neighbors had found effective to exploit natural local and seasonal concentrations of fish. Continually rising human numbers and disproportionately more who bought food on the market drove total fish consumption up. At the same time, however, more people and the environmental impacts of their greater agricultural and industrial activity detectably affected aquatic habitats, limiting and changing the fish stocks Europeans were accustomed to eat. To emerging local and regional imbalances between demand and traditional supply medieval Europeans applied various cultural and social responses, from rising prices to privatization of fishing rights on the one hand and communal or even precocious governmental regulation—of seasons, gear, catch, and activities deemed harmful to the resource—on the other. We can see in particular the social construction of both self-regulating "commons" and of private property in fisheries.

In all we have so far mentioned medieval Europeans essentially treated nature as given and responded to disjunctions between demand and resource by adjusting human behavior. By sometime in the high Middle Ages (the twelfth and thirteenth centuries), this approach visibly left demand still unsatisfied. Europeans began to engage nature differently, looking to modify it or to tap new parts of it to meet human wants. People today are well-acquainted with this more aggressive alternative. Those who want more of what nature offers try to "improve" nature to better serve human desires or to discover "untapped" resources they can make "useful." We think of intensification channeling a greater share of natural productivity to what humans value, or of extending human activity across a frontier to spaces formerly ignored. With respect to aquatic resources, the first strategy is plain in creation of an artificial fish culture and the second in offshore expansion, each driven by distinctive segments of consumption demand. Both brought complex economic and environmental consequences.

The Case of the Carp: Aquaculture as Environmental Revolution

Twelfth-century western Europeans, almost certainly French, evolved ways of controlling water and the fish they put in it to get consistent and seasonally useful food. When the technology was joined, probably not much before the mid-1200s, to a naturally well-adapted fast-growing fish of exotic eastern European origin, the carp, the result was the dominant and most advanced artificial fish production system used from the Atlantic to the Urals well into the nineteenth century and, in some regions, still today. Aquaculture revolutionized local ecologies and human relations with them, forming and controlling synthetic habitats for the sake of a non-native animal and to the harm of some native varieties. Whether aimed at indirect subsistence or market sales, fish farms made ordinary people and nature alike submit in new ways to elite cultural preferences and powers.

Simple measures to get more use from fisheries were always commonplace.[11] Live storage, use of artificial watercourses, and haphazard stocking of wild fry belonged to the historical setting but they are not domesticated fish farming.

A Benchmark: Advanced Traditional European Fish Farming

The evolution toward aquaculture in medieval Europe is best understood by taking as a benchmark the methods being used at the close of the Middle Ages in East Central Europe, then recognized as leading in this field.[12] "State-of-the-art" management and techniques used by Czech, Polish, and Austrian fish farmers are described in unique contemporary didactic manuals. The *De piscinis* (*On Fishponds*) which future Moravian bishop Jan Dubravius composed in the 1530s for mining capitalist and landowner Jakob Fugger was printed in Wrocław in 1547, in Zürich in 1559, and by 1600 in both Polish and English translations.[13] Olbrycht Strumienski's 1573 book on ponds gained authority from the author's experience as a professional manager.[14] The publications are augmented by recently discovered manuscript instructions of slightly earlier date in Czech and German.[15] What the didactic literature advanced as normal and best practice is corroborated by managerial records of actual enterprises at the time. There are surveys of Polish royal estates near Kraków in 1564;[16] complete operating accounts from pond enterprises in the southwestern Polish duchy of Oświęcim-Zator from the 1510s to the 1530s;[17] rich operational archives from the Bohemian magnate families Rožmberk and Pernštejn;[18] and widespread if piecemeal records from lesser Polish, Czech, Austrian, and Franconian noble and monastic producers.[19] All these sources show the same integrated technology operating aquatic "agro-ecosystems."

At the end of the Middle Ages the best-reputed fish farmers in Europe reared mainly common carp (*Cyprinus carpio*), with a side crop of pike and occasional lesser numbers of bream[20] or crucian carp.[21] Cold-water fishes such as trout and grayling lacked importance. Cultural methods emphasized full control over water and fish through each stage in the production cycle.

Continual production required coordinated management of many ponds, each with facilities to adjust the inflow, outflow, and level of water, from fully filled to partly empty or entirely dry, without undesired entry or exit of fish. Experts laid out ponds according to the terrain, preferring broad to deep ones even if the former required longer dams. Engineering skill predetermined the area to be flooded. In valleys a sequence of ponds, ideally with bypass channels, could feed one another downslope. Internal timber frames braced earthen dams and dikes, and stone or wicker armored exposed surfaces. Wood or metal grates and gates controlled passage through sluiceways of wood or stone. Large aquaculture enterprises had use for tanks only a few dozen meters square and for regulated lakes up to four hundred hectares—like Bezdrev pond at Hluboká in southern Bohemia, built in 1492.

The fish were bred from selected brood stock, though not by artificial means,[22] and moved as a uniform age group (year class, cohort) through a sequence of special-purpose ponds. As the water in a shallow and vegetated spawning or fry pond warmed enough in late spring the pond master introduced chosen ripe adult carp to emit and fertilize eggs in the natural way, then removed the parents. After a summer in the protected and warm fry pond, the carp larvae had become tiny fish by the time they were moved en masse to deeper waters for the winter. The cohort of young carp went for another year or two to a growth pond where they would reach "stocker" size of a hand's span, before being set into a much larger, recently refilled, and otherwise uninhabited fattening pond to get big enough to harvest.

Carp eat mainly small invertebrates they find on water plants or grub out of the bottom sediment and crush with teeth in their throats. Nutrient-rich waters offer more food, so managers promoted growth by adding manure or channeling runoff from animal or human wastes into their ponds. They also learned to concentrate the carp biomass into harvestable adults by introducing piscivorous pike to eat any fry bred by precociously mature members of the class. Depending on local temperature and climatic conditions, the carp reached table size (forty to sixty centimeters and about a kilogram in weight) by year four, five, or six. By 1550 Polish records suggest that two distinctive quick-growing domestic races of carp, "mirror carp" with a few large scales and entirely scaleless "leather carp," had emerged.

Demand-focused harvest took place in weekly stages during Lent when the market was at its annual peak, sometimes even where this meant breaking ice to get nets into the pond; or in a single push in the fall, when the crop not needed for Advent consumption was held in live storage for the Lenten season. In cool weather carp survive some days wrapped in damp cloths or straw for overland transport; portable tanks and boats with live wells were also sometimes available. Water in the fattening pond, which had been designed with large warm, shallow, biologically productive areas at the periphery and a much smaller deep central channel, was lowered to concentrate the fish. When all were netted out the flow of water was, if possible, entirely diverted and the pond site left to dry. Best practice called for the bottom to be plowed up, seeded with oats or barley, and then the grain used to pasture livestock through the summer. This fallow season or year readied the fattening pond for another cohort of young carp.

Productive operation of a late medieval fishpond enterprise thus called for integrated management of some tens or dozens of ponds and, exclusive of breeding stock, some half-dozen year classes of carp. In the first half of the sixteenth century these methods sustained, for instance, annual shipments of some fifteen hundred tonnes of live fish from the ponds of Upper Silesia (chiefly Oświęcim-Zator) to Kraków.[23] But the integrated system was absent from East Central Europe before the mid-fourteenth century and remained rare until enterprises using it proliferated in southern Bohemia, Moravia, and Poland during the late fifteenth century. Knowing the mature system's diagnostic characteristics—multiple ponds, active manipulation of water levels, and both breeding stock and year classes of fish—we can trace its beginnings elsewhere. Earlier technical development further to the west reached a climax with introduction of the exotic carp to practices first evolved with pond fishes native to Atlantic drainages.

Early Signs of Development

Seen from historical distance, the process of creating traditional European aquaculture involved transition from live storage to long-term rearing of fish in protected waters, natural or artificial, plus design and construction of physical facilities capable of manipulating the water and the fish for consistent growth and production. The present state of research suggests that new kinds of hydraulic engineering began in central France during the late eleventh and early twelfth centuries.[24]

Certain disputes over water rights and flooding indicate the construction and rhythmic operation of pond systems without earlier counterparts elsewhere. Dams and excavations for multi-pond projects were undertaken about

1080 by St. Vincent of Mans and other lay and religious lordships in Maine and Poitou, such as those of Laval at Marennes.[25] Monks at Ronceray complained in 1141 to Count Geoffrey IV of Anjou and canons of Bourges in 1160 to Count Étienne of Sancerré because the counts' new ponds had flooded properties of the religious.[26] Complex pond systems were begun by Cistercians at Châalis in the Valois after 1160 and by Burgundian nobles and monasteries between 1184 and 1206.[27] Some charters hint how these facilities were run. A pond built by St.-Père of Chartres flooded land of a neighbor named Aucher: sometime between 1101 and 1129 they came to agree that when the pond was filled, water and fish belonged to the abbey, but when it was dry, land and crop were Aucher's. Because some ponds on a Norman watercourse belonged to a knight, Robert de Buat, and others to the monks of La Trappe, in 1215 they arranged to coordinate the draining and later refilling.[28] By 1263 ponds owned by the knight Pierre de Palluau at Ograis in Berry were being drained and dried every three years,[29] while new constructions along the eastern edge of the Paris basin in Champagne came equipped with a full set of sluice gates and drains.[30]

Some of the same kind of landowners were also handling fish with forethought. Duke William of Normandy and St.-Benoit-sur-Loire agreed in 1067 to share the fish taken regularly from one of the ponds at St.-Jacques de Beuvron.[31] Surviving scraps of a financial account kept for the count of Champagne in 1217 indicate systematic stocking of his ponds,[32] and the 1235 ordinance on fiefs of King Louis IX assumed that fishponds had a five-year production cycle.[33] The fish were probably bream, the variety most consumed at La Charité-sur-Loire at this time.[34] In England, where after 1066 an ethnically and culturally French elite built and managed pond systems wholly like those of their continental cousins, the king's own fishponds commonly supplied "fat mother bream" (*bremas matrices et grassas*) and pike for stocking his other ponds and those of his favorites.[35] A late thirteenth-century guide to estate management, *Fleta,* explicitly recommended bream and perch for ponds.[36]

In the twelfth century carp were unknown west of the Rhine.[37] The native range of this heat-loving species covered the lower Danube and Black Sea drainages of southeastern Europe. Millennia of premedieval bone finds and verbal records are confined to that region. Until recently the westernmost fully wild population of carp still inhabited the Danube between Vienna and Bratislava.[38] As the carp is adapted to quiet weedy waters, which it rapidly colonizes, but ill-equipped to move upstream against fast, cold currents, its extension of range northwestward during the early Middle Ages may have had fortuitous help from store ponds and mill dams of humans. Was a time of climatic warming also influential? A stepwise progression of archaeological remains and written references put carp into the upper Danube basin by the late eleventh century and nearby tributaries of the upper Rhine by the mid-twelfth. At the confluence of Rhine and Main the sharp-eyed Hildegard of Bingen (1098–1179) knew the fish well.[39]

Carp almost explosively entered the French historical record in middle decades of the thirteenth century—and were there for the first time associated with ponds. Fragmentary financial accounts from 1258/59 have local officials of the count of Champagne stocking carp by the hundreds and thousands ("and six big pike") in some of the ponds at the Marne-side seigneurie of Igny-le-Jard, an area still full of diversions and standing water.[40] At almost the same time the king's governor of Paris, Etienne Boileau (1260/61–69) , issued regulations for marketing carp (*carpes, cuerpiaus*) among other fishes.[41]

A third genre of source, scholastic encyclopedias, confirms new contemporary awareness of carp in the French heartland. The chief authors, Thomas of Cantimpré, Vincent of Beauvais, and Albert the Great, respectively sons of the Low Countries, northern France, and the upper Rhine Valley, spent much of their careers in Paris, where their Aristotelian observations on natural history were compiled between the mid-1240s and early 1260s and then circulated widely.[42] All three describe carp as a fish of still waters skilled at evading nets. Thomas and after him Vincent appealed to popular knowledge ("Opinio vulgi est . . .") in telling how this fish spawned in water weeds, its young liked warm shallows, and sickness brought by stagnant August conditions was cured by fresh river water. Albert added that carp spawn in certain waters and grow large in others, the best having clay bottoms seeded to wheat and plowed over before flooding.[43] With that entry into well-known reference books, after the 1260s mere learned reference to carp is no longer evidence for local presence of the fish.

From the Paris basin to Burgundy and from Hainaut to the middle Loire, by the turn of the thirteenth century carp seem almost ubiquitous—carp growing in ponds, little carp for stocking, big carp for breeding, carp in elite kitchens and on elite tables.[44] Well before the Black Death of 1348–51 marked a secular turn in Europe's economic climate, clerks for great and lesser lords of northern France were surveying and keeping account books for carp-rearing enterprises essentially identical to Czech or Polish ones two hundred years later.[45]

English pond managers kept fine accounts in the thirteenth century, too, but they never once mentioned carp. The species is unrecorded in the British Isles before the mid-fifteenth century, and acknowledged as a rare and recent introduction there well into the sixteenth (which also gives the earliest known remains of carp).[46] Delayed entry of carp to England contrasts with Anglo-Norman introduction of fallow deer and rabbits.[47] A relative lag crossing the Channel fits other evidence that the fish had failed to reach northwestern France before the 1204–6 collapse of the Angevin realm split Anglo-Norman estates into different lordships and social networks. England received the pond technology in the twelfth century but not the carp in the thirteenth.

Diffusion

Well-evidenced signposts mark the subsequent spread of the whole integrated aquaculture system (multiple ponds used in rotation to rear chiefly carp) from northern France across other regions of transalpine Europe.[48] One vector went east from ducal Burgundy into the Franche-Comté—where all the familiar features occur in production accounts dating 1338–83[49]—and down the Rhône (or up the Loire) into Forez and Dombes. The latter regions experienced a double wave of pond construction. Early initiatives in the 1230s were emulated on a large scale only fifty and more years later, when lords invested in pond systems as one form of market-linked agricultural improvement. More flooding after the mid-fourteenth century sought rather to get some use out of abandoned farmland. Large-scale market-oriented carp production by a wide range of social groups—though the best operational records are, of course, for the nineteen large and more small ponds of the Count of Forez—became central to the regional economy.[50] Movement northward from the French- to the Flemish-speaking Netherlands, in contrast, is so far better traced through the fourteenth-century appearance of carp in customs schedules and food remains.[51]

Those tracking aquaculture eastward into central Europe might first stop at Heilsbronn, a Cistercian monastery on the Schwabach, a tributary of the Main. This estate can stand for the whole belt of ponds which developed in Upper Franconia. During the 1260s Heilsbronn, already more than a century old, began acquiring first natural, then artificial ponds in its general neighborhood. By the 1340s the monks were buying and rearing stocker carp for a multipond enterprise which spread over thirty kilometers from the convent and still mainly served their own consumption needs. In the late fifteenth and sixteenth centuries, however, and eventually as a state-managed foundation, Heilsbronn's large output supplied urban consumers, especially in prosperous Nürnberg about twenty-five kilometers away.[52]

Upper Franconia borders the Czech lands where a first great surge of pond construction for aquaculture occurred under the auspices of Emperor Charles IV (1347–78), who also had family and personal ties to France. Charles even reputedly urged pond building on grounds the country needed abundance of fish and humidity ("ut regnum nostrum piscibus et vaporibus abundaret"). Dwarfing all known earlier hydraulic works there, eighty-seven new projects can be documented before the Hussite uprising in 1418.[53] Though favoring hill country, which made for cheaper short dams but less productive deep ponds, at least some late fourteenth-century Czech pond masters were building the oldest large ponds (one hundred hectares and more) which still survive, installing elaborate sluices and valves, populating ponds with age-class stockers, and selecting fish for breeding. Their output went to towns in Bohemia, Germany, and Austria.[54] Czech politics calmed after 1450—when the Rožmberk estate at Třeboň already had three large ponds covering seven hundred hectares and seventeen smaller ones—giving opportunity for more large engineering works in both southern Bohemia and the flat lands of the upper Labe (Elbe) basin of the east. Then the Pernštejns, for instance, improved their Hluboká estate with the huge Bezdrev pond and Štěpánek Netolický ensured water supplies for Třeboň, which he managed from 1516 to 1530, by diverting a river through a forty-five-kilometer artificial canal, the Zlatá Stoka ("Golden Drain").[55]

Fish culture facilities were being built in central Silesia, then a dependency of the crown of Bohemia, even during the difficult 1430s and 1440s,[56] and by the 1450s they also appeared in the principalities of Oświęcim-Zator, at the low divide between the upper Odra and Wisła basins where Bohemian and Polish influence intersected. As observed, carp from here supplied Kraków in the early sixteenth century. That territory also probably transmitted aquaculture techniques still further further east and north in Poland to the zone between Kraków and Sandomierz and the region of Kalisz-Sieradz, where production grew after 1550.[57]

Demand, Power, and Impact

Distinctive in spatial, social, and economic terms, the particular segment of demand which medieval aquaculture served shaped its character and structured its impacts. The fish farms provided live fresh fish for inland elites. When the duke of Burgundy was in his titular home, his pond masters delivered thousands of carp to his castles. When the ducal family went to Bruges in coastal Flanders, they ate fresh seafood, but after a two-day journey inland, there were the pond fish back on the menu again.[58] Thirteenth-century English ponds delivered fresh bream and pike to the king and other magnates and so did those of the Polish king in 1420.[59] When the product went on the market it was not cheap, even for fish. At Namur in 1356 a hundred carp for a wedding feast came at twice the price of a cow.[60] During the Council of Konstanz (1414–18) a pound of carp cost the same as four pounds of beef or twenty loaves of bread.[61] A like ratio applied in Paris at the end of the sixteenth century, when noble households contracting in advance for bulk food deliveries paid three and a half times more for foot-long carp than for a pound of beef.[62]

Medieval fish farming acquired strong regional features. It belonged especially to interior Europe. From upper Poitou and the great arc of the middle Loire all the way to central Poland, zones noteworthy for fish culture arose inland of the roughly 150 kilometers preindustrial technologies could safely carry fresh sea fish. Resource endowments and consumption centers played further locational roles. Where large natural lakes offered ample and tasty competition, few fishpond enterprises endured.[63] Nor was all inland Christendom equally inundated. Areas distinguished for pond culture have characteristically impermeable soils and abundant water supplies. In Franconia, Burgundy, and Lorraine springs seep from discontinuities between porous and solid strata on valley slopes, while southern Bohemia, Dombes, and Sologne rely on surface runoff and locally disordered drainage.[64] Natural potholes and intermittent wetlands made good sites for ponds. But plainly, as in the case of medieval commercial viticulture,[65] concentrated demand for freshwater fish affected those suitable landscapes most accessible to it. A quintessential great medieval household, the pope's at Avignon, sent special agents up the Rhône and Saône to buy and ship the output of Burgundian ponds.[66] Good-sized towns concentrated a well-off clientele: Lyon bought the fish of Dombes; Nürnberg, Bamberg, and other south German towns were supplied by Franconian ponds but also by those of southern Bohemia—insofar as the latter had surplus from the demand of local elites and Vienna. Greatest consumer of all, Paris drew on the Seine basin (Valois, Champagne) and the Loire (Forez, Sologne). Medieval fish culture is associated with special resource endowments accessible to inland concentrations of wealth.[67]

Aquaculture was shaped and spread in medieval Europe by exercise of elite power, the same agent then contesting customary access to wild fisheries. Fishponds were private property. Most belonged to landowners. They were large investments. Assembling the land for a fattening pond at Massileux in Forez cost a thousand livres and as much more to build it. Work for a pond made at Chomutov in western Bohemia in 1408/9 cost two or three times what the owner could expect annually from an entire agricultural village.[68] The standard production system promised no returns for two to five years, and thereafter also continuing costs of operation and maintenance.[69]

Pond construction for lords threatened others' interests. In a litany of power, tropes of conflict and compensation everywhere run through detailed records. Late twelfth-century Abbot Samson of Bury St. Edmunds stirred up resentment in and outside the abbey when he raised the the dam of his fishpond at Babwell and the higher water damaged meadows, pastures, orchards, and arable of neighbors and other monks alike. The count of Beaumont, whose estate was on the Théve, a river in the Valois, complained in 1204/8 when the Cistercians of Châalis diverted water upstream at Commeles. In Champagne, other monks claimed compensation for lands lost to a new pond of Henri, seigneur de Hans, in 1302 and themselves paid Aaliz de Joinville, dame de Beaufort, not to flood their meadow in 1336.[70] A customary *droit d'inonder* (*d'inondation*) allowed lords in southeastern France arbitrarily to build dams on their own land and compensate neighbors afterwards for what the pond flooded, "for the construction of such a pond provides to the builder and the community more benefit than the inconvenience of neighbors whose properties are drowned in the flooding of the water." Involuntary victims otherwise kept only the crop from years when the pond was dry and the right to collect shoreline vegetation.[71] In Upper Saxony the abbey of Altzelle was continually at odds with its small-town neighbors in Rosswein, whose lands Altzelle flooded; with a convent at Nimbschen, whose woodland Altzelle drowned by raising a pond in 1495; and with another nearby town, Grimma, whose own pond washed out the abbey's mill.[72] Building ponds at the expense of subject peasants was one way lords wielded authority in East Central Europe around 1500. To take but one example from a court book of the Kraków cathedral chapter, villagers at Choczianowice who complained in 1502 that the lords's fishpond "frequently damages their fields and floods their meadows," blamed the local administrator for having raised the dam and, in preparation for another pond, cut down the local woodlot. Protest was vain.[73] Peasants undoubtedly lost meadows and sometimes arable when their lords turned land into water.

Depending on economic structures around them, medieval fishpond enterprises operated in the customary economy, the market economy, or in East Central Europe, the newly coercive economy of neo-serfdom. Whatever the setting, fish culture called for high technical skill and environmental knowledge on the part of managers, and for heavy labor. Workers were turned out to harvest fish from the ponds of Ebrach, a monastery in Upper Franconia, as a thoroughly old-fashioned obligation on certain peasant tenements, limited to the specific task and with equally customary remuneration of meals and two fish. Early fourteenth-century tenants on the count of Bar's Lorraine estates did compulsory service carting fry to stock ponds.[74]

Skilled monastic pond masters are curiously hard to identify, but a long catalog could be assembled of specialized professional engineers and managers, some employed on limited-term contract, others as long-time, high-ranking salaried servants, with the title of "pond master" or "fish master." We already encountered Štěpánek Netolický building the Zlatá Stoka at Třeboň in 1516. During his fifteen-year tenure he designed and supervised construction of nine large and thirty-seven small ponds, traveled to advise the archbishop of Salzburg and the count of Salm, and for one tough engineering problem called in as consultant his counterpart from the Poděbrady estate. At a time when contract pond masters in Bohemia were getting three grosz a day the Rožmberks were paying Netolický twenty-four hundred grosz a year. His retirement house still stands in a suburban district of Třeboň.[75] But this kind of expert can be tracked back just about as far as detailed records of fish culture go. Hugh Colard was stocking ponds for the count of Champagne in 1258/59 and "Master Nicholas the Fisherman" toured southern England for at least eighteen years (1244–62) for sake of the bishop of Winchester's ponds. In the 1340s Reynaud le Tarroillon had charge of work on many of the duke's ponds in Burgundy.[76]

Reynaud and his counterparts in the Sologne hired wage laborers for construction and fish harvesting.[77] Czech pond builders worked for a wage, too: one grosz a day in the 1300s and as much as two grosz a century later. But by that time "robot," forced labor service, was becoming a norm in East Central Europe, including for heavy fishpond work. A survey of the Kraków chapter estates in 1531 made this plain for villagers of Dłotów: "furthermore, the aforesaid men are bound to work at the fishing in the fishpond, however often it is demanded of them, and likewise on the repair of the dam of the fishpond and hauling the fish to the manor house at Pabianice, however often it is demanded of them."[78]

The product of fish culture might likewise be delivered in kind to the lord's household—so functioning precisely as a more dependable form of the indirect subsistence fisheries familiar in the earlier Middle Ages—or sold by auction or prior contract to merchants from town. In 1356, for instance, one pond near Namur shipped 3,700 carp to the count's table, and the royal ponds at Oświęcim in 1522 put 32,640 carp and 6,045 pike into their master's kitchen.[79] Monks at Châalis and at Heilbronn ate their fish, too. But commercial sales are well documented in fourteenth-century Burgundy, supplying the papal court and towns alike. One sale at l'Eperviére-sur-Saône handled 3,400 carp and 190 pike, representing about 60 percent of the harvest for 1353. In 1534 the Polish king's Oświęcim ponds sold 26,100 carp.[80] Dubravius had Vilém of Pernštejn (1449–1518) brag that he made more from sale of fish (alone) than the entire wealth of ancient Roman senators. Land tax assessments from Lower Austria a generation later do indicate some lordships getting 15–25 percent of annual cash returns from their sales of cultured carp.[81]

Thus, technically advanced aquaculture could meet elite demand for fresh fish in late medieval Europe through old-fashioned seigneurial self-sufficiency or up-to-date marketing. In either case, however, this response to wants backed up by power and wealth set off localized and small-scale but numerous and widespread "ecological revolutions."[82] Fish culture transformed ecosystems, landscapes, and the way people lived in them. There were winners, losers, and changes in human experience.

Medieval fish culture purposely multiplied a kind of aquatic habitat rare until then in Europe, namely warm, nutrient-rich (eutrophic), still waters. Ponds sustained ecosystems unlike those of former running streams or seasonal wetlands. Stillwater organisms—rooted water plants, midges, mosquito larvae, perch, pike, roach, tench, bream—gained living space; and notably those of running water—gravel-loving invertebrates, gudgeon, grayling, trout, barbel, various migratory fishes—lost. The scale of transformation is impressive, for the fish culture regions acquired large numbers of ponds. Careful local experts concur on the order of magnitude: twenty-five thousand ponds in Bohemia and twenty-two thousand in upper Franconia, twenty-five thousand hectares of ponded surface in upper Silesia and forty thousand hectares in central France.[83] With ample evidence of modern drainage of medieval ponds in areas as far apart as France and Poland, it is hard to dispute Robert Delatouche's guess that the historic province of Maine, for instance, contained at the end of the Middle Ages twenty-five times as many ponds as it did five hundred years later.[84]

Principal intended beneficiary of this transformation of aquatic habitat was the carp. Fish culture meant the large-scale human introduction of an exotic animal to extensive areas of western Europe where it had not formerly lived. Just the written records which survive from the thirteenth and fourteenth centuries itemize hundreds of thousands of small carp going into ponds there. As objects of controlled monoculture these fish were becoming domesticates. Select domesticated races of carp certainly had emerged in Europe by the sixteenth century. Yet more telling, physiological features symptomatic of domestication remain in all wild-living European carp populations west of the middle Danube.[85] In other words, the "wild" carp of central and western Europe are feral descendants of domesticates. Carp escaped and proliferated into wild habitats so promptly that astute medieval observers soon recognized them in nature. Late thirteenth-century French fisheries regulations and the first Polish writer on fishes, Stefan Falimirz in 1534, grouped carp among wild fishes of the river as well as cultivated ponds.[86]

Together and separately ponds and carp worked change on the larger environment. Some of this was on purpose: like other "agro-ecosystems," aquaculture meant to control and simplify environmental variables so that biomass concentrated in one desired species. But then this exotic animal "went native" with what are now predictable—if historically undocumentable—consequences for the aquatic ecosystems receiving it.[87] Biologists watching the carp spread into new continents during the late nineteenth century found it unusually quick to colonize temperate stillwater environments—which medieval economic development was proliferating—and to disrupt relatively simple fish populations which were already under stress from overfishing and environmental change—like the native cold-water fishes of medieval Europe. In wild ecosystems carp behavior harmed native species by accelerating changes in habitat and biomass.[88] Had Philip the Fair, for instance, owned a freshwater research station, historians could now hope for more than just increasing references to carp.

The environmental impact of the ponds is more visible. Medieval records let modern scholars see the eastern edge of the Paris basin (Champagne-Lorraine) and the valley lands of Berry and Sologne being transformed from woods and wetlands to open arable and ponds.[89] Shaded springs and trickling streams became sunny or mist-shrouded expanses of water and reeds. Charles IV himself alleged (p. 11 above) a humidifying effect on microclimates. Later observers would deplore the insalubrious climate of the pond-filled regions. Dombes and Forez in particular had by the early nineteenth century well-deserved malarial reputations, though the disease was only later found to spread not by "bad air" but by mosquitos bred in standing water.[90]

Lakes and ponds are, moreover, ephemeral landscape features always fated to fill with sediment and revert to dry land. Without continual maintenance dams crumbled, sluices decayed, vegetation encroached, and many artificial fishponds ceased to be functional stillwater ecosystems.[91] Pond masters also knew that the water from densely stocked ponds inhibited growth of fishes downstream, and tried to design their water systems to avoid this affecting their crop. Even more than agriculture in general, aquaculture was dubiously sustainable.

Long-term social effects were destabilizing, too. Besides the professional pond masters, certainly some ordinary rural people did find opportunities in the fishpond industry by servicing large producers or small local markets. Especially village notables in Silesia, Poland, and south Germany crop up in possession of one or two small ponds each. On big estates this could add up. When Himmelskron Abbey was secularized in 1547, eighty-two ponds were inventoried on its peasant tenements. Twenty villages in the Pszczyń district of Silesia had in 1536 seventy-one peasant ponds with capacity to produce almost a hundred thousand fry.[92]

But for understandable reasons peasant resistance to aquaculture did not abate. Rather, the spread of fish farming intensified peasants' long-standing opposition to privatized natural fisheries. The twelfth-century writer Wace was sure that Norman peasant rebels back in 997 had wanted not the generic water rights reported in his source but in particular "to seize the fish from the ponds."[93] Against theft of fish from an artificial pond customary and written laws made special provisions—*Sachsenspiegel,* for instance, set fines ten times higher than for poaching in natural waters[94]—and lords posted special guards, especially when water was lowered for harvest. Complaints about illicit and clandestine removal of fish from ponds are a commonplace of court records.[95] In Dombes, Jean-Claude Schmitt has argued, thirteenth-century peasant anger against widespread flooding sublimated into legends where evil emanated from the lord's pond. In 1388 and 1440, however, rebellious peasants publicly broke the dams.[96] Collective destruction and fishing of ponds followed by common public meals of the lord's fish are well known from the German Peasant War—where the social content of these acts is obvious—but also from feuds between Staffordshire factions in the 1530s.[97]

Finally, thousands of small ecological revolutions colored cultural style and expectation. The magnate offering "fish from our own ponds" asserted social power with at least tacit environmental implications: people and nature were turned to private purpose. Sellers and buyers of pond-reared fish dealt in a standardized commercial product. Dubravius discusses regional customs of selling carp by measure (volume) or by count, favoring the latter to avoid dispute "where carp of equal size are sold."[98] As the Moravian described, so the Polish and the Austrian pond masters behaved, selling their carp and pike in units of sixty (*kop, schock*).[99] Earlier French fish farmers counted by hundreds.

Fish farms served an elite segment of medieval demand for fish. In medieval aquaculture fish and landscape were intentionally transformed to fit human ends. Manipulating the environment gained intended results and others not intended as well. What nature failed to provide was remedied by altering nature, in this instance by establishing artificial ecosystems to domesticate a useful animal and introduce an exotic life form into still-natural surroundings. Neither environmentally nor socially neutral, the aquaculture response—intensified development—to rising European demand for fish modeled an expectation that fisheries were subject to human authority and invention. Again, an East Central European source articulates what cryptic western charters and account books can only suggest had been going on since the high Middle Ages: an inspection team touring lordships of the Kraków cathedral chapter in 1533 was instructed: "if there are no fish in a pond, ask who took them out."[100] Fish are present or absent by human agency. It is the fundamental assumption of the managerial mind.

Over the Horizon toward Abundance and "Tragedy"

While medieval Europeans inland constructed new fisheries at home, coast dwellers were reaching out into new populations of marine fishes. The internal frontier of intensification was more than matched by external frontiers of "untapped resources." Rather than making new ecosystems, mariners and merchants found new ecosystems and pulled them into the task of satisfying western Christendom's hunger for fish. Like aquaculture, the new marine fisheries were powered by demand—but theirs was a less elite and potentially much larger demand. Like inland ponds, coastal fishing evolved from traditional to novel methods—but the coasts offered more than one possibility for innovation and the evolutions were as often institutional as technological. Like interior farmers and consumers, fishers and eaters of sea fishes learned different expectations—but "over the horizon" fewer lessons were in manipulating a familiar nature and more in beating out other humans to take a piece from a nature ever less connected to what fishers and consumers called home.

Carp culture made a single narrative. Regional marine ecosystems presented distinct and unequal opportunities to increase output. Coastal artisans extended fishing efforts offshore into untouched stocks, species, and ecosystems and extracted products they could preserve for distant consumer markets. Market-driven expansion generated complex organizations, conflicts, and a view of limitless further frontiers over the horizon.

New Marine Fisheries

From the south coast of the Baltic all the way around to the English Channel seasonal spawning concentrations of pelagic herring (*Clupea harengus*) offered rich food within sight of land. Estuarine stocks could be taken in fixed traps and weirs and those along open shorelines from small boats with light seines and drift nets. In about 1200 Danish chronicler Saxo bragged that the fish along the Scanian coast were so abundant that they blocked shipping and could be caught by hand.[101] Herring are a distinctly oily fish, especially when preparing to spawn, and they spoil in twenty-four hours. When promptly dusted with salt, smoked, or packed in simple salt brine, they become "powdered," "red," or "pickled" herring,[102] an inexpensive food, portable in baskets, kegs, or jugs, and palatable enough for several months—especially during the cold season after the autumn spawning. By the mid-twelfth century coastal artisans from Picardy to Pomerania were tapping the silvery billions to feed themselves, their neighbors, and nearby inland populations, especially in the fast-growing towns.

The twelfth-century herring industry had resulted from more than a century of parallel development by artisan fishers with access to local consumers and stocks of fish. A herring fair was operating at Fécamp by 1088 and inland Poles already thought Pomerania a land of fresh fishes. That was the likely origin of herring bones found in eleventh-twelfth century layers at Wrocław, and definitely where Silesian nuns about 1200 sent small riverboats for their fish.[103] Towns on the Flemish coast gained papal dispensation in 1169/70 to fish on Sundays while the herring run was on.[104] Already some decades earlier the count of Boulogne had given twenty thousand herrings a year to Cluny,[105] and in the early 1160s Alan of Lille called herring " the most common of fish, [which] by his wide availability relieved the hunger of the poor."[106] Present but not numerous in food remains from ninth–tenth century London and Norwich, at the latter town herring rise to three-quarters of fish remains from the twelfth century. Along the southern Baltic shore, herring bones are common in coastal settlements and less so—though not absent—inland.[107]

Subsequent thirteenth-century expansion centered on easily accessible schools, encouraging regional concentration of fishing, processing, and marketing activities. Two areas gained special repute, the southern North Sea region associated with Great Yarmouth in Norfolk, and the waters off southwestern Scania, where both sides of the Öresund were then Danish. In the North Sea large autumn schools of herring along the coast of East Anglia attracted fishers and buyers from England, Flanders, and France. Peak years of the herring fair at Yarmouth about 1340 drew some five hundred ships and fragmentary customs accounts show annual exports of seven hundred to eighteen hundred lasts (at twelve thousand herrings the last).[108] Much of the recorded catch went to Flanders. From what fishers of Calais scooped from the same waters they smoked ten million herrings in 1321 and exported twenty million. Quantities caught, packed, and shipped by Yarmouth and other southern centers fell noticeably after the 1360s.[109]

Following immemorial local subsistence use Scanian herring quickly commercialized after 1200 under the influence of merchants from north German towns. Amply provided with salt from Lüneburg, cooperage, and hungry buyers in central Europe, what would become the Hanse towns gained territorial concessions on the Skanör peninsula. During the herring run of August to October merchants on the beach bought herrings caught just offshore by five to seven thousand small boats of Danish peasants exploiting a gap in agricultural routine, and paid Danish and German women to gut, salt, and pack them. After sale at the September Skanör herring fair, the product was shipped south. The peak Scanian pack came in the 1360s to 1380s, annually surpassing eight thousand lasts or, by various modern estimates, ranging between ten and twenty-five thousand tonnes. Burgundian noble councillor Philippe de Mézières saw the area in 1389 and marveled: "the herring makes its passage through the strait from one sea to the other in marvellously great number . . . so great one could cut through them with a sword. . . . the bounty of God . . . furnishing an abundance of herrings by which all Germany, France, and England and many other countries are fed in Lent, for poor Christians can have a herring who cannot afford a big fish."[110] Whether from overfishing or the lowered salinity of a cooling Baltic, around 1416–25 the Scanian schools failed and neither they nor the trade ever regained strength.[111]

It is plausible to infer that by about 1300 effective mercantile organization had taken the herring industry of northwestern Europe as far as its traditional techniques could go. Coastal concentrations of fish could be exploited effectively with a modicum of operating capital, much of it going for salt, and supply inexpensive Lenten protein to large regional concentrations of humans. Dependence on a few spawning stocks left the industry susceptible to natural environmental fluctuations affecting the year-to-year presence of the schools. From the mid-thirteenth century continued pressures of demand drove herring prices rapidly upward. Those at Calais doubled from 1268 to 1300 and doubled again by 1342. Short-term changes in supply superimposed year-to-year price variations exceeding 25 percent.[112] The bottleneck was time, the need for quick processing which anchored this fishery within sight of the beach, and the limited durability of its product, which hindered access to consumers well inland. Pegolotti warned Italian merchant travelers to be sure they got herring only from the most recent pack.[113]

Technical and economic innovation at the end of the fourteenth century transformed the herring fishery.[114] Fishers and fish merchants from the northern Netherlands, hitherto minor actors in the herring fishery of the southern North Sea, worked out procedures for onboard curing of herring. They gutted the fish right away but incompletely, leaving the pyloric caecae (now known to contain an enzyme that retards spoilage), and packed them with salt. The product, *kaakharing,* had better eating and keeping quality. Hauling salt required a larger ship, soon built for the purpose as a "herring buss," but then a larger crew could stay at sea catching and packing herring for a week or more. The tether to shore undone, Dutch herring fishers could depart the contentious English shores for unexploited spawning stocks in the central and northern North Sea, and eventually off Shetland and Iceland. Higher salinity fed these fish to larger size than in the Baltic. Ostend alone landed twenty-four hundred tonnes in 1467.[115]

The Dutch backed up technical superiority with investments—salt, nets, ships, barrels—and government policy. "The greatest impetus to the use of all the superior methods was the presence of a market for the preserved herring and a market that had potential for growth."[116] When the herring were unloaded, some went directly to urban consumers. At Deventer, where next to no eating of herring is visible in food waste from the the ninth through fourteenth centuries, in the 1400s and 1500s herring contributed 15 percent of the recovered fishbones.[117] More of the catch was washed and layered in barrels with salt for an extremely durable product, edible for a year and more. Price advantages on the Antwerp market offered incentives to substitute Dutch for Hanseatic herrings, so by the mid-fifteenth century Holland and Zeeland were supplying the shores of the North Sea, the Rhineland, and, after failure of the Scanian spawning stock, even the Baltic. The Dutch remained Europe's master herringers long into the seventeenth century.[118]

European herring fisheries went from broad use of local coastal stocks in the twelfth century through dependence on Scanian and southern North Sea schools to exploitation of distant-water populations by the end of the Middle Ages. The evolution was closely linked to the market which called forth greater commercial complexity and levels of investment. Before following the barrels of repacked salt herrings on their journeys inland, there are other offshore opportunities to explore.

Codfishes (gadids) and their biological or culinary associates make a story with perhaps the collective weight of the herring and certainly greater geographic range. Atlantic ("northern") cod (*Gadus morhua*) is a common demersal[119] predator in cold waters of the continental shelf. Across the North Atlantic from the Baltic to North Carolina, the one species has many breeding populations (races) with behavioral adaptations to particular environments. Local stocks are ubiquitous in near-shore habitats, but the largest fish—to 1.5 meters and forty kilograms—remain in deeper waters of the shelf down to six hundred meters. Traditionally caught with baited hook and line, cod has a long history of human use. In Arctic Norway middens of cod bones go back to the Bronze Age.

Two salient features of cod in the north are not immediately visible:[120] (1) Adult members of the large Arctic population migrate each winter in a southwesterly direction along northern Norway into spawning areas south of Lofoten;[121] and (2) no salt is there needed to preserve cod. With heads removed, split to the tail, and hung on racks in the Arctic wind and sun the white flesh "becomes as dry as wood," as an Italian castaway in Norway described the imperishable "stockfish" (Norse *skreið*).[122] Stockfish makers discard the head and forward vertebrae and ship off to the consumer a bundle of dried slabs still connected to the rear part of the backbone. Hence production and consumption sites have distinctive patterns of fish remains. Since prehistory stockfish were central to subsistence strategies in northern coastal Norway and this potential store of surplus food likely played a role in the creation of Viking-age chieftancies there. Dried cod served local subsistence in post-settlement Iceland, too. When available, local cod, fresh, dry-salted, or salted, was also eaten along European coasts further south.[123]

Development of Norse society and the cultural integration of elites into Europe during the eleventh and twelfth centuries—a process symbolized after the 1030s by official Christianity—modified demand for northern goods in, at first, the Norwegian west (Bergen) and south (Oslofjord).[124] To the furs and hides of marine mammals which the Vikings sought, their descendants added fish. Early twelfth-century King Eystein, brother of a crusader to Jerusalem, imposed a tax in kind on northern fishing and soon thereafter signs appear of English and north German demand for Norwegian cod. Norwegians themselves carried stockfish to England well into the thirteenth century, but Hanseatic merchants came to Bergen for it. Present purposes are not served by giving details of how the Hanse established economic hegemony in later medieval Norway. But it is germane to observe that the Germans more than once violently drove their English rivals from Bergen and had their return outlawed. But Norwegian monarchs did sustain sullen German and English acquiescence in the Bergen staple, the rule barring foreigners from trading beyond that city to Lofoten a thousand kilometers north or, into the fifteenth century, to Iceland a thousand kilometers west. Norse stockfish went to European tables through Bergen.

When it comes to counting stockfish production and thus gauging the scale of the Norwegian cod fishery, medieval records are mute. Knowing that the value of stockfish entering late fourteenth--century Lübeck each year was one quarter that of the herring[125] simply omits too much. Extrapolations by Norwegian scholars make out something in the order of two to four thousand tonnes of *skreið* to provide half the value of Norway's exports, with half or more of that weight going to England.

Structural constants and shifts in the fishery are more informative.[126] Early trips by northerners to Bergen gave way in the 1200s to Bergen merchants dealing in north-country harbors. Only after Norway's mid-fourteenth century loss of economic and political strength could Hanseatic merchants successfully circumvent or defy Norwegian protectionism and deal directly with northern processors of the fish. Most business still occured in Bergen, whence men of Lofoten sailed each May, but late fourteenth- and fifteenth-century sources document numbers of illegal ventures to the far north. In both venues merchants advanced goods to northerners on credit for future delivery of fish, so establishing at least the threat of debt bondage in an enterprise as unstable as Arctic fishing.

Stockfish delivered to Bergen in summertime had dried in the subfreezing overnight temperatures of an Arctic spring. Men who were otherwise peasants fished through the late Arctic night and brief dawn of February and March. The inshore stock of smaller cods which had served the original subsistence fishery could not sustain commercial demand. Fishing zones left the coasts near the farms and moved offshore of southeastern Lofoten, where the big migratory cod arrived to spawn.

Northern regions better survived Norway's fierce late medieval economic and demographic crisis than did the south. Deteriorating climatic and economic conditions for northern agriculture and for a mixed subsistence strategy highlighted the continued strength of demand for fish. As the grain received in Bergen for a given weight of stockfish doubled and redoubled between 1350 and 1500, settlements in the far north moved toward the coast. Some places and men came to specialize in fishing. Land rents, which medieval Norwegians had calculated in weights of butter, were now—the first evidence is from 1432—expressed in weights of *skreið*.[127]

Icelanders, mid-Atlantic cousins to Norwegians, fished cod increasingly for subsistence but not for trade before their independent commonwealth acquiesced to Norwegian suzerainty in 1264. Export markets were attracting effort by 1300, when first isolated cryptic references, then direct evidence, establish Icelandic *skreið* as strong competition to those of Arctic Norway. As dried fish supplanted woolen homespun as the island's export staple, the fishery challenged pastoralism as an axis of Icelandic culture. At least for trading purposes Icelanders also gave up their traditional standard of value, a unit of cloth, and adopted a set quantity of dried fish. The price of *skreið* in Iceland rose by 25 percent during the first half of the fourteenth century and, after a plateau, by another 70 percent in the fifteenth. Such prospects of gain help explain why four of five animal remains in Icelandic archaeological sites from the fifteenth and sixteenth centuries are of fish, mostly head and anterior skeletal elements of cods.[128]

Distant resources were pulled into the service of European demand for fish and so distant societies were pulled in as well. Norwegian and Icelandic scholars advance the stockfish trade and export production as the chief economic elements integrating these peripheries into the developed and expanded Europe of the later Middle Ages. An equally valid interpretation is that Europe was reaching out, and this point of view reveals long-term impacts on a wider scale. King Håkon V acknowledged the driving force of European demand in the famine year 1316, when he required any merchant seeking dried fish or other goods from Norway to arrive with grain.[129] Just after 1400 English traders, their search for stockfish at Bergen thwarted by German competition, reached Iceland. So did English fishers, who in 1415 explained that failure of fisheries in home waters sent them to the abundant resources off the island.[130] In tacit, sometimes open, defiance of the Bergen staple and the express commands of Danish and English monarchs alike, both trade and fishery grew. English buyers offered twice what *skreið* directed to Bergen might get. English fishers arrived in late spring and stayed to midsummer, curing their catch on board and ashore. After 1420 some wintered over. There are grounds to think the English fished deeper water with longer lines and more hooks than did the natives.

Not all welcomed the English. Icelandic annals and Danish diplomatic correspondence indict English mariners for stealing fish, looting, violence, and even the murder of the Danish governor. Landowners fearing loss of labor banned full-time fishing as well as the overwintering foreigners who might offer such work. Meanwhile Hanseatic merchants and seamen also wanted what Icelandic waters offered and they boasted better official relations. A naval war ensued in the 1470s and 1480s, with armed fleets from Hamburg confronting English escort vessels. Icelandic sources record eight armed clashes between Germans and English between 1486 and 1532. English traders, many from Bristol, reduced operations. English fishers, many from the east coast, hung on into the 1520s, even though in 1500 Iceland declared English long-liners outlaw, freely open to attack.[131] In England and in Germany fishing communities had customs protecting and allocating the resources of home waters. Over the horizon the wealth of nature belonged to the one who could seize it.

South and west of the British Isles other offshore opportunities were opened in the later Middle Ages. They were less tied than the northern commercial fisheries to a single species[132]—though the close cod relative, hake (*Merluccius merluccius*) had a major role—and engaged the attention of more various coastal communities around the English Channel and Bay of Biscay. The new fisheries built on prior experience with inshore and estuarine resources. All participants need not be surveyed here, nor have all been so far studied by historians.

Along the English south coast from Exeter to Southampton increased landings of hake and cod are visible in the written and archaeological record of the fourteenth century.[133] Hake especially was here an offshore species caught during the March–June period in the western approaches to the Channel south and west of Cornwall. The same waters and species also became a target for Breton fishers, who then began to leave sight of land.[134] The Bretons sailed for a day, fished night and day for hake, then returned. In 1427 papal permission for Sunday departures legitimized boats making two voyages a week. Sheds and racks for drying hake were a lucrative seigneurial monopoly—papal purchasing agents shipped thousands of Breton hake to Avignon in the 1370s and 1380s—but the fishers themselves salted or brined mackerel and sardines. Only toward the end of the fifteenth century did Bretons make longer voyages toward the Cornish coast or elsewhere to catch cod, which were until then absent from the markets they served.

On another western tip of the European mainland, Spanish Galicia, an abrupt coastal topography and a narrow (twenty-to-thirty-kilometer) continental shelf put deep water unusually near shore. Earlier medieval inshore subsistence fisheries blossomed after Spanish Christians gained control of the south and opened regular sea routes to the Mediterranean.[135] Large-scale fishing in the later thirteenth century established a base for new coastal towns—La Coruña, Pontevedra, etc.—to integrate economic activities in each Galician estuary. Many aspects—not least the target species and ways of preservation—are reminiscent of stockfish and herring industries in the north. Hake (*pixota*) were caught offshore on baited lines from small boats. One royal charter made very plain that these were to be dried, not salted, for commerce. Sardines (*Sardina pilchardus*)—a pelagic plankton feeder in the same family as herring—frequent western Iberian estuaries and inshore waters during summer and autumn. Skilled crews hired by town-based companies took them in costly encirclement nets and delivered the catch to processing plants ashore, where women removed the heads and viscera before stacking the fish for dry salting (*sardina de pila*) or racking them to smoke before brining (*sardina arencada*). Full-time sardine workers for half the year might in the other season fish hake or conger but also tend fields and vineyards. Annual shipments of Galician sardines to Valencia alone in the 1490s were twice the size of Dieppe's best fifteenth-century herring catch; at least comparable quantities were going to Portugal (itself a sardine producer and exporter), to Andalucia, and to Barcelona.

Subsistence or artisanal fisheries with potential for expansion in the high and later Middle Ages were not necessarily all that far "offshore," at least from the viewpoint of the fishers. We can but mention two northern analogs—dried pike (called *strekfuss*) around the eastern Baltic[136] and salmon exports from Scotland[137]—and look more closely at one in the Mediterranean, tuna trapping. The latter industry, which swelled along the northern coast of Sicily after the thirteenth century, had—for all its use or revival of an ancient capture technique—much in common with the other new marine fisheries. The *tonnara* was a semipermanent installation at a chosen coastal site where a long leader fence perpendicular to shore diverted migrating schools of tuna (*Thunnus thynnus*) into successive impounding chambers, ending in the "death chamber" where men raised the net floor and took the fish with barbed harpoons. Documented enterprises rose from perhaps a dozen in the thirteenth century to about thirty in the fifteenth, when the fishery expanded westwards from Palermo to Trapani. Trapping rights belonged to the crown but individual sites were leased to entrepreneurs who selected a skilled, quasi-hereditary labor chief, the *rais,* to assemble and operate the catching crew. The team of skilled and unskilled sea-workers formed a share association for the duration of the tuna season, mid-April to late June. Shoreward of each *tonnara* salaried and wage laborers butchered the tunas and brined the meat. Investors from Genoa, Palermo, or other Sicilian towns financed the enterprise by advance purchase of its catch or its product, *tonnina,* pickled and barreled tuna meat of various grades. A *tonnara* killed one to two thousand fish a year, which processed out to the same number of barrels, about forty to eighty tonnes of *tonnina.* In peacetime from the early fourteenth century on, Palermo annually shipped thousands of barrels to Naples and Tuscany—where Pegolotti knew the fish in the 1320s—to Rome and Liguria, and by the fifteenth century, "Spanish style" *tonnina* to Seville.[138] As the best meat was deboned and the bones were ground for other uses, archaeology cannot directly verify the written record of this fishery.

Markets, Ecosystems, Expectations

The newly opened or expanded frontier fisheries of the later Middle Ages established distinctive relations between markets and ecosystems. Unlike the traditional local catch, fish from "over the horizon" rarely reached the consumer fresh.[139] Most of the output from the new marine fisheries was preserved for consumers distant in time and in place from the animals and ecosystems on which they fed. This had complex consequences.

The big commercial fisheries preserved fish with two major technologies, simply conceivable as "wet" and "dry."[140] Drying gadids or pike easily made a long-lasting product but the flesh of oily fishes—herring, sardine, tuna—quickly turns inedibly rancid on exposure to air. They need fast anaerobic treatment. The dominant method, wet brining, required economical access to copious supplies of salt. About 1400 it took a tonne of salt to pack every five tonnes of Scanian herring and more for the Dutch method; five hundred barrels of *tonnina* required 15.5 cubic meters of salt.[141] Further, the natural drying of fish has a strict climatic precondition of extreme low humidity, whether from freezing cold or blazing heat, which confined its economic application to far northern or southern regions. Elsewhere even non-oily fishes had to be dried under cover and with salt to help absorb their fluids. Such dry salting gave a less durable product and left the processor dependent on the salt trade.

In towns of any appreciable size, salted, dried, and brined fishes were the stock in trade of "salt-fishmongers" or "herringers" (*harengier, allecistae, śledzowniki*), retail merchants distinct from purveyors of fresh local fish.[142] While herring or sardines might be eaten straight from the container, other preserved fish also demanded special preparation. Eberhard, chef to Duke Henry "the Rich" of Bavaria-Landshut (1404–50) advised making "a good stockfish" by boiling it like veal and letting it stand in the liquid for half a week before removing and washing the fish and then warming it in a buttered pan with ginger and saffron. A Parisian householder of the 1390s alternated soaking the stockfish and beating it with a mallet.[143] But then medieval physicians and dietitians, Eberhard among them, condemned all preserved fish on medical grounds and medieval elites thought them a gruesome penance. Bishop Matthias von Rammung of Speyer left no doubt in his 1470 kitchen ordinance: "Herring shall not be considered a fish dish, [for] one can in no way make them acceptable."[144] Salt herring and dried cod were not high-class food.

Economic features of the offshore fisheries derived from their technologies and the scale of demand to which they responded. Large-scale commercial fishing was an economic dynamo. Offshore resources called for longer voyages, larger ships, and crews with new capabilities. It took the sturdier but more expensive "dogger" design, for instance, to project English fishing activity out of eastern coastal waters as far as Iceland[145] and the *buss* to exploit herring out in the North Sea. One result was greater output. Achieving the necessary regional mobilization of labor, skill, and capital had further strong positive feedbacks for growth and development. At herring-rich Dieppe in the 1460s the smell of fish and the saltworks filled the air, disgusting the members of a passing Czech embassy.[146] Successful fishing multiplied production of raw materials such as salt, of equipment (ships, nets), and of packaging (barrels). Preparation and distribution drew more capital investment, not only for Dutch herring but also earlier at Yarmouth and in Sicilian tuna traps.[147]

Forward linkages most plainly demonstrate the mass-market and long-distance commercial distribution of this commodity. Preserved fish was sold in bulk, "by the last and by the hundred" as some English accounts put it.

Indeed the standard measure for Galician sardines was the "thousand" by count, and actual shipments of sardines and herrings alike were, as already noticed, in many, many thousands. Some thirty million herrings are estimated to have entered Paris in 1321. Merchants from elsewhere could retail salt fish in Kraków for three days—though they could await a fish day—and thereafter sell only by the cask or keg. Standard barrels for *tonnina* held forty-five and sixty kilograms.[148]

The mass had social substance, too, with herrings acknowledged a fish of "the poor" by Alan of Lille in the mid-twelfth century and Philippe de Mézières at the end of the fourteenth. Portuguese King João II (1481–95) had the same view of sardines.[149] Actual rations suggest these men were correct. In 1390 the almoner to the king of France bought seventy-eight thousand herrings on the Paris market and distributed them to hospices and poor households. Police guards at Tours in 1480 were fed herring, dried hake, and ray, while their employers preferred carp, pike, and lamprey. Sardines were the only fish eaten by late medieval Portuguese workers and peasants. Recipients of charity dinners in Valladolid chewed hake and sardines.[150]

Rare records of legitimately comparable values give a quantitative sense. Clerks in twelfth-century Flanders assessed ten herrings as equal to eight eels or one plaice. Salt herring entering Paris in 1321 paid the lowest duty among fishes, half that on fresh herring or mackerel and one-fifth that on whiting.[151] When the bishop of Coventry bought food in 1461, herrings at four to the penny compared favorably to 1½d. for an eel; a stockfish at 3d. gave more for the money than a fresh tench at 6d. or a pike at 8–12d. Market regulations at Frankfurt am Main in 1487 priced a pound of herrings the same as two-thirds of a pound of salt cod, one-third of a pound of salmon, and a pound of beef.[152] No fish on medieval markets provided cheap calories, but preserved varieties did offer the most inexpensive fish people could buy.

These inexpensive bulk commodities were carried long distances to faraway consumers. Herring from Pomerania at nearest, and after 1300 more normally from Scania or the North Sea, were always the most common, even typical, fish on the market in Kraków. Herring were the only marine fish bought—in Lenten accounts from 1396, 1400, and 1401—by Kaspar von Schlandersberg, younger son of a gentry family in south Tirol, about as far from a live herring as anywhere in western Europe.[153] Among the best customers at Yarmouth were traders to Gascony.[154] "*Stofix,*" an elderly Parisian husband explained in the 1390s to his new young wife, came from the "frontiers of the sea" (*es marces de la mer*) and were dried so they kept a very long time.[155] Stockfish were the one fish of the sea in the oldest German-language cookbook known, the "Buoch von guoter spise" which a Würzburger compiled in 1350.[156] Traders in the English West Country contracted to ship many tons of hake and other fishes into Castile.[157] The first known consignment of fishes from the Atlantic to southern Italy in 1426—sixty dozen dried hake and 3,550 sardines[158]—pointed straight back to Galicia. But a verbal clue takes the linkage back further still. *Bacalao* itself—Spanish, Portuguese, and Italian (*baccalà*) for a dried gadid—was a term borrowed from Flemish (*bakkeljaw*) where it was formed by metathesis from *kabeljaw,* known since the mid-twelfth century as the lowland Germanic (Flemish/Dutch, Low German) name for a cod.[159] Though these fishes became staples of local cuisines, they never came from local waters.

As symbolic of the new offshore fisheries as the prosperous stench of Dieppe, then, must be those headless cod skeletons in Paris, while their heads moldered in heaps beside the Norwegian Sea. Quite without modern technologies the medieval fish trades separated the death of these animals from their ultimate consumption and disposal. Long before the nineteenth century medieval offshore fisheries, pressed by mass demand for the least expensive fish, distanced consumers from the animal and the producing ecosystem. At the same time, requirements for bulk preservation and marketing pushed the largest offshore fisheries toward dependence on abundant stocks of single species. This made them susceptible to both depletion and effects of natural environmental fluctuations, like the breakdown of the Scanian herrings and a possible collapse of southern coastal cod stocks in 1517.[160]

Market-driven expansion propelled offshore fisheries into new structures, some institutional, others mental. Elements of capitalist organization supplanted artisanal structures: offshore fishing evolved trading companies, not guilds of fishers. Patterns of mercantile investment have already been noticed. This not only provided resources for greater production, it tied producers more closely to fishing as an occupation. In Norway and Denmark specialized full-time fishers first appeared under Hanseatic influence in the fourteenth century.[161] Professionals worked the Sicilian tuna traps, too. The common mechanism for changed relationships was advance purchase of the catch, which capitalized a seasonal venture at the risk of debt bondage. Fishers were tied to distant markets and market values. At the same time advance purchase again confirms the eagerness of frontier traders—at Yarmouth, for a time in Lofoten, in Aberdeen, and in Palermo—to obtain fish for sale to consumers across Europe.

At a larger scale this binding to the market is called economic integration, which the Middle Ages experienced not as "globalization" but as "Europeanization." A central role for the fishery is especially argued for

North Atlantic communities—Norway, Iceland, other island societies—which joined western Christendom during this period and whose economic connection entailed exchange of staple resources for European manufactures.[162] Perhaps the clearest mark of economic integration is when it works to disadvantage, as when better and cheaper Dutch herring around 1400 pushed the Hanse's Scanian product off the Flemish market or when Norwegian cod prices collapsed after 1567 under competition from cheap barreled salt fish.[163] Dutch herring are, of course, a tale of seizing the mercantile center, not joining the periphery. Galician and Sicilian fisheries also deserve consideration in this larger perspective, and not merely as engines of local enterprise.[164]

The other side of the story is competition among user groups over rights and access to frontier resources. In struggles among market-driven fishers and merchants the "tragedy of the commons" was scripted offshore. Maritime rivalries among Norman, Breton, Gascon, Basque, Galician, and Portuguese fishers formed part of the prelude to Anglo-French wars of the 1290s. The North Sea was an arena for struggles between Norman, Picard, Flemish, Dutch, and English herring boats.[165] Anglo-Hanse conflict seems to have been especially intense, first in Norway and later in waters around Iceland, where native fishers also became engaged.[166] Anticipating Anglo-Icelandic relations of the 1970s, this might be called "Cod War I." Nor was the Mediterranean a zone of peace, even among Christian fishers. Mainland fishers and men from the island of Lerins long contested the intervening strait, while in 1469 boats from Antibes waged a "miniature naval battle" against their rivals from Cannes.[167] In contrast, for all the capitalist quality of the *tonnara,* participants in that fishery undertook to limit their take by using some locations only in alternate years.[168]

Could one attribute some of that difference to a separation even of fishers themselves in the distant-water fisheries, from the producing ecosystem which they exploited? The expectation is born offshore: the resource is superabundant, it needs only to be taken. This was the plain view of the English petitioners of 1415 who, having left depleted home waters, found "a great plenty of fish" off Iceland's shores.[169] Indeed to the extent they and other newcomers did penetrate depths beyond reach of less well-equipped Icelanders, they could have found something like pristine stocks. Squeezed again out of Iceland, where did Bristol interests go? Farther offshore. Just one step beyond Iceland are the banks of Newfoundland.[170] Hear (a Milanese agent report) what Bristol's hired navigator, Zuan Caboto, and his English shipmates said they had found in 1497: "they affirm that the sea is covered with fish which are caught not merely with nets but with baskets, a stone being attached to make the basket sink in the water, and this I heard the said Master Zoanne relate. And said Englishmen, his companions, say that they will fetch so many fish that this kingdom will have no more need of Iceland, from which country there comes a very great store of fish which are called stockfish."[171] There again was the unlimited abundance Philippe de Mézières and before him Saxo had seen at the peak of the Scanian herring fishery and in its earliest days. Over the horizon the frontier abounds with what is needed to satisfy European demand.

New fisheries, inland and frontier, formed a double response to the widening gap between medieval European demand for fish and the supply traditional artisans could pull from natural local fish populations and habitats. The response to elite inland demand for fresh fish was aquaculture, where Europeans learned to manipulate and humanize aquatic ecosystems and animals. The response to mass demand for inexpensive fish was frontier fisheries, where Europeans learned to push further, to invest more, and to fight other users for the biggest piece of a distant inexhaustible abundance.

NOTES

1. Jean Desse and Nathalie Desse-Berset, "Pêches locales, côtières ou lointaines: Le poisson au menu des parisiens du Grand Louvre, du 14ème au 18ème siècle," *Anthropozoologica* 16 (1992): 119–26; Miriam Sternberg, "L'approvisionnement de Paris en poisson au 16ème siècle: que disent les sources bibliographiques? Que peut on espérer des données ichtyofauniques des Jardins du Carrousel?" ibid., pp. 127–30.

2. Marie-Christine Marinval-Vigne, "Consommation d'animaux sauvages en milieu ecclésial à Orléans au XVIe siècle: Données archéozoologiques et livres de cuisine," in *Exploitation des animaux sauvages à travers le temps,* ed. Jean Desse and Frédérique Audoin-Rouzeau (Juan-les-Pins, 1993), pp. 474–83.

3. Frédérique Audoin-Rouzeau, *Ossements animaux du Moyen-Age au monastère de La Charité-sur-Loire* (Paris, 1986), pp. 146–47.

4. Wim Van Neer and Anton Ervynck, "Food Rules and Status: Patterns of Fish Consumption in a Monastic Community (Ename, Belgium)," *Archaeofauna* 5 (1996): 155–64.

5. Theo de Jong, "Huisdieren, jachtwild, vissen en weekdieren: Een weerspiegeling van gevarieerde maaltijden," in *Het Kasteel van Eindhoven: Archeologie, ecologie en geschiedenis van een heerlijke woning 1420–1676,* ed. Nico Arts (Eindhoven, 1992), pp. 214–31; and "Fish Consumption at Eindhoven

Castle: Archaeological Remains versus Historical Sources," in *Fish Exploitation in the Past: Proceedings of the 7th Meeting of the I.C.A.Z. Fish Remains Working Group,* ed. Wim Van Neer, Annales du Musée Royal de l'Afrique Central, Sciences Zoologiques, vol. 274 (Tervuren, 1994), pp. 129–38.

6. Dirk Heinrich, "Fish Remains of Two Medieval Castles and of an Urban Context: A Comparison," ibid., pp. 211–16.

7. Elisa Ferreira Priegue, *Galicia en el comercio maritimo medieval* (Santiago, 1988), pp. 667–72, 728–39.

8. Luis Martínez García, "La asistencia material en los hospitales de Burgos a fines de la edad media," in *Manger et boire au Moyen Age,* I. Colloque de Nice, 1982, Publications de la Faculté des lettres et sciences humaines de Nice, 1st ser., nos. 27–28 (Paris, 1984), pp. 352–55; D. Menjot, "Notes sur le marché de l'alimentation et la consommation alimentaire a Murcie a la fin du Moyen-Age," ibid., pp. 202–3; A. Rucquoi, "Alimentation des riches, alimentation des pauvres dans une ville castillane au XVe siècle," ibid., p. 302; Bernard Vincent, "Consommation alimentaire en Andalousie orientale (Les achats de l'Hôpital Royal de Guadix, 1581–1582)," *Annales: Economies, sociétés, civilisations* 30, nos. 2–3 (1975): 449–53.

9. Henri Bresc, "La pêche et les madragues dans la Sicile médiévale," in *L'Homme méditerranéen et la mer,* ed. Micheline Galley & L. L. Sebai (Tunis, 1985), pp. 168–69.

10. Some of the following appeared in my "Economic Development and Aquatic Ecosystems in Medieval Europe," *American Historical Review* 101 (1996): 631–69, and is elaborated in my forthcoming book *The Catch: Medieval European Fisheries and the Antecedents of Today's World Fisheries Crisis: An Essay in Environmental History.*

11. See details in my chapter, "Medieval Fishing," in *Working with Water in Medieval Europe: Technology and Resource-use,* ed. Paolo Squatriti (Leiden, 2000), pp. 331–93.

12. The subject, poorly known elsewhere, has received much attention, often repetitive, from regional writers on aquaculture and history: Jiří Andreska, *Vývoj rybářství,* Zemědělské muzeum Ohrada Hluboká n/Vlt., Průvodce expozici (Prague, [1977]), (not paginated); Jiří Andreska, "Development of Fish-Pond Culture in Bohemia," in *The Fishing Culture of the World: Studies in Ethnology, Cultural Ecology and Folklore,* ed. Bela Gunda, vol. 1 (Budapest, 1984), pp. 77–89; and Jiří Andreska, *Rybářství a jeho tradice* (Prague, 1987), pp. 32–51; Herbert Knittler, *Nutzen—Renten—Erträge: Struktur und Entwicklung frühneuzeitlicher Feudaleinkommmen in Niederösterreich,* Sozial- und wirtschaftshistorische Studien, vol. 19 (Vienna, 1989), pp. 146–81; Stanislaw Brzozowski and Mieczyslaw Tobiasz, "Z dziejów rybactwa małopolskiego," *Studia i Materialy z Dziejów Nauki Polskiej,* ser. B: *Historia*

nauk biologicznych, vol. 9 (Warsaw, 1964), pp. 11–37; Stefan Górzyński, *Zarys historii rybołówstwa w dawnej Polsce* (Warsaw, 1964), pp. 41–50, 59–60; Rudolf Hurt, *Dějiny rybnikařství na Moravě a ve Slezsku,* 2 vols. (Ostrava, 1960), 1:53–219; Aleksander Nyrek, *Gospodarka rybna na Górnym Śląsku od połowy XVI do połowy XIX wieku* (Wrocław, 1966), pp. 89–154; Wojciech Szczygielski, *Gospodarka stawowa na ziemach południowo-zachodniej Rzeczypospolitej w XVI–XVIII w.: Studium z dziejów postępu w dawnym gospodarstwie wiejskim,* Prace Instytutu Historycznego Uniwersytetu Lódzkiego, no. 11 (Lódz, 1965), and *Z dziejów gospodarki rybnej w Polsce w XVI–XVIII wieku,* Instytut Historii Kultury Materialnej Polskiej Akademii Nauk, Studia z Dziejów Gospodarstwa Wiejskiego, vol. 9, no. 2 (Warsaw, 1969). Compare Brian K. Roberts, "Fish Culture in Sixteenth-Century Poland," *Agricultural History Review* 16 (1968): 161–63.

13. Jan Dubravius, *De piscinis ad Antonium Fuggerum* (Vratislaviae: Andreas Vinglerus excudebat, 1547); also published as *De Piscinis et Piscium qui in eis aluntur naturis libri quinque . . ad illustrem virum Antonium Fuggerum. Item Xenocratis de alimento ex aquatilibus Graece & Latine nun primum aeditus: cum scholijs Conradi Gesneri* ([Zürich:], 1559), which further appeared in 1596, 1623, and 1671; the modern edition is ed. Anežka Schmidtová in *Sborník filologický ČSAV,* 1, suppl. 1 (1953): 11–45. The English translation, *A New Book of Good Husbandry* (London, 1599), is available as a facsimile reprint in *Three Books on Fishing (1599–1659),* intro. J. Milton French (Gainesville, Fla., 1962). A Polish translation by Andrzej Proge was *Janusza Dubrawiusza O Rybnikach y Rybach* (Kraków: w drukarnia Wojciecha Siekielowica [ca. 1600]).

14. Olbrycht Strumienski, *O sprawie, sypaniu, wymierzaniu i rybnieniu stawów* (Kraków, 1573; 2d ed., Kraków, 1897). For both printed handbooks the source analysis and review by Stefan Inglot and Aleksander Nyrek, "Jana Dubraviusa i Olbrychta Strumienskiego dziala o gospodarce rybnej Studium porówznawcze," in *Studia z dziejów polskich i czechosłowackich,* ed. Ewa and Karol Maleczyński, vol. 1 (Wrocław, 1960), pp. 249–82, remains authoritative.

15. Andreska, *Rybářství,* pp. 49–51, treats economic instructions written 1525 by Vojtěch z Pernštejna, "Instrukce rybní pro panství Potštýnské a Litické," and a 1540 agricultural manual by Jan Brtvína z Ploskovic with chapters on pond culture. I am preparing to edit two texts hitherto unknown to scholars: Trier, Stadtbibliothek, Hs 608 (1954) [1958/1422 4°], fols. 1–38, "Friedrich von Flersheim, Fischbuch 1530," and Vienna, Österreichische Nationalbibliothek, Codex vindobonensis 13103, "Vischpuech," internally dated to 1542.

16. Jan Malecki, ed., *Lustracja województwa krakowskiego 1564,* 2 vols., Lustracje Dóbr Królewskich, XVI–XVIII wieku, Małopolska (Warsaw, 1962–

64), 2:225–26 et passim.

17. Roman Rybarski, *Gospodarstwo Księstwa Oświęcimskiego w XVI wieku,* Polska Akademia Umiejętności: Rozprawy Wydziału Historyczno-Filozoficznego, 2d ser., vol. 43, no. 2 (Kraków, 1931), pp. 56–84.

18. Josef Šusta, *Fünf Jahrhunderte der Teichwirtschaft zu Wittingau: Ein Beitrag zur Geschichte der Fischzucht mit besonderer Berücksichtigung der Gegenwart* (Stettin, 1898), pp. 1–29.

19. Besides works in note 12 above see Willy Cnopf, *Die Entwicklung und wirtschaftliche Bedeutung der Teichwirtschaft in Mittelfranken* (Nürnberg, 1927), also published as an article with same title in *Archiv für Fischereigeschichte* 11 (1927): 26–82.

20. *Abramis brama* is a large but slow-growing cyprinid (member of the carp family) native to still waters across Europe north of the Alps and Pyrenees.

21. *Carassius carassius* is a slow-growing middle-sized cyprinid closely akin to goldfish (*Carassius auratus*). This Central European native tolerates both low oxygen and cold winter conditions better than carp or bream.

22. Unlike trout or salmon, for instance, cyprinids do not naturally pair up and make nests for their eggs, but groups of ripe females and males gather in vegetated shallows and emit eggs and milt simultaneously, with much splashing and mixing, after which the adults leave the shallow nursery habitat.

23. Nyrek, *Gospodarka rybna,* pp. 29–38, thinks the Polish capital thus received about 50–90 percent of Upper Silesian output.

24. Paul Benoit and Monique Wabont, "Mittelalterliche Wasserversorgung in Frankreich: Eine Fallstudie: Die Zisterzienser," in *Die Wasserversorgung im Mittelalter,* ed. Klaus Grewe, for the Frontinusgesellschaft, vol. 4 of *Geschichte der Wasserversorgung* (Mainz, 1991), pp. 189–96; G. de Gislain, "Le rôle des étangs dans l'alimentation médiévale," in *Manger* 1 (1984): 89. Writers on ponds and fish culture in medieval France are commonly unaware of their counterparts in late medieval and early modern East Central Europe.

Reasoning from silence leaves discomfort. It is good to have the lack of solid evidence for early medieval fish culture confirmed by the likes of Heinrich Fichtenau, *Living in the Tenth Century: Mentalities and Social Orders,* tr. Patrick J. Geary (Chicago, 1991), p. 281. Likewise the rich records of Cluny at its height of wealth and prestige up to the 1150s refer to catching and temporarily storing fish from the Saône and its tributaries, but have no trace of rearing them in fishponds. See Ulrich of Cluny, "Consuetudines Cluniacensis," (*PL,* vol. 149, col. 703); administrative instructions, surveys, and accounts in Auguste Bernard and Alexandre Bruel, eds., *Recueil des chartes de l'abbaye de Cluny,* 6 vols. (Paris, 1876–1903), nos. 3789, 3790, 4132, and 4143; Georges Duby, "Le budget de l'abbaye de Cluny entre 1080 et 1155: Économie domaniale et économie monétaire," *Annales: Économies, sociétés, civilisations* 7 (1952): 155–71, and "Un inventaire des profits de la seigneurie clunisienne a la mort de Pierre le Vénérable," *Studia Anselmiana* 40 (1956): 129–40 (both reprinted in his *Hommes et structures du moyen âge: Recueil d'articles* [Paris, 1973], pp. 61–82, 87–101).

25. Raymond Delatouche, "Le poisson d'eau douce dans l'alimentation médiévale: L'exemple du Maine," *Bulletin philologique et historique (jusqu'à 1610) du comité des travaux historiques et scientifiques,* 1967, vol. 1 (Paris, 1969), pp. 174–75, and works there cited; Roland Sanfaçon, *Défrichements, peuplement et institutions seigneuriales en Haut-Poitou du Xe au XIIIe siècle* (Québec, 1967), pp. 26, 85–89; George T. Beech, *A Rural Society in Medieval France: The Gâtine of Poitou in the Eleventh and Twelfth Centuries* (Baltimore, 1964), pp. 38, 106.

26. Gislain, "Le rôle des étangs," p. 95, n. 5; Guy Devailly, *Le Berry du Xe siècle au milieu du XIIIe: Étude politique, religieuse, sociale et économique* (Paris, 1973), p. 298, and compare pp. 295–96.

27. François Blary, *Le domaine de Châalis, XIIe–XIVe siècles: Approches archéologiques des établissements agricoles et industriels d'une abbaye cistercienne,* Comité des Travaux Historiques et Scientifiques, Mémoires de la Section d'Archéologie et d'Histoire de l'Art, vol. 3 (Paris, 1989), pp. 31–40; Jean Richard, "Le commerce du poisson en Bourgogne et les étangs de la région Autunoise," *Mémoires de la Société Éduenne* 44 (1983): 181–97, and "Les étangs et le commerce du poisson en Bourgogne," in *L'eau en Bourgogne,* Cahiers du Centre d'Etudes Regionales de Bourgogne, no. 3 (Dijon, 1986), p. 99.

28. Roger Grand and Raymond Delatouche, *L'agriculture au moyen âge de la fin de l'empire romain au XVIe siècle,* vol. 3 of *L'Agriculture à travers les âges,* (Paris, 1950), p. 541.

29. Devailly, *Le Berry,* pp. 556–57.

30. Walter Maas, *Les moines-défricheurs: Etudes sur les transformations du paysage au Moyen-Age aux confins de la Champagne et de la Lorraine* (Moulins, 1944), pp. 74–75.

31. Gislain, "Le rôle des étangs," p. 95, n. 6.

32. Felix Bourquelot, "Fragments de Comptes du XIIIe siècle," *Bibliothèque de l'École des Chartes* 24 (1863): 66–68, where the scribe did not say what kind of fish were stocked.

33. Eusèbe J. de Laurière et al., eds., *Ordonnances des roys de France de la troisième race recueillés par ordre chronologique,* 21 vols. (Paris, 1723–1849), 1:55–56.

34. Audoin, *Ossements animaux,* p. 147, discusses fish remains from the floor of an eleventh- to twelfth-century refectory.

35. J. M. Steane, "The Royal Fishponds of Medieval England," in *Medieval Fish, Fisheries and Fishponds,* ed. Michael Aston, BAR British series, vol. 182 (Oxford, 1988), pp. 39–68, and notably p. 45; J. McDonnell, *Inland Fisheries in Medieval Yorkshire, 1066–1300,* University of York, Borthwick Institute of Historical Research, Borthwick Papers, no. 60 (York, 1981), pp. 9–12; Edward Roberts, "The Bishop of Winchester's Fishponds in Hampshire, 1150–1400: Their Development, Function and Management," *Proceedings of the Hampshire Field Club and Archaeological Society* 42 (1986): 125–38.

36. *Fleta,* ed. H. G. Richardson and G. O. Sayles, Selden Society, vol. 72 (London, 1955), II.73, p. 247. Contra Grand and Delatouche, *L'agriculture,* p. 544, the text has no carp.

37. A systematic survey and interpretation of the material and verbal evidence is Richard C. Hoffmann, "Remains and Verbal Evidence of Carp (*Cyprinus carpio*) in Medieval Europe," in Van Neer, *Fish Expoitation,* pp. 139–50, later slightly revised and augmented in my "Environmental Change and the Culture of Common Carp in Medieval Europe," *Guelph Ichthyology Reviews* 3 (1995): 57–85.

38. Eugene K. Balon, "The Common Carp, *Cyprinus carpio*: Its Wild Origin, Domestication in Aquaculture, and Selection as Colored Nishikigoi," *Guelph Ichthyology Reviews* 3 (1995): 1–55.

39. Hildegard of Bingen, *Physica* [also known as *Liber simplicis medicinae*],V.11, as "Subtilitatum diversarum naturarum creaturarum libri novem" in *Sanctae Hildegardis abbatissae opera omnia* in *PL,* vol. 197 (Paris, 1882), col. 1277.

40. Bourquelot, "Fragments," p. 71.

41. R. de Lespinasse and F. Bonnardot, eds., *Les métiers et corporations de la ville de Paris: Le Livre des métiers d'Etienne Boileau* (Paris, 1879), pp. 213, 216. The same rules also appear in a royal ordinance for marketing fish in Paris, which Laurière et al., *Ordonnances,* 2:583–86, gives from a confirmation of 1320, after (p. 575) expressing doubt about the 1254/58 date given that text by Nicholas Delamare, *Traité de la police,* vol. 3 (Paris, 1719), pp. 298–302. Delamare, however, had access to manuscript sources later lost: see Caroline Bourlet, "L'approvisionnement de Paris en poisson de mer aux XIVe et XVe siècles, d'après les sources normatives," *Franco-British Studies* 20 (1995): 6–7, and Bernadette Auzary-Schmaltz, "Les Contentieux en matière d'approvisionnement, d'après les registres du Parlement de Paris (XIVe–XVe siècles)," ibid., p. 60. Philip IV's 1289 fisheries ordinance set

the value for carp at two per denier: H. Duplés-Agier, "Ordonnances inédites de Philippe-le-Bel et de Philippe-le-Long sur la police de la péche fluviale," *Bibliothèque de l'Ecole des Chartes* 14 (1852): 51.

42. Comparable lists of fishes assembled by learned men in this same region during the twelfth and early thirteenth centuries say nothing of carp: Gui of Bazoche, "Epistola 23," in *Liber epistularum Guidonis de Basochis,* ed. Herbert Adolfsson, Acta Universitatis Stockholmiensis: Studia Latina Stockholmiensis, vol. 18 (Stockholm, 1969), pp. 89–99; Alexander Neckham, *De natura rerum libri duo: With the Poem of the Same Author De laudibus divinae sapientiae,* ed. Thomas Wright, Chronicles and Memorials of Great Britain and Ireland during the Middle Ages (Rolls Series), vol. 34 (London, 1863), II.22–46, pp. 142–58. Nor did carp occur among the thirty-three fishes mustered for allegorical war in the French "Battle of Lent and Carnival," assembled in Picardy or Normandy at the start of the thirteenth century: Grégoire Lozinski, ed., *La Bataille de Caresme et de Charnage* (Paris, 1933), p. 121.

43. Thomas of Cantimpré [Thomas Catimpratensis], *Liber de natura rerum,* VII.23, *Editio princeps secundum codices manuscriptos,* pt. 1: *Text,* ed. H. Boese (Berlin: 1973), pp. 258–59; Vincent of Beauvais [Vincentius Bellovacensis], *Speculum naturale,* XVII.40, vol. 1 of his *Speculum quadruplex sive speculum maius,* 4 vols. (Duaci: Baltazaris Belleri, 1624; repr. Graz, 1964–65), col. 1274; Albert the Great [Albertus Magnus], *De animalibus,* XXIV.26, ed. Hermann Stadler as *De Animalibus libri XXVI nach der Cölner Handschrift,* Beiträge zur Geschichte der Philosophie des Mittelalters, vols. 15–16 (Münster, 1916–20), pp. 1525–26. Intellectual context is provided by Christian Hünemörder, "Die Geschichte der Fischbücher von Aristoteles bis zum Ende des 17. Jahrhunderts," *Deutsches Schiffahrtsarchiv* 1 (1975): 188–93.

44. Carp recipes and menus are to be found in the oldest thirteenth-century form of what would come to be called the "Viandier" of Taillevent (Terence Scully, ed. and tr., *The Viandier of Taillevent: An Edition of all Extant Manuscripts* [Ottawa, 1988], see pp. 1–31 and notably recipe 102); in the "Enseignements" of ca. 1300 (in Lozinski, *Bataille,* pp. 185–86), and in the literary satire on court customs written in Paris in 1316, the "Roman du Comte d'Anjou," line 1132 (ibid., pp. 192–93; see commentary by A. Planche, "La table comme signe de la classe: Le témoignage du Roman du Comte d'Anjou [1316]," in *Manger* 1 [1984]: 253–54).

45. Some examples: D. D. Brouwers, ed., *Cens et rentes du Comté de Namur au XIIIe siècle,* vols. 1 and 2, pts. 1 and 2 of *L'administration et les finances du Comté de Namur du XIIIe au XVe siècle, Sources,* Documents inédits relatifs à l'histoire de la Province de Namur publiés par ordre du Conseil Provincial (Namur, 1910–26), vol. 2, pt. 2, pp. 433–34; Isabella Guérin, *La vie rurale*

en Sologne aux XIVe et XVe siècles, École pratique des hautes études, VIe Section, Centre de recherches historiques, Les hommes et la terre, no. 5 (Paris, 1960), pp. 131–42; Richard C. Hoffmann, "*Carpes pour le duc . . . :* the operation of fish ponds at La Perriére-sur-Saône, Burgundy, 1338–52," *Archaeofauna* (Madrid) 4 (1995): 33–45; Richard, "Le commerce du poisson," pp. 188–90; G. Sivery, *Structures agraires et vie rurale dans le Hainaut à la fin du moyen âge* (Lille, 1977), p. 104, and works there cited; Léo Verriest, ed., *Corpus des records de coutumes et des lois de chefs-lieux de l'ancien Comté de Hainaut,* Sources de l'histoire du droit rurale (Mons, 1947), pp. 4–5, and *Le polyptyque illustré dit "Veil Rentier" de messire Jehan de Pamele-Audenarde (vers 1275)* (Brussels, 1950).

46. Thomas H. Turner, ed., *Manners and Household Expenses of England in the Thirteenth and Fifteenth Centuries Illustrated by Original Records* (London, 1841), pp. 560–64 (management records for the Duke of Norfolk's fishponds, 1462–72). That firm reference antedates the carp served at King Richard III's coronation banquet in 1483 (Anne F. Sutton and P. W. Hammond, eds., *The Coronation of Richard III: The Extant Documents* [Gloucester, 1983], pp. 294–95, 300) and the ones mentioned in the "Treatyse of fysshynge wyth an angle" printed by Wynkyn de Worde in the 1496 second "Book of St. Albans," but the piece of the Treatyse which might have mentioned carp is missing from the surviving midcentury manuscript, now Yale Beinecke Library MS 171: see John D. McDonald, *The Origins of Angling* (Garden City, N.Y., 1963), pp. 214–15. The so-called "Ancient Cookery" which the Society of Antiquaries published from Royal Society MS Arundel no. 344 in *A collection of Ordinances and Regulations for the Government of the Royal Household, made in divers reigns. Also Receipts in ancient Cookery* (London, 1790), pp. 423–76, said to be a copy done shortly after 1399 of an older text from the English royal household, offers the choice "baken breme or carpe" in one menu for a fish day (p. 449); failing modern scholarly verification of the manuscript's date and originality, this easily interpolated reference must be thought dubious, especially because carp is not in the two fish-rich recipe collections which are convincingly linked to the court of Richard II: Constance B. Hieatt and Sharon Butler, eds., *Curye on Inglysch: English Culinary Manuscripts of the Fourteenth Century (Including the Forme of Cury),* Early English Text Society, 2d ser., vol. 8 (London, 1985), pp. 19–30, 81–91, 93–145. Nor are carp found among fish remains from twelfth- to fourteenth-century St. Albans (Dale Serjeantson, personal communication) or other contemporary and likely English sites; the earliest so far known come from an early sixteenth-century midden in Surrey: Andrea Bullock, "Evidence for Fish Exploitation from Tudor Deposits Excavated at Little Pickle, Surrey/England, in 1989," in *Archaeo-Ichthyological Studies: Papers*

Presented at the 6th Meeting of the I.C.A.Z. Fish Remains Working Group, Schleswig, ed. Dirk Heinrich, *Offa* 51, suppl. (Neumünster, 1994), pp. 267–75. Factual errors mar the interesting discussion in Christopher K. Currie, "The Early History of the Carp and its Economic Significance in England," *Agricultural History Review* 39 (1991): 97–107.

47. Annie Grant, "Food, Status and Religion in England in the Middle Ages: An Archaeozoological Perspective," in *L'animal dans l'alimentation humaine: Les critères de choix,* ed. Liliane Bodson, Actes du colloque international de Liège, 26–29 novembre 1986, *Anthropozoologica,* suppl. 2 (Paris, 1988), pp. 141–43.

48. But not, significantly, into Spain (where carp was a sixteenth-century introduction), nor into Italy. Italian cookbooks of the fourteenth century lack carp, and the first occurs in the mid-fifteenth-century "Libro de arte coquinaria" by the Lombard Maestro Martino, in *Arte della cucina: Libri di recetti, testi sopra lo scalco, il trinciante e i vini dal XIV al XIX secolo,* ed. Emilio Faccioli, 2 vols. (Milan, 1966), 1:19–204, notably p. 194, and *Anonimo meridionale: Due libri di cucina,* ed. Ingemar Boström, Acta universitatis Stockholmensis, Romanicia Stockholmiensia, vol. 11 (Stockholm, 1985); or, marginally, in a menu of the Savoyard Chiquart from 1420 (Terence S. Scully, ed., "*Du fait de cuysine* par Maistre Chiquart 1420 [Ms. S 103 de la bibliothèque Supersaxo, à la Bibliothèque cantonale du Valais, à Sion]," *Vallesia* 40 [1985]: 135).

49. Pierre Gresser and J. Hintzy, "Les étangs du domaine comtal en Franche-Comté d'après les comptes de gruerie du XIVe siècle," in *Société d'émulation du Jura: Travaux 1975–1976* (Dôle, 1978), pp. 129–56.

50. On the origins of ponds in the Dombes, writings by Paul Perceveaux, "Essai sur l'origine des Etangs de Dombes," *Bulletin de la Société des naturalists et archéologues de l'Ain* (1962): 81–90, and "Structures et relations économiques en Dombes à la fin du Moyen Age," *Cahiers d'histoire* 12 (1967): 339–57, supersede that by M.-C. Guigue, *Essai sur les causes de la dépopulation de la Dombes et l'origine de ses étangs* (Trévoux, 1908), while the long-lasting pond-based society there is peculiarly accessible in Wilhelm Egloff, *Le paysan dombiste: Etude sur la vie, les travaux des champs et le parler d'un village de la Dombes: Versailleux (Ain),* Société de Publications Romanes et Françaises sous la direction de Mario Roques, vol. 20 (Paris, 1937), pp. 84–94. Conditions in the Forez are treated in Vincent Durand, "De l'établissement des étangs dans la plaine du Forez," *Bulletin de la Diana* 5 (1889–90): 101–8, J. de Fréminville, "Les comptes du maître des étangs de Forez de 1400 à 1430," *Bulletin de la Diana* 7 (1893–94): 5–20; and Etienne Fournial, *Les villes et l'économie d'échange en Forez aux XIIIe et XIVe siècles* (Paris, 1967), pp. 687–90; and become an intriguing case study for Fernand

Braudel in *The Identity of France,* vol. 1: *History and Environment,* tr. Siân Reynolds (New York, 1989), pp. 204–6.

51. Frans van Mieris, ed., *Groot charterboek der graaven van Holland, van Zeeland en Heeren van Vriesland,* 4 vols. (Leyden, 1753–56), 2:656; compare finds reported in D. C. Brinkhuizen, "Preliminary Notes on Fish Remains from Archeological Sites in the Netherlands," *Palaeohistoria* 21 (1979): 83–90, and "Visresten uit twee middeleeuwse vindplaatsen te Leeuwarden," in *Planten- Vis- en Vogelresten uit vroeghistorisch Leeuwarden* (Leeuwarden, 1983), pp. 19–20; Theo de Jong, "Tolbrugstraat, Breda, de botresten: Resultaten van het archeologische botrestenonderzoek" (Department of Archeology, Municipality of Breda, 1988); Anton Ervynck and Wim Van Neer, "A Preliminary Survey of Fish Remains in Medieval Castles, Abbeys and Towns of Flanders (Belgium)," in Heinrich, *Archaeo-Ichthyological Studies,* p. 304; M. Seeman, "Monnickendam: The Faunal Remains," in *Medemblik and Monnickendam: Aspects of Medieval Urbanization in Northern Holland,* ed. H. A. Heidinga and H. H. van Regteren Altena, Cingula, vol. 11 (Amsterdam, 1989), pp. 125–34.

52. Findings of Alfred Heidacher, *Die Entstehungs- und Wirtschaftsgeschichte des Klosters Heilsbronn bis zum Ende des 15. Jahrhunderts: Gründung, Gründer, Wirtschafts- und Verfassungeschichte* (Bonn, 1955), pp. 118–19, need to be adjusted in light of management information from 1338 to 1374 in Staatsarchiv Nürnberg, Klosterverwalteramt Heilsbronn, Rechnungen, vol. 1, and the chronology of land acquisition revealed by charters published in Günther Schuhmann and Gerhard Hirschmann, eds., *Urkundenregesten des Zisterzienserklosters Heilsbronn,* pt. 1: *1132–1287,* Veröffentlichungen der Gesellschaft für fränkische Geschichte, 3d ser., vol. 3 (Würzburg, 1957). Cnopf, *Teichwirtschaft,* pp. 26–82, treats sixteenth-century operations of the seventy-one ponds Heilsbronn then controlled.

53. František Graus, *Dějiny venkovského lidu v Čechách v době předhusitské,* 2 vols. (Prague, 1953–57), 2:32–35, 345–46, 483–86; Šusta, *Teichwirtschaft,* p. 2, provides the statement of Charles.

54. The general accounts of Andreska, *Vývoj rybářství,* and "Development of Fish-Pond Culture," pp. 77–80, and Adolf Kalný, "Stará práva rybolovu na tekoucích vodách bývalého třeboňského panství," in Ceskoslovenská akademie zemědělská, *Současnost, perspektivy rozvoje a tradice československého rybářství* (Prague, 1978), pp. 67–74, need balance with the well-grounded historical evidence in Kateřina Charvátová, "Manorial Farms of Cistercian Abbeys of Mediaeval Bohemia," in *Historia i kultura cystersów w dawnej polsce i ich europejskie związki,* ed. Jerzy Strzelczyk (Poznań, 1987), p. 133; Augustin Neumann, *Prameny k dějinám duchovenstva v době předhusitské a Husově* (Olomouc, 1926), no. 20, pp. 130–33; Josef Šusta, *Purkrabské účty panství Novohradského z let 1390–91,* Historický archiv, vol. 35 (Prague, 1909), p.

34; and Josef Hemmerle, ed., *Die Deutschordens-Ballei Böhmen in ihren Rechnungsbüchern 1382–1411,* Quellen und Studien zur Geschichte des Deutschen Ordens, vol. 22 (Bonn, 1967), pp. 139, 154. Note that the signs of advanced fish culture appearing on Teutonic Order estates in Bohemia after the mid-1380s (compare ibid., pp. 46–71) and before 1410 coincide with construction of carp ponds at the grand master's palace in Marienburg, Prussia, in or just before 1396: compare A. Sielmann, ed., "Die Reste des Marienburger Konventsbuch aus den Jahren 1395–1398," *Zeitschrift des Westpreußischen Geschichtsvereins* 60 (1920): 70.

55. Šusta, *Teichwirtschaft,* p. 4. The chronology in Lower Austria paralleled that in Bohemia: Knittler, *Nutzen—Renten—Erträge,* pp. 152–58.

56. Richard C. Hoffmann, *Land, Liberties, and Lordship in a Late Medieval Countryside: Agrarian Structures and Change in the Duchy of Wrocław* (Philadelphia, 1989), pp. 365–67.

57. Szczygielski, *Gospodarka stawowa,* pp. 21–23.

58. P. Beck, "L'approvisionnement en Bourgogne ducale aux XIVème et XVème siecles," in *Manger* 1 (1984): 175–76; Hoffmann, "*Carpes pour le duc . . .*"; Monique Sommé, "L'alimentation quotidienne à la Cour de Bourgogne au milieu du XVe siècle," in *Bulletin philologique et historique (jusqu'à 1610) du Comité des travaux historiques et scientifiques,* 93e Congrès national des Sociétés savantes, Tours, 1968, vol. 1: *Les problèmes de l'alimentation* (Paris, 1971), pp. 103–17.

59. Steane, "Royal Fishponds," pp. 49–50; Roberts, "Fish Culture"; McDonnell, *Inland Fisheries*; C. J. Bond, "Monastic Fisheries," in Aston, *Medieval Fish,* pp. 92–95; Christopher K. Currie, "The Role of Fishponds in the Monastic Economy," in *The Archaeology of Rural Monasteries,* ed. Roberta Gilchrist and Harold Mytum, BAR British Series, vol. 203 (Oxford, 1989), pp. 153–60; compare Franciszek Piekosiński, ed., *Rachunki dworu króla Władysława Jagiełły i królowej Jadwigi z lat 1388 do 1420,* Monumenta medii aevi historica res gestas poloniae illustrantia, vol. 15 (Kraków, 1896), pp. 545 et passim.

60. Joseph Balon, "La pêche et le commerce du poisson dans le comté de Namur au moyen âge," *Namurcum* 19 (1942): 31.

61. Herbert Hitzbleck, "Die Bedeutung des Fisches für die Ernährungswirtschaft Mitteleuropas in vorindustrieller Zeit unter besonderer Berücksichtigung Niedersachsens" (Diss., Göttingen, 1971), pp. 105–8.

62. Pierre Couperie, "Les marchés de pourvoierie: Viands et poissons chez les Grands au XVIIe siècle," in *Pour une histoire de l'alimentation,* ed. Jean-Jacques Hémardinquer (Paris, 1970), pp. 245–46.

63. Artificial ponds built near Tegernsee in 1455 were soon abandoned because the brethren found the fish "too mossy" beside their familiar white-

fish and trout: Johann N. Kißlinger, *Chronik der Pfarrei Egern am Tegernsee,* Oberbayerisches Archiv für vaterländische Geschichte, vol. 52, pt. 3 (Munich, 1907), p. 96.

64. Devailly, *Berry,* pp. 62–65; Guérin, *Sologne,* pp. 131–35; Perceveaux, "Etangs de Dombes," pp. 81–82, and "Structures et relations économiques en Dombes," pp. 339–40; Cnopf, *Teichwirtschaft,* pp. 23–26.

65. The great wine producers of medieval Europe were located, not in the best areas for the vine, but as close as possible (in terms of bulk transport) to thirsty northern consumers.

66. Richard, "Le commerce du poisson," p. 191; Richard, "Les étangs et le commerce"; and Jean Richard, "Transports par eau et péages de Chalon a Avignon: A propos des fournitures de poisson a la cour des papes," *Actes du 112e Congrès national des Sociétés savantes, Lyon, 1987: Section d'histoire médiévale et de philologie* (Paris, 1989), pp. 40–44.

67. The contrasting English experience validates the diagnosis. In the twelfth and thirteenth centuries state-of-the-art fish culture enterprises were installed on manors supplying royal, magnate, and monastic household needs but most facilities fell into disuse and were abandoned during the fourteenth century, when consistent commercial supplies of fresh marine fish became competitive throughout the country. Records of English fish culture thereafter hint at self-indulgence, especially by early Tudor churchmen (C. F. Hickling, "Prior More's Fishponds," *Medieval Archaeology* 15 [1971]: 118–23; critique in Currie, "The Role of Fishponds," pp. 158–59), and eventually at Elizabethan enthusiasm for imitation and experiment (as shown in translation of Dubravius and in John Taverner's *Certain Experiments Concerning Fish and Frvite: Practised by Iohn Taverner Gentleman, and by him published for the benefit of others* (London: William Ponsonby; facsimile reprint, The English Experience, no. 75, Amsterdam, 1968).

68. Fournial, *Les villes en Forez,* p. 691; Hemmerle, *Die Deutschordens-Ballei Böhmen,* p. 93.

69. Jean Mercier, *La seigneurie de Jaucourt en Champagne méridionale (approche socio-économique)* (Bar-sur-Aube, 1993), pp. 77–80, closely follows expenses for upkeep of ponds in the late fourteenth century.

70. *The Chronicle of Jocelin of Brakelond Concerning the Acts of Samson Abbot of the Monastery of St. Edmund,* ed. and tr. H. E. Butler (London, 1949), pp. 130–31; Blary, *Le domaine de Châalis,* p. 56; Maas, *Les moines défricheurs,* p. 75.

71. "dummodo etiam ex constructione talis stagni afferatur majus commodum construi facienti et reipublicae, quam sit incommoditas vicinorum quorum proprietates ex inundatione aquae submerguntur"; Durand, "De l'établissement des étangs," pp. 105–7, quoting the "Decisiones Gratiano-

politanae," question 91, from the Dauphiné. Perceveaux, "Structures et relations économiques," p. 348, describes the "Coutume de Villars" of Dombes as parallel to customs in Forez and Nivernais. Compare Georges Helmlinger, *La Dombes: Les hommes et l'eau* (Bourge-en-Bresse, 1990), pp. 25–26.

72. Eduard Beyer, *Das Cistercienser-Stift und Kloster Alt-Zelle in dem Bisthum Meissen* (Dresden, 1855), pp. 176, 183–84, 420, 422, 703–4.

73. Boleslaw Ulanowski, ed., *Księgi sądowe wiejskie,* 2 vols., Starodawne prawa polskiego pomniki, vols. 11–12 (Kraków, 1921), 2:430–32. Other instances of lords' fishponds harming peasant resources are ibid., pp. 418, 421, 425, 469–72; Górzyński, *Zarys historii rybołowstwa,* pp. 48–49; and, Hurt, *Dęjiny rybnikářství,* pp. 104–8.

74. Hildegard Weiss, *Die Zisterzienserabtei Ebrach: Eine Untersuchung zur Grundherrschaft, Gerichtsherrschaft und Dorfgemeinde im fränkischen Raum* (Stuttgart, 1962), pp. 77–78; Hubert Collin, "Les ressources alimentaires en Lorraine pendant la première partie du XIVe siècle," in *Bulletin philologique et historique (jusqu'à 1610) du Comité des travaux historiques et scientifiques,* 93e Congrès national des Sociétés savantes, Tours, 1968, vol. 1: *Les problèmes de l'alimentation* (Paris, 1971), pp. 64–65.

75. Andreska, "Development of Fish-Pond Culture," and *Rybářství,* pp. 39–49; Šusta, *Teichwirtschaft,* pp. 2–3.

76. Bourquelot, "Fragments," p. 73; Roberts, "The Bishop of Winchester's Fishponds," pp. 130–35; Hoffmann, "*Carpes pour le duc . . .*" Another managerial career can be traced at Namur in the 1350s: Balon, "La pêche et le commerce," pp. 28–31.

77. Guérin, *Sologne,* pp. 135–36.

78. Ulanowski, *Księgi sądowe wiejskie,* 2:472: "praeterea fassi sunt homines praedicti teneri ad piscandum in piscina, quoties illis mandatur, similiter ad reformandum piscinae aggerem et ad deferendum pisces ad curiam Pabyanicze, quoties illis mandatur. . . ." Górzyński, *Zarys historii rybołowstwa,* p. 50; Hurt, *Dějiny rybnikářství,* pp. 124–28; and Alois Míka, "Ceské rybnikářství a problém počatků původní akumulace kapitálu v českých zemích," *Československý časopis historický* 2 (1954): 262–72, think this normal practice. The same applied at Altzelle: Beyer, *Alt-Zelle,* pp. 428–29.

79. Balon, "La pêche et le commerce," p. 30; Rybarski, *Gospodarstwo,* p. 75. Like magnitudes come from ducal Burgundy (Hoffmann, "*Carpes pour le duc . . .*"; Beck, "L'approvisionnement"); while English deliveries were more often in the hundreds (Roberts, "The Bishop of Winchester's Fishponds"; Steane, "The Royal Fishponds," pp. 49–50; McDonnell, *Inland Fisheries,* pp. 19–24).

80. Richard, "Le commerce du poisson," "Les étangs et le commerce," and "Transports par eau"; Gresser and Hintzy, "Les étangs du domaine comtal"; Guérin, *Sologne,* p. 137; Rybarski, *Gospodarstwo,* p. 76.

81. Dubravius, *De piscinis,* ed. Schmidtová, pp. 23–24; Knittler, *Nutzen—Renten—Erträge,* pp. 162–66. Compare the estimates in Nyrek, *Gospodarka rybna,* pp. 40–43 , or Szczygielski, *Gospodarka stawowa,* pp. 41–47, 220–47, but note that in this region counts of fish sold (in contrast to those stocked) occur only in later records.

82. I choose the term advisedly to accentuate the scope and ramifications of "transformations in relations of humans with nonhuman nature" commensurate with the theoretical model advanced by Carolyn Merchant, "The Theoretical Structure of Ecological Revolutions," *Environmental Review* 11 (1987): 265–74.

83. Andreska, "Development of Fish-Pond Culture"; Nyrek, *Gospodarka rybna,* pp. 42–43; Robert-Henri Bautier, *The Economic Development of Medieval Europe* (London, 1971), pp. 198–99; Cnopf, *Teichwirtschaft,* p. 22.

84. Delatouche, "Le poisson d'eau douce," pp. 173–75; Szczygielski, *Gospodarka stawowa,* p. 19.

85. Balon, "The Common Carp"; and, from a different perspective, R. Boddeke, *Vissen en vissen: Ervaringen van een visserijbioloog en sportvisser* (Amsterdam, 1971), pp. 124–30.

86. Lespinasse and Bonnardot, *Les métiers et corporations,* pp. 212–18; Józef Rostafinski, *Średniowieczna historia naturalna w Polsce: Symbola ad historia naturalna medii aevi,* Munera saecularis universitatis Cracoviensis, vols. 7–8 (Kraków, 1900), 1:70–71.

87. "Successful establishment of an exotic species must necessarily precipitate changes in the physical and biological characteristics of the aquatic ecosystem receiving the introduction. . . . That no effects should result from such perturbations strains one's confidence in ecological principles." Jeffrey N. Taylor, Walter R. Courtenay, Jr., and James A. McCann, "Known Impacts of Exotic Fishes in the Continental United States," in *Distribution, Biology, and Management of Exotic Fishes,* ed. Walter R. Courtenay, Jr., and Jay R. Stauffer, Jr. (Baltimore, 1984), pp. 323, 352.

88. Alwyne Wheeler, *The Fishes of the British Isles and North-West Europe* (East Lansing, Mich., 1969), pp. 178–79, summarizes carp biology, and more is in Otto Heuschmann, "Die Weissfische (Cyprinidae)," in *Handbuch der Binnenfischerei Mitteleuropas,* ed. R. Demoll and H. N. Maier, vol. 3 (Stuttgart, 1957), pp. 53–63. R. L. Welcomme, "International Transfers of Inland Fish Species," in Courtenay and Stauffer, *Exotic Fishes,* pp. 33–36, reviews exotic fish introductions; Taylor, Courtenay, and McCann, "Known

Impacts," classify ecological effects (pp. 324–26) and then catalog the known impacts of carp (pp. 335, 336, 342, 345, 349).

89. Maas, *Les moines-défricheurs,* p. 78, sees a "symbiosis"; Devailly, *Le Berry,* p. 567, eschews progressive claims.

90. Grand and Delatouche, *L'agriculture au Moyen-Age,* p. 540; Perceveaux, "Etangs de Dombes" and "Structures économiques en Dombes"; Durand, "Etablissement des étangs"; Braudel, *Identity of France,* pp. 204–6.

91. Steane, "The Royal Fishponds," p. 50.

92. Ulrich Schmidt, "Himmelkron: Grundbesitz und wirtschaftliche Verhältnisse des ehemaligen Zisterzienserinnenklosters," *Archiv für Geschichte von Oberfranken* 48 (1968): 51; Nyrek, *Gospodarka rybna,* p. 55. Fournial, *Les villes en Forez,* pp. 689–90, found a handful of peasant-owned ponds marketing fish in late fourteenth-century Forez.

93. "Es viviers prendre les peissuns." Wace, *Roman de rou,* book 2, line 891, in *Maistre Wace's Roman de Rou et des ducs de Normandie,* ed. Hugo Andresen, 2 vols. (Heilbronn, 1877–79), 2:64.

94. *Sachsenspiegel Landrecht,* II.28.1–2 , ed. Karl A. Eckhardt, MGH Fontes iuris germanici antiqui, n.s., vol. 1, pt. 1, 2d ed. (Göttingen, 1955), p. 157. Like provisions in the Charter of Beaumont used as customary law in Lorraine are noted in Collin, "Les ressources alimentaires," p. 43.

95. Gislain, "Le rôle des étangs," p. 92; McDonnell, *Inland Fisheries, p.* 18; M. Aston and C. J. Bond, "Worcestershire Fishponds," in Aston, *Medieval Fish,* pp. 440, 442; R. F. Hartley, "Medieval Fisheries and Fishponds in Leicestershire (including Rutland)," ibid., p. 294; Górzyński, *Zarys historii rybołowstwa,* pp. 49–50.

96. Jean-Claude Schmitt, *The Holy Greyhound: Guinefort, Healer of Children since the Thirteenth Century,* tr. Martin Thom, Cambridge Studies in Oral and Literate Culture, vol. 6 (Cambridge, 1983), pp. 164–65.

97. Hermann Heimpel, "Fischerei und Bauernkrieg," in *Festschrift Percy Ernst Schramm zu seinem siebzigsten Geburtstag,* ed. P. Classen, 2 vols. (Wiesbaden, 1964), 1:353–72; Currie, "Early History of the Carp," pp. 102–3. In the tension-ridden England of 1376 sixty people attacked a park of Evesham Abbey near Ombersley in Worcestershire and took one hundred shillings worth of fish from the pond (Christopher C. Dyer, "The Consumption of Fresh-water Fish in Medieval England," in Aston, *Medieval Fish,* p. 35).

98. "Statim emptor et sine lite cum uenditore transigit, ubi pares statura Cyprini uenundatur," Dubravius, *De piscinis,* V.7, ed. Schmidtová, p. 65.

99. Rybarski, *Gospodarstwo,* p. 75; Malecki, *Lustracja województwa krakowskiego,* 2:241.

100. Ulanowski, *Księgi sądowe wiejskie,* 2:481: "quodsi in eis pisces modo non sunt, inquiratur diligenter, quis eos expiscatus fuerit."

101. Saxo Grammaticus, *Saxonis Gesta danorum,* preface, 2.4, ed. J. Olrik and H. Ræder, 2 vols. (Copenhagen, 1931), 1:6: "Ab huius ortivo latere occasivum Scaniae media pelagi dissicit interruptio, opimam praedae magnitudinem quotannis piscantium retibus adigere soliti. Tanta siquidem sinus omnis piscium frequentia repleri consuevit, ut interdum impacta navigia vix remigii conamen eripiat nec iam praeda artis instrumento, sed simplici manus officio capiatur." Also tr. Peter Fisher in *The History of the Danes,* ed. Hilda Davidson, 2 vols. (Cambridge, 1980), 1:7.

102. Herring, no Mediterranean fish, received the Middle Latin name *{h}allec,* from a Roman word for fermented fish sauce and ultimately from the Indo-European root for "salt" (*hal*). Medieval authors constantly associated herring with salt: William of Hirsau's sign language did "herring" as "fish" + "salt" (*PL,* vol. 150, col. 442); Hildegard of Bingen found it the only fish—she considered 35 varieties—healthier to eat salted than fresh (*Physica,* V.22, *PL,* vol. 197, col. 1280). Many local studies treat medieval herring fishing, but no scholar has examined its whole history. Charles L. Cutting, *Fish Saving: A History of Fish Processing from Ancient to Modern Times* (London, 1955), pp. 53–73, offers an anecdotal overview from a British perspective, sometimes with what have become obsolete references; Michel Mollat du Jourdin, *Europe and the Sea,* tr. Teresa L. Fagan (Cambridge, Mass., 1993), pp. 145–46, offers working hypotheses.

103. Marian Rulewicz, *Rybołówstwo Gdanska na tle Ośrodków Miejskich Pomorza od IX do XIII Wieku,* Gdańsk Wczesnośredniowieczny, vol. 10. Gdańskie Towarzystwo Naukose, Wydział i Nauk Społecznych i Humanistycznych: Prace Komisji Archeologicznej, no. 11 (Wrocław, 1994), pp. 68–88, 342–47; Zofia Kozikowska, "Ryby w pokarmie średniowiecznych (X–XIV w.) mieszkanców Wrocławia na Ostrowie Tumskim jako wskaźnik gatunków łowionych w wodach danych okolic lub dobierających tam drogą handlu," *Acta Universitatis Wratislaviensis,* vol. 223 (Prace zoologiczne, vol. 4) (Wrocław, 1974), pp. 3–14; Heinrich Appelt and Winfried Irgang, eds., *Schlesisches Urkundenbuch,* 4 vols (Vienna, 1963–88), nos. 123, 140.

104. Jean-Claude Hocquet, "Les pêcheries médiévales," in *Histoire des pêches maritimes en France,* ed. Michel Mollat (Toulouse, 1987), pp. 39–44; Anne-Dominique Kapferer, *Fracas et murmures: Le bruit de l'eau dans un Moyen-Age picard et boulonnais* (Amiens, 1991), pp. 93–96.

105. Auguste Bernard and Alexandre Bruel, eds., *Recueil des chartes de l'abbaye de Cluny,* 6 vols. (Paris, 1876–1903), no. 4132, p. 481, predating 1148; Josephus-Maria Canivez, ed., *Statuta capitulorum generalium ordinis Cisterciensis ab anno 1116 ad annum 1786,* 8 vols., Bibliothèque de la Revue d'histoire ecclésiastique, vols. 9–14b (Louvain, 1933–41), vol. 1, *Ab anno 1116 ad annum 1220,* anno 1210, no. 10.

106. *The Plaint of Nature,* tr. James J. Sheridan (Toronto, 1980), prose I, pp. 94–98.

107. Alison Locker, "The Fish Bones," in Robert Cowie and Robert L. Whytehead, "Two Middle Saxon Occupation Sites: Excavations at Jubilee Hall and 21–22 Maiden Lane," *Transactions of the London and Middlesex Archaeological Society* 39 (1988): 149–50; "Fish Bones," in Robert L.Whytehead et al., "Excavations at the Peabody Site, Chandos Place, and the National Gallery," ibid., 40 (1989): 148–51; "Fish Bones," in Brian S. Ayers, *Excavations at Fishergate, Norwich, 1985,* East Anglian Archaeology Report no. 68 (Norwich, 1994), pp. 42–44. Compare Norbert Benecke, "Zur frühmittelalterlichen Heringsfischerei im südlichen Ostseeraum: Ein archäologischer Beitrag," *Zeitschrift für Archäologie* 16 (1982): 283–90, an overview of fifty-eight sites, with the later findings Benecke reported in his "Die Fischreste aus den slawischen Inselsiedelungen in der Lieps und im Tollensesee," in *Lieps: Eine slawische Siedlungskammer am Südende des Tollensesees,* ed. V. Schmidt, Beiträge zur Ur-und Frühgeschichte der Bezirke Rostock, Schwerin und Neubrandenburg, vol. 16 (Berlin, 1984), pp. 118–21, and "Die Fischreste aus der Mecklenburg, Kr. Wismar," in *Die Mecklenburg: Eine Hauptburg der Obodriten,* ed. P. Donat (Berlin, 1984), pp. 183–87. The small, soft bones of herring pose special zooarchaeological difficulties, and little should be made of their apparent absence from unsieved sites (compare Norbert Benecke, "Untersuchungen zum Einfluß der Bergungsmethode auf die Qualität von Tierknochen," *Ausgrabungen und Fünde* 30 (1985): 260–65, and Ervynck and Van Neer, "A preliminary survey," esp. pp. 304–5.

108. Defined in *Fleta,* II.12, ed. Richardson and Sayles, p. 119.

109. A. Saul, "The Herring Industry at Great Yarmouth c. 1280–1400," *Norfolk Archaeology* 38 (1981): 33–43; A. Saul, "Great Yarmouth and the Hundred Years War in the Fourteenth Century," *Bulletin of the Institute of Historical Research* 52 (1979): 105–15; Hocquet, "Les pêches médiévales," pp. 48, 53–56, 79–86.

110. Philippe de Mézières, *Le songe d'un vieil pèlerin,* ed. and tr. G. W. Coopland, 2 vols. (Cambridge, 1969), 1:129–30 (tr. at 1:249–50).

111. Philippe Dollinger, *The German Hansa,* rev. ed., tr. D. S. Ault and S. H. Steinberg (London, 1970), pp. 37–39, 239–41, can be updated by Arnved Nedkvitne's contributions to *Medieval Scandinavia: An Encyclopedia,* ed. Phillip Pulsiano et al. (New York, 1993), namely "Fishing, Whaling, and Seal Hunting" (pp. 194–96) and "Trade" (pp. 649–53), both with extensive bibliographies.

112. Hocquet, "Les pêcheries médiévales," pp. 53–56.

113. Francesco Balducci Pegolotti, *La pratica della mercatura,* ed. Allan Evans (Cambridge, Mass., 1936), p. 380.

114. I follow the analysis of Richard W. Unger, "The Netherlands Herring Fishery in the Late Middle Ages: The False Legend of William Beukels of Biervliet," *Viator* 9 (1978): 335–56, and "Dutch Herring, Technology, and International Trade in the Seventeenth Century," *Journal of Economic History* 40 (1980): 253–80, with recent refinements by John H. Munro, "Patterns of Trade, Money, and Credit," in *Handbook of European History 1400–1600: Late Middle Ages, Renaissance and Reformation,* ed. Thomas A. Brady, Jr., Heiko A. Oberman, and James D. Tracy, vol. 1: *Structures and Assertions* (Leiden, 1994), pp. 161–63.

115. Michel Mollat, *Le commerce maritime normand à la fin du moyen age: Etude d'histoire économique et sociale* (Paris, 1952), p. 314n.

116. Unger, "Dutch Herring," p. 256.

117. G. F. IJzereef and F. Laarman, "The Animal Remains from Deventer (8th–19th Centuries AD)," *ROB: Berichten van de Rijksdienst voor het Oudheidkundig Bodemonderzoek* 36 (1986): 435–36, analyze 10,028 fish vertebrae recovered by sieving.

118. Others could learn the technology and exploit northern schools, too. In peacetime (1460s to 1480s) the yearly catch at Dieppe oscillated by about 25 percent around a four-hundred-tonne median (Mollat, *Le commerce maritime,* pp. 313–17, 598–99). Barreled herring were an important innovation in late fourteenth-century Scotland which from the mid-fifteenth century sustained a significant export trade: Elizabeth Gemmill and Nicholas Mayhew, *Changing Values in Scotland: A Study of Prices, Money, and Weights and Measures* (Cambridge, 1995), pp. 317–23.

119. A fish oriented to the bottom rather than the surface (pelagic), but not necessarily feeding on the bottom.

120. Much of what follows is summarized from Przemysław Urbańczyk, *Medieval Arctic Norway* (Warsaw, 1992), pp. 29–34, which assembles material and debates from scattered Norwegian-language books and papers, many of them unpublished.

121. Like movements occur along Iceland, Greenland, and Newfoundland.

122. In conditions more humid than the far north but still cold, cod dried under cover with some salt ("dry-salting") becomes a product resembling stockfish, called "klippfisk," "rundfisch," or the like. Ling (*Molva molva*), torsk (*Brosme brosme*), haddock (*Melanogrammus aeglefinus*), and pollacks (*Pollachius* species) are often taken with cod and handled the same way (compare Cutting, *Fish Saving,* pp. 118–22). Other important gadids (cod

relatives) such as whiting (*Merlangius merlangus*) or hake (*Merluccius merluccius*) are accessible in different habitats.

123. So not all cod eating is stockfish, nor evidence of long-distance trade in fish products. For differential use of local cod populations see Dirk Heinrich, "Fang und Konsum von Kabeljau im Mittelalter," *Offa* 43 (1986): 87–93, abridged as "Fishing and Consumption of Cod (Gadus morhua Linnaeus. 1758) in the Middle Ages," in *Fish and Archaeology,* ed. D. C. Brinkhuizen and A. T. Clason, BAR International Series, no. 294 (Oxford, 1986), pp. 42–52, and Dirk Heinrich, "Fischreste als archäologische Quellengattung—Probleme und Ergebnisse," *Archäologische Informationen* 12, no. 2 (1989): 174–75.

124. Much of what follows is a précis, with emphasis on particular relevant features, of an interpretation detailed in Reidar Bertelsen , "A North-East Atlantic Perspective," *Acta Archaeologica* 61 (1990) 22–26; Urbańczyk, *Medieval Arctic Norway,* pp. 42–78 et passim; Nedkvitne, "Fishing, Whaling, and Seal Hunting," and "Trade"; E. Ebel, "Fischerei und Fischereimethoden: Literarisches zur Fischerei im Norden," in *Reallexikon der Germanischen Altertumskunde,* ed. Heinrich Beck et al., vol. 9 (Berlin, 1994), pp. 143–44; and Sophia Perdikaris, "Scaly Heads and Tales: Detecting Commercialization in Early Fisheries," *Archaeofauna* 5 (1996): 21–33.

125. Dollinger, *The German Hansa,* p. 242.

126. Notably developed in Urbańczyk, *Medieval Arctic Norway,* pp. 132–47.

127. Ibid., pp. 240–61.

128. Bruce S. Gelsinger, *Icelandic Enterprise: Commerce and Economy in the Middle Ages* (Columbia, S.C., 1981), esp. pp. 181–94, and Kirsten Hastrup, *Nature and Policy in Iceland 1400–1800: An Anthropological Analysis of History and Mentality* (Oxford, 1990), esp. pp. 66–75, 235–75, respectively provide classic economic and cultural approaches, while Thomas Amorosi, "Icelandic Archaeofauna: A Preliminary Review," *Acta Archaeologica* 61 (1990); 276–81, details the dramatic shift in animal remains. Ebel, "Fischerei und Fischereimethoden," pp. 140–43, summarizes.

129. Urbańczyk, *Medieval Arctic Norway,* p. 141.

130. *Rotuli Parliamentorum . . . tempore Henrici R. V,* vol. 4 ([Westminster], n.d.), pp. 79–80; Geoffrey J. Marcus, *The Conquest of the North Atlantic* (New York, 1980), p. 131. A classic of commercial history is Eleonora Carus-Wilson, "The Iceland Trade," in *Studies in English Trade in the Fifteenth Century,* ed. Eileen Power and Michael M. Postan (London, 1933), pp. 155–82.

131. Sigurdur A. Magnússon, *Northern Sphinx: Iceland and the Icelanders from the Settlement to the Present* (Montreal, 1977), pp. 106–7; Marcus, *Conquest of the North Atlantic,* pp. 125–54; Hastrup, *Nature and Policy in Iceland,* p. 73.

132. More species, each with less biomass, characteristically distinguish warm southern from cold nutrient-rich northern waters.

133. Maryanne Kowaleski, *Local Markets and Regional Trade in Medieval Exeter* (Cambridge, 1995), pp. 307–12; M. Wilkinson, "The Fish Remains," in Mark Maltby, *Faunal Studies on Urban Sites: The Animal Bones from Exeter 1971–1975*, Exeter Archaeological Reports, vol. 2 (Sheffield, 1979), pp. 74–81; Jennie Coy. "Medieval Records versus Excavation Results: Examples from Southern England," *Archaeofauna* 5 (1996): 55–63.

134. Henri Touchard, *Le Commerce maritime breton a la fin du Moyen age* (Paris, 1967), pp. 58–61.

135. Ferreira Priegue, *Galicia en el comercio maritimo,* pp. 130–56, 345–46.

136. Local discussions and references to eating, trading, and producing *strekfuss* include Horst A. Willam, "Die Fischerei des Deutschen Ordens in Preussen bis zu Dietrich von Altenburg," *Jahrbuch der Albertus-Universität zu Königsberg/Pr.* 11 (1961): 148–49; Jürgen Sarnowsky, *Die Wirtschaftsführung des Deutschen Ordens in Preußen (1382–1454),* Veröffentlichungen aus den Archiven Preussischer Kulturbesitz, vol. 34 (Cologne, 1993), pp. 130–31, 625–26, 814–25; Franciszek Piekosinski and Józef Szujski, eds., *Najstarsze księgi i rachunki miasta Krakowa od r. 1300 do 1400,* Monumenta medii aevi historica res gestas Poloniae ilustrantia, vol. 2 (Kraków, 1878), pp. 235–41; Piekosinski, *Rachunki dworu króla Władyslawa Jagiełły,* pp. 22, 310 et passim; Sabine Sten, "Trading with Fish in Medieval Sweden: Some Examples from Archaeological Bone Finds," *Archaeofauna* 4 (1995): 65–66; Nils Friberg, *Stockholm i bottniska farvatten: Stockholms bottniska handelsfält under senmedel tiden och Gustav Vasa* (Uppsala, 1983), pp. 196–99.

137. Gemmill and Mayhew, *Changing Values in Scotland,* pp. 303–17.

138. Bresc, "La pêche et les madragues"; Henri Bresc, *Un monde méditerranéen: Economie et société en Sicilie 1300–1450* (Rome, 1986), pp. 264–72, and "La pêche et les madragues"; Pegolotti, *La pratica della mercatura,* pp. 181, 205.

139. One notable exception was England, small enough for carts from the coast to reach the interior with still-edible seafood, especially in cooler seasons: Christopher C. Dyer, "Les régimes alimentaires en Angleterre, XIIIème–XVème siècles," in *Manger* 2 (1984): 263–74. Medieval Europe's largest consumption market, Paris, just at the 150–kilometer threshold where freshness could not be guaranteed, built up an elaborate structure of packhorse stages, brokers, and *poissoniers de mer* to hasten the catch from Norman ports to consumers' tables (Bourlet, "L'approvisionnement de Paris," and Auzary-Schmaltz, "Les contentieux," pp. 59–67, with sources there cited).

140. C. L. Cutting, "Historical Aspects of Fish," in *Fish as Food,* vol. 2: *Nutrition, Sanitation, and Utilization,* ed. Georg Borgstrom (New York, 1962), pp. 8–16; Mollat, *Europe and the Sea,* pp. 63–65; Helmut Hundsbichler, "Nahrung," in Harry Kühnel et al., *Alltag im Spätmittelalter* (Graz, 1984), p. 203. For simplicity's sake I omit the less long-lasting smoked fish, acid pickled fish, or fish in gelatine, which also entered at least local medieval markets.

141. Bresc, *Un monde méditerranéen,* p. 269.

142. Hercule Géraud, *Paris sous Philippe le Bel d'après les documents originaux, et notament d'après un manuscrit contenant le role de la taille imposée sur les habitants de Paris en 1292* (Paris, 1837), p. 516; Lespinasse and Bonnardot, *Les métiers et corporations,* pp. 218–22; Laurière et al., *Ordonnances,* 2:575–82 (compare Delamare, *Traité de la police,* as discussed in note 41 above); René de Lespinasse, ed., *Les métiers et corporations de la ville de Paris,* vol. 1, *XIV–XVIIIe siècle: Ordinances Générales, Métiers de l'alimentation* (Paris, 1886), pp. 13–19, 409–26; Franciszek Piekosinski, ed., *Kodeks dyplomatyczne miasta Krakowa (1257–1506),* Monumenta medii aevi historica res gesta Poloniae illustrantia, vols. 5 and 7 (Kraków, 1879–82; repr. New York, 1965), nos. 262, 299; Jan Ptasnik, *Cracovia artificum 1300–1500* (Kraków, 1917), nos. 201–2, 218, 230.

143. Anita Feyl, "Das Kochbuch des Eberhard von Landshut (erste Hälfte des 15. Jhs.)," *Ostbairische Grenzmarken* 5 (1961): 353–64 (recipe 23); *Le Ménagier de Paris,* ed. Georgine E. Brereton and Janet M. Ferrier (Oxford, 1981), p. 237 (para. 194).

144. "Heringe sollen nit fur vische geachtet werden, man mochte es dan nit mol gebessern." Gerhard Fouquet, "'Wie die kuchenspise sin solle': Essen und Trinken am Hof des Speyerer Bischofs Matthias von Rammung (1464–1478)," *Pfälzer Heimat* 39 (1988): 27. Cutting, *Fish Saving,* pp. 4, 27, 30–32, reiterates the general point.

145. P. Heath, "North Sea Fishing in the Fifteenth Century: The Scarborough Fleet," *Northern History* 3 (1968): 59–60.

146. Quoted in Mollat, *Le commerce maritime,* p. 314, n. 251 from A. H. Wratislaw, *Diary of An Embassy from King George of Bohemia to King Louis XI of France in the Year of Grace 1464* (London, 1871), p. 45.

147. Unger, "The Netherlands Herring Fishery" and "Dutch Herring"; Mollat, *Europe and the Sea*; and Bresc, *Un monde méditerranéen,* extensively consider these backward linkages.

148. C. J. Given-Wilson, "Purveyance for the Royal Household, 1362–1413," *Bulletin of the Institute for Historical Research* 56 (1983): 147–50; Ferreira Priegue, *Galicia en el comercio maritimo,* p. 147; Bourlet, "L'approvisionnement," p. 6; Piekosiński, *Kodeks dyplomatyczne Krakowa,* no. 336, p. 459; Bresc, *Un monde méditerranéen,* p. 269.

149. Alan of Lille, *The Plaint of Nature,* pp. 173–74; Philippe de Mézières, *Le Songe d'un vieil pèlerin,* pp. 129–30; Maria Helena da Cruz Coelho, "Apontamentos sobre a comida e a bebida do campesinato Coimbrão em tempos medievos," *Revista de historia económica y social* (Lisbon) 12 (1983): 95.

150. L.-C. Douët-d'Arcq, *Comptes de l'Hotel des rois de France au XIVe siècle* (Paris, 1865), p. 266 (as cited by Coopland in Philippe de Mézières, *Le Songe d'un vieil pèlerin,* p. 130n); Bernard Chevalier, "Alimentation et niveau de vie a Tours à la fin du XVe siècle," in *Bulletin philologique et historique (jusqu'à 1610) du Comité des travaux historiques et scientifiques,* 93e Congrès national des Sociétés savantes, Tours, 1968, vol. 1: *Les problèmes de l'alimentation* (Paris, 1971), p. 145. More respectable poor in France eat herring in André Endrès, "Alimentation d'assistance à l'Hôtel-Dieu de Meaux en 1527," ibid., pp. 209–56, and Claude Hohl, "Alimentation et consommation a l'Hôtel-Dieu de Paris aux XVe–XVIe siècles," ibid., pp. 181–208. Iberian counterparts are in A. H. de Oliveira Marques and Jopo Pedro Ferro, "L'alimentation au Portugal du moyen age au XVIIIe siècle," in *La sociabilité à table: Commensalité et convivialité à travers les âges,* ed. Martin Aurell, Olivier Dumoulin, and Françoise Thélamon, Publications de l'Université de Rouen, vol. 178 (Rouen, 1992), p. 286; Rucquoi, "Alimentation des riches," p. 302.

151. Raymond Delatouche, "Importance relative des divers produits agricoles dans le 'Gros Brief' de Flandre pour 1187," in *Bulletin philologique et historique (jusqu'à 1610) du Comité des travaux historiques et scientifiques,* 93e Congrès national des Sociétés savantes, Tours, 1968, vol. 1: *Les problèmes de l'alimentation* (Paris, 1971), p. 30; Hocquet, "Les pêcheries médiévales," pp. 53–56; François Maillard, "Tarifs des 'coutumes' pour la vente du poisson du mer à Paris au début du XIVe siècle," *Bulletin philologique et historique* (1966): 247–58.

152. Dyer, "Consumption of Fresh-Water Fish," p. 31; Hitzbleck, *Die Bedeutung des Fisches,* p. 106. At Cheb (Eger) on the border between Upper Franconia and Bohemia, in 1465 the only fish cheaper than herring were dried bleak, and only small cyprinids could be bought fresh for less than the price of a stockfish (ibid., p. 107).

153. Francis W. Carter, *Trade and Urban Development in Poland: An Economic Geography of Cracow, from Its Origins to 1795,* Cambridge Studies in Historical Geography, vol. 20 (Cambridge, 1994), pp. 134–39; E. von Ottenthal, ed., "Die ältesten Rechnungsbücher der Herren von Schlandersberg," *Mitteilungen des Instituts für Österreichische Geschichtsforschung* 2 (1881): 603. Hemmerle, *Die Deutschordens-Ballei Böhmen,* pp. 110, 114–16, 155 et passim, has many herrings in Bohemia while 306,000 of them were paying tolls at Chalon sur Saône in 1406 (Beck, "L'approvisionnement," p. 175). The heartlands of aquaculture never lacked for preserved fish.

154. Saul, "Herring Industry," p. 36.

155. *Ménagier,* p. 327 (para. 194).

156. Hans Hajek, ed., *Daz buoch von guoter spise: Aus der Würzburg-Münchener Handschrift,* Texte des späten Mittelalters, vol. 8, no. 20 (Berlin, 1958), p. 22.

157. Wendy Childs, *Anglo-Castilian Trade in the Later Middle Ages* (Manchester, 1978), p. 99.

158. Bresc, "La pêche et les madragues," p. 169, n. 7: *merluza, sardina roia* (smoked sardines), and *sardina arrencada.*

159. *Etymologisches Wörterbuch des Deutschen,* ed. Wolfgang Pfeifer, 2d ed. (Berlin, 1993), s.v. *kabeljau*; Charles Du Fresne, Sieur Du Cange, et al., *Glossarium mediae et infimae latinitatis* (Paris, 1840–50), s.v. *cabellauwus*; Walther von Wartburg, *Französisches etymologisches Wörterbuch,* vol. 2 (Basel, 1971), s.v. *Kabeljauw—cabillaud*; Real Academia Española, *Diccionario de autoridades* (1726; facs. repr. Madrid, 1964), s.v. *bacallao.*

160. Ferreira Priegue, *Galicia en el comercio maritimo,* p. 148, and sources there cited.

161. Ebel, "Fischerei und Fischereimethoden."

162. Bertelsen, "North-East Atlantic Perspective"; Urbańczyk, *Medieval Arctic Norway.*

163. Urbańczyk, *Medieval Arctic Norway,* p. 261.

164. A germ of the interpretation is already present when Bresc, "La pêche et les madragues," p. 168, briefly compares the *tonnara* with the Sicilian sugarcane industry as forms of "precocious capitalism."

165. Saul, "Herring Industry," pp. 38–40; Mollat, *Europe and the Sea,* pp. 76, 148–52.

166. Besides works mentioned in note 131 above, see Dollinger, *The German Hansa,* pp. 243, 429–30; Terrence H. Lloyd, *England and the German Hanse, 1157–1611: A Study of Their Trade and Commercial Diplomacy* (Cambridge, 1991), pp. 137–40, 190, 238–39, 260; John D. Fudge, *Cargoes, Embargoes and Emissaries: The Commercial and Political Interaction of England and the German Hanse 1450–1510* (Toronto, 1995), pp. 25–72, 90–91, 118–19, 164–70.

167. Roger Aubenas, *Le droit de pêche de l'abbaye de Lérins,* Equipe des Historiens Cannois, fascic. no. 2 (Cannes, 1953), pp. 8–9, with more in Louis Stouff, *Ravitaillement et alimentation en Provence aux XIVe et XVe siècles* (Paris, 1970), pp. 201–2.

168. Bresc, *Un monde méditerranéen,* pp. 264–70.

169. *Rot. Parl.,* 4:79: "ils ount trovez lieux, l'ou graunt plentie de tiel maner pesson est prise sur les costes de Island. . . ."

170. Mollat, *Europe and the Sea,* pp. 104, 146–47, emphasizes both the particular (compare Marcus, *Conquest of the North Atlantic,* pp. 164–73) and the general direct and intimate connection between European offshore fishing and its extension to Newfoundland. David H. Sacks, *The Widening Gate: Bristol and the Atlantic Economy, 1450–1700* (Berkeley, Calif., 1991), pp. 34–36, situates the men who backed those voyages.

171. "et affirmanno che quello mare è coperto de pessi, li quali se prendenno non solo cum la rete ma cum le ciste, essendoli alligato uno saxo ad cio che la cista se impozi in l'aqua, et questo io l'ho oldito narrare al ditto messer Zoanne. Et ditti inglesi suoi compagni dicono che portaranno tanti pessi che questo regno non havera piu besogno de Islanda, del quale paese vene una grandissima mercantia de pessi che si chiamanno stochfissi." James A. Williamson, *The Cabot Voyages and Bristol Discovery under Henry VII,* Hakluyt Society, 2d ser., vol. 120 (Cambridge, 1962), document 24 (Raimondo de Raimondo de Soncino, letter to the Duke of Milan, December 18, 1497.

For discussion of "Everything We Know about John Cabot" see Peter E. Pope, *The Many Landfalls of John Cabot* (Toronto, 1997), pp. 11–42, who takes pains (p. 13) to "start with the following facts. First of all, he wasn't John Cabot, of course. On the other hand, he wasn't Jean and he wasn't even Giovanni, as biographers like to call him. He was Zuan Caboto. Although a citizen of Venice, he was not born there but naturalized circa 1472." It is a salutary reminder of the *European* quality of frontier enterprise at the end of the Middle Ages.

2

Cattle-Grazing in the Southeastern United States, 1670–1949

An Economic and Social Adaptation

John Solomon Otto

In 1895, Frederic Remington, the famed artist of the American West, published an article on the "Cracker Cowboys of Florida" in *Harper's Magazine.* After years of painting cowboys in the West, he finally visited Florida to view its "cracker cowboys" or "cattle-hunters," as they called themselves. He found that the cracker cowboys were "new birds," whose lifestyle and landscape differed dramatically from their Western counterparts. Unlike the latter, who rode horned saddles and used saddle ropes ("lariats" and "lassos") to handle cattle, the Florida cattle-hunters rode hornless saddles, had no saddle ropes, and used long whips and "cur-dogs" to handle cattle. In contrast to the Western cowboys who herded bulky "longhorn cattle," Florida cattle-hunters tended "half-starved and horribly emaciated" little animals that were the size of burros. Unlike the West, with its rolling grasslands and deserts, the Florida landscape was "flat and sandy" with "straight pine timber," "scrub palmettoes," and "tough wiregrass." And unlike the West, where the cattle ranchers were fencing in the open range with barbed wire, the Florida pinewoods and savannas were still unfenced range.[1]

During his visit, Remington became acquainted with Florida cattle grazing, which he contrasted unfavorably with western ranching. Returning to his first love, Remington continued to paint western ranching and history until his death in 1909. His depictions of western cowboy life became icons that were widely distributed in the printed media, and helped create our century-long fascination with cowboy life.[2]

In the years since Remington's death, scholars have exhaustively researched western cattle-ranching life, exploring the origins, lifeways, and values of the cowboys.[3] In contrast to the attention that has been lavished on the cowboys, only a handful of studies have been published on the Florida cattle-hunters and their kinsmen in Georgia and South Carolina.[4] Despite this scholarly neglect, cattle grazing was once an important economic and social adaptation in the southeastern United States. Beginning in colonial South Carolina around 1670, open-range cattle grazing survived until 1949 in the state of Florida (see figure 1).

Cattle arrived with the first settlers of Carolina Colony in 1670. Although the Lords Proprietors, the founders of Carolina Colony, wanted "Planters there and not Graziers," they supplied their colonists, who hailed from the British West Indies and the British Isles, with breeding cattle from Virginia, New York, and Bermuda. Despite the cougars and wolves which lurked in the Carolina woods, the imported cattle proved prolific. A visitor to Carolina in 1682 noted the "great encrease [*sic*] of the cattle, which was "rather to be admired than believed." Only "six or seven years past, the country was almost destitute of Cows . . . and now they have many thousand Head." This remarkable increase of cattle provided a surplus of animals for home consumption as well as for export overseas. By 1682, Carolinians were regularly exporting barrels of salt beef to the British West Indies. Thanks to the West Indian trade, many Carolina colonists turned to commercial cattle raising.[5]

Cattle grazing permitted Euroamerican agriculturalists to adapt to the southeastern coastal flatwoods—an ecologically diverse region which contained hardwood forests, riverain swamps, treeless savannas, and grassy pinewoods (see figure 1). The forests and swamps yielded excellent harvests, but the returns from crops planted on the infertile savanna and pinewood soils were meager. The colonists therefore treated these grassy wastelands as unfenced grazing lands for livestock that transformed useless grass and browse into useful meat and hides. Finding expanses of pinewoods and savannas in the coastal flatwoods of Carolina, and later also of Georgia and Florida, Euroamerican agriculturalists successfully adapted to these infertile environments by grazing cattle herds and harvesting yearly "crops" of scrub beeves (see figure 2).[6]

Carolinians raised cattle on stock farms that were commonly known as "penns" or "cowpens." Penns were farmsteads with stock pens, dwellings, and fields, sitting amidst expanses of unfenced grazing land, or open range. Carolina laws required colonists to enclose their fields with "sufficient Fences," so that cattle could graze on any unfenced lands without fear of trespass. The finest grazing lands in coastal Carolina were "the Savannas (or grassy plains) of severall [*sic*] magnitudes clear of trees." In addition to the treeless savannas, the pinewoods, which contained carpets of wild grasses, provided a grazing ground for cattle in the warm months. During the brief winters, when frosts killed the grasses, Carolinians burned the pinewoods

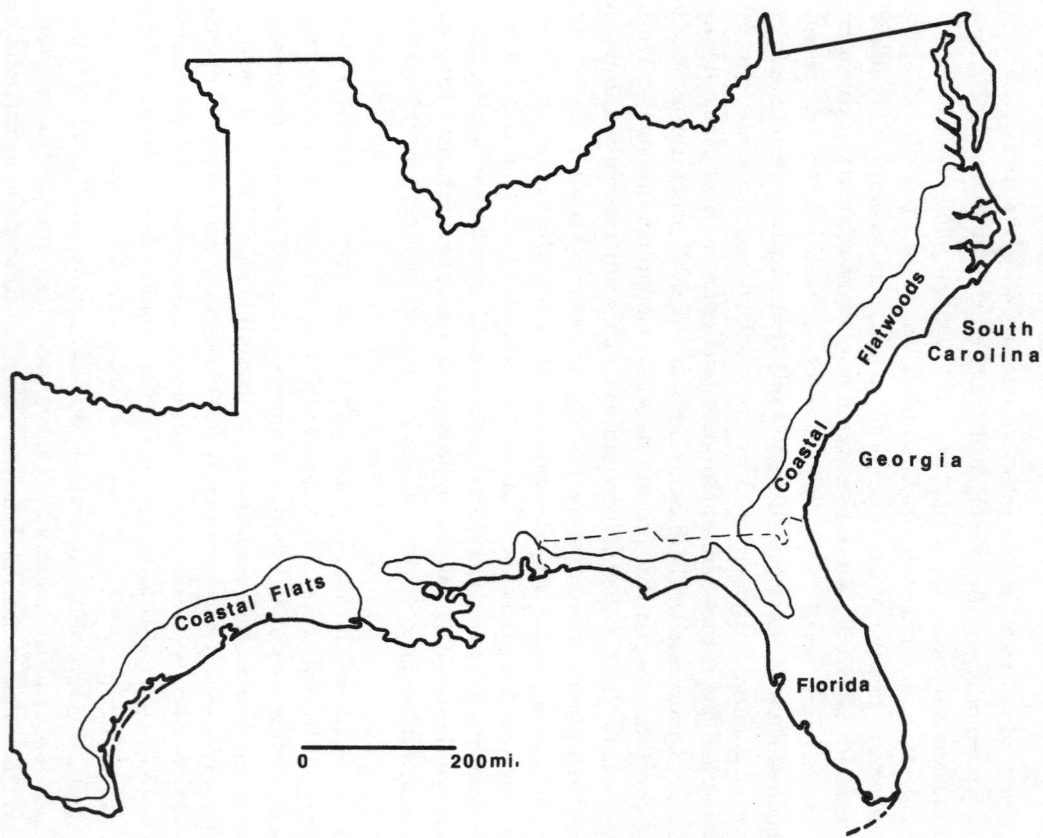

1. The Southeastern United States.

2. The Southeastern Coastal Flatwoods Environment: Lake Kissimmee State Park, Florida.

and savannas in order to remove "great Quantities of the dry Russet Grass" and expose the fresh grass for their cattle. In the winter months, cattle also grazed on the evergreen canes which grew in the lowlands. Since cattle grazed throughout the year on the open range, Carolina laws obligated owners to identify their stock with distinctive "ear[marks] and burnt marks [brands]." To prevent cattle from straying on the unfenced range, Carolinians relied on herdsmen—white servants and black slaves—to pen their stock nightly in order to "keep them from running wild." Nightly penning not only tamed the cattle but also manured the soil within pens, creating ideal garden spots. One Carolinian advised his neighbors to "dig in their cowpens immediately after the removal of their pens" and plant crops. In addition to supplying manure, cattle also furnished beef and hides for market. Each fall, mounted herdsmen collected the range steers, driving beeves to Charleston for processing. Carolinian merchants exported barrels of salt beef to the West Indies and shipped cattle hides to New England.[7]

Cattle grazing provided Carolinian colonists with exportable cattle products, while they searched for cash crops to export. From 1670 to the 1690s, Carolinians experimented with such possible export crops as tobacco, citrus, ginger, wine grapes and sugarcane. But in the 1690s, they discovered that the crop best suited to Carolina's environment was African rice.[8]

Exporting rice as well as cattle products, Carolina proved economically viable, but the fledgling British colony faced a potential threat from the neighboring Spanish colony of LaFlorida. Founded in 1565, LaFlorida was a sparsely settled colony that stretched from coastal Georgia to the Apalachicola River, whose inhabitants included Hispanic soldiers and civilians as well as Christianized Indians. Most of the colony's Hispanos lived near the presidio (fortress town) of St. Augustine, whereas the Christian Timucua, Apalachee, and Guale Indians resided in the hinterland—in northern Florida, western Florida, and coastal Georgia respectively. The Franciscans who had converted the Indians during the seventeenth century had also introduced domestic animals, including Spanish cattle. Although the Guale raised few animals, the Timucua and the Apalachee Indians grazed cattle in old fields, savannas, and pinewoods. In addition to the mission herds, there were over thirty cattle farms, or grazing haciendas, in northern Florida, granted by Spanish governors of Florida to Hispanic civilians who were encouraged to raise beef for the St. Augustine presidio. A typical hacienda was a circular tract about eight leagues (twenty miles) across, which supported a herd of about two hundred cattle. A couple of Indian or African herdsmen managed the cattle that grazed on the unfenced haciendas throughout the year. During the winter, herdsmen burned the unfenced range in order to freshen the grass. In the spring, they collected the range cattle, inventoried the stock, and branded the calves. During the fall, they rounded up the herds in order to select the beef cattle. Though many beeves were driven to St. Augustine, where they were slaughtered to feed the presidio's soldiers and civilians, Florida's *hacendados* (cattle owners) found it more profitable to export dried beef, tallow, and cattle hides to Cuba. Furnishing cattle products for export and providing beef for the St. Augustine presidio, cattle raising on haciendas and missions helped Spain maintain its foothold in LaFlorida.[9]

Hoping to isolate Spanish LaFlorida, Carolinian colonists forged trading alliances with the Yamasee, Creeks, and other southeastern tribes. Trading firearms for deerskins, Carolinians encouraged their Indian trading partners to raid the Christian Indians of LaFlorida. By 1700, the Yamasee had driven the Guale Indians south of the St. Marys River and had occupied the Georgia coast. When Queen Anne's War (1702–13) began, Governor James Moore of Carolina called on Carolina's Indian allies and led a mixed force of Indians and Carolinians into LaFlorida. Though they failed to capture the St. Augustine presidio, Moore's forces ravaged the missions and haciendas of Florida's Timucua Province, returning to Carolina with stolen cattle and Indian captives. In 1704, Moore led another expedition of Carolinians and Creek Indians into Florida's Apalachee Province, where they destroyed haciendas and missions. Returning to Carolina with cattle and captives, Moore's raiders sold Florida stock and Indian slaves in the Charleston markets. The Spanish cattle from Florida augmented the Carolinian herds of "native" cattle, which had descended from north European stock, and the Florida Indian captives in turn swelled the ranks of Carolina's slave population. Following the destruction of Florida's missions and haciendas, the Spanish retreated into the St. Augustine presidio, and the Carolina colony enjoyed a brief respite from conflict.[10]

During this time of relative peace, Carolina's cattle graziers and rice planters prospered, exporting cattle products to British colonies and shipping rice to Europe. Carolinian cattle graziers relied on a mixed labor force of African slaves, Indian captives, and British servants to tend their stock, which grazed on the unfenced pinewoods and savannas of coastal Carolina, or the "low country." The low country, however, also became a producer of "dry rice," raised in upland fields cleared from hardwood forests, and "wet rice," grown in lowland fields reclaimed from river swamps. Since African slaves, who introduced rice to Carolina, proved indispensable for growing the crop, Carolinians imported hundreds of African captives to work in the rice fields, and rice planting soon surpassed cattle grazing in economic importance. By 1712, Carolina exported 12,727 barrels of rice to Britain, valued at forty thousand pounds, as against only 1,963 barrels of salt beef, worth less than four thousand pounds, to the West Indies. Although rice planting became the leading agricultural pursuit, cattle grazing did not disappear. Many Carolinian planters raised cattle as well as rice, and the term "planter" became the "common Denomination" for those who followed "Tillage or Grasing [*sic*]."[11]

Cattle grazing and rice planting proved profitable pursuits, but these agricultural adaptations required large amounts of land. Searching for fresh pasturage and new fields, graziers and planters pressed southward along coastal Carolina. By 1715, Carolinian agriculturalists were living as far south as the Edisto River, near the boundary of the Yamasee Indian reservation, set aside by the Carolina colonial assembly in 1707 to serve as a hunting ground for the Yamasee who supplied the colony with deerskins, where Carolinians were forbidden to graze their stock. Despite this prohibition, Carolinian cattle strayed into the Yamasee reservation, competing with deer for woodland forage. As deer populations declined, the Yamasee found it increasingly difficult to collect enough deerskins to meet their trade debts. The indebted and frustrated Yamasee Indians lashed out at the Carolinians in the spring of 1715, killing hundreds of colonists, slaves, and cattle. The surviving Carolinians retaliated against the Indian attackers, and the Yamasee retreated to Spanish St. Augustine, where they settled in pueblos outside the fortress.[12]

The Yamasee revolt plunged the Carolinian economy into depression. By 1717, Carolina was exporting only 790 barrels of salt meat and 8,300 barrels of rice. Angered by the Proprietors' inability to protect the colony, Carolinians toppled the proprietary government in 1719. Yet the Yamasee, who were based at St. Augustine, continued to raid Carolina, and the agricultural economy continued to suffer. But in 1730, the British Crown paid off the Lords Proprietors, and Carolina became a royal colony. With the return of political stability, Carolinians finally mastered the Yamasee Indian raiders, and the colonists rebuilt the rice and cattle industries.[13]

Importing thousands of African captives to work on rice plantations and cattle farms, Carolina counted nearly forty thousand black slaves by 1750, and the colony had a black majority. Though most Carolina slaves labored in the rice industry, many worked on cowpens. A typical cowpen crew consisted of a white supervisor, or "cowpen-keeper," and several black cattle-hunters. This crew tended the hundreds of cattle that grazed in the unfenced pinewoods and savannas. In the fall, the cowpen workers collected the marketable beeves, driving the animals overland to Charleston for butchering. By the mid-eighteenth century, Carolina was still exporting about two thousand barrels of salt beef yearly to the West Indies.[14]

Since cattle grazing remained economically important, herds and cowpens were advertised for sale in local newspapers. Such an advertisement in a Charleston newspaper offered to sell a "Stock of very fine, tame, well-governed Cattle, containing in the whole about 2,000 head." These animals grazed on "as good a range for cattle as any in the Southern parts of the province, having a large cane swamp . . . for a winter's range and a most extensive and plentiful summer's range." In addition to the cattle herds, there was a "Settlement suitable for a a Cow pen, with necessary buildings thereon, and a good pasture inclosed [*sic*], together with about 50 head of Horses and Mares, and a small stock of Hogs, some provisions, and many necessary Cowpen and plantation utensils." The advertisement also mentioned that a "very careful white man," who had "managed the stock for some years," was available for employment "on easy terms."[15]

By the mid-eighteenth century, such cowpens dotted the countryside of South Carolina, and the colony contained an estimated seventy thousand cattle. Since every animal needed at least fifteen acres of open range in order to find enough forage during the year, these cattle required over a million acres of unfenced land. South Carolina was so overstocked with cattle that cowpen keepers led their herds into the neighboring colonies of North Carolina, which became a royal colony in 1730, and Georgia. North Carolina was a colony of small planters who resented the intrusion of these South Carolinian graziers. In 1766, the North Carolina colonial assembly passed an act "to prevent the inhabitants from South Carolina driving their Stocks of cattle from thence to range and feed in this Province," unless they purchased "a sufficient quantity of land for feeding the said Cattle." Carolinian graziers, however, found a friendlier legal environment in Georgia, where the colonial assembly adopted herding laws resembling those of South Carolina. Georgia required planters to fence in their fields, so marked cattle were free to graze on unfenced lands at no charge.[16]

Founded in 1732, Georgia began its history as a philanthropic experiment. James Oglethorpe and other British philanthropists obtained Crown lands from King George II in order to establish a refuge colony for British paupers and European Protestants. To protect the morality of Georgia's colonists, the philanthropic Trustees forbade the introduction of African slaves. Arriving in the new colony, Georgia' s colonists turned to farming and grazing in order to make a living. Cattle grazing proved a labor-saving agricultural adaptation in the new colony, where hands were in short supply. Obtaining mixed-breed cattle from South Carolina, Georgians established cowpens near the early settlements. Though these cowpens furnished beef and milk for the colonists, Georgia exported little salt beef before 1750. But in that year, Georgia's Trustees repealed the ban against slavery. This permitted slaveowning Carolinian rice planters and cattle graziers to enter Georgia. Reverting to the British Crown in 1753, Georgia became a rice-planting and cattle-grazing colony which resembled neighboring South Carolina.[17]

When the naturalist John Bartram visited Georgia in 1765, he observed the colony's cowpens, which were "the greatest curiosity this country affords." Traveling southward from Savannah, Bartram and his companion, Mr. Galphin, an Indian trader and cowpen owner, passed through "much poor piney ground," arriving at Galphin's "first cowpen." This was a "kind of house, or hut, near a good spring, in which four or five negroes" and "one white man" resided. They looked after "a number of cattle of various kinds" that occupied "a range of country of six or ten miles round." The "chief employment of these herdsmen" was "to tend the calving of the cows" and "to bring those to the pen that stand most in need of assistance." Bartram also noted that several Georgians had "two or three of these cowpens at ten miles distance from each other," and the cattle grazed in "distinct herds," feeding "both winter and summer in their respective walks."[18]

Although cowpens were found from Georgia to the Carolinas by the 1760s, cattle raising was also practiced in the new British colony of East Florida. In 1763, at the conclusion of the Great War for the Empire (1754–63), Spain ceded LaFlorida to Britain, which created the colonies of East and West Florida. Evacuating the Hispanic residents of St. Augustine, Spain left East Florida to the British colonists and to the Seminoles ("wild ones")—the descendants of Creek Indians who had settled in northern Florida after the destruction of the Spanish missions and tamed some of the wild Spanish cattle and horses that they had found there. By 1763, when Florida reverted to Britain, the Seminoles were already grazing cattle and horses in the savannas and pinewoods. Two years later, at the Picolata Congress, the Seminoles ceded their lands east of the St. Johns River to the British, and withdrew to the country west of the St. Johns, where they pursued hunting, gardening, and livestock raising.[19] Passing through East Florida in 1774, the naturalist William Bartram described the Seminole livestock raisers in the Alachua area: "Soon after sun-rise, a party of Indians on horseback, appeared upon the savannas, to collect together several herds of cattle, which they drove along near our camp, toward the town" of Cuscowilla.[20]

In 1775, when the Revolutionary War (1775–83) began, East and West Florida remained loyal to Britain, but Georgia and South Carolina joined the rebellion. Divided loyalties in the southeastern colonies resulted in bitter warfare between loyalists and rebels. The rebels raided East Florida, and the loyalists in turn attacked Georgia. During these raids, loyalist banditti frequently stole slaves and cattle from outlying Georgia cowpens.[21]

When the Revolutionary War ended in 1783, Britain ceded the Floridas back to Spain, but many British loyalists remained in the Spanish Floridas. Among these were cattle graziers, who supplied beef to the Spanish garrisons. The cattle herds of the British Floridians, however, were raided by Georgian cattle thieves and Seminole Indians. Forming posses, Floridians tried to track down their stolen cattle.[22]

Given the unrest in the Floridas, Spanish authorities were unable to prevent Georgians and Carolinians, whom they called "crackers," from settling without permission in East Florida. "Cracker," a curious term which meant a "boaster" among other things, was the nickname of the poor southern backwoodsmen. Don Vicente Manuel de Zespedes, the Spanish governor of Florida, warned his superiors about these "cracker" interlopers in his letter of 1790: "These 'crackers' are nomadic like Arabs and are distinguished from savages [Seminoles] only in their color, language, and the superiority of their depraved cunning and untrustworthiness." Although Spanish soldiers patrolled the border, dozens of these "cracker" families had entered East Florida, where they survived by hunting, farming, and grazing. If the governor tried to compel them to leave, he "would be required to use force to detain these families, to tear down their shacks, and to destroy their crops." After Governor Zespedes left office in 1790, his warnings went unheeded. "Cracker" families continued to enter East Florida, where they squatted on Spanish land and grazed livestock. Claiming the wild cattle which wandered in the pinewoods, they often clashed with the Seminole Indians over the use of grazing lands.[23]

The weakly defended Spanish Floridas not only attracted southeastern "cracker" graziers, but also lured runaway African slaves from the southeastern states. The runaways sought sanctuary with the Seminole Indians, who treated them as valuable auxiliaries. Black Seminoles settled in their own villages, serving with the Indians as scouts, interpreters, and fighters.[24]

When the War of 1812 began, the Seminoles and their black auxiliaries raided into Georgia, easily defeating the local militiamen. These successful raids, however, provoked a retaliatory invasion by American militiamen in 1813. After destroying the Seminoles' villages and driving off their cattle, the invaders left the Indians on the brink of starvation. This punishing raid ended the major campaigns for the rest of the war. But after the war ended in 1814, border troubles erupted again. In 1816, General Andrew Jackson led an American expedition into the Spanish Floridas in order to attack a troublesome fort that was manned by black Seminoles. After American soldiers destroyed the fort, the Seminoles renewed their raids on Georgia. Two years later, Jackson led another foray into the Floridas, planning to punish the Seminole raiders. Defeating the Seminole forces west of the Suwanee River, Jackson tried to overrun the Spanish Floridas. Jackson's attempted conquest of the Floridas embarrassed the American government, and President James Monroe ordered Jackson to withdraw. But in 1819, Spain and the United States negotiated the Adams-Onis Treaty, which peacefully transferred the Spanish Floridas to the Americans. When the treaty was ratified by the U.S. Senate in 1821, the Floridas became a territory of the United States. And two years later, the Treaty of Moultrie Creek ended the hostilities between the United States and the Seminole Nation. Although the treaty set aside a reservation for the Seminoles in central and southern Florida, this did not end the tensions between them and Florida settlers.[25]

Hoping to attract settlers to Florida, the territorial legislature adopted livestock laws resembling those passed in colonial Carolina and Georgia. A Florida law of 1823 required planters to enclose their crops with fences, so that cattle would be free to graze on unfenced lands. Five years later, the Florida territorial legislature passed a law requiring graziers to identify their range cattle with earmarks and brands. Finding familiar livestock laws, "cracker" cattle graziers from Georgia and Carolina entered Florida, and these newcomers pressed upon the Seminole reservation.[26] When the graziers' cattle trespassed on Seminole lands, the enraged Indians destroyed the stray animals. It was "the killing of the cattle of the 'crackers'" that was "the most fruitful source of disputes" between the Seminole Indians and the Florida settlers.[27]

Complaints about lost cattle contributed to the Treaty of Payne's Landing in 1834. This draconian agreement called for the removal of the Seminole Indians to a new reservation in what is now Oklahoma. When the Seminoles resisted, the United States became involved in a bloody war that lasted from 1835 to 1842. By war's end, most of the Seminoles had been forcibly removed from Florida, and most of the Indians' cattle had passed into the hands of Florida settlers.[28]

Since only a remnant of the Seminole Nation remained in southern Florida, the federal government claimed most of the old Seminole reservation, adding these lands to Florida's public domain. Hoping to attract settlers to Florida's public lands, the U.S. Congress passed the Armed Occupation Act in 1842, offering 160 acres of public land to any armed settler who would live on his homestead for five years. Though the act expired the next year, Florida's public lands continued to attract settlers. In 1841, the U.S. Congress had passed a Preemption Act, which allowed settlers to locate on the public domain, earn some cash, and purchase up to 160 acres of land for only $1.25 an acre. And in 1847, Congress passed a Military Bounty Act, allowing veterans of America's wars to claim homesteads on the public domain. Taking advantage of these measures, numerous cattle-grazing families from the southeastern states located in Florida after the Seminole War ended in 1842. Acquiring homesteads from the federal government, families grazed their livestock on the unclaimed public lands of southern Florida.[29]

Deacon E. I. Wiggins of the Mt. Enon Primitive Baptist Church recalled his family's move to the wilds of southern Florida after the conclusion of the Seminole War:

> After the war was over and the Indians were at peace again, Father, Uncle Rabon, Uncle John Raulerson, and Uncle John Thomas decided to move to the frontier in South Florida. So, in 1844, they gathered up all of their stock and drove them down to South Florida and stopped them about twenty five miles east of old Ft. Brook[e] , where Tampa now stands. . . . Now, after they got there, they turned their cattle loose on the range and rode over the country for a few days looking for the best location. . . . Each one drove out to his selected place for settlement and unloaded the families and furniture out in the wild woods without a shelter or tent to keep them dry. The forest was full of wild beast[s] such as Bear, Panther, Wolves and other smaller animals, but they went to work and soon had up shelters and log cabins. Then, they commenced clearing and fencing in land, and they penned their cattle on the land at night in order to protect them from the wild beast[s], while it made their land rich. They were soon having nice

gardens and plenty of sweet potatoes and [dry] rice. . . . They soon had good size [*sic*] fields and [were] making plenty of corn and sugar cane. . . . People kept moving down from Georgia, Alabama and other places from year to year. Being mostly stockmen, they settled pretty far apart—from five to ten to twenty and forty miles apart. So, every year they kept adding more settlers to those little settlements. . . .[30]

Settling in Florida, cattle graziers, or "cowmen" as they called themselves, found wild Spanish cattle, which they incorporated into their mixed-breed herds. By mixing Iberian with north European stock, cowmen created the famous "Florida scrub" cattle. Florida scrubs survived on coarse native forage during the year, they tolerated the heat of southern Florida, and they developed immunity to such endemic stock diseases as tick-borne fevers. Despite their hardiness, scrub cattle were small and scrawny, gaining weight during the warm months when forage was plentiful and losing weight during the cool months when food was scarce. Scrub bulls weighed only six hundred or so pounds, cows averaged about five hundred pounds, and full-grown steers weighed about the same (figure 3).[31] Weighing only a few hundred pounds, scrub cattle "could cover range so sparse that heavier blooded stock literally walked themselves to death trying to find enough to eat."[32]

3. Florida Scrub Steer, Lake Kissimmee State Park, Florida.

Grazing on the unfenced public lands of southern Florida, scrubs lived in herds which stayed within a familiar territory. A typical herd, or "bunch," of scrub cattle included about one hundred bulls, cows, calves and steers, which ranged within two thousand or so acres. This herd territory generally included tracts of pinewoods, hardwood forests, and savannas. During most of the year, cattle grazed in the pinewoods, which contained an understory of native grasses. But in the winter, when forage was scarce in pinewoods, cattle browsed in hardwood tracts, eating the long Spanish moss that hung from the trees. Cattle also foraged in the treeless savannas which often contained stands of evergreen canes. And in the late winter, cowmen burned the pinewoods in order to remove the frost-killed grass and to expose the fresh growth. Within a few weeks, the burned pinewoods yielded a carpet of fresh grass.

When grass was abundant in the spring, most scrub cows dropped their calves. Since the Florida woods harbored wolves and cougars which preyed on stock, cowmen collected newborn calves and mother cows, penning them within rail enclosures on their farmsteads (figure 4). During the day, cows ventured out to graze in the pinewoods and savannas, but at night, they returned to the safety of the pens to nurse their calves. At this time, the graziers' wives and daughters sparingly milked the cows to get milk for household use. By summer, the milk supply began to dwindle, and cowmen released the cows and calves, permitting them to rejoin the herds on the pinewoods range.

4. Reconstructed Cowpen, Lake Kissimmee State Park, Florida.

After releasing their stock, cowmen used the old cowpens as garden sites, since the manure and urine from the penned cattle had enriched the sandy soils. Cowmen plowed up the fertilized cowpen sites and planted such garden crops as sweet potatoes and corn; some of the corn went to fatten grass-fed beeves for home consumption. Although cowpen gardening put roasted potatoes, hominy grits, and fresh beef on the tables, it was cattle grazing that earned the money.

Since cattle grazed on the unfenced range during the year, the greatest expense in cattle grazing was to collect the range stock for processing. Each fall, cowmen gathered their range cattle for marking, branding, and marketing. During these roundups, or "cow-hunts," the graziers' dogs, or "cur-dogs," proved indispensable. Dogs chased cattle out of the brush into the open pinewoods and savannas, where mounted herdsmen, armed with long "blacksnake" whips, drove the animals into temporary pens for handling. Floridians claimed that their nickname of "crackers" came from the cracking sound of these rawhide whips. After penning the range stock, cowmen processed the yearly crop of calves. They "mammied-up" the calves, matching unbranded calves with their branded mothers. Cowmen then branded the calves' flanks and marked their ears with distinctive cuts. Each cowman had his own brand and earmarks, which he registered at the county courthouse. After castrating most male calves to create the next generation of steers, cowmen released the calves and breeding animals on the unfenced range. They kept the steers, which were older than three years, for marketing, and mounted on horseback and accompanied by cur-dogs, they drove the beeves along overland trails to coastal markets.

Though some Florida cowmen relied on black slave cow-hunters, many cowmen owned no slave laborers and had to depend on each other. Though cowmen lived on scattered farmsteads, the cattle-grazing families formed rural neighborhoods, or "settlements," whose members were linked by kinship, marriage, and friendship. Frequently, a crossroads church and cemetery was the focus for a settlement, bringing the scattered families together for preachings, weddings, and funerals. The scattered families in a settlement also came together to help their neighbors clear land, build houses, construct pens, collect range cattle, and drive beeves to market.[33]

During antebellum times, Florida cowmen usually drove beef steers to the distant markets of Savannah, Georgia, and Charleston, South Carolina. Since such trail drives took weeks to complete, cowmen searched for a more convenient market for their beeves. Fortunately, James McKay, a Scottish-born merchant who lived in Tampa, Florida, opened the Cuban cattle market in the late 1850s. Cuba was Spain's leading plantation colony in the New World, producing such export crops as sugar, coffee, and tobacco on so large a scale that it had to import foodstuffs. Among the essential foods which Cuba imported were thousands of tons of dried beef from South America. Turning to Florida as a source of fresh beef, Cuba began importing live steers from James McKay in Tampa. Acquiring vessels to ship Florida cattle to Cuba, McKay bought steers from local cowmen, paying them with Spanish gold coins, and Tampa surpassed Savannah and Charleston as the preferred market for Florida cattle. By 1860, McKay was shipping four hundred steers a month from Tampa to Havana.[34]

Exporting beeves to Cuba and driving steers to the southeastern ports, Florida cowmen prospered in the years between 1845, the year that Florida became the twenty-seventh state to join the Union, and 1861, the year that Florida became the third state to secede from the Union. They lived modestly in pine-log cabins that were equipped with homemade furniture and store-bought utensils. Despite this humble lifestyle, many grazing families owned thousands of dollars worth of property.[35] The probate inventories of deceased cowmen reveal their hidden wealth. As an example, when Eli Hart, a backwoods Florida grazier, died in 1859, he left an estate valued at $3,729.00. Although his inventory listed such modest possessions as cowhide-bottomed chairs, moss-filled mattresses, and iron cooking pots, he owned a sizable estate in slaves and livestock. In addition to holding a slave woman and her two children who were valued at $1,800.00, Hart owned a bay horse worth $150.00, sixty hogs that were valued at $150.00, and three hundred head of cattle which were worth $1,350.00.[36]

At the outbreak of the Civil War, Florida was a sparsely settled state, which counted only 140,000 inhabitants but contained nearly 400,000 cattle.[37] Ideally, Florida held enough cattle to supply steers for the Cuban trade as well as to furnish beeves for hungry Confederate soldiers. In reality, the Union naval blockade stifled Florida' s export trade to Cuba, and the lack of rail transportation hindered shipments to the Confederate armies.

When the Civil War began in April 1861, the Union navy imposed a blockade on the Southern coastline, intercepting any ships carrying war contraband to or from Confederate ports. Union vessels prowled Florida's coastline, and federal forces captured most of the major ports in Florida. Though blockade-running vessels carried some Florida cotton to the Bahamas and James McKay's ships smuggled some Florida beeves to Cuba, Florida's export trade languished during the Civil War years.[38]

Despite the blockade, Florida sent thousands of beef cattle to neighboring Confederate states. The state's cattle shipments, however, were limited by inadequate rail transportation. Florida possessed three major railways, but none of these connected with the Georgia rail system. The Florida-based Pensacola & Georgia Railway and the Georgia-based Atlantic & Gulf

Railroad attempted to build a connecting link from Live Oak, Florida, to Lawton, Georgia, but this connector was not operable until March 1865. Lacking such a link, Floridians had to drive cattle overland to Georgia railheads for shipment to the Confederate armies.[39]

When the war began, the Confederate Commissariat turned to Florida as a source of beef for the Rebel armies. In the fall of 1861, the Commissariat contracted with Jacob Summerlin, a leading cowman in southern Florida, to deliver twenty-five thousand beeves to Confederate soldiers in the southeastern states. Hiring local drovers, Summerlin collected cattle in southern Florida and trailed them to the railheads in southern Georgia. When Summerlin's beef deliveries proved inadequate, the Commissariat appointed Major Pleasant White as chief commissary of Florida in June 1863, and White in turn chose Captain James McKay as head of the commissary district of southern Florida. McKay's drovers were responsible for gathering cattle in southern Florida and delivering them to White's depots in northern Florida. In all, McKay, Summerlin, and others furnished an estimated thirty thousand beeves to the Confederate Commissariat in 1863.

Union commanders were determined to stop the flow of Florida beef to the Rebel armies. Although the Union invasion of northern Florida was turned back at Olustee in February 1864, Union forces seized Fort Myers in southern Florida. From this base, they raided into the interior of southern Florida, returning with hundreds of stolen cattle. At Punta Rassa, Union soldiers built a wharf to load these cattle for shipment to the Union naval base at Key West. In response, McKay called for the creation of local militia units, or "cow cavalrymen," to protect, collect, and drive Florida cattle. Enrolling local cowmen into militia companies, Confederate commissary officers continued to collect and deliver Florida beeves to the Rebel armies. Despite the best efforts of the cow cavalrymen, poor transportation and Union raids limited the delivery of Florida cattle. During the crucial year of 1864, Florida supplied only twenty-five thousand beeves to the Confederate war effort. And during the winter of 1865, few Florida cattle reached the beleaguered Confederate soldiers in the southeastern states.[40]

When the Civil War ended in the spring of 1865, the surviving Florida cowmen had to rebuild the cattle industry after four years of destruction, neglect, and loss of income due to the fact that the Confederacy had paid for their cattle in devalued currency and worthless promissory notes. Returning to his farmstead after the war ended, F. C. M. Boggess, a former cow cavalryman, found that "his house and all his fences" had been burned by Unionists. Boggess, however, had "a nice stock of cattle" in the woods, though "his calves had not been marked" during the war. As soon as he "could split rails," Boggess constructed a shed to house his family, and built cowpens to hold his stock. He then "went to hunting and marking and branding" his half-wild cattle.[41]

While cowmen were rebuilding their herds, McKay was reviving the Cuban cattle trade. In December 1865, when George Thompson, an inspector for the Freedmen's Bureau, was touring southwestern Florida, he came upon McKay's steamer, the *George Marvin,* near Punta Gorda, loading a cargo of scrub cattle for shipment to Cuba. Moving on to Tampa in January 1866, Thompson again saw the *George Marvin,* loading another 250 head of cattle for the Havana market. Thompson looked askance at these scrawny Florida animals and vowed that he "would as soon think of eating a piece of dog as such meat at home." Yet James McKay assured Thompson "that in June or July after the cattle have a chance to renew their flesh," it was "as tender and delicious as any beef in the country." Whether fat or lean, Florida scrub cattle fetched premium prices in Cuba. McKay purchased "the cattle of the Stock Raisers at from 10 to 15 dollars (in gold) per head," and he sold "them in Havanna [*sic*] at from 17 to 27 dollars in gold."[42]

The revival of the Cuban cattle trade proved a boon to Florida's postbellum cowmen. Since it cost only a couple of dollars to raise a scrub steer for market on the open range, and since that same steer fetched up to one Spanish doubloon or fifteen dollars on the Cuban market, cattle grazing proved highly profitable. Spanish doubloons were so plentiful in southern Florida that cowmen stashed sacks of gold between the logs of their cabins, and their babies played with home-made rattles containing the coins.[43] It was estimated that stock raising returned "a gain of 35 per cent. upon the principal." Despite such profits, the raising of scrub cattle was not without "drawbacks," and they were "very serious ones."[44]

The herds of scrub cattle, which ran "at large in the 'range' many miles from the residence of the owner," were "constantly being butchered, driven off, sold, and appropriated by hordes of base and unprincipled scoundrels."[45] Cattle thieves operated with impunity on the open range: they branded unmarked calves, altered brands on older cattle, and even lopped off the ears of earmarked stock. Wealthier cowmen had to hire armed guards to protect their range herds from cattle thieves. If they caught thieves in the act, the guards administered summary justice.[46]

Cattle stealing was not the only drawback to cattle grazing. There was also a shortage of agricultural labor after the Civil War. The war had not only emancipated the black slave cow-hunters, but it had also taken a heavy toll of the white working class. Fortunately, stock raising required far less labor than most types of agriculture.[47] It was possible for "one man with one horse" to manage a thousand head of cattle "the whole year round, except at branding and gathering times." At such times, cowmen had to rely on hired help to gather stock, or a few stockowners joined together to collect cattle.[48] Although wealthier cowmen such as Jake Summerlin, who claimed twenty thousand cattle, hired poor white herdsmen to help him tend his animals, humbler cowmen called on their neighbors for help in collecting, branding, and driving stock.[49]

Despite labor shortages and cattle thefts, Florida cowmen prospered after the Civil War, thanks to the Cuban cattle trade. During the Ten Year Rebellion (1868–78), when the Cubans tried to overthrow Spanish colonial rule, most of the cattle on the island were destroyed, and Cuba imported breeding stock as well as beef cattle from nearby Florida. In the years between 1870 and 1880, Florida exported 166,000 cattle, which were valued at $2,447,000.[50]

During these heady times, cowmen drove thousands of scrub beeves each year to Tampa, Punta Gorda, and Punta Rassa for shipment to Cuba. Returning with saddlebags full of gold coins, they purchased more cattle for their herds. Amassing thousands of cattle, several cowmen earned the moniker of "cattle kings." They included Jake Summerlin, the bespectacled "King of the Crackers"; Francis A. Hendry, who served several terms in the Florida state legislature; Howell T. Lykes, the son-in-law of James McKay; and " Judge" Ziba King, a storekeeper who became the wealthiest cowman in the state.[51]

Yet even the wealthiest cowmen continued to live in humble log houses, which were surrounded by ramshackle outbuildings, luxuriant gardens, and straggling groves. Viewing such homesteads from the roadways, many postbellum visitors to Florida dismissed graziers as "poor white 'crackers.'"[52] A British traveler, who stopped at the homestead of Dr. Huff, a wealthy cowman, was shocked by Huff's backwoods lifestyle:

> Dr. Huff's farm consisted of a few acres planted with corn, tomatoes, melons, and sweet potatoes, all in a flourishing condition, while round the house were a few orange-trees, guavas, and bananas. . . . Though [he was] a large owner of stock, fresh butter was unknown in the house, as is the case throughout Florida. . . . The only meat ever seen at table, except what [game] our guns provided, was tough bacon, which was served thrice a day, at breakfast, dinner, and supper. This with a large dish of hominy, with grease poured over it, was actually the only food procurable or thought necessary. . . .[53]

More famous than any cattle king, however, was Morgan Bonaparte ("Bone") Mizell, the archetypal Florida cow-hunter. Though Mizell acquired several thousand cattle during his lifetime, he always lost them through mismanagement. The hard-drinking Mizell spent most of his career as a cow-hunter, working for Judge Ziba King, the Parker brothers, Eli Morgan, and other wealthy cowmen. Tending cattle all his life, Mizell never got married, never owned a home, and rarely slept in a bed. In 1895, when Bone Mizell was already a living legend, Frederic Remington, who was preparing his article on the "Cracker Cowboys of Florida," immortalized Bone in one of the illustrations, "A Cracker Cowboy," which depicted the weather-beaten Mizell astride a bony Florida horse and accompanied by his cur-dog.[54]

In his article on "cracker cowboys," Remington emphasized the sensational side of Florida cattle grazing—range wars, gunfights, and cattle stealing. He even included a witty exchange between a "Judge Zuigg" (a thinly disguised Judge Ziba King) and a "Zorn Zuidden" (a thinly disguised Bone Mizell) about the legality of cattle stealing. Zorn Zuidden, Judge Zuigg's former employee, had been accused of stealing the judge's cattle. When brought before the judge, Zuidden righteously declared: "Now, see here, old man Zuigg, what do you want fer to go and git me arrested fer? I have stole thousands of cattle and put your mark and brand on 'em, and jes because I have stole a couple of hundred from you, you go and have me indicted. You jes better go and get that whole deal nol prossed" (from *nolle prosequi,* "refusal to proceed" in a court case).[55]

Although Remington stressed the more colorful aspects of Florida cattle grazing, he did make one terse comment about the marketing of Florida cattle: "There used to be a trade with Cuba, but now that has gone; and beyond the supplying of Key West and the small fringe of settlements they have no market."[56] Remington's assertion was somewhat exaggerated, but the Cuban cattle boom was short-lived. Having imported breeding stock as well as beef cattle from Florida, the island rebuilt its cattle herds after the failure of the Ten Year Rebellion. By 1881, it was possible for Cuba to reduce cattle imports, and the colonial government imposed a heavy duty on foreign livestock. These duties dampened Florida's cattle trade with Cuba, so cowmen began supplying beef cattle to Key West, Florida, which had became a cigar-making and sponge-fishing center. Cowmen continued to drive beef cattle to Punta Rassa for shipment to Key West and Cuba, but exports fell to a fraction of former levels. By 1890, annual shipments from Punta Rassa averaged only ten thousand head of cattle. When Frederic Remington visited Florida in 1895, the cattle trade was indeed depressed.[57]

But in 1895, the Cubans renewed their rebellion against Spanish rule. In an effort to starve out the rebels, the Spanish governor general ravaged the Cuban countryside, destroying crops and livestock. Since most of Cuba's cattle were killed in the fighting, the Spanish colonial government had to import foreign stock. In 1897, the governor revoked the import duties, allowing free entry for Florida beeves. Florida cattle, which had been selling for as low as five and seven dollars per head, soared in value to eleven and thirteen dollars a head. Encouraged by rising prices, cowmen and cow-hunters spent the fall of 1897 "rounding up stock and selecting from herds those ready for the slaughter pens" of Cuba. Then, crews of cow-hunters drove the animals to Punta Rassa for shipment to Cuba.[58]

Although there were predictions that the shipment of Florida cattle to Cuba would continue for years, the United States went to war with Spain in 1898, and the Americans handily defeated the Spanish forces. After the island obtained full independence in 1902, it turned to Venezuela rather than Florida as a source of cattle. Beefy Venezuelan cattle displaced scrawny Florida steers in the Havana markets.[59]

Losing their Cuban trade once again, Florida cowmen turned to the American market, supplying beeves to slaughterhouses and packing plants in Tampa and other cities in Florida. Yet they found little demand for their small steers elsewhere. The Chicago meatpacking companies, which dominated the American beef market, preferred meaty western steers to the lean Florida scrubs. In order to win a share of the American market, Florida cowmen had to improve the quality of their beef cattle.

Generations of random breeding on Florida's open range had yielded scrub cattle that were prized more for their hardiness than for their beefiness. The beef quality of scrub cattle, however, could be improved by cross-breeding cows with blooded bulls. Cross-breed steers possessed the hardiness of the cows and the beefiness of the bulls, thus commanding a higher market price than lean scrub steers. Yet whenever cowmen tried to mate blooded bulls with scrub cows, the bulls soon died from tick-borne fever—an endemic stock disease that rarely killed immune scrub cattle. As long as tick fever was prevalent in Florida, there was little hope of improving scrub cattle by cross-breeding with blooded stock from tick-free areas.

Tick fever could be eliminated only by destroying the ticks which carried the fever-producing pathogens. Researchers discovered that dipping tick-infested cattle in vats of arsenic solution killed the attached ticks before they dropped to the ground to lay their eggs. By repeatedly dipping cattle to disrupt the tick life cycle, it proved possible to eventually eradicate the ticks within an area. Planning to eliminate tick fever and thus permit the improvement of Florida cattle, the state legislature adopted a compulsory tick eradication program in 1923. The state legislature divided Florida into a series of quarantine zones. Using barbed-wire fences to isolate the zones, state agents established local dipping vats, where tick-infested range cattle were periodically dipped in arsenic solution to kill the parasites. After a couple of years of cattle dipping, a quarantine zone was declared free of ticks and tick fever. Starting on the Georgia-Florida border in 1923, the dipping program and quarantine zones progressed southward. By 1944, the state of Florida was virtually free of fever ticks.[60]

Following the eradication of the fever tick, cowmen introduced blooded bulls to improve their scrub cattle. Brahman bulls were most favored, since they showed the greatest resistance to the heat and humidity of Florida. After breeding Brahman bulls with scrub cows, cattle owners then mated the mixed-breed cows with Hereford and Angus bulls to further improve the beef quality of their stock. With such mixed-breed steers, Florida cowmen finally won a share of the American beef market: by 1950, Florida ranked twelfth among beef-producing states.[61]

Though mixed-breed cattle survived on the open range, they thrived on improved pastures. After World War I (1914–1918), Florida cowmen began experimenting with artificial pastures, clearing pinewoods and planting exotic grasses. Grazing on these pastures, mixed-breed cattle recorded substantial weight gains, reaching market size long before mixed-breed animals that grazed on unimproved pinewoods range.[62]

Despite such successful experiments with artificial pastures, little could be done to improve the Florida range as long as most of the land was still owned by the state, or by private timber companies, to which Florida had sold much of its public pinewoods land during the early twentieth century in hopes of raising revenues for the state government. When the Great Depression (1929–40) hit Florida, however, many timber companies declared bankruptcy and abandoned thousands of acres of tax-delinquent land which reverted to the state. Holding huge amounts of tax-forfeited timber land that earned no revenue, the state government sold it to the highest bidders—usually, in fact, for the amount of the delinquent taxes. As a result, wealthier cowmen were able to acquire large tracts of grazing land at little cost.

After World War II, the wealthier cowmen continued buying more timberland, thus reducing the amount of open-range grazing land that was available to poorer cowmen. In addition, truck farmers and citrus growers expanded their fields and groves at the expense of the remaining open range. Finally, suburban areas infringed on range lands, when thousands of retired northerners congregated in tract housing developments after the war.

Open-range grazing was not compatible with modern suburban life and agribusiness. Range cattle trampled vegetable fields, invaded citrus groves, and strayed onto public highways and streets where they collided with cars and trucks. Complaints about traffic accidents and stray cattle prompted the Florida legislature to pass a fencing law in 1949, requiring all stock-owners to fence in their cattle. Owners who allowed their livestock to wander into roads, fields, and groves were punished with stiff fines or prison terms. This tough law closed the last of the open range in Florida, and ended nearly three centuries of cattle grazing in the southeastern United States.[63]

Although open-range grazing disappeared in Florida by the mid-twentieth century, this long-lived economic adaptation left a social legacy. The cattle-grazing economy was associated with a distinctive folk culture that was shared by the graziers and their families, and which did not entirely disappear. Many of Florida's contemporary cattle ranchers are the descendants of old-time "cracker" cowmen, they still call on their neighbors for occasional help, and some even worship in the country churches where their ancestors are buried. Yet modern cattle ranchers bristle whenever outsiders call them "crackers"—a pejorative term that Florida newcomers hurl at native-born Floridians. But to the old-time cowmen, "cracker" was an honorific term and not a pejorative. As the descendant of one antebellum cowman told the author:[64] "A lot of people would just say, 'I'm a Florida Cracker!' And he was proud of it. . . . I think it comes from cracking cow-whips. Whenever the people drove their cattle into market, they'd come in cracking their cow-whips, and somebody would say, 'Here comes those Crackers!'"

NOTES

I wish to thank Professors William C. Jordan and Mary Henninger-Voss for inviting me to present a paper for the "Cattle Culture" conference at the Shelby Cullom Davis Center on January 24, 1997, and I also thank the faculty and students of the Princeton University History Department for their cogent comments on my paper.

1. Frederic Remington, "Cracker Cowboys of Florida, " *Harper's New Monthly Magazine* 91 (1895): 339–45.

2. L. Logan, "The Geographical Imagination of Frederic Remington: The Invention of the Cowboy West," *Journal of Historical Geography* 18 (1992): 75–90.

3. For example, see J. Frank Dobie, *Tales of Old-Time Texas* (Boston, Mass., 1956); William W. Savage, Jr., *Cowboy Life* (Norman, Okla., 1975); David Dary, *Cowboy Culture* (New York, 1981); Richard W. Slatta, *Cowboys of the Americas* (New Haven, Conn., 1993).

4. See, for example, W. Theodore Mealor, Jr., and Merle C. Prunty, "Open-Range Ranching in Southern Florida," *Annals of the Association of American Geographers* 66 (1976): 360–76; Mart A. Stewart, "Whether Wast, Deodand or Stray: Cattle Culture and the Environment in Early Georgia," *Agricultural History* 65 (1991): 1–28; Gary S. Dunbar, "Colonial Carolina Cowpens," ibid., 35 (1961): 125–31.

5. J. S. Otto, *The Southern Frontiers (1670–1860)* (Westport, Conn., 1989), p. 30.

6. The coastal flatwoods environment is described in Joe A. Edmisten, "The Ecology of the Florida Pine Flatwoods" (Ph.D. diss., University of Florida, 1963), pp. 1–13, 28, 92–94; and the grazing potential of pinewoods and savannas is discussed in Robert S. Campbell and Wesley Keller, eds., *Range Resources of the Southeastern United States,* American Society of Agronomy, Special Publication no. 21 (Madison, Wis., 1973), pp. 4–15.

7. The cattle industry in late seventeenth century Carolina is described in J. S. Otto, "Livestock-Raising in Early South Carolina (1670–1700): Prelude to the Rice Plantation Economy," *Agricultural History* 61 (1987): 13–24; and the British, Hispanic, and African antecedents of Carolinian cattle grazing are discussed in J. S. Otto and N. E. Anderson, "The Origins of Southern Cattle-Grazing: A Problem in West Indian History," *Journal of Caribbean History* 2 (1987): 138–53.

8. Otto, *The Southern Frontiers (1670–1860)*, p. 35. The origins of the Carolinian rice industry are discussed in Peter H. Wood, "'It Was a Negro Taught Them': A New Look at African Labor in Early South Carolina," *Journal of Asian and African Studies* 9 (1974): 160–79; and Judith A. Carney, "From Hands to Tutors: African Expertise in the South Carolina Rice Economy," *Agricultural History* 67 (1993): 1–30.

9. See Charles W. Arnade, "Cattle Raising in Spanish Florida,1513–1763," *Agricultural History* 35 (1961): 116, 118–19, 122; Brian G. Boniface, "A Historical Geography of Spanish Florida, circa 1700" (M.A. thesis, University of Georgia, 1971), pp. 52, 85, 93, 117, 123, 126–27, 138, 140–41, 144, 150–51, 203; Amy T. Bushnell, "The Menendez Marquez Cattle Barony at LaChua and the Determinants of Economic Expansion in Seventeenth Century Florida," *Florida Historical Quarterly* 56 (1978): 407–8, 410, 417–18, 423–24; Amy T. Bushnell, *Situado and Sabana: Spain's Support System for the Presidio and Mission Provinces of Florida,* American Museum of Natural History, Anthropological Papers, no. 74 (Athens, Ga., 1994), pp. 17–18, 139, 141, 174–75.

10. Mark F. Boyd et al., *Here They Once Stood* (Gainesville, Fla., 1951), pp. 8–10, 19, 48–50; Boniface, "Historical Geography of Spanish Florida," pp. 42, 44–45; Amy T. Bushnell, "Patricio de Hinachuba," *American Indian Culture and Research Journal* 3 (1979): 1–16.

11. Otto, *The Southern Frontiers (1607–1860),* pp. 34–35.

12. See Richard L. Haan, "The 'Trade Do's Not Flourish as Formerly': The Ecological Origins of the Yamasee War of 1715," *Ethnohistory* 28 (1981): 341–58; John H. Hann, "St. Augustine's Fallout from the Yamasee War," *Florida Historical Quarterly* 68 (1989): 180–200.

13. Otto, *The Southern Frontiers (1607–1860),* pp. 37–38.

14. See Dunbar, "Colonial Carolina Cowpens," p. 126; Terry G. Jordan, *Trails to Texas: Southern Roots of Western Cattle Ranching* (Lincoln, Neb., 1981), pp. 28–29; Russell R. Menard, "Slave Demography in the Lowcountry, 1670–1740," *South Carolina Historical Magazine* 96 (1995): 283, 294.

15. Advertisement in *South Carolina Gazette* (Charleston, S.C.), July 18, 1768; newspaper on microfilm at The Newberry Library, Chicago.

16. See Walter Clark, ed., *The State Records of North Carolina,* vol. 23 (Goldsboro, N.C., 1904), pp. 676–77; Allan D. Candler and Lucien Knight, eds., *The Colonial Records of the State of Georgia,* vol. 18 (Atlanta, Ga., 1910), pp. 73–75; Otto, *The Southern Frontiers (1607–1860),* p. 42.

17. Milton L. Ready, "An Economic History of Colonial Georgia, 1732–1754" (Ph.D. diss., University of Georgia, 1970), pp. 101–3; Stewart, "Whether Wast, Deodand or Stray," pp. 1–11; Mart A. Stewart, *"What Nature Suffers to Groe": Life, Labor and Landscape on the Georgia Coast, 1680–1920* (Athens, Ga., 1996), pp. 72–77, 86, 91, 94.

18. "[John] Bartram's Observations in a Journey through Georgia," *The Gentleman's Magazine* 37 (1767): 168.

19. "An Exhortation to Gentlemen of Small Fortunes to Settle in East Florida," ibid., 37 (1767): 21–22; James M. Covington, "Migration of the Seminoles into Florida, 1700–1820," *Florida Historical Quarterly* 46 (1968): 344–49; Brent R. Weisman, *Like Beads on a String: A Culture History of the Seminole Indians in Northern Peninsular Florida* (Tuscaloosa, Ala., 1989), pp. 37–48.

20. William Bartram, *Travels through North & South Carolina, Georgia, East & West Florida* (Phildadelphia, 1791), p. 190.

21. Joseph B. Lockey, "The Florida Banditti, 1783," *Florida Historical Quarterly* 24 (1945): 87–88; Joe A. Akerman, Jr., *Florida Cowman: A History of Florida Cattle Raising* (Kissimmee, Fla., 1976), pp. 24–25.

22. Susan R. Parker, "Men Without God or King: Rural Settlers of East Florida, 1784–1790," *Florida Historical Quarterly* 69 (1990): 135, 140, 149–51.

23. James A. Lewis, "Cracker—Spanish Florida Style," ibid., 63 (1984): 185–204. See also Marion F. Shambaugh, "The Development of Agriculture during the Second Spanish Period" (M.A. thesis, University of Florida, 1953), pp. 47–48, 50, 68; and J. S. Otto, "Cracker: The History of a Southeastern Ethnic, Economic and Racial Epithet," *Names* 35 (1987): 28–30.

24. Shambaugh, "The Development of Agriculture During the Second Spanish Period," p. 56; Weisman, *Like Beads on a String,* pp. 75, 174.

25. Philip C. Brooks, *Diplomacy and the Borderlands: The Adams-Onis Treaty of 1819,* University of California Publications in History, vol. 24 (Berkeley, Calif., 1939), pp. 1, 30–34, 71, 170, 191; John K. Mahon, *History of the Second Seminole War, 1835–1842* (Gainesville, Fla., 1967), pp. 20–50.

26. Leslie A. Thompson, *A Manual or Digest of the Statute Law of the State of Florida* (Boston, Mass., 1847), pp. 134, 419.

27. An American Orson, "The Seminoles," *The Knickerbocker* 7 (1836): 453.

28. Mahon, *History of the Second Seminole War, 1835–1842,* passim.

29. J. S. Otto, "Hillsborough County (1850): A Community in the South Florida Flatwoods," *Florida Historical Quarterly* 62 (1983): 182–84.

30. Deacon E. I. Wiggins, *A History of the Mt. Enon Association, Holding to the Doctrine of Absolute Predestination* (Tampa, Fla., [1921]), pp. 4–5.

31. John E. Rouse, *The Criollo: Spanish Cattle in the Americas* (Norman, Okla., 1977), p. 186; Joe G. Warner, *Biscuits and 'Taters: A History of Cattle Ranching in Manatee County* (St. Petersburg, Fla., 1980), pp. 5–6.

32. Stetson Kennedy, *Palmetto Country* (New York, 1942), p. 216.

33. See J. S. Otto, "Open-Range Cattle-Herding in Southern Florida, " *Florida Historical Quarterly* 65 (1987): 317–26, for a detailed discussion of antebellum cattle-raising practices.

34. The antebellum cattle trade in Florida is discussed in J. S. Otto, "Florida's Cattle-Ranching Frontier: Hillsborough County (1860)," ibid., 63 (1984): 77–78, and his "Florida's Cattle-Ranching Frontier: Manatee and Brevard Counties (1860)," ibid., 64 (1985): 55, 60.

35. The southeastern cattle-grazing lifestyle is described in J. S. Otto, "Reconsidering the Florida 'Cracker,'" *Journal of Regional Cultures* 4 (1984): 8–9.

36. "Old Guardianship—Inventory and Settlement," Records on microfilm at the Circuit Probate Division, Clerk of Circuit Court, Hillsborough County Courthouse, Tampa, Fla.

37. "A Few Facts from Florida," in U.S. Department of Agriculture, *Report of the U.S. Department of Agriculture for the Year 1871* (Washington, D.C., 1872), pp. 161, 167.

38. Alice Strickland, "Blockade Runners," *Florida Historical Quarterly* 36 (1957): 85–93; Robert A. Taylor, *Rebel Storehouse: Florida in the Confederate Economy* (Tuscaloosa, Ala., 1995), pp. 30–38, 96.

39. J. S. Otto, *Southern Agriculture during the Civil War Era (1860–1880)* (Westport, Conn., 1994), p. 37.

40. Taylor, *Rebel Storehouse,* pp. 91–92, 99–130.

41. F. C. M. Boggess, *A Veteran of Four Wars: The Autobiography of F. C. M. Boggess* (Arcadia, Fla., 1900), p. 74.

42. "Journal of Geo. F. Thompson, as Inspector, Bureau of Freedmen, Refugees and Abandoned Lands, on a Tour of Central Florida and the lower West Coast—Dec. 1865," pp. 79, 102, 107, 115, 117–18; Manuscript Collection, Box 24, P. K. Yonge Library of Florida History, Gainesville, Fla.

43. Warner, *Biscuits and 'Taters,* pp. 50, 57, 65.

44. Dennis Eagan, *The Florida Settler, or Immigrant's Guide* (Tallahassee, Fla., 1873), p. 139.

45. Ibid., pp. 138–139.

46. See Akerman, *Florida Cowman,* pp. 209–16.

47. See *Florida: Its Climate, Soil and Productions* (Jacksonville, Fla., 1869), pp. 52, 61.

48. "Cattle Breeding in Monroe County," *The Semi-Tropical* 2 (1876): 203.

49. "Journal of Geo. F. Thompson," pp. 80–81; "Letter from South Florida," *Southern Cultivator* 29 (1871): 299; "Cattle Breeding in Monroe County," p. 202.

50. "Florida," in U.S. Census Office, *Report on the Productions of Agriculture as Returned at the Tenth Census* (June 1, 1880) (Washington, D.C., 1883), p. 1101; Lawrence E. Will, "King of the Crackers," *Tequesta* 26 (1966): 35.

51. See F. Trench Townshend, *Wild Life in Florida with a Visit to Cuba* (London, 1875), pp. 144–46; J. F. Bartholf and F. C. M. Boggess, *South Florida* (Jacksonville, Fla., 1881), pp. 29, 39; Will, "King of the Crackers," pp. 33–34.

52. For example, see J. S. Bradford, "Crackers," *Lippincott's Magazine* 6 (1870): 457–63; George M. Barbour, *Florida for Tourists, Invalids and Settlers* (New York, 1882), pp. 53–55; "Florida 'Crackers,'" *Littell's Living Age* 159 (1883): 625–28.

53. Townshend, *Wild Life in Florida,* pp. 55–57.

54. See Remington, "Cracker Cowboys of Florida," p. 340; Jim Bob Tinsley, *Florida Cow Hunter: The Life and Times of Bone Mizell* (Orlando, Fla., 1990), frontispiece, pp. 1, 4–8, 32–38, 70–71.

55. Remington, "Cracker Cowboys of Florida," pp. 340–42.

56. Ibid., p. 342.

57. Charles L. Norton, *A Handbook of Florida* (New York, 1890), pp. 21, 51; Tinsley, *Florida Cow Hunter,* pp. 50–51.

58. "Cattle-Raising in South Florida," *Florida Times-Union and Citizen* (Jacksonville, Fla.), December 19, 1897; newspaper on microfilm at the University of Florida Library, Gainesville.

59. William T. Mealor, Jr., "The Open-Range Ranch in South Florida and Its Contemporary Successors" (Ph.D. diss., University of Georgia, 1972), p. 271; Mealor and Prunty, "Open-Range Ranching in Southern Florida," pp. 367–68.

60. Mealor, "The Open-Range Ranch in South Florida," pp. 59–61, 75–76; Mealor and Prunty, "Open-Range Ranching in Southern Florida," pp. 368–71; Akerman, *Florida Cowman,* pp. 234–43.

61. See H. O. Lathrop, "Distribution and Development of the Beef Cattle Industry of Florida," *Journal of Geography* 50 (1951): 133–44.

62. Robert S. Rummell, "Beef Cattle Production and Range Practices in South Florida," *Journal of Range Management* 10 (1957): 74–75; Rouse, *The Criollo,* p. 187; Mealor and Prunty, "Open-Range Ranching in Southern Florida," pp. 374–76; Akerman, *Florida Cowman,* pp. 247–51.

63. "Cattlemen Must Make New Cattle Law Work," *The Florida Cattleman* 13 (1949): 6, 30; Mealor and Prunty, "Open-Range Ranching in Southern Florida," pp. 371–74; Akerman, *Florida Cowman,* pp. 253–55.

64. Otto, "Reconsidering the Florida 'Cracker,'" pp. 12–13.

3

Beef in a Box
Killing Cattle on the High Plains

Ken C. Erickson

I grew up as a cattleman, I like cattle. I have no problems when they are cattle and I have no problems when they are thoroughly beef. The transformation bothers me.
—Packing company owner, 1989

Non vi si pensa quanto sangue costa.
(No one thinks of how much blood it costs.)
—Dante, *Paradiso*

A Tour

The two yardmen at "Branding Iron Beef" (the pseudonym of a huge meat factory on the High Plains) pushed a group of cattle from the holding pen and up into the chute.[1] Unlike in the feed yards these cattle came from, there wasn't much dust. Underfoot was rugged concrete, not High Plains dirt. These two men, monolingual Spanish speakers, call themselves *arrieros*, stock drivers. The English version of their union contract calls them *yardmen*. Walking along beside the chute, they encourage the wary animals in soft Spanish. Loud noises and rough prodding excite the livestock and can cause bruises, or worse. The *arrieros* may nudge them a bit with their electric cattle prods but they don't touch the shock button very often.

Pushed farther up the metal-roofed chute to the second-floor knockbox, the cattle face a set of one-way gates, hinged at the top, to keep them from backing out. They usually resist the noise—and, allegedly, the smell—emanating from the killfloor below. The chute narrows, and the cattle arrive at the knockbox. Beside the chute, in a narrow isle overlooking the killfloor below, stands the *matador*, the knocker. He directs the pneumatic stun gun by its twin handles, a safety feature to prevent accidental discharge. Stun guns have been known to kill knockers as effectively as they stun cattle. The gun, about the size and shape of a large orbital sander, is tethered to the ceiling by a spring-loaded cable. The knocker sways it over the animal's skull, moving it as the animal tries to back out of the box.

On that afternoon, I watched the *matador* with a group of new-hire men and women who were taking their first tour of the Branding Iron beef plant. All the slaughter workers we watched were men, and all were Mexicans. The tour group was mostly Spanish speaking, except for one woman, Tammy, a newcomer about to start her first job away from her home in a small town in Oklahoma. We had all gathered a few moments before, outside the building on the asphalt, next to a metal stairway that climbed to the knockbox. Our tour guide and instructor had described what we would see, mostly in Spanish, but with some details in English for Tammy. He then let us climb the stairs and crowd into the doorway to watch the routine.

At a work station just below the *matador* and his bovine target, the shackler, occasionally within view in his hockey mask and helmet, reached for a length of chain attached to a trolley. He was ready to take a quick wrap around the left hind hoof when it would drop through the slot in the floor of the knockbox. That happened, as it should, about the time the stun gun made its muffled pop, the sound of a steel rod banging into the animal's skull, cracking it and rendering the animal limp. Limp, but still breathing.

Beyond the knocker, down on the killfloor, this industrialized transformation of flesh into food continued. The killfloor is humid, warm, and loud. There is the smell of blood and manure but none of the acrid stench most people associate with packing plants. That smell, the one that reaches people driving by on the highway, comes from rendering, not from slaughter. Don Stull's description of the killfloor at the largest packing plant in the world, an easy drive from our new-hire vantage point above the knockbox, matches the Branding Iron killfloor in most respects. There, as in any packinghouse, the business of shackling the hind hoof of a cattle is

> easier said than done, since its legs may be jerking from the stun. As each animal (referred to in both the singular and the plural as "cattle") is knocked and shackled, it falls forward and down onto a mechanized overhead trolley and swings out onto the floor, hanging upside down from its left hind hoof.
>
> After traveling the full width of the floor, the suspended animal makes a 90 degree turn to meet the sticker. Holding a knife with a four-inch blade, he deftly makes an incision in the hide at the base of

> the throat, careful not to pierce the throat itself. The can-hanger then hooks a stainless-steel cylinder under the animal's open throat. The bleeder quickly severs an artery and blood gushes into the cans, spilling over to drench the workers and run through the grated floor beneath their feet. The three stand shoulder to shoulder like gladiators, wearing hockey helmets and face masks for protection from flailing hoofs and their own knives, which are sometimes knocked into their faces. The bleeder deals the fatal blow, but dying is a process. Signs of life such as muscle twitches and jerking limbs may continue well down the line.[2]

At Branding Iron, death is represented in technical terms. The steps are marked out on job descriptions and on factory blueprints, distancing the moment of death from easy capture so that the fatal event is not a moment but a set of stages. This is a calculated death. The transition from live cattle to meat is never marked by a single blow, or even by a single industrial tool in a packing plant. Determining exactly when the killing happens, and identifying which job actually does the deed, is frustrated by the minute division of tasks along the way from animal to meat.[3] Noëlie Vialles, in her ethnographic study of the abbatoirs of the Ardour in France, shows how this serves cultural ends, masking culpability.[4] The industry benefits, too. Hiding the moment of death is important to an embattled, marginally profitable industry, one that is so often looking over its shoulder at its vocal critics: independent cattle feeders and producers, labor unions, workplace health and safety inspectors, animal and consumer rights advocates. And dividing up the fatal task does more than hide the moment of animal death: it is important for efficient production. Creating several tasks out of dispatching a cattle makes it easier to train new workers, limits the number of tools each must use, and allows quick repetition of the steps. After slaughter, meat factories add even more shop-floor complexity as they break down meat processing into smaller and smaller steps, which are carried out by immigrants from Mexico and South Texas, former refugees from Vietnam and Laos, Euroamericans, and African Americans from anywhere that jobs are scarce. It is an old story, but one rarely told outside the new packing towns of the High Plains.[5]

Death may be represented as an event, but in fact it is a process, one that is marked by ritual when it happens to humans.[6] Throughout most of human history, animal slaughter has also been marked by ritual, elaborated and conducted in special ways, and generally formulated, in Leviticus or among East African pastoralists, as sacrifice. An anthropological review of the patterns of animal sacrifice is beyond the scope of this essay but it is clear that most cattle-raising humans outside of late capitalism attended to both the symbolic and the economic importance of cattle when they slaughtered them. Evans Pritchard, in his work on the Nuer, reported that animals were never killed just for meat, nor were they killed without taking into account the empirical limits on herd productivity.[7] The process of animal death in industrial settings, on the other hand, is new and markedly different because, among other things, the rituals of modern beef production are not about the kinds of social reproduction entailed in small-scale beef slaughter and redistribution.

What follows is barely a dip into the stream of cattle and humans that flow through the enormous new meat factories on the High Plains. A historical summary of the beef-packing industry highlights how the job of killing and cutting cattle in the slaughterhouse has not changed very much in the past century. Some aspects of work in the slaughter and processing divisions of Branding Iron Beef are described to see how work practice is ritualized and symbolized given the dangerous nature of slaughter operations and the rationalized business of processing. Because immigrants work at the boundary between animal life and death, our data on the plant's workforce are examined to provide a preliminary look at how ethnicity and gender, language and culture are deployed in the factory. Finally, some preliminary findings about individual reactions to work with routinized animal death and meat cutting are offered. These findings provide a basis for assessing the prospects for better understanding the nature of the slaughterhouse workplace, and of rethinking the problems that routinized animal death and routinized meat processing create for both industry and worker.[8]

Learning about Meat

The packing industry does not like to see itself in print or on television. At this writing, the major beef packers do not have sites on the World Wide Web, either. What researchers have been able to learn has sometimes been learned covertly or from interviews outside the plant.[9] Our research team had a public role in the packing community and shared an epistemological commitment to overt, collaborative research practice.[10] But gaining access to the shop floor wasn't easy, even after the Ford Foundation funded a two-year study of Garden City, Kansas, one of the new packing towns. Early in 1987, the foundation announced its national Changing Relations Project, intended to guide public discussion of what promised to be an especially immigrant-rich 1990 United States census, one that would reflect the greatest growth in U.S. new immigrant population since the great wave of newcomers at the turn of the century.[11] The Garden City ethnographic team owed its existence to the expansion of one large beef plant and the construction of another, the IBP plant, arguably the largest of its kind in the world. The immigrant workers recruited for these plants changed the cultural texture of the region and made it an object of anthropological study.

Don Stull, the local project's principal investigator, gained entry to IBP by taking repeated official tours and by enrolling in a beef-grading course offered by the local community college. He participated in the course practica, held in the plant's coolers, where beef carcasses were chilled and graded before they were cut, shrink-wrapped, and boxed.[12] Another team member, Michael Broadway, analyzed regional and national data to track the geographic and economic trajectory of the globalized meat industry.[13] The team interviewed packers, managers, and their families, and for a short time, I was retained by a Nebraska packer to look into its astronomical new-hire turnover rate—more than 40 percent of the workforce per month.[14] (Managers there weren't surprised by the high turnover, but the owner wanted to sell the plant and was bringing in consultants to give the appearance of a problem under study and under control.) Working in Nebraska provided participant observation data about the processing floor to add to our interview, tour, and cooler information. An opportunity for more time in a plant came later.

In 1993, after the end of the Garden City study, three researchers, Don Stull, Miguel Giner, and myself, formed another ethnographic team and were awarded a research contract to study the shop floor of another town's packinghouse—not the world's largest but very nearly so—here called "Branding Iron Beef." Workers there, most of them Mexicans, had walked out in defiance of both their union representatives and Branding Iron's local management. We were hired by a joint union and management committee (but paid by the corporation) to look into "cultural issues" in the plant that were thought to be related to the wildcat strike. To this end, we sought to discover how managers, supervisors, and workers dealt with one another and with the official corporate culture. We used team ethnographic methods—participant and non-participant observation and interviews with managers, supervisors, community members, and workers.[15] We emerged from the plant with a better understanding of how the meatpacking industry makes some demands that workers of any language or culture resist.[16]

Unpacking Meatpacking's History

The U.S. packing industry originated, so the story goes, when pork processing became concentrated along the rivers of Ohio, Indiana, Illinois, and

Kentucky.[17] Then as now, livestock processing facilities were located with access to raw materials and transportation in mind. "In those days . . . pork was salted down and packed in barrels, and hence arose the name 'the packing industry'."[18]

Strategies by western railhead packers to control eastern markets went hand in hand with cost cutting, accomplished through deskilling—dividing slaughterhouse work into smaller and smaller jobs and offering less and less money in return for doing them. There were also some genuine technological innovations in the preservation and transportation of meat that made possible the industry's first period of growth.

The development of the refrigerated rail car and its promotion by Phillip D. Armour ensured the packers' control of eastern markets.[19] The development of these cars, the use of a moving chain to move the carcasses along, and later, the development of mechanical refrigeration in the slaughterhouse, completed the industry's first round of technological innovations.[20] The resulting plants were technically simple. Cattle were hung on hooks suspended from trolleys that rode along metal overhead rails pulled by a moving chain, exactly as they are today at Branding Iron. The plants themselves were huge, multistory affairs, built to take advantage of gravity. Cattle and hogs were driven up switchbacks and long, inclined ramps to meet their fate on the top floors. The knocker, always male, wielded a sledgehammer that stunned the animal before it was hoisted up and hung from the trolley, there to be pulled along, in the early days by gravity, and later on by a powered chain, as it was bled, skinned, gutted, and split. This was, and is, the killfloor.

Inedible parts of the animal were dropped through holes in the killfloor to tanks and troughs below, much as they are in today's boxed beef plants.[21] In the 1880s Armour added a motorized chain to move the carcasses along. "This probably was the first attempt in industry to control by power the flow of raw product during a manufacturing process."[22] Henry Ford claimed to have taken the idea for continuous-flow production from the packinghouse.[23]

After the introduction of the powered chain, the output of the axe-wielding splitters, (again, the job has the same name today), more than doubled in ten years. By then, four splitters "were handling twelve hundred at a lower hourly [wage] rate. This meant an average of thirty animals per man hour; by 1900, the average was up to thirty-five."[24] These days, the carcass no longer pauses at the splitter's station, and motorized saws have replaced the axe. Branding Iron kills as many as 325 head every hour, on two eight-hour shifts, using splitters on moving platforms that rise up, then lower and move along with the carcass. The semiautomated side puller, which strips the hide from the ribs, and the down puller, which completes the operation before dropping the hide down a hole, also appear to be recent innovations. In any event, all the big plants have them now; industry photos of older and smaller plants don't reveal them.[25]

Profitability depended—and depends—on control of markets, transportation, and cheap, deskilled labor. By the turn of the century, the all-around butcher had been replaced "by a killing gang of 157 men divided into 78 different trades, each man performing the same minute operation a thousand times during a full workday."[26]

Labor's response was long and complex. In 1904 in Chicago, twenty-eight thousand packinghouse workers walked off their jobs and into the streets. The walkout linked workers of all ethnic backgrounds. Union broadsides were printed in six languages, churches provided relief, and the women, said one settlement house worker, "will be the last to give up."[27] But there was plenty of surplus labor; while the unions made some gains, the packers hired nonunion workers, and the process of deskilling continued. After that, in the age of corporate liberalism, efforts to control labor power included both the stick and the carrot, as Carver's manual for packinghouse personnel managers demonstrates, with its advice for starting health clubs and citizenship classes for workers.[28]

The rise of the early packers—Cudahay, Armour, and Swift—depended on a ready supply of labor, on production processes designed in accord with Fredrick W. Taylor's scientific management principles, and on a few new tools like the refrigerated rail car and the moving chain.[29] The recent restructuring in the industry, and the rise of the "Big Three"—IBP, Monfort, and Excel—has mirrored the first period of rapid growth in the industry by its grounding in shipping-related technical change and continued Taylorized management.

Boxed beef, the latest technical innovation, was but one element of the global restructuring in which capital got the upper hand over wages and regulations during the global recession of the 1970s. Broadway writes that IBP was responsible for developing boxed beef; he sets 1967 as the date of its introduction.[30] By globalizing its control of labor markets, it took advantage of nonunion workers, often women and immigrants, to provide the labor power for intensified, low-cost production.[31]

Boxed beef does for Branding Iron what refrigerated rail cars did for the members of the original Beef Trust. It provides marketing advantages, extends the life of a perishable product, and makes shipping easier. The early refrigerated rail cars, with carcass beef hanging on overhead rails, were always tipping over (Armour 1907). Semitrailers hauling "swinging beef" down the interstate had the same problem. Boxed beef put an end to large-scale shipments of swinging beef; but it did not put an end to old fashioned, Taylorized work.

THE RITES OF BOXED BEEF AT BRANDING IRON

En d'autres termes, la mort aux yeux des primitifs, est une initiation.
(In other words, death in the eyes of primitive people is an initiation.)
—Robert Hertz, 1923

Branding Iron, built in the early 1980s, has lower injury and turnover statistics than many other producers of boxed beef.[32] Branding Iron is unionized, and is considered to be among the more progressive of the late twentieth-century beef industry giants. As new workers learn during their tour, cattle move from the yards, up the ramp, into the knockbox, and out onto the slaughter floor. After skinning, gutting, and splitting, the carcass halves are graded, weighed, and shunted off to the coolers, and the transition from living flesh to meat is complete. The carcasses wait overnight (or over the weekend) before emerging on the other side of the plant, in the processing division, a refrigerated shop floor with its own managers, surrounded by its own cafeteria, locker room, and maintenance room. There, the swinging beef is transformed again, this time into boxed beef.

To produce Branding Iron boxed beef, one cattle is transformed into seven heavy cardboard boxes 22.5 inches long by 17.5 inches wide by 10.5 inches deep. Together, the seven boxes (a magic number in the Occident) weigh about 427 pounds.[33] Boxed beef is the unmarked category in the corporate lexicon. Neither hamburger, nor organ meats, nor carcass beef—all produced at Branding Iron—are included in the official boxed-beef calculus. The manly arts of boxed beef production focus on these seven boxes.[34] Boxed beef production is a site of contest and negotiation, a place where the unquestioned givens of human relations, life and death, are both acted out and called into question every day.[35]

Slaughter and Danger

In the slaughter division, death is in the foreground, and it engages all the senses. Paul Stoller has forcefully reminded ethnographers that privileging the eyes over the other senses impoverishes ethnographic understanding.[36] The overwhelming sense impressions of the killfloor—loud machinery, unfamiliar smells, humidity, dramatic death—force attention to more than the visual, especially for the first-time visitor. The killfloor makes a striking contrast with the clean, well-lit and businesslike offices that occupy the second story of the adjoining building, where managers and clerical staff monitor the daily operation of the plant. This is the place where visitors are greeted, where there are few hints of the radically different environment just a few yards away.

The office is the official space belonging to the engineers and the managers, clean and dry; the space belonging to slaughter workers is down below, bloody and humid. But managers and engineers move with ease between those worlds by making a simple trip up or down the stairs—whereas slaughter workers rarely go upstairs. The killfloor comes easily into the manager's consciousness with a simple shift of conversation in the upstairs meeting room, from balance sheets, perhaps, to the death-defying near misses that happen with some regularity.

It is the routinization of animal death that stands out on the killfloor. The corporation is ambivalent about sanctioning the development of individual or group symbolic accommodation to the anxieties that arise in the process. Physical risk is not less real in the processing division, where the bloodless cuts of meat that pass through processing are just that: meat, a fait acompli. But in the slaughter division, there are arguably more symbols of old-time butcher craft, more ways of representing what it means to be a butcher. Signs of ritual in the usual sense only appear at the edges of production, where new workers are made into butchers, because production methods throughout the plant tightly control workers' physical movements. Individual oblations, if any are offered by the slaughter worker, are internal. The only ritual in the actual, moment-by-moment knife work on the killfloor lies in the reenactment, cattle after cattle, of the rationalized production system.

In a sense, individual atonement for killing is balanced in slaughter by the sometimes real, sometimes merely talked-up danger and pollution surrounding death, blood, and living matter out of place.[37] Real danger is present in hanging an undead cattle from an overhead trolley and working beneath it. Cattle can kick; the contest in the blood pit can at times be a real contest between man and beast, more immediate than the general contest between workers and the deathless activity of meat production. Unlike in the "trap," where cattle are dispatched in the abbatoirs of the Ardour, cattle dropping out of the Branding Iron knockbox are just a few yards from the flayers and the gutters. Commotion in the blood pit can put everyone at risk. Here, in industrial slaughtering, "risk reintroduces an appearance of fairness, which offsets the facile treachery of the 'trap'."[38]

There were always at least two men in the blood pit at Branding Iron. On one occasion, the shackler missed a cattle's leg, and the animal fell from the trolley to lie senseless in the pit. One of the stickers had to leave his station to maneuver an overhead cable into position, hook a leg, and hoist the animal back up so the shackler could reattach the animal to a trolley. The slippery floor, and the fact that the chain did not stop, made this difficult. The cattle kept coming. The sticker explained this to me in the nurse's station just outside the killfloor, where he was re-wrapping his knee a few days after his fall.

Occasionally, the knocker also misses, and a cattle drops down the knockbox ramp and onto the slaughterhouse floor. There is a fence and an iron gate at the head end of the blood pit, beneath the knockbox, a physical reminder that something alive, something that might need to be fenced in, just might come rolling down the incline. A cattle might need to be captured, sentient and breathing, behind the fence. Though it happens only very rarely, quick thinking, real reaction to a real danger, not routinized repetition, may be called for here. Here in the slaughter division, coexisting with (and partially explained by) the presence of occasional danger, are symbols of craft skill that are not so prominent in processing. For example, while many slaughter workers wear a rubber apron, none wear white frocks, the aseptic, scientific, and anonymous garb of the processing worker. Maintenance men and slaughter workers wear blue shirts clearly embroidered with their names, displaying both individuality and the dress of a tradesperson. These workers are not denied this display of identity.

Processing and Identity

The processing division, where chilled carcasses are quartered, cut, trimmed, ground, and boxed, is the more recently mechanized and, in some respects, more rationalized side of the plant. Processing, the relatively bloodless domain that was formerly closer to the domestic sphere, part of the grocery store or specialty butcher's work, is the newer element in the industrial pattern. It is noisy here, and as in slaughter, easy conversation is replaced by shouting. Danger from falling equipment, from falling sides or quarters of beef, from slips and falls, and from repetitive motion disorders like carpal tunnel syndrome is very real here, and the statistics demonstrate it: the new processing divisions are loud and dangerous places to work.[39]

The processing division presents workers with different constraints, a different environment through which to make individual meaning and symbolically mark group identities and personhood (not machinehood) in the meat plant. Ethnicity and gender, if they are visibly expressed at all in the processing division, have to peek out from under white frocks. Workers personalize their nametags on their hardhats, and sometimes playfully paint one another's frocks with graffiti drawn with a juicy piece of trimmed meat. They all receive a laundered, generic white frock at the beginning of the shift, and a second, clean one after lunch. Although manicures and nail polish are hidden beneath white gloves, mascara and blue eye shadow may be seen beneath the brims of some hardhats. U.S. and other national flag stickers are fairly common features—usually on men's hats. A bandanna emblazoned with the Mexican flag is sometimes visible beneath the back of a white hardhat, flashing a red, green, and white reminder of the wearer's national identity, although rock and roll band logos are seen just as frequently.

The business of offal production (it is pronounced as it smells), including tripe, edible intestines, and organ meats, and the hamburger trim and grind lines are often, just as in Swift's and Armour's day, the province of women and new immigrant packers.

"Reducing the level of skill required" is a phrase found over and over again in trade publications touting the latest new tool in processing. The industry not only continues to routinize its activities, but tries to take the skill out of them. These efforts must continue to be unsuccessful because the natural variability in the raw material and the cultural variability produced by everyday work practice among interdependent workers on a line demands a certain level of physical and interpersonal skill. This is true in all present-day technical work.[40] Current experiments with automated beef-boning machines are but another verse in the old deskilling song. The New Zealand Meat Industry Research Institute hopes its new automatic beef-boning machine will "reduce labor requirements, both in the number of workers and in the level of skill required to run the machines."[41] Branding Iron has no such automatic boning machines: it has knife workers. The men and women who run the knives at Branding Iron appear to be mostly Mexican immigrants, and indeed, the plant's documents support this.

Branding Iron's Workers

According to the plant's employment records, about 68 percent of the workers are Hispanic, as against 23 percent white, 8 percent Asian, and 2 percent African American (see table 1). The so-called "Hispanic" ethnic category, fabricated by the U.S. government for census and labor reporting purposes, is not particularly helpful for understanding the cultures on the shop floor. It masks important national, linguistic, and cultural differences among Spanish-surnamed workers. Many are new immigrants—Mexicans, Tejanos, Chicanos, and a few South and Central Americans—while others are established residents, long-time packers who usually refer to themselves

TABLE 1. Gender and Ethnicity at Branding Iron

	Men[a]	Women[a]	Total[a]
Native American	1 (0)	2 (0)	3 (0)
African American	31 (0)	7 (2)	38 (2)
Asian	114 (6)	58 (13)	172 (8)
White	400 (22)	113 (25)	513 (23)
Hispanic	1,263 (70)	279 (61)	1,542 (68)
Total	1,809	459	2,268

Source: "Branding Iron" employment records, 1993.

[a]Figures in parentheses show each ethnic group as a percentage of male, female, and all employees.

as "Mexican Americans." But the catchall Hispanic category does reveal a general pattern: that Spanish-surnamed workers dominate the Branding Iron workforce.

Packinghouses have recently made special efforts to recruit women to work on the line; some managers and owners suggest they are more "stable," a common term of praise in managerial discourse about low-wage jobs.[42] Indeed, IBP's new Lexington, Nebraska, plant includes a day care center in an effort to attract women workers.[43] It is mainly men who do the killing and the heavy knife work at Branding Iron. Women mostly do the so-called finer, domesticated work: in the clerical department, in offal, trim, and hamburger. In general, the plants rely on the labor power of immigrants and women, but give most of the responsibility (and reward) for managing the plant to Euroamerican men.

Ethnicity and gender are not spread evenly through Branding Iron's job hierarchy. This mirrors the meat industry as a whole. Managers throughout the meat and poultry industry are overwhelmingly native-born, middle-aged men. Most are Anglo-Americans with limited knowledge of other national languages or cultures.

By contrast, most "production workers in the industry are immigrants, predominantly from Mexico, South America and Asia."[44] This is in line with the industry's history. Over seventy years ago, Arthur Carver, a personnel manager for the venerable Swift & Company, wrote:

> The labor element engaged in the meat packing industry is made up of representatives of many nationalities. The nature of the work to be done in many departments of a packing house is such that the necessary degree of skill may be acquired in a very short time. The industry has consequently been for years an attractive one to immigrants who have had no special training in any kind of skilled work but are confronted with the necessity of earning a living. The result is that,

> even with the restrictions which have been placed upon immigration within recent years, many nationalities are represented in the working force of almost every meat packing establishment.[45]

Carver's assessment still rings true today. Note his attention to the "necessary degree of skill." What he had in mind was a low level of skill. He went on to describe a plant in which immigrant workers made up 43 percent of the plant's total, 35 percent lower than the "nonwhite" percentage at the Branding Iron plant. If Carver's plant was representative, it indicates that there is nothing new in the intensified labor recruiting that the boxed beef packers have employed in recent years to staff their new factories.

We found that young, inexperienced men dominated the work force at Branding Iron and many of them were recent Mexican immigrants. Length of employment for all hourly workers is 3.5 years, but overall averages are misleading. An analysis by Stull found two distinct patterns of tenure for hourly workers—one among old-line packers, and one among recent immigrants. While 55 percent of hourly workers had been hired since 1992 and averaged only a bit more than one year of service, the other 45 percent averaged 6.3 years. Among these were many old-line packers who had been at the plant since it opened.

When the newcomers and old-line packers are examined more closely, Hispanics emerge as the most recent employees of all. Using surname as an indicator of ethnicity, hire dates and ethnicity of hourly production workers before and after 1990 were examined to measure long- versus short-term employment. Only 29 percent of the 2,030 line workers employed at Branding Iron in March 1994 were hired before 1990. Of these 592 long-time workers, 57 percent were Hispanic and 43 percent were non-Hispanic. Among the newer workers (those hired since 1990), 80 percent were Hispanic. The newer the worker, the more likely he or she is to be listed among the Hispanic names in the plant's records.

We wondered how many Hispanics were new immigrants, so Stull explored Spanish given names as a proxy measure for recency of immigration. Among all hourly production workers with Spanish surnames, 84 percent had Spanish given names as well, suggesting that they are likely to be new immigrants or members of families who recently immigrated to the United States. Non-Hispanics tended to work on first shift, which is generally preferred by workers to the second, or night shift. And long-term workers were more likely to hold positions that demanded less stamina. Visitors find precious few long-term workers among the "industrial athletes" (so one admiring manager called them) at the shackler's station or among the stickers in the slaughter division's blood pit.

Hispanic hourly workers are thus more likely to work at physically demanding jobs that are subject to higher rates of injury and turnover and are also nearest to animal death, shackler and sticker among them. These workers are more likely to be monolingual speakers of Spanish and to lack suitable job alternatives in the community. This means that they are likely to leave the plant for varying intervals of rest and rehabilitation in Mexico, only to return at some point. In fact, 20 percent of hourly employees hired by Branding Iron since 1992 were rehires. Employees who leave the plant for extended periods risk loosing their accumulated seniority, and if they are rehired, they get the positions that others do not want.

Beef Plant Languages and Cultures

To the Mexicans the American cowboy owed his vocation . . . He obtained from Mexican sources all the tools of his trade, all the technic of his craft, the very words by which he designated his utensils, the very animals with which he dealt.

—*Philip Ashton Rollins*, 1979

A sociolinguistic look at the packing plant would explore the cultural gaps that language and culture leave behind when distinctive cultural groups come together in a single arena. [46] The boundaries between Branding Iron's cultures are hardly distinct, and the plant itself, like any industrial shop floor, creates a powerful cultural pull that limits and directs the expression of individual linguistic and cultural identity in many ways.

The high proportion of Mexicans in our new-hire tour of the killfloor offers one point of departure. After all, the Anglophone cowboy (and cowgirl) owe most of their cattle-working technology and lexicon to their Mexican *colegas*.[47] In the plant, many Mexican words for jobs on the slaughter side do not require affixes, hyphens, or additional context for them to denote the cattle-related job at hand. The American English words, on the other hand, rely on the beef plant context to point to a beef plant job. The Mexican lexicon seems pre-adapted to cattle, to killing cattle, and to work in the beef plant.

The English term *knocker*, used in this and all meat plants, has to be said in context for it to mean "one who kills cattle." *Matador* (not used in Spanish for "killer of humans," for which *asesino* is generally reserved) means killer of cattle in both bullring and beef plant.[48] This term is used by Spanish speakers in the plant. They do not code-switch to the packinghouse-English word. The English word *sticker*, with a bovine-related meaning only in the context of the packing plant, becomes *picador* in Spanish, intimately associated with *el corrido* and always used to describe the man who bleeds the animals. No code switching is needed for this lexical item. Even the cattle driver, the *arriero*, is a more specific word than the rather generic *yardman* of the union contract.

Among *norteños*—men from Northern Mexico—and probably among working-class men and women who come to the High Plains from anywhere in the Mexican Republic, echoes of ranching are always within earshot. *Norteño* music, the story songs of the *corrido* tradition, and the *charreada* (the Mexican rodeo), are part of after-work recreation in or near all the High Plains packinghouse towns.[49] All are full of cattle and ranching imagery. Male Mexican friends in the packing town call one another *buey*, "steer," using the name as a generic pronoun for reference and address, with some affection for a close friend and with more than a hint of malice (and emasculation) for an enemy. And there are complex historical associations between the packing industry and Mexico. During the Mexican revolution, Pancho Villa himself owned and operated a beef plant in an effort to break the hold that Texas packers had on the marketing of Mexican cattle.[50] Linguistic and cultural ties to the cattle industry are part and parcel of the equipment that Mexican packers bring to the killfloor.

Stereotypical *machismo*, Mexican male bravado, is also supposedly part of the equipment that Mexican immigrants bring to Branding Iron. But Mexicans have not cornered the market on beef plant *machismo*. The "war stories" of the old-line packers are retold with the same manly bravado. And there are very macho women on the killfloor, working at physically demanding jobs, sometimes running pneumatic "round knives" that flay the cattle, or pulling their "count" on the gut table. Many can and do participate in sexually explicit verbal dueling with their male colleagues. But except for someone wandering away from her usual job station, no women were seen at the front end of the killfloor line.

The young Mexican workers who proudly did the heavy and grisly work of hanging severed heads on hooks grinned and struck bloody muscleman poses as we watched them do their work. But physical wear and tear takes its toll even on the most macho. Industry-wide workman's compensation claims bear this out (as do Branding Iron's own injury rates).[51] A scan up and down the heavy boning lines in the processing department—the chuck line and the break line where the work is hardest and the risk of repetitive motion injury high—suggests that Mexicans might be found there more often than their overall numbers in the plant might suggest. Their presence on a particular line reflects the personnel department's notions of "tough, hardworking Mexicans," the kind of racialized stereotype we sometimes heard from supervisors and managers alike. At the same time, the numbers also reflect worker agency. Mexicans seek positions under Spanish-speaking supervisors, some of them highly educated immigrants, whose interpersonal and training skills they appreciate. And many Mexicans desire the higher wages paid for the heavy boning jobs. For those with family ties in Mexico, a job that pays well until a shoulder or a wrist blows out is an option that seems to make sense. Multiple household incomes are more often available to Vietnamese families than appear to be the case for the many single Mexican men in the plant. For a single Mexican man, the option of a lower-stress job is less attractive in the absence of the financial and emotional support of kin and fictive kin found in Southeast Asian households.[52]

Euroamericans in the plant, like their Mexican coworkers, also exhibit a range of cultural backgrounds: city people, first- and second-generation immigrants, sons of packinghouse executives, daughters of Oklahoma dirt farmers, and women from working-class families in the region's cities. These established residents also participate in the pan-American cowboy culture.[53] Men and women alike participate in the manly talk and long-suffering tolerance that is exhibited along with vocal complaints about the difficulty of working with short-crewed lines or stressed supervisors. What they sometimes share with one another that the Mexicans tend not to share is a history of labor organizing in old-line packing plants .[54]

And what about the Southeast Asians who make up about 8 percent of the work force? Plenty of them work heavy boning jobs on the processing line. They do turn up on the killfloor but they may be equipped to view the bloody business of slaughter in a somewhat different light. One wonders about the older Vietnamese men, who often experienced long-term incarceration during and after the struggle in their homeland. None of these men are on the killfloor, but are found working lighter jobs elsewhere in the plant. And the "Asian" category masks a certain ethnic diversity. There are Lao workers here, whose cultural and linguistic traditions are rooted in a rural, Buddhist homeland that lacks both the urbanization and Roman Catholicism found to a greater extent in many parts of Vietnam.[55]

In keeping with their strong Buddhist tradition,[56] Lao workers at Branding Iron, unlike many of their Vietnamese coworkers, seem to share a concern about the taking of animal life, even though many (probably most) of them will eat meat. A feast that I attended with Lao/Thai speakers (and with my former High Plains Lao meatpacker host) in northeast Thailand, just across the Mekong River from Laos, included poultry; and household consumption of meat among Lao in U.S. and Canadian packing towns is probably quite a bit higher than in the homeland.[57] But none of the three people, including two family members, for whom the feast was held played a part in killing the birds. Lao workers on the High Plains are generally not vegetarians, but they restrict their participation in the beef plant to working with meat after someone else has done the killing, or they work in other areas of the plant—the laundry or the front office.

When the immigrants and refugees arrived to work in the High Plains plants in the 1980s, established resident folklore was full of hard-drinking and tough-fisted Anglo packers, knife-throwing Vietnamese, hard-working but larcenous Mexicans, and mysteriously dangerous Laotians. A few locally well-publicized incidents of violence fed the impression that any of these would be good candidates for work on the packinghouse killfloor. This sort of stereotyping sometimes guides worker placement in the plant. But initial placement becomes an issue for negotiation when experienced packers come to work. New hires, after passing their probation, may also "bid out" to other jobs in either slaughter or processing. So when non-packinghouse cultural forces operate, they do so to the extent that they mesh with the bloody, hypermasculine shop-floor comportment that is exhibited by nearly everyone in the plant, mediated by the rationalized structure of line work.

It seems as though Mexican immigrants and tough, Anglo packers fit the killfloor and the heavy boning jobs in processing because manly stoicism and athleticism fit perfectly with the context of industrial beef production. Such a macho stance allows a direct confrontation with the realities of industrial animal death. It also draws energy away from collective action, something that the sociologist Paul Willis and others demonstrate is part of all industrial machismo.[58]

TABLE 2. Ethnicity by Slaughter and Other Departments

	Slaughter[a]	Other Departments[a]	Total[b]
Anglo and Native American[c]	136 (45)	169 (55)	305 (14)
Hispanic	521 (33)	1,079 (67)	1,600 (75)
Vietnamese	39 (26)	109 (74)	148 (7)
African American	8 (21)	30 (79)	38 (2)
Laotian	1 (2)	42 (98)	43 (2)
Total	705 (33)	1,429 (67)	2,134 (100)

Source: July 1994 employment listing.

[a] Figures in parentheses are of the percentage of each ethnic group employed in Slaughter and in other departments, respectively.

[b] Figures in parentheses are percentages of the total workforce of Slaughter and other departments that is provided by each ethnic group.

[c] Native Americans make up less than 1 percent of the workforce in the departments shown here, and are aggregated with the Anglo employees.

So how are workers distributed between slaughter and processing at Branding Iron? A preliminary tabulation of one day's personnel listing (table 2) offers a beginning answer.

The "Slaughter" column in table 2 includes most of the task groups that fall under the jurisdiction of Branding Iron's slaughter superintendent: yards, slaughter, slaughter maintenance, and offal. Thus, jobs like washing intestines or boxing and freezing organ meats are grouped with the killing, bleeding, skinning, and gutting jobs of slaughter. On the other hand, rendering and coolers were grouped under "Other Departments" along with all jobs that are under the supervision of the processing superintendent: boxing and warehouse jobs are also lumped here under that heading. Since Lao surnames are so distinctive, they were pulled from the "Asian" list and shown separately, with interesting results.

The table shows that there are more Anglos than there should be in slaughter. They are overrepresented there, much as would be expected from the aggregate personnel data. Among them are many old-line packers. The number of Hispanics in the plant, and our present inability to accurately distinguish the new immigrants from the established resident Mexican Americans makes it difficult to know if there are patterns beneath the relatively unsurprising distribution of Hispanics between slaughter and other departments at Branding Iron. The harder jobs are uniformly the high-turnover jobs; among the hardest are those at the head end of the slaughter and processing divisions, where we always saw more new immigrant Mexicans than any other kind of worker. There are only a few African Americans. One set of numbers stands out: there are almost no Lao in slaughter.

That Lao don't work in slaughter does not do much to change the shop floor. That Mexicans do work in slaughter helps reproduce high shop-floor turnover, injury patterns, and weak collective action. They often work there as sojourners without local kin support and certainly without state-sponsored health or welfare safety-net support. The differences in placement among all the plant's workers are only interpretable with reference to the overall structure and habitus of the Branding Iron workplace, the way it participates in reproducing culturally interpretable ways of talking about and acting on language and cultural difference. We have much to learn about how individuals and groups deal with the business of large-scale animal death. But we know that the industry has its own scientific ways of dispatching animals, its own set of practices and words, its own rituals. And while our research data are sketchy, individual responses to industrial slaughter can be seen to take several forms.

INDIVIDUAL RESPONSES TO ROUTINIZED ANIMAL DEATH

At a structural level, plants are organized to remove individual culpability from the complete process of cattle killing and cutting. No worker in the plant functions as a generalist. As the initial description of the killfloor suggested, the plant—the aggregate and not the individual worker—does the killing. This is a principal finding of Nöelie Vialles, whose study of thirty-two abbatoirs in southwest France documents how "the action of each man is left indecisive; the key moment, the crucial point, is reduced to the harmless suspension of a body that had slumped to the floor."[59] At Branding Iron, workers (and supervisors) we have seen and worked with either repress the moment of killing ("get used to it" and tolerate it in silence), or they displace it with macho poses and locker-room bravado. If they have managed to stay around the industry long enough, they make some combination of these a part of their professional identity as butchers.

Managers include former all-round butchers; during their morning meetings, they apply a rationalized language of production to the work of killing cattle, one that puts the object of the work out of management discourse. "They plan to kill 5300 today. When reporting they rarely say head or cattle, just numbers. While one could argue that this is merely instrumental behavior which saves time and avoids the obvious, one could also argue that it is subconscious avoidance of openly acknowledging that they are in the business of killing."[60] So even the cattle are hidden in some of manager's talk; the killing is pulled out of its context, and applied to an invisible subject. The managers in this instance keep the sentient being out of the picture. This is a rationalized "getting used to it."

For workers, "getting used to it" takes place only after an impressive first encounter, one to which all new hires are exposed. This is ritual initiation—a rite of passage that makes packers out of new hires, and to which all of them are subjected—not just the minority of new hires who will end up working in slaughter.

> So we head into slaughter, right into where [the cattle] come out of the knock and make the corner to the *pateros* where they make the first cuts while/after they are bled. I stood on one end and watched the six faces [of the new workers], minus one guy who had worked there before who spent most of his time keeping us waiting while he said hello to all his friends: he was in Mexico for 6 months. They were looking up, all in a line. What a picture these new hires were, watch-

> ing the cattle coming by, fresh killed and drooling. We were supposed to be looking for emergency exits but what it turned out to be was kind of a *rite de passage*, getting splashed with blood as we walked along the line, many amazed smiles from the new hires, looks of "yucch!" from the Vietnamese woman, looks of admiration by the new hires for the people handling the knives.[61]

In fact, the Vietnamese woman ended up working on the killfloor. The wage scale in slaughter is very slightly higher for new workers and she found a job stamping carcasses with an ink marker. In time, like all meat packers, she "got used to it," and so did we.[62] The workers (and eventually, corporate ethnographers) get along with the death by replacing initial revulsion (or was it amazement?) with boredom. We researchers have not asked workers about this enough. I know I wanted to fit in, to be a participant observer, to be as macho (as detached) as many of the workers seemed to be while on the job. "I tracked through slaughter, amazed at how quiet and almost serene it seemed, in a surreal way. The brightly lit carcasses gently swinging along; workers seemed to be in slow motion. Not at all as hectic as it seemed at first blush. Once you get past the blood and gore the place does get routine; people's faces sure show it."[63] One month earlier, Don Stull made a similar comment in his fieldnotes: "I had very good views of many of the jobs from my various vantage points—the gutters from the leggers stand, etc. . . . I can tell that I'm getting used to the killfloor—I don't really watch as intently as I once did and the whole process no longer holds me in awe—that doesn't mean I understand it that much better, however."[64]

Often, workers acted as though they positively valued the violence of the killfloor. The Mexican head hangers were one example. Tammy, the young new hire who was away from home for the first time, was another. She sought in slaughter an object and an outlet for her own aggression. She said she actually liked the knockbox, and wanted to work there. She said, "There are some men she'd like to shove in there. . . . she wondered if a girl had ever done that job."[65] And old-line packers retold with pride their experience of the bad old days when sledgehammers were still used to stun cattle, and when they were able to gain control of the full range of butcher-craft skills.[66]

Vialles suggests that the industrial slaughterhouse has done damage to the "logic" surrounding the transformation from "animal to edible," replacing craft work with rituals of factory efficiency. Ritual action is often found at the borders between categories like new hire and packer, live animal and bloodless meat.

The modern meat plant makes a ritual out of killing cattle in one way. It is highly redundant: repeated over, and over, and over again. Repetition alone, however, makes for a factory, not a ritual in the usual sense. Ritual affects something in the physical, social, or metaphysical world(s).[67] Ritual does something about social or natural categories, or following Malinowski, bridges the gaps between the relatively predictable domain of reliable technical or social action and the not so predictable domain beyond the reach of such action.[68] So ritual, at least from one of several possible theoretical standpoints, plays a role in regulating social life, or in buffering, directing, or controlling contests about it. No matter how ritual is conceived from a social scientific standpoint, a rational workplace like Branding Iron should put ritual aside. Ritual belongs to human things, not to machines. The kind of rationality that ritual entails is generally represented as running very much in opposition to "scientific" rationality.

Nevertheless, we have been told that there is a Muslim ritual specialist employed by one nearby meat plant to bless the killfloor so that organ meats can be shipped to observant Muslim countries—ritual in the service of "rational" economic ends. Other than that, ritual in the meat factory is found off the line, sometimes sanctioned by the corporation, sometimes not. It emerges in the break room, where women workers (more often than men) celebrate birthdays, births, and other life transitions. It appears where workplace anniversaries are celebrated in cooperation with the corporation, just as in other factory and workplace settings. The "official" new-hire training done in a Branding Iron classroom is a rite of passage sanctioned by Branding Iron, complete with little rituals of obeisance—obligatory signings to record that workers have been told about this chemical hazard or that worker's compensation policy.

Unofficial—and genuinely transformative—rites of passage were visible when skillful supervisors set aside extra space and time apart from the official training. Official and unofficial ritual space was blended when trainers created a convivial, jocular moment in which to complete a progress report that the corporation required in order to move a new hire out of probation and into regular, and more highly paid, full-time status. These ritual moments make new hires into packers, packers with all the (wo)manly stoicism and macho bravado that adheres to the killing of cattle and the cutting of meat. These rituals are not about animal death per se because Branding Iron does not do the work of animal sacrifice. Branding Iron does the work of animal commodification.

Ritual is not completely absent but who controls and transmits the butcher craft knowledge on the shop floor, whose version of the knowledge counts, is a contested field. Our work at Branding Iron provided a glimpse at how the industry re-marks the line between animal life and animal death with mechanized routines. But workers respond as they must: they learn from their colleagues the skills not taught in the official training, or they learn them from skillful supervisors on and off the line. This suggests that the flexibility of capitalist industrial production in adapting and adopting cultural practices from among its workers is an unplanned result of the blending of humans and the production line, a flexibility that draws on worker symbolic resources while it uses workers' physical labor.

Ethnographers of the shop floor have discovered that they cannot conceive of capitalist production in deterministic terms. The work at Branding Iron can and does adapt to local conditions, local cultures, local languages, local histories that make sense out of (or resist) factory work.[69] Places like Branding Iron call into question the idea of local cultures. There are many languages spoken here, many meanings to be shaped by and to reshape meatpacking. True, the shop floor is a place that demands a shared set of worker skills. A shared task makes a shared identity as a packer possible; having the skills means having the embodied knowledge to do the work, a knowledge that can transcend language barriers.[70] But by devaluing skill, the corporation prevents the formation of the shared ground that is necessary for communication and collective action across linguistic boundaries.

The industry strengthens and perpetuates individual worker difference when it hides worker skill. At the same time, the exigencies of this kind of meat production, with standardized uniforms and standardized motions, tend to erase the differences, making the knife work robotic. Violent undertones in packer hypermasculinity both respond to and feed this corporate practice, providing "proof" that these polyglot workers are anything but rational, thinking actors capable of making good decisions about how to kill and cut cattle. Employing worker reactions in this way is a flexible, well-established, and largely unconsious way of doing the business of meat.

The industrial border between live animal and bloodless meat was drawn over one hundred years ago in the packinghouses of the Midwest. There, unionized butchers marched through the streets with icons of their craft stitched to their banners, and tools slung over shoulders and tucked into their belts.[71] They developed traditions that transmitted and celebrated butcher skills. The butchers at Branding Iron, like their old-line butcher counterparts in contemporary France, decry the loss of these skills. They don't believe that carcass washes and disinfecting machines will replace careful knife work; they believe that the heavy chain mail safety armor causes injuries that would never happen if skillful butchers ran their knives unencumbered by protective clothing designed for "unskilled" workers. These technical innovations, however, represent a corporate effort to attempt to erase both the skill and the danger of the packinghouse.

CONCLUSION

You gotta kill cows. That's the main thing.
—Packinghouse supervisor, 1994

We have much to learn about the interplay among languages and cultures in factory settings and also about the emotional consequences of industrial slaughter.[72] Stull has mentioned the dead-cow dreams that packer's wives have told him about, dreams that awaken their spouses. The husbands don't talk about it. Gandhi tells of having a similar dream after eating meat for the first time.[73] And we have learned that on the High Plains, the smell of death is quite real. Like High Plains residents, our research team came to call it "the smell of money." That metaphor masks the deadly origin of the smell. Broken hands and wrists come out of those plants, along with the meat, the money, and the smells.

This brief excursion into the Slaughter and Processing Divisions at Branding Iron demonstrates the flexibility of the meat industry in adapting and adopting both macho displacement and numbed repression, in having room for Lao workers and Mexican *matadors*, old-line packers and young women from Oklahoma. The turnover in this industry is still running at 5 percent a month or more. This cost is partly born by the packing companies, who pay for it with product quality problems and with expensive worker compensation claims. The communities that house the plants also pay for it when they must deal with the social dysfunction that constant turnover among low-wage, young, mostly single male workers generates.

Rethinking this workplace calls for more than a technical solution. It calls for a recognition of the symbolic balance that is so easily upset when animal blood is spilled on an industrial scale. The problem is more than one of better training and slower lines. The problem is both in the worker reaction to the shop floor, and in the ways in which continued Taylorization robs workers of the ability to add meaning to their work, to build identities out of something more than macho toughness. The process of rethinking the packinghouse will have to engage packers who know how to construct places for the ritual celebration and transmission of craft skills. It will engage managers who know how to lobby the corporation for new, quieter spatial arrangements in which workers can talk about work while they work, where they can be more than robots. This still leaves room for risk and danger, for meat cutters without knives become machines, and slaughter workers who kill without exposing themselves to danger face a greater risk: becoming *asesinos* instead of *matadors*.

NOTES

The packers at Branding Iron Beef have my thanks for allowing us to be part of their shop floor. I am particularly indebted to Donald D. Stull and the rest of the Ford Gang for setting a context in which team ethnography could take place. Without their fieldnotes and insights, furthermore, no exploration of the packinghouse could hope to paint more than a dim picture. I am grateful also to the 1997 Fellows of the Davis Center for their comments on an earlier draft of this article. Workplace ethnographers and friends, Eduardo Davel of the Ecole des Hautes Etudes Commerciales de Montréal, and Julian Orr of Xerox PARC, Palo Alto, shared bibliographic insights and important thoughts about the organization of work—both have provided (and will provide) collegial criticism. And I thank the educator and cattleman's daughter, Janice Mowry, for sharing her insights.

1. Sources of epigraphs are as follows: Michael Broadway, "Interview Transcripts: Owner of Major U.S. Packing Firm" (Garden City, Kansas, Changing Relations Project Files, Department of Anthropology, University of Kansas, Lawrence, Kansas, 1989); and James M. Saslow, *The Poetry of Michelangelo: An Annotated Translation* (New Haven, Conn., 1991), p. A40 (the quotation is given as it appears on the upright of the cross in a *Pietà* by Michelangelo).

2. Donald Stull, "Knock 'em Dead: Work on the Killfloor of a Modern Beefpacking Plant," in *Newcomers in the Workplace: Immigrants and the Restructuring of the U.S Economy*, ed. Louise Lamphere, Alex Stepick, and Gullermo Grenier (Philadelphia, 1994), pp. 44–77.

3. Ibid.

4. Noëlie Vialles, *Animal to Edible* (Cambridge, 1994).

5. Only a few stories in the national press have focused on the meat plant. An exception won a Pulitzer: Tony Horwitz, "Nine to Nowhere: The Grim Side of '90s Growth Jobs," *Wall Street Journal*, December 1, 1994.

6. Human death and its rites mark a transition from one realm, from one status, to another; Robert Hertz, "Contribution à une étude sur la représentation collective de la mort," in *Mélanges de sociologie religieuse et folklore* (Paris, 1928), p. 82.

7. Raymond Firth, "Offerings and Sacrifice: Problems of Organization," *Journal of the Royal Anthropological Institute* 43 (1963): 12–24.

8. The significance of meat as a gustatory category is explored in Nick Fiddes, *Meat: A Natural Symbol* (London, 1991).

9. There is a folk tradition in the plants about journalists and social scientists who enter as workers without revealing their investigatory intentions, and some researchers have to rely on interviews when the management is hostile to their studies, as did Carol Andreas in *Meatpackers and Beef Barons* (Niwot, Colo., 1994).

10. Donald D. Stull and Jean Schensul, eds., *Collaborative Research and Social Change: Applied Anthropology in Action* (Boulder, Colo., 1989); Ken C. Erickson and Donald D. Stull, *Doing Team Ethnography: Warnings and Advice* (Beverly Hills, Calif., 1997).

11. Six researchers—Donald D. Stull of the University of Kansas; Janet Benson of Kansas State University; Michael J. Broadway of Wichita State University; Ken C. Erickson of the Kansas Department of Social and Rehabilitation Services; and Mark Grey and Arthur Campa, both of the University of Colorado—made up the Garden City, Kansas, research team. The other sites were Chicago, Houston, Miami, Philadelphia, and Los Angeles: Louise Lamphere, ed., *Structuring Diversity: Ethnographic Perspectives on the New Immigration* (Chicago, 1994).

12. Donald D. Stull, Michael J. Broadway, and Ken C. Erickson, "The Price of a Good Steak," ibid., pp. 35–64.

13. See Michael Broadway, "Beef Stew: Cattle, Immigrants, and Established Residents in a Kansas Beefpacking Town," in Lamphere et al., *Newcomers in the Workplace*, pp. 24–43; Michael Broadway, "From City to Countryside: Recent Changes in the Structure and Location of the Meat-and Fish-Processing Industries," in *Any Way You Cut It: Meat Packing and Small Town America*, ed. Donald D. Stull, Michael J. Broadway, and David Griffith (Lawrence, Kans., 1995), pp. 17–40; Michael Broadway, "Following the Leader: IBP and the Restructuring of Canada's Meatpacking Industry," *Culture and Agriculture* 18 (1996): 3–8.

14. Ken Erickson, "Guys in White Hats: Short Term Participant Observation on the Packing House Floor," in Lamphere et al., *Newcomers in the Workplace*, pp. 78–98.

15. Participant observation was our primary method. We began in February 1994, when we were given a tour of the plant. We participated in new-hire training and divided the ethnographic tasks so that when we were all on site, one of us could attend the "morning management meeting" while another would be in the plant during second shift, from late afternoon until midnight. We did not live in the community but commuted

from our homes in eastern Kansas, staying in a motel near the plant for seventy-five total days on site, until we completed a final report to union and management one year after our entry.

16. Donald Stull, Ken Erickson, and Miguel Giner, "Management and Multiculturalism," *Meat&Poultry* 42, no. 4 (1996): 44–50; Ken Erickson, "Muscle Power and Meat: Rewriting a Story of Progress," in *The Story of Progress*, ed. Gøsta Arvastson and Mats Lindqvist, Studia Ethnologica Upsaliensia, no. 17 (Uppsala, 1996), pp. 63–75.

17. James Barrett, *Work and Community in the Jungle: Chicago's Packinghouse Workers—1894–1922* (Urbana, Ill., 1987), p. 14.

18. L. D. H. Weld, "The Packing Industry: Its History and General Economics," in *The Packing Industry: A Series of Lectures Given under the Joint Auspices of the School of Commerce and Administration of the University of Chicago and the Institute of American Meat Packers* (Chicago, 1923), p. 70.

19. J. Ogden Armour, *The Packers, the Private Car Lines, and the People* (Philadelphia, 1906).

20. Arthur Cushman, "The Packing Plant and Its Equipment," in *The Packing Industry*, p. 100.

21. Jimmy Skaggs, *Prime Cut: Livestock Raising and Meatpacking in the United States, 1607–1983* (College Station, Tex., 1986).

22. Cushman, "The Packing Plant and Its Equipment," p. 127.

23. Henry Ford, with Samuel Crowther, *My Life and Work* (Garden City, N.Y., 1922), p. 81.

24. Barrett, *Work and Community in the Jungle*, p. 27.

25. Stull, "Knock 'em Dead," pp. 59, 63.

26. Barrett, *Work and Community in the Jungle*, p. 24.

27. Ibid., p. 176.

28. Arthur H. Carver, *Personnel and Labor Problems in the Packing Industry* (Chicago, 1928).

29. Braverman notes that Taylor documented (but did not cause) key changes in modern wage work. For Taylor, deskilling was not so much a reduction in skill as a new allocation of work and reward, a redistribution of control over work, from generalist master craftsmen–foremen to technocrats, planners, managers, and engineers. See Harry Braverman, *Labor and Monopoly Capital: The Degradation of Work in the 20th Century* (New York, 1974); C. Bertrand Thompson, *The Taylor System of Scientific Management* (Chicago, 1917; repr. Easton, Me., 1975).

30. Broadway, "Beef Stew," p. 30.

31. M. J. Piore and Charles Sable, *The Second Industrial Divide* (New York, 1984).

32. The epigraph is from Hertz, "Contribution à une étude sur la représentation collective de la mort," p. 86.

33. "Branding Iron Beef," *Your Beef Guide for the Eighties* (Garden City, Kans., 1985) ("Branding Iron" Files, Department of Anthropology, University of Kansas, Lawrence, Kansas).

34. The annual cycle of work in the plant is still linked to a ritual cycle of feasting. Thanksgiving and Christmas, in the United States at least, are not periods of high beef consumption; production slows significantly during these times. Summertime barbecues (especially, the packers will say, the Fourth of July) make for overtime hours and frequent weekend work in the plants.

35. Burkard Sievers, *Work, Death, and Life Itself: Essays on Management and Organization* (Berlin, 1994).

36. Paul Stoller, *The Taste of Ethnographic Things: The Senses in Anthropology* (Philadelphia, 1989).

37. Mary Douglas, *Purity and Danger: An Analysis of Concepts of Pollution and Taboo* (New York, 1966).

38. Vialles, *Animal to Edible*, p. 124.

39. Lourdes Gouvea and Donald D. Stull. "Dances with Cows: Beefpacking's Impact on Garden City, Kansas and Lexington, Nebraska," in Stull et al., *Any Way You Cut It*, pp. 85–107.

40. Julian E. Orr, *Talking About Machines: An Ethnography of a Modern Job* (Ithaca, N.Y., 1996).

41. Stephen Judd, "On the Cutting Edge," *Meat&Poultry* 40, no. 5 (1994): 32.

42. David Griffith, *Jones's Minimal: Low-Wage Labor in the United States* (Albany, N.Y., 1993); Robert Hackenberg, David Griffith, Donald Stull, and Lourdes Gouvea, "Creating a Disposable Labor Force," *Aspen Institute Quarterly* 5, no. 12 (1993): 78–101; Aiwa Ong, "The Gender and Labor Politics of Postmodernity," *Annual Review of Anthropology* 20 (1991): 279–309.

43. Gouvea and Stull, "Dances with Cows."

44. Keith Nunez, "Who Are We?" *Meat&Poultry* 40, no. 7 (1994): 46–47.

45. Carver, *Personnel and Labor Problems in the Packing Industry*, p. 20.

46. The epigraph is from Philip Ashton Rollins, *The Cowboy: An Unconventional History of Civilization on the Old-Time Cattle Range* (Albuquerque, N.Mex., 1979), p. 1.

47. Will James, *Cowboys North and South* (New York, 1928).

48. In Argentina, *matadero* refers to the person in charge of the slaughterhouse; Anna Maria Martinez de Sanchez, *La vida cotidiana en Córdoba: El abasto de carne 1783–1810* (Córdoba, Argentina, 1995).

49. Americo Paredes, "*With his Pistol in his Hand*": *A Border Ballad and Its Hero* (Austin, Tex., 1958).

50. Manuel Machado, *The North Mexican Cattle Industry, 1910–1975* (College Station, Tex., 1981).

51. The most recent published data on industry injury rates is in Donald Stull and Michael Broadway, "Killing Them Softly," in Stull et al., *Any Way You Cut It*, pp. 61–84. The pattern is the same in Canada according to Joel Novek, "The Labour Process and Workplace Injuries in the Canadian Meat Packing Industry," *Canadian Review of Sociology and Anthropology* 29 (1989): 17–37.

52. For information on Southeast Asian households, see Janet Benson, "The Effects of Packinghouse Work on Southeast Asian Refugee Families," in Lamphere et al., *Newcomers in the Workplace*, pp. 99–128.

53. James, *Cowboys North and South*; William Savage, *The Cowboy Hero: His Image in American History and Culture* (Norman, Okla., 1979).

54. Hardy Green, *On Strike at Hormel : The Struggle for a Democratic Labor Movement* (Philadelphia, 1990). Recently at Branding Iron, an upstart union challenged the powerful UFCW with an election, an effort that linked some old-line and new-immigrant packers across language and cultural boundaries.

55. Neil Jamieson, *Understanding Vietnam* (Berkeley, Calif., 1993); Dông Tiên, *Nuóc Tôi, Dân Tôi* (San Jose, Calif., 1992).

56. Buddhist tradition can mean many things. For local varieties of Buddhism in Tai-speaking regions, see Stanley Tambiah, *Buddhism and the Spirit Cults of Northeast Thailand* (Cambridge, 1970); David W. Haines, ed., *Refugees as Immigrants Cambodians, Laotians, and Vietnamese in America* (Totowa, N.J., 1989).

57. Thao Nhouy Abhay, "Le Baci," in *Présence du Royaume Lao*, ed. René de Berval, *France-Asia* 118–20 (1956) (special issue): 817–20.

58. Paul Willis, "Shop Floor Culture: Masculinity and the Wage Form," in *Working Class Culture: Studies in History and Theory*, ed. John Clarke, Charles Cricher, and Richard Johnson (London, 1979), pp. 195–97; Aiwa Ong, "The Gender and Labor Politics of Postmodernity"; Phillippe Bourgois, "The Political Economy of Resistance and Self-Destruction in the Crack Economy: An Ethnographic Perspective," in *The Anthropology of Lower Income Urban Enclaves: The Case of East Harlem*, ed. Judith Friedenberg, Annals of the New York Academy of Sciences, no. 749, (New York, 1995), pp. 97–118.

59. Vialles, *Animal to Edible*, p. 46.

60. Donald Stull, "Branding Iron Fieldnotes" (Department of Anthropology, University of Kansas, Lawrence, June 9, 1994), p. 1.

61. Ken Erickson, "Branding Iron Fieldnotes" (Department of Anthropology, University of Kansas, Lawrence, June 21, 1994), p. 3.

62. Fiddis, *Meat: A Natural Symbol.*

63. Ken Erickson, "Branding Iron Fieldnotes" (Department of Anthropology, University of Kansas, Lawrence, September 22, 1994), p. 3.

64. Donald Stull, "Branding Iron Fieldnotes" (Department of Anthropology, University of Kansas, Lawrence, August 18, 1994), p. 14.

65. Ken Erickson, "Branding Iron Fieldnotes" (Department of Anthropology, University of Kansas, Lawrence, July 27 and 28, 1994), p. 5.

66. Donald Stull, "Branding Iron Fieldnotes" (Department of Anthropology, University of Kansas, Lawrence, August 9, 1994), p. 6.

67. Edmund Leach, "Ritualization in Man in Relation to Conceptual and Social Development," in *Reader in Comparative Religion: An Anthropological Approach*, ed. William Lessa and Evon Vogt (New York, 1972), pp. 333–37; Catherine Bell, *Ritual Theory, Ritual Practice* (New York, 1992).

68. Bronislaw Malinowski, *Magic, Science and Religion, and Other Essays* (New York, 1948).

69. Ong, "The Gender and Labor Politics of Postmodernity"; Laetitia Cairoli, "Factory as Home and Family: Female Workers in the Moroccan Garment Industry," *Human Organization* 57 (1998): 181–89.

70. Rajendra Singh, Jayant Lele, and Gita Martohardjono, "Communication in a Multilingual Society: Some Missed Opportunities," *Language and Society* 17 (1988): 43–45.

71. Barrett, *Work and Community in the Jungle.*

72. The epigraph is taken from Ken Erickson, "Branding Iron Fieldnotes" (Department of Anthropology, University of Kansas, Lawrence, June 24, 1994), p. 4.

73. M. Gandhi, *Diet and Diet Reform*, (Ahmedabad, 1949), cited in Fiddis, *Meat: A Natural Symbol*, p. 67.

Part Two

ON THE BORDERLANDS OF EXPANSION AND EMPIRE: THE HUNT

4

The Wild and the Tamed

Indians, Euroamericans, and the Destruction of the Bison

Andrew C. Isenberg

Between October 1867 and April 1868, a special United States–Indian Peace Commission negotiated a series of treaties with several nomadic Indian societies of the western plains. The commissioners intended to end the sporadic warfare between Indians and Euroamericans that had persisted since the last years of the Civil War. At the root of the violence in the western plains was a contest over the use of natural resources: Euroamericans seeking access to grazing lands and gold mines in Colorado and Montana interfered with the Indians' use of their primary resource, the bison. To resolve the conflict, the commission conceded to the Indians rights to hunt bison in the northern and southern plains. In the first round of negotiations, in the valley of the Medicine Lodge Creek, the Comanches, Kiowas, Southern Cheyennes, and Southern Arapahos ceded the territory north of the Arkansas River while retaining the right to hunt bison south of it. In the second round, at Fort Laramie, the Sioux, Northern Cheyenne, and Northern Arapahos ceded the territory south of the Platte River and reserved undisturbed use of the hunting territory north of it, west of the Missouri River, and east of the Bighorn Mountains.[1]

Although the Indians surrendered the region between the Platte and the Arkansas rivers, they secured their most important objective, the right to hunt bison in designated territories. Yet the seemingly straightforward treaties were only one part of the contradictory regulation of resource use in the plains. The federal peace commissioners could not speak for the State of Texas, which had retained its public lands when it had entered the Union in 1845; after the Civil War, millions of West Texas longhorns began to move northward into the bison's range.[2] While the Treaties of Medicine Lodge prohibited Euroamericans from settling in the Indians' hunting territory, they did not expressly forbid Euroamerican hunters from pursuing bison there.[3] After 1870, Euroamerican hide hunters poured into the region south of the Arkansas and within a decade had nearly exterminated the herds.

The Treaty of Fort Laramie was likewise an imperfect guarantee of the northern plains nomads' unimpeded use of the bison. Although it explicitly banned Euroamerican hunters from the territory north of the Platte, the law also allowed scientific exploration, land grants, and mining claims—the legal apparatus of Euroamerican settlement—to go forward.[4] In 1864, the federal government had set aside land in the Dakota and Montana territories for the construction of the Northern Pacific Railroad. The treaty did not specify a northern boundary of the Indians' hunting territory, virtually ensuring conflict in 1873 when the Northern Pacific reached the Missouri River. Moreover, in 1874, George Armstrong Custer led a scientific expedition to the Black Hills in the center of the area reserved for the Sioux. Custer's report of an abundance of gold in the Black Hills sparked a rush of Euroamerican miners into the Sioux territory. They were followed by Euroamerican hide hunters who decimated the herds in the northern plains.

The Euroamericans who intruded upon Indian hunting lands in the 1870s and early 1880s slaughtered millions of bison. Yet by 1867, most Indian and Euroamerican signatories to the treaties of Medicine Lodge and Fort Laramie already knew that the bison were diminishing rapidly. In the northern plains, the most obvious cause of the demise was the Indian-Euroamerican commerce in bison robes, the animal's thick winter coat. At the time of the treaty negotiations, Indian hunters' industriousness had driven the robe trade into decline. The trade had fallen from 110,000 robes annually in 1849 to 50,000 in 1859.[5] Fort Pierre on the Missouri River, the center of the northern plains robe trade in 1849 when it collected 75,000 robes, accumulated only 19,000 in 1857.[6]

Beginning in the 1840s, the emigration of California-bound Euroamericans through the plains also contributed to the diminution of the bison. The emigrants' livestock ruined the bison's forage by trampling or consuming the grasses in a wide swath along the Platte River, which paralleled much of the Oregon-California Trail. In 1853 alone, emigrants herded over one hundred thousand cattle and forty-eight thousand sheep along the south side of the Platte. Before 1860, emigrants probably drove over half a million cattle and an equal number of sheep to the Pacific coast.[7] Similar destruction occurred in the southern plains along the Santa Fe Trail. The degradation of the bison's habitat destroyed or displaced large numbers of the species.[8]

In addition to the nomads' commercial predation and the degradation of river valley habitats by Euroamerican emigrants, the volatile, semi-arid climate of the grasslands undermined the bison population at mid-century. In the first forty years of the nineteenth century, precipitation in the plains was above average.[9] Beginning in 1846 and continuing for the next decade, however, rainfall was below average. A Kiowa pictographic calendar recorded a drought in the southern plains in 1855.[10] Lone Dog, the keeper of the Yanctonai Sioux pictographic winter count, noted drought conditions in the northern plains in 1865–66.[11] According to the records of Euroamerican amateur meteorologists, the central plains suffered severe drought in 1861-63 on a par with the infamous "dust bowl" drought of the 1930s.[12] Ecological and economic pressures thus combined to diminish the bison in the middle of the nineteenth century.

The diminution of the bison did not escape the notice of William Tecumseh Sherman, a member of the Peace Commission. He wrote to his brother, John Sherman, a U.S. senator from Ohio, on June 17, 1868, that "the commission for present peace had to concede a right to hunt buffaloes as long as they last, and this may lead to collisions, but it will not be long before all the buffaloes are extinct near and between the railroads."[13] The language of the treaties Sherman helped to negotiate was temporizing: the Indians "yet reserve the right to hunt . . . so long as the buffalo range . . . in such numbers as to justify the chase." Sherman was ultimately proved correct: by 1883, the bison was nearly extinct, and the plains nomads had submitted to the reservation system.

Sherman's Peace Commission was the forward arm of federal government policy in the plains, which in turn reflected dominant cultural notions of the rightness of the Euroamerican conquest of the West. Most Euroamerican soldiers, emigrants, hunters, and ranchers in the plains shared a belief in the inevitability of Euroamerican triumph over Indians and the wilderness. They believed that civilized Euroamericans and domesticated livestock were destined to replaced savage Indians and wild bison. Some Euroamericans in the 1870s drew on opposing cultural traditions and argued that a truly civilized society would pursue a policy of humanity towards both Indians and the bison, but ideas that championed competition and conquest prevailed. The army and the federal government thus institutionalized cultural notions that channeled the transforming forces of the American industrial economy into the bison range. Together with inherent environmental pressures, those forces pushed the bison to the brink of extinction.

While the dynamic western plains environment contributed to the destruction of the bison, a spasm of industrial expansion was the primary cause of the bison's near-extinction. Euroamericans hunted bison to satisfy an increasing demand for animal hides in industrializing America. Heavy leather belts were the sinews of nineteenth-century industrial production:

mills relied on them to power their machinery.[14] In the 1870s, however, the tanning industry was in crisis. In 1850, Adirondack tanneries paid an average of $2.20 per hide. By 1870, the average price had nearly doubled. Because the demands of the tanning industry outstripped the domestic supply, 70 percent of the hides tanned in the United States in 1850 were imported from Latin America, primarily from the Argentine pampas.[15] Continuing demand prompted the industry's interest in the bison as a relatively low-cost source of hides.

Exploitation of the bison required a heavy investment of industrial capital, energy, and knowledge. In 1870, tanners perfected the chemical process of transforming bison hides into leather by soaking them in a strong lime solution.[16] The result was a porous but highly elastic leather ideally suited to industrial belting. Toward the end of the Civil War, American munitions makers developed accurate, large-bore rifles suited to bison hunting.[17] The extension of railroads to the plains permitted hide hunters to ship their products to the east. The Union Pacific, which joined the Central Pacific in Utah in 1869, traversed the plains through Nebraska; the Kansas Pacific reached Denver in 1870; and the Atchison, Topeka, and Santa Fe reached Dodge City, Kansas, in 1872. Altogether, the hunting of the bison in the 1870s and early 1880s was unquestionably the work of an industrial society. The western plains in the 1870s were a remote extension of the global industrial market and an object of its demand for natural resources.

The destruction of the bison in the plains to fuel the demand for hides was part of a broad pattern of environmental degradation in industrializing America. The tanning industry consumed not only animal hides but trees and lime in enormous quantities; it returned unused portions of those resources to local environments in the form of pollution. The tanning industry initially concentrated amidst dense forests of eastern hemlock in northeastern Pennsylvania and the Adirondack region of New York, because the bark of hemlock is rich in tannin, the chemical essential to the production of leather. The industry's unsustainable consumption of tanbark depleted accessible supplies of hemlock in the Adirondacks by the mid-1880s and a few decades later in Pennsylvania. The *New York Times* attributed the failure of Shaw and Brothers, the largest tanning company in the United States, in 1883, to the firm's exhaustion of local tanbark supplies.[18] Deforestation had widespread environmental as well as economic consequences, ranging from the destruction of local wildlife habitats to increased erosion and runoff of soil into streams.[19] Tanneries, which discharged large amounts of organic pollution into watercourses, including particles of lime and animal hair and flesh, were also a leading contributor to the noxious condition of rivers in the industrializing Northeast.[20] Late nineteenth-century tanneries were thus an environmental malignancy that consumed bison, razed forests, and fouled rivers.

The industrial system that brought Euroamerican hide hunters to the western plains wrought extensive environmental change. Yet Euroamerican hide hunters in the plains also necessarily adapted themselves to the constraints of the western plains environment and the ecology of the bison. Because the bison spent most of the year fragmented into small foraging groups, Euroamericans dispersed in search of bison in small hunting outfits, typically consisting of a shooter, a cook, and three or four skinners, which staked out camps along the rivers where the bison sought water. In the early 1870s, outfits were camped every few miles along the Smoky Hill, Arkansas, Cimarron, Canadian, Washita, and Red rivers.

Although they preferred to be called "buffalo runners," most hide hunters did not pursue bison from horseback like the plains nomads. Instead, they employed a method known as the "still hunt." At dawn, a hunter slipped up to the downwind side of a group of bison and concealed himself several hundred yards away in a wallow or behind the rise of a bluff. Using a telescopic sight to scan the herd, he located the lead cow of the group and killed her with his first shot, which he aimed below the animal's shoulder to strike her lungs and the bottom of her heart. The remaining animals could hear the report of the rifle but neither see nor smell their enemy. Nor could they rely on their leader to alert them to the danger. As the lead animal of the group lay dead or dying, dropping first to one foreleg, then the other, and finally collapsing, the hunter picked off several more of the bewildered animals. Such a "stand" varied in productivity, from as few as ten to as many as fifty hides in one morning's work if the hunter took care to kill the wariest bison before they smelled blood and led the herd out of range. The hunter's labors ended before noon, when the skinners took over. Ordinarily, they removed hides until sundown.[21]

The impact of the hide hunters upon the herds in the southern plains was devastating. In 1872, Col. Richard Irving Dodge wrote that the area near Dodge City, the center of the hide trade in the early 1870s, was thick with bison. By the fall of 1873, however, "where there were myriads of buffalo the year before, there were now myriads of carcasses. The air was foul with a sickening stench, and the vast plain which only a short twelve months before teemed with animal life, was a dead, solitary desert." Even those bison who avoided bullets felt the impact of the hunters. Dodge wrote that "Every drink of water, every mouthful of grass is at the expense of life, and the miserable animals continually are harassed, are driven into localities far from their haunts, anywhere to avoid the unceasing pursuit."[22] George Lemmon, a Nebraska rancher, wrote that after 1875, bison rendered no tallow and only stringy meat, "as hide hunters had the few remaining animals almost run to death."[23] As with the passenger pigeon—millions of which market hunters in eastern North America slaughtered in the 1870s—the bison's communal breeding system relied on the assembly of large groups in the summer. Reproductive success likely declined with group size in the 1870s, as unceasing predation prevented the congregation of the herds in the rutting season, upsetting the bison's patterns of migration and reproduction and thus inhibiting a recovery of the bison population.[24]

How many bison did the hide hunters slaughter in the 1870s? The productivity of the outfits varied. W. S. Glenn, who hunted bison in Texas in 1876–77, recalled that "a remarkably good hunter would kill seventy-five to a hundred in a day, an average hunter about fifty, and a common one twenty-five, some hardly enough to run a camp. It was just like in any other business. A good skinner would skin from sixty to seventy-five, an average man from thirty to forty, and a common one from fifteen to twenty-five."[25] Charlie Hart's outfit stacked up 2,003 hides in six weeks, an average of about fifty hides a day.[26] Bat Masterson's crew, which was camped close to Fort Dodge and, unlike most outfits, took the time to take the meat from the carcasses, averaged only fifteen hides a day.[27]

Colonel Dodge estimated that the number of hides shipped on the Santa Fe, Kansas Pacific, and Union Pacific railroads between 1872 and 1874 was 1,378,359. His figures were largely guesswork because only the Santa Fe provided him with a statement of the number shipped, and Dodge simply assumed that the other two lines each shipped a number equal to the Santa Fe's 459,453. Yet the Santa Fe's accounting of the hides it shipped in this period—even if accurate—was no reflection of the actual number of bison killed. Dodge noted that in the early years of the slaughter hunters were especially wasteful. Poor hunters wounded two or three bison for every one they killed; the crippled animals later fell to wolves. Skinners ruined hides as they flayed them, or failed to stretch and stake them to dry properly. When skinners failed to finish their work before sundown, wolves often tore the unskinned carcasses to pieces by the next morning. Dodge estimated that in 1872 every hide shipped to market represented five dead bison. Eventually, as the outfits learned their craft, they became more efficient: in 1873 every hide represented only two dead bison. Altogether, Dodge estimated that hide hunters slaughtered 3,158,730 bison between 1872 and 1874, an average of just over one million animals each year.[28]

If the hide hunters' estimates of their own productivity were correct, the impact of commercial hide hunting was still more devastating than Dodge guessed. Glenn estimated that a "common" hunter killed between fifteen and twenty-five bison a day.[29] The Kansas hunter Frank Mayer set twenty-five hides as his "regular quota."[30] An outfit that accumulated that many hides a day could amass 2,250 hides in three months. Such a figure was not unreasonable: Charlie Hart and his outfit compiled 3,361 hides in three months.[31] In one year, a hide hunter known as "Kentuck" brought in 3,700 hides.[32] J. W. Woody claimed to have killed 3,200 bison in the winter of 1876–77.[33] Thomas Linton of Troy, Kansas, killed 3,000 in 1872; Zack Light killed 2,300 in the winter of 1873–74; Joe McCombs killed 4,900 in Texas in the winter of 1877–78; William "Doc" Carver killed 5,700 in 1875; Orlando Brown killed 5,855 bison during a two-month span in 1876—indeed, the nearly constant report of his .50-caliber rifle during that time rendered him deaf in one ear.[34] While a competent hunter could have a single productive season—or inflate his estimates to appear more successful before his peers—more common, perhaps, was Josiah Wright Mooar's tally. He estimated that he killed twenty thousand bison in a career as a hunter between 1870 and 1879—twenty per day if he hunted three months of every year.[35] The Newton *Kansan* reported in 1872 that at least one thousand hide hunters roamed the southern plains.[36] If that number were in the grasslands for three months of the year, they could have easily supplied tanneries with over two million hides every year. Waste and the disruption of the bison's patterns of subsistence and reproduction could have raised the hide hunters' capacity for destruction to over four million bison a year in the first years of the slaughter.

By what means could so many hides have been transported from the southern plains every year, if not on the Santa Fe, Kansas Pacific, and Union Pacific railroad lines? While railroads were the primary means of shipping bison hides to tanneries, Dodge's exclusive focus on the railroads obscured more traditional means of transportation. Many hunting outfits hauled their hides by the cartload not to railway stations in Kansas, where dealers paid lower prices, but to markets in Texas, Colorado, and New Mexico. The hunter James Cator, for instance, hauled his hides by wagon to dealers in Las Vegas, New Mexico, and Granada, Colorado.[37] Dodge's estimate of the number of hides carried by the railroad was but a fraction of the number that were collected at temporary hide markets in the southern plains, such as Buffalo Gap in Taylor County and Hide Town in Scurry County, Texas. Like many mining towns in the Mountain West, these places flourished only briefly before the local resource became scarce. At one hide town, a skinner recalled, "there were acres of ground covered with hides."[38] At another, Fort Griffin, Texas, a government freighter compared the hide yards to "a great lumber yard. The hides were stacked tier upon tier in rows."[39] Far from the Kansas railroads, the hides accumulated in the Texas Panhandle found their way out of the southern plains by the cartload.

The hide hunters' slaughter could not be sustained by a southern plains herd that probably numbered between ten and fifteen million in the late 1860s. A herd of that size produced perhaps three million calves a year, although wolves, accidents, drought, and competition from other grazers probably killed one-half of that number in their first year.[40] Indeed, in addition to the hunters, bison suffered from environmental pressures. In 1871–72, a severe winter followed by drought destroyed, by one estimate, two hundred thousand head of cattle in western Texas. While the loss of cattle reduced the competition for the bison's range, the blizzard and drought probably killed large numbers of bison as well. In October 1873, a grassfire spread for one hundred miles along the north side of the Kansas Pacific railroad, destroying millions of acres of forage.[41] Texas fever, a disease endemic to Texas longhorns, may have spread to bison in the 1870s as the West Texas cattle population expanded to the north.[42] By the end of the 1870s, the combination of human and environmental pressures had reduced the southern plains bison to a few hundred stragglers.

After the near-extermination of the bison in the southern plains, the destruction moved to the north. Just as the extension of the Santa Fe Railroad to Dodge City spurred the expansion of commercial hide hunting in the southern plains, the extension of the Northern Pacific Railroad to Miles City, Montana Territory, in 1881 made possible the slaughter in the northern plains. Yet the northern hide hunters did not equal the destructiveness of their colleagues to the south. A *New York Times* correspondent in Montana estimated that the hunters brought in only ten thousand hides in the winter of 1880.[43] The naturalist William Hornaday thought that the slaughter was greater. He estimated that between 1881 and 1883, hide hunters in the north shipped nearly three hundred thousand hides east. Yet in 1884, the Northern Pacific carried only three hundred hides. Although hunters accumulated fewer hides, something certainly destroyed the northern plains bison: a rancher who traveled one thousand miles across the region in the early 1880s was, by his own account, "never out of sight of a dead buffalo and never in sight of a live one."[44]

In the northern plains, environmental pressures—both inherent in the grassland and anthropogenic—conspired with hunters to reduce the number of bison to a few stragglers. The demise of the northern plains bison resulted primarily from the combination of drought and ranches. The dominant shortgrasses of the western plains are well-adapted to the drought-prone climate. By concentrating roots at the top of the soil, and keeping leafage minimal, shortgrass species conserve their moisture and take advantage of scarce rainfall.[45] Nonetheless, acute drought can devastate the shortgrasses upon which grazing animals such as the bison depend. Yet when drought shrivels shortgrasses in the western plains, it also opens niches for them in the ordinarily more humid grasslands to the east. The mixed-grass plains—so called because they lie between the semi-arid, shortgrass, western plains and the humid, tallgrass, eastern prairies—are far enough from the Rocky Mountains rain shadow that their dominant species, such as little bluestem, can grow to a height of one to two feet. During severe drought, however, little bluestem is reduced to a tiny proportion of the mixed-grass community, while shortgrass species, which thrive in relatively dry conditions, invade and dominate.[46] Thus in the historic plains drought did not so much destroy the bison's range as temporarily shifted it to the east.

So long as bison were able to move eastward in search of forage, they could mitigate the impact of drought. Yet by the late 1860s, they faced an obstacle. Lone Dog's winter count recorded the arrival of Texas cattle in the region north of the Platte River in 1868–69.[47] In succeeding years, bison increasingly faced competition for rangeland from domestic livestock. Between 1874 and 1880, the number of cattle in Wyoming increased from ninety thousand to over five hundred thousand.[48] By 1883, over five hundred thousand cattle grazed in eastern Montana.[49] Between 1874 and 1890, the cattle population in southern Alberta rose to one hundred thousand.[50] Domestic cattle not only encroached upon the bison's range, but, moving in from the south and east, they occupied the region to which bison migrated during periods of drought. Drought struck the northern plains in the 1870s and early 1880s: climatologists recorded droughts in eastern Wyoming and northeastern Colorado in 1873, 1879, and 1882. The handful of stations in Montana recorded low summer precipitation in 1877–78 and 1881–83.[51] During his two-year sojourn in the plains beginning in 1882, Theodore Roosevelt killed only a few bison for trophies, but his ranch on the North Dakota–Montana border was one of many that likely prevented northern plains bison from migrating to ranges in the mixed-grass plains to escape drought. Moreover, by occupying the bison's range, cattle made a recovery of the bison population unlikely.

Inherent environmental causes of bison mortality—drought, blizzard, and grassfire—also took their toll on the northern plains bison in the 1870s and early 1880s. When Romanzo Bunn settled in Dakota Territory in 1880, he discovered the bleached skeletons of hundreds of bison in his half-section. The skeletons were largely intact, and Bunn found no broken bones or bullet-marked ribs among the remains. Bunn attributed the deaths to blizzard, such as the one he endured in his first winter in the northern plains in 1880–81. The blizzard that winter covered the plains with a foot of snow; drifts were twenty feet high in some places.[52] Grassfire in Alberta in 1878 destroyed many bison and drove survivors south into the Missouri

River valley, toward the bullets of hide hunters and the competition of domestic livestock.

In the western plains, anthropogenic and environmental causes of bison mortality worked in concert. When Euroamerican hunters and ranchers added their pressures to the inherent causes of bison mortality in the dynamic plains, the population collapsed almost completely in only a decade and a half. When Hornaday surveyed the bison population in January 1889, he discovered only twenty-five in the Texas Panhandle, twenty in the foothills of Colorado, ten between the Yellowstone and Missouri rivers, twenty-six near the Bighorn Mountains, and two hundred in Yellowstone National Park.[53] By the mid-1890s, market hunters had reduced the Yellowstone bison to twenty-five.[54] In just a few years, two dynamic forces—the plains environment and the American industrial economy—had combined to nearly obliterate the millions of bison that had inhabited the grasslands.

As the scope of the hide hunters' destruction of the bison became clear, Euroamericans in the eastern United States clashed over the wisdom of permitting the slaughter in the West to continue. Those who opposed the destruction drew on two reform efforts in the postwar United States: animal protection and Indian humanitarianism. The advocates of the slaughter appealed to an emerging belief that the extinction of the bison and the subjugation of Indians, however brutal, were not only the just deserts of savages and wild beasts but necessary to open the Great Plains to Euroamericans and their domesticated livestock. The triumph of those beliefs permitted the rapid destruction of the bison in the 1870s and early 1880s.

An enthusiasm for animal protection was the most important obstacle to the destruction of the bison. American efforts to prevent cruelty to animals began in the 1820s. Following the example of the Royal Society for the Prevention of Cruelty to Animals, founded in England in 1824, eighteen states enacted laws regulating the treatment of horses, sheep, and cattle between 1828 and 1861. Prominent antebellum reformers endorsed kindness to animals; Charles Lowell preached a sermon on the topic in Boston in 1837. Yet animal protection became a bona fide reform movement in the United States only after the Civil War. Henry Bergh, the son of a wealthy immigrant shipbuilder, founded the American Society for the Prevention of Cruelty to Animals in New York in 1866. Pennsylvania and Massachusetts SPCAs followed in 1867 and 1868, respectively, organized in both cases by wealthy, reform-minded women. By 1874, SPCA chapters were located in thirty of the largest cities in the Northeast and Upper Midwest.

The concentration of animal protection societies in the centers of American industry was not coincidental. Nineteenth-century animal protection was largely a reaction to industrialization. Nearly all reformers were middle-class city dwellers. Critical and fearful of urban mechanization and poverty, they focused their efforts largely on farm animals which, they believed, were symbolic of rural innocence. The prevention of the cruel treatment of animals would, they hoped, counteract the aggressive, competitive amorality of the marketplace. According to the historian James Turner, the advocates of animal protection "were conformists, comfortable, at heart happy with their up-to-date industrial world." Sensitive to the ill effects of industrialism, however, they "wanted to protest the sordid acquisitiveness of Victorian capitalism, but not too loudly."[55]

Anticruelty was also a decidedly feminizing movement, in its sizable female constituency and its invocation of a domestic ethic of gentleness. Caroline Earle White in Philadelphia and Emily Appleton in Boston were early organizers of SPCAs in their cities, although both were formally excluded from the societies they helped to found. In 1869, White became the president of the Women's Branch of the Pennsylvania SPCA. In a short time, the Women's Branch rivaled the original organization in its effectiveness, energy, and fund-raising ability.[56] Anticruelty was, above all, an explicit effort to extend the ideology of feminine compassion to men and boys who were presumed to be responsible for most of the cruelties practiced upon animals. Pet keeping and stories of the cruel treatment of innocent animals were intended to instill in boys an ethic of kindness. Reformers were particularly concerned about boyhood hunting, which, they feared, established a taste for killing.[57]

While anticruelty tracts criticized hunting, they focused their attentions largely on urban and suburban animals: pets, carriage horses, and stray dogs. The hunting of wild animals was, at first, not a significant part of their mission. But the destruction of the bison excited Bergh to action. In the early 1870s, he received a number of letters describing the destruction of the bison. "We speak on behalf of the buffalo, antelope, and various wild game of the western prairies," a group of women from Freeport, Illinois, recently returned from a trip through the plains, wrote to Bergh in December 1872. "Visiting the sickening scenes of slaughter, we find the Plains thickly strewn with carcasses of buffalo, deer, and antelope."[58] High-ranking army officers stationed in the plains also wrote to Bergh objecting to the slaughter. General William Hazen and Lieutenant Colonel A. G. Brackett protested to Bergh, who included Hazen's letter in a report to ASPCA members and forwarded Brackett's letter to the *New York Times*. He sent copies of both letters to Congressional supporters, hoping to spur the passage of legislation to regulate the hide hunters.

Brackett's letter in particular revealed how the SPCA's efforts on behalf of the bison sought to extend the feminine, domestic rhetoric of anticruelty—

previously applied only to farm animals, pets, songbirds, and some urban draft animals—to a wild animal. Toward that end, Brackett took care to present the bison as both "harmless and defenseless": a gentle, nearly domesticated animal. "[T]here is as much honor and danger in killing a Texas steer as there is in killing a buffalo," wrote Brackett. "It would be equally as good sport," he argued, "to ride into a herd of tame cattle and commence shooting indiscriminately."[59] Brackett's repeated analogies to domestic animals, which were protected in many states by anticruelty legislation, and his conclusion that the killing of bison was not simply needless but "cruel," revealed his debt to the arguments of the anticruelty advocates.

Indian humanitarianism, like animal protection, was a relatively new passion in the 1870s. The 1870s were liminal years in United States Indian policy, marking a transformation from a policy of separation to one of assimilation. From the beginning of the nineteenth century until the Civil War, policymakers sought to confine Indian populations to reservations in order to segregate them from Euroamerican settlers. Indian policy in the 1870s was, in one sense, a continuation of this program. "Our civilization is ever aggressive, while the savage nature is tenacious of traditional customs and rights," wrote Columbus Delano, the secretary of the interior, in 1873. "This condition of things calls loudly for more efficient efforts to separate Indians from whites by placing them on suitable reservations."[60] Yet in the years after the Civil War, the United States also embarked on a so-called "peace policy," a cooperative effort between the federal government and Protestant missionaries to Christianize and educate reservation Indians. The peace policy foreshadowed the ambitious, deeply flawed program of Indian assimilation that began in the 1880s and continued into the twentieth century.[61]

To a considerable extent, Indian humanitarianism emerged from the United States' experience in the Civil War. Tired of bloodshed, the editors of the *New York Times* condemned an army attack on a Piegan winter camp in 1870 as a "sickening slaughter"—the same terms the Freeport women used to describe the killing of bison. To subdue Indians by violence imperiled "our standing before the world as a Christian nation." The heightened interest in racial justice for recently emancipated slaves also influenced Indian humanitarianism. "We have long been doing justice to the negro," the *Times* wrote in 1870, on the occasion of the Oglala leader Red Cloud's visit to Washington. "Is it not almost time to see what we can do for the Indian?"[62]

Yet the humanitarianism of the peace policy was not a complete departure from the history of hostility between Indians and Euroamericans. "What we must do if we mean to save the remnant of the Indians," the *Times* editorialized in 1870, "is to gather them all in a small district which we can really police and protect, and there teach them the arts of civilized life." What the *Times* euphemistically termed "gathering" was, in plain words, the application of military force. Indeed, the editors endorsed military efforts to confine the plains nomads to reservations; they approved of the 1876 campaign to subdue the "hostile" Sioux led by Crazy Horse and Sitting Bull. Only reservation Indians deserved material support and Christian education. Those nomads who continued to resist the pressure to go to reservations, the *Times* concluded, remained "utterly out of the reach of every humanizing influence." Following the Sioux defeat of Custer at the Little Big Horn in June 1876, this already straitened channel of humanity was closed. In the wake of Custer's death, the *Times* counseled that the Sioux deserved "condign" punishment.[63]

Although all humanitarians wanted Indians of the plains eventually to renounce nomadic bison hunting in favor of agriculture, some strongly disapproved of the slaughter of the bison. They believed that the destruction of the bison was counterproductive; it merely drove desperate Indians to acts of violence. The editors of *Harper's Weekly* contended in 1874 that "the indiscriminate slaughter of the buffalo . . . has been the direct occasion of many Indian wars. Deprived of one of their chief means of subsistence through the agency of white men, the tribes naturally take revenge by making raids on white settlements and carrying off stocks."[64] *Harper's* asked its readers to take an unprecedented step: to extend the sentimentality and morality of animal protection and Indian humanitarianism from peaceful reservation Indians to off-reservation hunters, and from domestic to wild animals. More typical was the view of the *Nation*: while attributing Indian violence to the demise of the bison, it urged that nonreservation Indians be "hunted down," and subjected to a redoubled effort at cultural assimilation.[65] The wild Indians of the plains required taming.

These contradictions in the reformers' approach to Indians would eventually be the undoing of the effort to save the bison. That effort began promisingly in 1874 when Greenburg Lafayette Fort, a Republican from central Illinois, introduced a bill in the House of Representatives making it "unlawful for any person who is not an Indian to kill, wound, or in any manner destroy any female buffalo, of any age, found at large within any of the Territories of the United States." Fort was a Union Army veteran and a career politician whose four terms in the House between 1873 and 1881 were distinguished by his penchant for "waving the bloody shirt"—a postwar term for the appeal to wartime loyalties—and his idiosyncratic, proto-Populist reformism (he was an early advocate of the coinage of silver). While Fort styled himself a reformer, many of his colleagues saw him as no more than a crank. His oftentimes inane motions, amendments, and objections ("I object to everything that is out of order," he once said in debate) provoked the exasperation of the chamber.[66]

As a member of the Committee on the Territories, Fort regularly encountered the policies of Columbus Delano, the secretary of the interior between 1870 and 1875. While Fort routinely seized upon popular causes to further his appeal, Delano's career was an embattled one, characterized by his unflinching advocacy of unpopular positions. A former member of Congress from Ohio, Delano had been a proponent of radical Reconstruction. As commissioner of internal revenue from 1869 to 1870 and thereafter as secretary of the interior, he headed two of the most venal offices of the federal government in a period distinguished for its extensive public corruption. While Delano was probably an honest man—a rare commodity in President Ulysses Grant's Cabinet—he was unmindful of his subordinates' corruption and too unwilling or uninterested to root it out. The press and backbenchers such as Fort subjected Delano to a barrage of criticism for his failure in this respect. In 1875, the House investigated Delano's supervision of contracts to supply Indian reservations with beef, and the press clamored for the secretary's resignation.[67]

Fort's 1874 bison bill brought him into direct opposition with Delano. The Illinois representative's idiosyncratic brand of reformism and symbolic politics garnered initial widespread, if shallow, support for his legislation. The 1874 bill and another, identical one Fort introduced in 1876 offered Fort's reform-minded colleagues in the House the opportunity to castigate the secretary of the interior on another matter related to Indian subsistence. Reformers in both parties supported the bills. In 1874, when Republicans were in the majority in the House, and 1876, when Democrats were in the majority, the Committee on the Territories unanimously approved the bills and a majority of the House supported the legislation; a majority of the Senate voted for the 1874 bill.[68]

The bills drew support because they marshaled the arguments of both animal protection and Indian humanitarianism. Fort made his sympathies plain in 1876, when he reintroduced his bill, saying, "For my part I favor the society which has in view the prevention of cruelty to animals." Most legislators who spoke in favor of the bills were northeastern or midwestern reformers. George Hoskins, a Republican from New York, perhaps taking his cue from the letter of Lieutenant-Colonel Brackett, which Fort read into the record during the course of the 1874 debate, compared Fort's bill to extant state game and animal protection laws. Fort, who preferred a hyperbolic style of debate, described the killing of bison as "wanton wickedness," a term usually found in the SPCA's tracts. One of Fort's reform-minded colleagues in the House, the Republican Joseph Hawley of Connecticut, brought the issue closer to home, alluding to both motherhood and rural innocence in describing a hide hunter as a man likely to "shoot down his mother's cow in the barn-yard." Invoking the high moral authority of the SPCA, the supporters of Fort's bill portrayed the killing of bison as a thoroughly sordid business.

Supporters of the bill further appealed to the arguments of Indian humanitarians. In effect, Fort's bills were attempts to restore the integrity of the treaties of Medicine Lodge and Fort Laramie by reserving to the plains nomads the undisturbed use of the remaining bison. In his opening address on his 1874 bill, Fort articulated the strongest argument of the humanitarians: "I am not in favor of civilizing the Indian by starving him to death." Fort's colleague, the Republican David Lowe of Kansas, chimed in with the observation that the destruction of the nomads' means of subsistence "will not do in this age of civilization and Christianity."[69]

Yet the advocates of Fort's bill were uncomfortable as partisans of the Indians. Lowe denied that his support of the bill was "simply a matter of sentiment in behalf of the Indian." In 1876, Fort declared that "I have no especial sentimentality in my bosom for the Indian; I have no especially friendly feeling for the savage." While on the floor of the House the question of animal protection was morally—or, at least, politically—unassailable, Indian policy was a nettlesome problem. In the course of the 1874 debates, the Ohio Republican James Garfield—who six years later would be elected president—echoed Secretary Delano to pose the question that ultimately derailed Fort's legislation. However cruel the actions of the hide hunters might be, Garfield pointed out, they were destroying the primary resource of the plains nomads and thereby forcing them to reservations. "If the barbarism of killing buffalo for mere wanton sport has any compensation in it," he asked, "perhaps it may be this is a compensation worthy of our consideration."

The effort to confine Indians of the plains to reservations posed a dilemma to Indian humanitarians. Forcing Indians to reservations and the aims of the peace policy were indeed complementary. Quite apart from the peace policy's hostility to Indian cultural traditions, it was not the entirely benign program that its appellation suggests. As Delano explained in 1873, the "so-called peace policy sought, first, to place the Indians upon reservations as rapidly as possible." Only on reservations were Indians treated to missionaries' government-sponsored advice and aid. Against nonreservation Indians, the federal government's policy was punitive. Delano wrote, "whenever it is found that any tribe or band of Indians persistently refuse to go upon a reservation and determine to continue their nomadic habits . . . then the policy contemplates the treatment of such tribe or band with all needed severity . . . thereby teaching them that it is better to follow the advice of the government." Delano saw the destruction of the bison as a means of pacification: "The rapid disappearance of game from the former hunting-grounds must operate largely in favor of our efforts to confine the Indians to smaller areas, and compel them to abandon their nomadic customs."[70]

Delano interpreted the treaties of Medicine Lodge and Fort Laramie as permitting the destruction of the bison by Euroamerican hide hunters. He applauded the results of their efforts, writing that he "would not seriously regret the total disappearance of the buffalo from our western prairies, in its effect on the Indians, regarding it rather as a means of hastening their sense of dependence upon the products of the soil and their own labors." He deplored the presence of hide hunters on the relatively small Indian reservations but noted that the Treaties of Medicine Lodge did not forbid them from the area south of the Arkansas River designated as the southern plains nomads' hunting territory. In the northern plains, he regretted the stipulation of the Treaty of Fort Laramie of 1868 granting the Sioux the exclusive right to hunt north of the Platte River. He therefore stated—quite unilaterally—that the number of bison ranging north of the Platte was insufficient to support the Sioux, and recommended in both his 1873 and 1874 reports that Congress abrogate that article of the treaty.[71] Congress did not act, so in 1875, Delano and Grant told a Sioux delegation led by Red Cloud and Spotted Tail that the region between the Platte and Niobrara rivers no longer contained enough bison to support the Sioux and must be ceded to the United States.[72]

Delano had many sympathizers in Congress. In 1874, the Missouri Republican Isaac Parker called his policy sound. In 1876, the Texas Democrat John Hancock explicitly rejected humanitarianism in favor of Delano's policies. "I hope, sir, there is no humanitarian sentimentality that would induce legislation for the protection of the buffalo," he said to Fort. "If the theory upon which the Government is now treating the Indians is a proper one, and I am inclined to believe it is the best, the sooner we get rid of the buffalo entirely the better it will be for the Indian and for the white man, too." While Fort's support was strong enough to ensure passage in both houses in 1874 and in the House in 1876, well-placed opponents defeated the bills. President Grant, who had made his support for Delano's policy plain to Indian delegates who visited Washington, killed the 1874 bill with a pocket veto.[73] In 1876, Fort's bill passed in the House but never came to a vote in the Senate Committee on the Territories.

Delano's policy of relying on the exhaustion of the herds in order to relegate Indians to reservations was entirely consistent with a midcentury image of the bison that was radically different from the SPCA's depiction of a gentle ruminant. Popular accounts of the bison described a violent species consistently losing its battles against human hunters and the forces of nature. This view of the bison pervaded an 1869 description in a popular journal. Theodore Davis, the author of the article, argued that "the buffalo is fast disappearing." To the pressure of Euroamerican hunters Davis added Indian hunters and environmental factors: blizzard, drowning, and the predation of wolves. The bison competed among themselves as well. Describing the contests between bulls to establish a rank order during the rutting season, Davis wrote: "The bulls fight viciously, and are attended during these combats by an admiring concourse of wolves, who are ever ready to come in at the death of either of the combatants, or will even take a chance in and finish any killing that has been imperfectly done." One of the illustrations accompanying the article showed a "Battle for Life": a bison fighting off an attack by a pack of wolves.[74] In the 1860s and 1870s, such scenes were a staple of popular depictions of the bison.[75]

The inevitability of the bison's extinction pervaded an 1876 study of the species by Joel Allen, a Harvard zoologist. Allen's book was a hybrid of neo-Lamarckian and natural selection theories and a combination of scientific and popular accounts of the bison. Allen came to his subject while analyzing the fossil remains of extinct species of bison; the extirpation of *Bison latifrons* and *Bison antiquus* foreshadowed the fate of *Bison americanus*. His study was largely a recitation of the bison's disappearance from one region of North America after another, owing to both Euroamerican and Indian hunters and environmental pressures. He called the species "sluggish"—in other words, unfit—and concluded that "the period of extinction will soon be reached."[76] Allen's views exemplified both the popular and academic understanding of Charles Darwin's theories of natural selection. Most Americans conflated Darwin's ideas with those of Herbert Spencer and neo-Lamarckian naturalists such as Edward Cope and Alpheus Hyatt to view natural selection as a process that weeded out the unfit in favor of ever more excellent forms of life: from bison to cattle, from Indians to Euroamericans.[77]

Lawmakers friendly to Delano's Indian policy embraced the idea that the destruction of the bison was an inevitable part of the advance of civilization. They were particularly attracted to the notion that domesticated cattle made a higher use of the range than bison. Representative Omar Conger of Michigan said in 1874 that "there is no law which a Congress can enact, that will stay the disappearance of these animals before civilization. They eat the grass. They trample upon the plains upon which our settlers desire to herd their cattle and their sheep. . . . They are as uncivilized as the Indian."[78] Hancock of Texas seconded that sentiment in 1876: the bison "are, at most, but game. Men have not been able to domesticate them so as to make them useful in any respect as a domestic animal. They take up as much room and consume as much provender as cattle and horses or any other character of useful domestic animals."[79] Relying on popular and dubiously scientific accounts of the bison's tenacity to contrast the bison with familiar domesticated species, Conger and Hancock revealed the tenuousness of the SPCA's depiction of the bison as nearly tame. They argued, in short, that the untamed bison was the resource of the savages; the bison's displacement by domestic cattle was a victory for civilization.

Ratification of Fort's bills may not have mattered, because most army officers in the plains were disinclined to enforce it. Military authorities in the plains were as enthusiastic as Delano about the disappearance of the bison. Colonel Dodge did not prevent hide hunters from entering the Indians' preserve south of the Arkansas River. In 1873, he reportedly told the hide hunter Josiah Wright Mooar, who had come to inquire about the propriety of hunting south of the Arkansas, "If I were a buffalo hunter, I would hunt where the buffaloes are."[80] When Sir William F. Butler, a British officer, confessed to Dodge that he had shot more than thirty bison on the North Platte, Dodge was gleeful. Butler wrote, "I could not but feel some qualms of conscience at the thought of the destruction of so much animal life, but Col. Dodge held different views. 'Kill every buffalo you can.' he said; 'every buffalo dead is an Indian gone.'"[81]

The army adhered to Delano's policy of allowing the exploitation of the herds to proceed unfettered. Even some hide hunters subscribed to this view of the slaughter. "The buffalo didn't fit in so well with the white man's encroaching civilization," argued Mayer. "So he had to go."[82] While he recognized that hide hunters deprived the plains nomads of their means of subsistence, John Cook reflected that the "ruthless slaughter" was "simply a case of the survival of the fittest. Too late to stop and moralize now."[83] Eastern reformers reluctantly agreed. In the Black Hills, the *New York Times* editorialized in 1875, "the red man will be driven out, and the white man will take possession. This is not justice, but it is destiny."[84] These rationalizations rested on the assumption of the innate superiority of Euroamericans and their land use strategies. The belief in Euroamerican progress assumed an inevitable advancement toward higher forms: from Indians and bison to Euroamericans and domestic livestock. For Delano, as for Sherman—who had been among the first to predict the demise of the bison—the replacement of the bison by domesticated livestock was not only an inevitability but a solution to their intractable problem of pacifying the western plains.

While hide hunters such as Mayer and Cook approved of the federal policy toward Indians and the bison, they were not its real beneficiaries.

Bison hunters often found themselves engaged in losing battles against hide dealers, freighters, and retailers. Exorbitant prices for supplies and a glutted market in hides in the mid-1870s broke many bison hunters. The destruction and competition unleashed on the bison not only deprived the plains nomads of their primary resource, but deprived many hide hunters of the rewards of their labors.

For a few years, the slaughter produced tremendous profits, but not necessarily for the hunters. Commercial hide hunting required a considerable investment, beginning with $400 to $650 for a good wagon and $125 for a .50-caliber Sharp or Remington rifle. Altogether, it cost Mayer $2,000 to outfit himself for the hunt.[85] The large initial investment meant that many independent hunters went broke. Others judged themselves fortunate just to break even.[86] Even after outfitting themselves, hunters, like miners in gold rush boomtowns, paid inflated prices for ordinary goods and services. Prices in Dodge City were scandalously inflated.[87]

The accounts of James Cator are instructive. While Cator did not keep an account book, he saved many of his receipts from his bison hunting career between 1874 and 1878. In his first year as a hunter, he almost certainly failed to break even. While his receipts—probably incomplete—show that he sold nearly eleven hundred dollars worth of hides to A. C. Myers and Charles Rath of Dodge City, he also purchased nearly twenty-nine hundred dollars worth of supplies. In 1875 and 1876, he looked to increase his profits by selling not only hides but meat, tongues, and the occasional wolf or coyote pelt. As prices dropped, Cator, like Bussell, hauled his hides to competing, far-flung dealers in search of the best price, selling his hides in New Mexico and Colorado.[88]

As hunters flooded the market with hides, the value of the product steadily declined. A tanner who experimented with bison hides in 1871 paid Josiah Wright Mooar $2.25 per hide. In 1872, before the rush of hunters into the southern plains had begun in earnest, Mooar sold his hides to a tannery in Pennsylvania for $3.50 per hide.[89] By 1874 and continuing until 1876, however, the price of a bull hide in the glutted market had fallen to $1.15, and that of cow and calf hides to sixty-five cents.[90] In December 1874, Dodge City dealers paid James Cator $2.15 for his bull hides. Two years later, their price had fallen to eighty cents per hide.[91]

The falling value of the hides drove some hunters to greater devastation. Some hunters maximized their profits by expanding their carnage to include the poisoning of wolves, foxes, and coyotes. The hunters stocked up on poison—usually strychnine—before setting out for a hunting excursion. At the end of a day's labor of shooting and skinning bison, they laced pieces of bison meat with the poison and left them for scavengers. Using this technique during a hunting trip in Kansas in 1879, Arthur Bill's outfit killed four hundred bison and fifty wolves in twelve days.[92] Hunters who had left poisoned meat for wolves were obliged to secure the pelts at dawn, before birds picked the wolf carcasses to pieces. Nevertheless, thousands of birds died from eating the poisoned entrails of the wolves.[93] As the number of bison declined, "wolf-trapping" was so remunerative that the bison hunter George Simpson claimed that some hunters concentrated on poisoning wolves during the winter.[94] This ancillary destruction signaled the decline of southern plains bison hunting. By 1879, Josiah Wright Mooar reported to his mother that he and his brother, John Wesley Mooar, had abandoned bison hunting: Josiah Wright was hauling supplies to the Colorado mines while John managed a ranch.[95] By 1880, Fort Griffin, Buffalo Gap, and Hide Town were nearly deserted. The bison hunters returned to work for the Santa Fe Railroad, went to the silver mines of Arizona, or moved north to pursue the bison in Montana and Wyoming.[96]

After the bison had been hunted out by 1883, the plains were strewn with their bones. A. M. Bede, a county judge from Fort Yates, North Dakota, recalled of his early days in the northern plains, "the country out here used to look like a charnel house with so many skulls staring at a man, and so many bones that newcomers felt nervous, and, in some cases, could hardly plow the land."[97] L. C. Fouquet of Kansas likewise called the bones "a nuisance to our breaking of the sod."[98] Poor homesteaders and Indians soon found a use for the bones, however, particularly during the drought years of the mid- to late 1880s. They scavenged the plains for bison bones, which they sold to bone dealers for delivery to sugar refineries or fertilizer plants. It took one hundred bleached skeletons to amass one ton of bones, which sold for four to twelve dollars. While the scavengers did not become wealthy, the bone trade sustained many poor families. An early settler in North Dakota recalled that his family "hauled 14 tons of buffalo bones. . . . I don't know how we would have lived if it had not been for the money we got that way."[99]

The bone trade was immense. In 1886, a resident of Dodge City saw a rick of bison bones a quarter of a mile long and as high as the bones could be thrown.[100] Dealers, such as the Northwestern Bone Syndicate of North Dakota, purchased thousands of tons each year, and railroad companies annually shipped nearly five thousand boxcars. The cargo was destined for the Michigan Carbon Works in Detroit and the Northwestern Fertilizer Company and Empire Carbon Works of East St. Louis, Illinois. By the mid-1880s, the Michigan Carbon Works produced five thousand tons of bone black, a pigment, and four thousand tons of bone ash, a fertilizer.[101] The most prominent resource of the plains had been ground down to its salable parts and incorporated into the American industrial economy.

The hide hunters' slaughter of the bison was a boon to Eastern tanneries, who had paid an average of $4.20 per hide in 1870, but only $3.40 per hide a decade later.[102] The destruction of the bison also accomplished the aims of the policymakers who sought to pacify the plains nomads. Once the hunters had eradicated the herds, the nomads had no recourse but to go to the reservations. Like poor Euroamerican farmers, the Indians gathered bison bones for sale to dealers. A reporter for *Harper's* asked in 1893, "Do the Indians make a living gathering these bones?" Yes, replied a railroad inspector, "but it is a mercy that they can't eat bones. We were never able to control the savages until their supply of meat was cut off. We have had no trouble worth speaking of since 1883, however."[103] On the reservation, the Indians' existence was meager. They subsisted in part on government beef. At the Red Cloud Agency in western Nebraska, Oglala Sioux hunters prepared for the delivery of cattle from the government as they had once prepared for a summer bison hunt. As the cattle were released into their corral, the Indians set after them on horseback, slaughtering the cattle in an imitation—culturally rich but economically impoverished—of the communal bison hunts of the past.[104] Government authorities permitted the communal hunts to continue until 1897, when, in an effort to stifle this form of cultural expression, they built a slaughterhouse and began distributing butchered meat to the Indians. The Oglalas responded by setting fire to the slaughterhouse one night.[105]

The costs of the near-extermination of the bison were not shared equally. The plains nomads suffered the loss of their primary resource and consequently of their lands and economic autonomy. The benefits of the destruction of the bison were likewise inequitably distributed. Hunters transformed millions of bison into profitable commodities, yet merchants and industrialists appropriated most of the wealth generated by the destruction of the herds. In the plains, the slaughter left in its wake not wealth but poverty and misery.

Euroamericans did not slaughter millions of bison between 1870 and 1883 because they believed nature provided an inexhaustible supply of commodities. Rather, they anticipated the extinction of the species. They regarded the disappearance of the herds as a triumph of both domestication over wildness and civilization over savagery, because the extermination of the bison removed the nomads' primary resource and cleared the plains for Euroamericans and their livestock.

The erosion of the nomads' hunting territory presaged the triumph—an inevitable triumph, in many contemporaries' minds—of Euroamericans. The engine of Euroamericans' advancement into the plains was the ability of an industrial society to destroy thoroughly the bison herds and thus deny their use to the nomads. While a few lawmakers saw the destruction as deplorable—a violation of the treaties, a provocation to Indian war, and a profligate waste—others saw it as a salient example of Euroamerican industriousness. The plains had remained a wilderness while in Indians' hands, yet hide hunters extracted extraordinary wealth from the grasslands, even though, when their scorched earth campaign was over, most of the hunters found themselves no wealthier than before.

NOTES

1. See Charles Kappler, ed., *Indian Affairs: Laws and Treaties*, vol. 2 (Washington, D.C., 1904), pp. 887–95, 977–89, 998–1005.

2. Terry G. Jordan, *North American Cattle-Ranching Frontiers: Origins, Diffusion, Differentiation* (Albuquerque, N.Mex., 1993), pp. 208–40.

3. Both Indian signatories and some bison hunters who attended the negotiations recalled that the hunters had promised not to hunt south of the Arkansas, but as the hunter Dick Bussell noted, "the hunters paid no attention to the treaty." Richard Bussell, interview with J. Evetts Haley, July 19, 1926, Canadian, Texas; Bussell, "Buffalo Hunting," December 27, 1929, Panhandle-Plains Historical Society, Canyon, Texas. The bison hunter Mark Huselby also recalled that "the whites had made an agreement to stay back, but they had begun to encroach upon the plains, which were the Indians' hunting grounds." Huselby, interview with Haley, June 18, 1925, Mobeetie, Texas, Panhandle-Plains Historical Society.

4. Hendrik Hartog pointed out to me the problem of multiple sources of legal authority in the nineteenth-century United States. See his article, ""Pigs and Positivism," *Wisconsin Law Review* (1985): 899–935.

5. William S. Hatton, *Annual Report of the Office of Indian Affairs*, 31st Cong., 1st Sess., 1849, S. ex. doc. 1 (Serial 550), p. 1074. The figure for 1859 is from Pierre Chouteau of the American Fur Company. Chouteau to Elias Marsh, May 31, 1859; Journal of Elias Marsh, M.D., St. Louis to Ft. Benton, May–August 1859. Journals and Diaries Collection, Missouri Historical Society, St. Louis.

6. Alexander Ramsey, *Annual Report of the Commissioner of Indian Affairs*, 1849, 31st Cong., 1st Sess. (Serial 550), p. 1022; Joel A. Allen, *The American Bisons, Living and Extinct* (Cambridge, Mass., 1876), p. 188. For the decline of Fort Pierre, see Henry A. Boller, *Among the Indians: Eight Years in the Far West, 1858–1866* (Chicago, 1959), p. 22.

7. John D. Unruh, *The Plains Across: The Overland Emigrants and the Trans-Mississippi West, 1840-60* (Urbana, Ill., 1979), p. 334.

8. For an analysis of the effect of emigrants on the bison, see Elliott West, *The Way to the West: Essays on the Central Plains* (Albuquerque, N.Mex., 1995), pp. 51–83. For other studies of the effect of the environment on the bison, see Rudolph W. Koucky, "The Buffalo Disaster of 1882," *North Dakota History* 50 (winter 1983): 23–30; Dan Flores, "Bison Ecology and Bison Diplomacy: The Southern Plains from 1800 to 1850," *Journal of American History* 78 (September 1991): 465–85; Richard White, "Animals and Enterprise," in *The Oxford History of the American West*, ed. Clyde Milner et al. (New York, 1994), pp. 247–49.

9. Merlin P. Lawson, *The Climate of the Great American Desert: Reconstruction of the Climate of the Western Interior United States, 1800–1850*, University of Nebraska Studies, n.s. 46 (Lincoln, 1974).

10. James Mooney, *Calendar History of the Kiowa Indians* (Washington, D.C., 1979), p. 300.

11. Garrick Mallery, "Pictographs of the North American Indians," *Fourth Annual Report of the Bureau of Ethnology, 1882–83* (Washington, D.C., 1886), p. 124.

12. Cary J. Mock, "Drought and Precipitation Fluctuations in the Great Plains during the Late Nineteenth Century," *Great Plains Research* 1 (February 1991): 26–57.

13. Rachel Thorndike Sherman, ed., *The Sherman Letters: Correspondence between General and Senator Sherman from 1837 to 1891* (New York, 1894), p. 320.

14. Lucius F. Ellsworth, *Craft to National Industry in the Nineteenth Century: A Case Study of the Transformation of the New York State Tanning Industry* (New York, 1975), p. 193.

15. Barbara McMartin, *Hides, Hemlocks, and Adirondack History: How the Tanning Industry Influenced the Region's Growth* (Utica, N.Y., 1992), pp. 8, 27, 88, 108; Victor S. Clark, *History of Manufactures in the United States, 1860–1914* (Washington, D.C., 1928), pp. 465–66.

16. Henry R. Proctor, *A Text-Book of Tanning: A Treatise on the Conversion of Skins into Leather, Both Practical and Theoretical* (New York, 1885), p. 144.

17. Lewis T. Nordyke, "King of the Buffalo Slayers," *Farm and Ranch* (April 1939): 7.

18. *New York Times*, September 16, 1892, quoted in McMartin, *Hides, Hemlocks*, p. 119. Gordon G. Whitney, *From Coastal Wilderness to Fruited Plain: A History of Environmental Change in Temperate North America, 1500 to the Present* (New York, 1994), pp. 186–88.

19. For a nineteenth-century assessment of the effects of deforestation, see George Perkins Marsh, *Man and Nature, Or, Physical Geography as Modified by Human Action*, ed. David Lowenthal (Cambridge, Mass., 1965), pp. 186–89.

20. Theodore Steinberg, *Nature Incorporated: Industrialization and the Waters of New England* (New York, 1991), pp. 206–9.

21. For descriptions of still hunting, see Homer W. Wheeler, *Buffalo Days: Forty Years in the Old West; the Personal Narrative of a Cattleman, Indian Fighter, and Army Officer* (Indianapolis, Ind., 1923), pp. 80–81; John R. Cook, *The Border and the Buffalo* (Austin, Tex., 1989), pp. 168–76; Richard Irving Dodge, *The Plains of North America and Their Inhabitants* (Newark, Del., 1989), pp. 151–53. See also T. S. Bugher, interview with Haley, July 17–18, 1925; J. W. Woody, interview with Haley, October 19, 1926, Panhandle-Plains Historical Society.

22. Dodge, *Plains of North America*, pp. 150–51.

23. Lemmon Papers, State Historical Society of Nebraska, Lincoln.

24. Douglas B. Bamforth, "Historical Documents and Bison Ecology on the Great Plains," *Plains Anthropologist* 32 (February 1987): 1–16. For the passenger pigeon, see H. Ronald Pulliam and Thomas Caraco, "Living in Groups: Is There an Optimal Group Size?" in *Behavioural Ecology: An Evolutionary Approach*, ed. J. R. Krebs and N. B. Davis, 2d ed. (Sunderland, Mass., 1984), p. 134; Jenny Price, "Flight Maps: Encounters with Nature in Modern American Culture" (Ph.D. diss., Yale University, 1998).

25. Rex W. Strickland, ed., "The Recollections of W. S. Glenn, Buffalo Hunter," *Panhandle-Plains Historical Review* 22 (1949): 25–31.

26. Cook, *Border and the Buffalo*, p. 171.

27. Henry H. Raymond, "Diary of a Dodge City Buffalo Hunter, 1872–1873," ed. Joseph W. Snell, *Kansas Historical Quarterly* 31 (winter 1965): 349–51.

28. Dodge, *Plains of North America,* p. 155.

29. Strickland, "Recollections of W. S. Glenn," pp. 25–31.

30. Frank Mayer and Charles B. Roth, *The Buffalo Harvest* (Denver, Colo., 1958), pp. 49–52.

31. Cook, *Border and the Buffalo*, p. 167.

32. Wheeler, *Buffalo Days*, p. 80.

33. Woody, interview with Haley, February 11, 1928, Panhandle-Plains Historical Society.

34. Wayne Gard, *The Great Buffalo Hunt* (Lincoln, Neb., 1959), pp. 128–29.

35. J. Wright Mooar, interview with J. Evetts Haley, November 25, 1927, Panhandle-Plains Historical Society.

36. *Kansan* (Newton), November 28, 1872.

37. Cator Family Papers, Panhandle-Plains Historical Society.

38. S. P. Merry, interview with J. Evetts Haley, Amarillo, Texas, August 21, 1926, Panhandle-Plains Historical Society.

39. John Chadbourne Irwin Papers, Southwest Collection, Texas Tech University, Lubbock, Texas.

40. Andrew C. Isenberg, "Social and Environmental Causes and Consequences of the Destruction of the Bison," *Revue Française d'Etudes Américaines* 70 (October 1996): 15–27.

41. "Snow on the Plains," *New York Times*, December 13, 1871; "Great Loss of Cattle in Texas," ibid., March 29, 1872; "Incidents in Frontier Experiences—How the Monarch of the Plains is Hunted," ibid., April 2, 1874.

42. The disease, which affects grazing animals in warm climates, is caused by parasites transmitted by certain ticks belonging to the *Boophilus* genus. The parasites enter a host animal's bloodstream and attack the red blood cells, causing high fever, anemia, jaundice, and weakness. Texas longhorns were immune to the disease caused by the ticks they carried, but when they were driven to market they spread the affliction to midwestern livestock. Like anthrax, which may have spread among Canadian bison in the 1820s and 1830s, Texas fever could spread without close contact between domestic and feral animals: the ticks could drop off their old hosts and attach themselves to new ones. Bison were susceptible to the disease. In 1905, preservationists established a bison refuge in Oklahoma where longhorns had formerly grazed. They feared that the bison would be infected; indeed, of the first fifteen bison introduced to the refuge, two died of Texas fever in the first year. T. S. Palmer, "Our National Herds of Buffalo," *Tenth Annual Report of the American Bison Society, 1915–1916* (New York, 1916), pp. 44–46.

43. "Montana's Indian Puzzle," *New York Times*, April 4, 1880.

44. Theodore Roosevelt, *Hunting Trips of a Ranchman* (New York, 1885).

45. James K, Detling, "Processes Controlling Blue Grama Production in the Shortgrass Prairie," in *Perspectives in Grassland Ecology*, ed. Norman R. French (New York, 1979), pp. 25–39; Victor Shelford, *Ecology of North America* (Urbana, Ill., 1963), p. 340; Philip L. Sims, J. S. Singh, and W. K. Lauenroth, "The Structure and Function of Ten Western North American Grasslands I: Abiotic and Vegetational Characteristics," *Journal of Ecology* 66 (March 1978): 270; Sims and Singh, "The Structure and Function of Ten Western North American Grasslands II: Intra-Seasonal Dynamics in Primary Producer Compartments," ibid., 66 (July 1978): 565; O. E. Sala and W. K. Lauenroth, "Small Rainfall Events: An Ecological Role in Semiarid Regions," *Oecologia* 53 (June 1982): 301–4.

46. Robert T. Coupeland, "Effects of Changes in Weather upon Grasslands in the Northern Plains," in *Grasslands*, ed. Howard B. Sprague (Washington, D.C., 1959), p. 293.

47. Garrick Mallery, "Picture-Writing of the American Indians," *Tenth Annual Report of the Bureau of American Ethnology* (Washington, D.C., 1889), pp. 285–86.

48. Ernest Staples Osgood, *The Day of the Cattleman* (Chicago, 1970), p. 53.

49. William Cronon, *Nature's Metropolis: Chicago and the Great West* (New York, 1991), p. 220.

50. See David H. Breen, *The Canadian Prairie West and the Ranching Frontier, 1874–1924* (Toronto, 1983), pp. 65–66.

51. Mock, "Drought and Precipitation Fluctuations," pp. 43–47.

52. Bunn, "The Tragedy of the Plains," *Forest and Stream*, October 29, 1904, pp. 360–61.

53. William T. Hornaday, *The Extermination of the American Bison, with a Sketch of its Discovery and Life History* (Washington, D.C., 1887), p. 513.

54. Isenberg, "The Returns of the Bison: Nostalgia, Profit, and Preservation," *Environmental History* 2 (April 1997): 179–96.

55. James Turner, *Reckoning with the Beast: Animals, Pain, and Humanity in the Victorian Mind* (Baltimore, 1980), pp. 31–57.

56. Ibid.

57. See Katherine Grier's essay in this volume, below, pp. 316–62.

58. Edward P. Buffett, "Bergh's War on Vested Cruelty," vol. 6; American Society for the Prevention of Cruelty to Animals Archives, New York. See also Zulma Steele, *Angel in Top Hat* (New York, 1942), p. 162.

59. "Buffalo Slaughter," *New York Times*, February 7, 1882.

60. Columbus Delano, *Annual Report of the Secretary of the Interior*, 1873, 43rd Cong., 1st Sess., House ex. doc. vol. 4, pt. 1, ix.

61. For nineteenth-century Indian assimilation, see Frederick E. Hoxie, *A Final Promise: The Campaign to Assimilate the Indians, 1880–1920* (Lincoln, Neb., 1984); Henry E. Fritz, *The Movement for Indian Assimilation, 1860-1890* (Philadelphia, 1963).

62. "The Slaughter of the Piegans," *New York Times*, February 24, 1870; "The Piegan Slaughter and Its Apologists," ibid., March 10, 1870; "The Montana Massacre—Col. Baker's Report," ibid., March 12, 1870; "The Last Appeal of Red Cloud," ibid., June 17, 1870.

63. "The Oldest of American Difficulties," ibid., May 22, 1870; "The Indian War," ibid., July 17, 1876.

64. *Harper's Weekly*, December 12, 1874, pp. 1022–23.

65. "Our Indian Wards," *The Nation*, July 13, 1876, pp. 20–22.

66. *Congressional Record*, 44th Cong., 1st Sess. (January 31, March 13, April 24, 1876), pp. 773, 1678, 1813; ibid., 45th Cong., 1st Sess. (November 23, 1877), pp. 622–31.

67. "Contracts for Indian Supplies and Transportation for the Fiscal Years Ending June 30, 1873, and June 30, 1874," House Report No. 778, 43rd Cong., 1st Sess. (Serial 1627). For Delano, see Daniel W. Delano, Jr., *Franklin Roosevelt and the Delano Influence* (Pittsburgh, 1946), pp. 101–33; Allan Nevins, *Hamilton Fish: The Inside History of the Grant Administration*, vol. 2 (New York, 1936).

68. In 1874, House Republicans held 192 seats, Democrats 92, and other parties or independents 14. The voice vote on Fort's bison bill was 132 ayes and noes not counted. In 1876, House Democrats held 169 seats, Republicans 109, and other parties and independents 14. The bison bill passed by a vote of 93 to 48. After tellers were ordered, the vote was 104 to 36. For the debates on the bison bills, see "Protection of Buffaloes," *Congressional Record*, March 10, 1874, pp. 2105–9; "Slaughter of Buffaloes," ibid., February 23, 1876, pp. 1237–41.

69. Ibid., March 10, 1874, pp. 2106–7.

70. Delano, *Annual Report of the Secretary of the Interior*, 1873, p. 5

71. Ibid., pp. vi–vii; *Annual Report of the Secretary of the Interior*, 1874, 43rd Cong., 2nd Sess., House ex. doc. no. 1, vol. 6, pt. 5, p. xi.

72. "The President's Last Interview with the Sioux," *New York Times*, June 3, 1875.

73. "Red Cloud: The Indians in Washington Visit the White House," ibid., May 29, 1872; "The Sioux and the White House," ibid., May 27, 1875.

74. Theodore R. Davis, "The Buffalo Range," *Harper's New Monthly Magazine* (January 1869): 147–63.

75. See Larry Barsness, *The Bison in Art: A Graphic Chronicle of the American Bison* (Fort Worth, Tex., 1977), pp. 62–63, 80–81.

76. Allen, *American Bisons*, pp. 35, 56, 180–81.

77. Peter J. Bowler, *Evolution: The History of an Idea*, rev. ed. (Berkeley, Calif., 1989), pp. 24, 258. For the neo-Lamarckians, see also Edward J. Pfeifer, "United States," in *The Comparative Reception of Darwinism*, ed. Thomas F. Glick (Chicago, 1988), pp. 168–206.

78. *Congressional Record*, March 10, 1874, p. 2107.

79. Ibid., February 23, 1876, p. 1239.

80. J. Wright Mooar, "Buffalo Days," as told to James Winford Hunt, *Holland's* (February 1933): 10, 44.

81. *Sir William Butler: An Autobiography* (New York, 1913), p. 97. For the question of the military and the bison, see David D. Smits, "The Frontier Army and the Destruction of the Buffalo: 1865–1883," *Western Historical Quarterly* 25 (1994): 313–38.

82. Mayer and Roth, *Buffalo Harvest*, pp. 27–28.

83. Cook, *Border and the Buffalo*, pp. 166–67.

84. "Taming the Savage," *New York Times*, April 15, 1875.

85. Mayer and Roth, *Buffalo Harvest*, pp. 49–52.

86. William E. Connelley, ed., "Life and Adventures of George W. Brown, Soldier, Pioneer, Scout, Plainsman, and Buffalo Hunter," *Collections of the Kansas State Historical Society* 17 (1926–28): 111, 123.

87. Wilson, "Details of a Buffalo Hunt of 1876 in Kansas," Kansas State Historical Society, Topeka.

88. Cator Family Papers, Panhandle-Plains Historical Society.

89. Odie B. Faulk, *Dodge City: The Most Western Town of All* (New York, 1977), pp. 32–33.

90. H. B. Lovett, interview with L. F. Sheffy, June 23, 1934, Pampa, Texas, Panhandle-Plains Historical Society.

91. Cator Family Papers, Panhandle-Plains Historical Society.

92. Arthur C. Bill, "The Buffalo Hunt," Kansas State Historical Society.

93. Frank J. North Papers, State Historical Society of Nebraska. For the poisoning of wolves, see also Wentin Wilson, "Details of a Buffalo Hunt," Kansas State Historical Society.

94. Simpson, interview with Haley, Canadian, Texas, July 18, 1926, Panhandle-Plains Historical Society.

95. John W. Mooar, Fort Griffin, to his mother, Pownal, Vermont, March 5, 1879, John W. Mooar Papers, Southwest Collection, Texas Tech University.

96. Frank Collinson Papers, Panhandle-Plains Historical Society.

97. Bede, Fort Yates, North Dakota, to Edmund Seymour, New York City, October 29, 1919, American Bison Society Papers, Conservation Collection, Western History Department, Denver Public Library, Box 275, File 14.

98. L. C. Fouquet, "Buffalo Days," *Collections of the Kansas Historical Society* 16 (1923–25): 347.

99. Quoted in LeRoy Barnett, "The Buffalo Bone Commerce on the Northern Plains," *North Dakota History* 39 (winter 1972): 23–41.

100. O. H. Simpson, Dodge City, Kansas, to the American Bison Society, January 19, 1934, American Bison Society Papers, Box 276, File 7. See also Joe Killough, interview with D. T. Leachman, December 16, 1945, Panhandle-Plains Historical Society.

101. Barnett, "Buffalo Bone Commerce," pp. 23–41.

102. McMartin, *Hides, Hemlocks*, p. 88.

103. Hamlin Russell, "The Story of the Buffalo," *Harper's New Monthly Magazine* (April 1893): 798. See also Mrs. Mae Tubb Dolcater, interview

with Morris Dolby, January 1, 1948, Panhandle-Plains Historical Society.

104. Collinson Papers, Panhandle-Plains Historical Society.

105. James R. Walker, "The No Ears, Short Man, and Iron Crow Winter Counts," in *Lakota Society*, ed. Raymond J. DeMallie (Lincoln, Neb., 1982), p. 153.

5

The Imperial Hunt in Colonial Kenya, c. 1880–1909

Edward I. Steinhart

Introduction

At the end of the nineteenth century, when the Scramble for Africa was at its most frenzied, the bearers of the scepters of empire were, as often as not, armed also with high-powered hunting rifles. In Kenya in particular, the coincidence of hunting and empire building was extraordinary. With the arrival of aristocratic settlers as permanent residents of Britain's premier "sportsmen's paradise," the fusion of Britain's dominion over Kenyan human and its wild life was given the gloss of elegance and sophistication which maintained the colony as a tourist destination for decades to come. But in the early years of imperial conquest and administration, a more rough-and-ready empire builder cum hunter dominated both Africans and the shooting fields. In the following pages, I will first examine how the British traditions of the Hunt were forged into a highly refined ideology of sportsmanship amounting almost to a code of gentlemen's honor and chivalry, an ideal rooted in the realities of class and privilege handed down and adapted from the European Middle Ages. Second, I will demonstrate how this ideology was imperfectly translated to the Kenyan bush by the first generations of professional hunters, imperial explorers, administrators, colonial police, and soldiers during the age of imperialism. If the picture that emerges lacks the romance of titled noblemen and royalty or the glamour of the later champagne safaris and their pursuit of adventure and big game, it is, nonetheless, the story of the first foundations of the Imperial Hunt, a discourse of domination over man and nature that lies at the heart of the imperial idea.

European Roots of the Imperial Hunt

The Europeans who began in the last half of the nineteenth century to pursue big game in what would become Kenya were the inheritors and carriers of an ancient tradition of hunting. Since the Middle Ages, this tradition had become a highly charged symbolic activity, steeped in ritual and social meaning beyond the simple pursuit of animals for food and hides. The Hunt, as this tradition has been called,[1] would clash in crucial ways with the practices of hunting by the diverse pastoral and farming communities that Europeans would encounter in Kenya in the colonial era. In the following pages, I hope to examine a few of the most salient elements of the European tradition as it manifested itself among those hunters among East Africa's imperial invaders who profoundly affected the transformation of hunting, in the process establishing a new, imperial hunting tradition, produced by the colonial encounter.

Unlike the African traditions of hunting found in Kenya, the European hunting heritage had become, since the late Middle Ages, a class-divided and contested arena for the symbolic expression of mastery over both nature and the lower orders of society.[2] Three elements of the tradition would directly affect the relations between African and European hunters in Kenya: the view of big game hunting as the exclusive prerogative of gentlemen, and the exclusion of peasants from the legal pursuit of game; the understanding of wildlife as a form of property whose ownership and use were controlled and determined by law; and finally, the symbolic uses of the hunt as reaffirmation and demonstration of the social hierarchy which gave meaning to the lives of the gentlemen hunters or sportsmen of the European tradition.[3]

Since classical antiquity, the hunting of large and dangerous animals within the cradle areas of Western culture has been seen as a manly exercise, its virtues used in the development of leadership skills and martial ability among rulers and warriors.[4] This is not to say that hunting was in any way restricted to the ruling classes of antiquity, merely that among the elite of warriors and princes it was those skills honed by hunting which were seen as giving value and meaning to the pursuit of game and which elevated it above the common practice of hunting for the pot.

During the Middle Ages, hunting in northern and western Europe would begin its transformation from a practical skill and subsistence-oriented economic activity to the ritualized sport of gentlemen and nobles, the Hunt. This can best be understood as part of a dialectical process by which the hunting of common folk was demeaned and eventually forbidden by the growing power of the landed classes and the warrior nobility and their appropriation of hunting resources (animals, forests, and technology) to their own exclusive use. Thus, hunting by peasants and commoners was gradually devalued and transformed into an illegal act against the king and his privileged nobles.

Poaching—the illegal taking of game, especially the king's deer—can be seen as the defining crime of the medieval social order. Poachers from the lower orders of society who took only small game for the pot might be tolerated or ignored by the king's gamekeepers and foresters. Those men of means—gentlemen without private estates or access to the privileges extended to the nobility by royal assent—who took the king's deer, enjoyed no such tolerance. The gentleman poacher became a figure of legend and folklore, with Robin Hood being only the best known of the breed. Poaching by gentlemen came to signify an act of defiance against a privileged order, a ritual of rebellion and declaration of independence.[5] In England especially, a hunting culture based on "poaching fraternities" of gentlemen hunters was engendered by the peculiar dialectic of medieval hunting and game regulation.[6]

As hunting became an exclusive privilege of the high nobility, it "took on an opposite significance among the aristocracy" to the ideas of rebellion and defiance: in those circles, hunting became associated with upper-class status, and hunting practices became encrusted with courtly ceremony. . . ."[7] Thus, the creation of a class of commoner and gentlemen poachers, who "associated [hunting] with freedom, feasting, and rebellion against authority,"[8] was the obverse of the emergence of an aristocracy of the chase, for whom the symbolic significance of hunting, as much as that of the nobleman's sword, was to designate high status as members of England's (and to some extent Europe's) warrior elite.

The association of hunting with military skills and leadership among the feudal nobility made deer hunting in particular a symbol of social superiority just as it changed other kinds of hunters into poachers. In late medieval England, "[h]unting was a preparation for war and was regarded as the best means, short of actual combat, for testing valour and developing the skills necessary for mounted combat. Hunting was less dangerous than duelling, which in any case, was forbidden by the laws of England. . . . "[9] The old connection between hunting and combat became, under medieval conditions of a ruling class of mounted warriors, an outward sign of nobility, proclaimed in custom and law.

During the early modern era from 1485 to the English Civil Wars (1640), "English game legislation became more and more socially restrictive."[10] Parliamentary game laws severely curtailed legal hunting by imposing high property qualifications, leaving many who had previously enjoyed the right to hunt to turn their resentments against aristocratic privilege. These laws "can also be viewed as part of the seignorial assault upon commons and wastes,"[11] and thus as part of the destruction of the independence of smallholders and peasants by the landed aristocracy.

What the parliamentary laws proclaimed was supported by the laws of hospitality. Only the wealthiest of individuals could afford to entertain in the royal manner, which normally "included opportunities for hunting"[12] in the private parks and estates of the nobility and in royal game parks. By 1600, only the great lords could offer such accommodation and "hunting was for great men only, 'not peasants.'"[13] The drama of struggle over the control of wildlife and the suppression of poaching was being moved onto the same stage where the central struggle for class hegemony within English society was being enacted.

The response of the peasantry and gentry to the aristocratic assault on their access to wildlife resources was to make "[p]oaching . . . a national pastime in Tudor and Stuart England."[14] A century-long conflict replaced the medieval game of gentlemanly poaching during the seventeenth century. This "long affray" of the poaching wars would continue with mounting intensity in the eighteenth century, and would last until the Great War brought English rural society into the twentieth century.[15]

Prior to this time, the authorities and the poachers had treated "illegal hunting . . . [as] a manly game . . . a game that reinforces male gender identity."[16] For the poachers, deer stealing was "an assertion of independence and an act of private rebellion."[17] For the gamekeepers and foresters "among the 'poaching fraternity,' the game [of controlling poaching] was an opportunity to display male virtuosity."[18] This sense of poaching as a boy's game played by men, which developed in the Middle Ages, would never completely die off and would influence many of the attitudes to poaching among the gamekeepers and settlers of colonial Kenya. But changes in English society in the modern era would bring about new, more severe attitudes that eventually came to overshadow the more playful views described here.

A new seriousness seems to have entered the long affray when the turmoil of civil war and revolution brought the landed gentry to a new social and political prominence in England in the eighteenth century.[19] The new squirearchy, less secure in its new status and privileges than the old nobility and perhaps recognizing new economic values in their land and its animal resources, would approach hunting as an attack on their property rights. Associated with a host of other crimes against property such as smuggling, arson, and even the blacking of faces to hide one's identity, poaching would be subject to a draconian code imposing capital punishment on those who continued to play the game. Not only was royal game, the king's deer, to be protected by the game laws, but even the snaring of hares on private land was subjected to the harshest of penalties after 1723.[20] This change in the legal treatment of poaching reflected a social transformation within English rural society that continued to affect ideas about hunting until the industrial transformation in the second half of the eighteenth century. By then hunting had all but ceased to be a manly game and a contest among gentlemen. In the words of Charles Chenevix Trench, a Kenya resident and amateur historian of English poaching: "In the early eighteenth century, . . . deer-stealing . . . was no longer 'a brave diversion' for persons of quality. The gentleman poacher, apart from an occasional eccentric, passed into history."[21]

A new, rougher element of society began to engage in illegal hunting while the gentry came increasingly to view poaching as a form of class warfare, an attack on their newly gained privilege to be opposed with ruthless determination. Allied now with the declining nobility in what had become their common attachment to hunting privilege, the gentry would shift the face of battle down the social scale. It would be the rural poor who would bear the brunt of the assault on their access to hunting opportunities and resources that accompanied the assault on the commonage and forests which had in the past supported peasant independence. Enclosures and game law restrictions were the two fronts of the same advance of the newly confirmed propertied classes in rural England against the lower orders of rural workers, peasants, and cottagers. The campaign against poaching was the battle standard beneath which marched the advancing power of the squirearchy.

Nowhere are the power and the values of the landed gentry more evident than in the emergence of foxhunting or "riding to hounds" as the preferred form of sport among the upper classes. During the late eighteenth century, foxhunting "came to be looked upon as one of the chief promoters . . . of unity, stability, harmony, and devotion to traditional deferential values" within English country districts.[22] Foxhunting would bring together in the field of sport all the classes of rural society in a visible ritual reenactment of the privileges and deference required by the rural elite: a dramatization both of mastery over the natural world and of the subordination of the social world to the local lords of the countryside. According to David Itzkowitz, foxhunting "meetings were, in fact, one of the major sources of contact among the various classes, but the relative differences in social station were never forgotten. . . . While hunting people boasted of the openness of the hunting field, there was no question that the values the field fostered were conservative and aristocratic."[23] In the ritualized Hunt, the costumed and rigorously stage-managed pursuit of a specially released fox by packs of trained hounds and gentlemen (and ladies) on horseback (as in the less formal and stagy, but still symbolically loaded "shooting party"), a profoundly antidemocratic and hierarchical impulse among "hunting people" was periodically reinforced by the ritual slaughter of a wild animal by sportsmen. Only the upper classes were permitted to ride to hounds and the assembled lower orders of rural society were compelled by custom and social power to bear witness. Combining the symbolic elements of medieval military authority with the very real economic powers of the landed classes, fox "[h]unting alone developed a mystique that raised it to the level of a national institution, that, to the minds of more than just hunting men, could be trifled with only at the risk of wide ranging social consequences."[24]

Such an important institution of English rural life as was the Hunt in the nineteenth century was a natural candidate for introduction into Kenya in the early twentieth century. Efforts were made to establish a hunting meet in the vicinity of Nairobi, where horses and dogs were bred by British settlers, and where silver-backed jackals substituted for foxes and African domestic servants and peasants stood in for their English counterparts. However, foxhunting was already becoming passé in England and never really took root among the transplanted English gentry and nobles of rural Kenya. Nonetheless, the breeding of both horses and dogs remained a popular and profitable business in Kenya for many years.[25]

The decline of English foxhunting began in the period of the 1870s and 1880s, largely as a consequence of the agricultural depression of that period and the chorus of complaints by nonhunting farmers against the damage caused to their property and their profits, leading to demands for the payment of compensation.[26] An increase in the poisoning of foxes and the growing use of wire fences were further evidence of efforts by nonhunting landholders to spoil the game for the hunters. This was only one aspect of a broader opposition to foxhunting which had emerged in English society among those branches of the middle class who saw it (but also hunting privilege in general) as a waste of time and valuable resources, a neglect of duty, and in what would become a key part of the modern anti-hunting sentiment, as a particularly obnoxious form of cruelty to animals.[27]

Behind this mounting attack on foxhunting as England's "national institution," the symbolic guardian of the order and social stability of traditional rural society, lay the profound transformation of English society and values engendered by industrialism and the rise to prominence of a new middle class. New attitudes towards nature were "linked to the growth of towns and the emergence of an industrial order in which animals became increasingly marginal to the processes of production."[28] Hunting of all kinds was significantly affected by the industrial and urban transformation and the commercialization of agriculture which led to the decline in foxhunting. From the early 1800s, an increasing intrusion into the countryside by members of the urban middle classes, including many women, especially after the railroads linked town to country, meant a gradual displacement of the landed gentry from their exclusive seats of privilege. By the early twentieth century, the figure of the foxhunter or country gentleman as a weekend sportsman was becoming an object of ridicule and contempt. The general decline of the aristocratic classes, whose fate was sealed by the First World War, left the sportsman within European culture as a symbol of the now decadent order of rural life.

Nowhere does this change in values become more evident than in the use by twentieth-century filmmakers of hunting and gentlemen hunters as symbols of the decadence and moral bankruptcy of the rural elite. Jean Renoir's classic *The Rules of the Game* is a poignant and humorous eulogy to the dying order represented as a weekend hunting party in rural France.[29] A more recent and stinging account of the anachronistic violence and moral decay of the English upper classes on the eve of World War I is Alan Bridges's *The Shooting Party,* a merciless caricature of bird-shooting hunters as archetypes of social decadence.[30]

The embourgeoisement of European culture not only drove the gentlemen hunters from their perches atop the rural social hierarchy. By the midnineteenth century, it also began to drive many of them out of Europe in search of both game and the kind of deferential rural society for which they pined. The decline of animal and bird populations in England which accompanied the enclosure and systematic agricultural exploitation of the countryside sent would-be hunters who could afford to pay for their sport in pursuit of happy hunting grounds elsewhere—first to the continent of Europe, especially Switzerland, and to the Indian subcontinent in pursuit of tigers and the hunting of wild pigs from horseback (*shikar*).[31] Frequently British and European sportsmen came to figure in the saga of the North American wilderness as the trans-Mississippi West briefly became a happy hunting ground for those seeking bison, bear, ram, moose, and elk. By midcentury, European gentlemen hunters had discovered and would soon virtually eliminate the wild game of southern Africa, aided in good measure by "Boer" hunters and settlers, who although not considered gentlemen, were politically independent and capable of defending their hunting rights.[32]

By the 1880s, African game trails were leading European sportsmen and professional hunters north of the Limpopo and Zambezi into the yet unexploited hunting grounds of East Africa. They would carry with them not just the desire to hunt big game, but also the banner of imperial conquest.[33] Most important, they would display in East Africa a very distinctive idea of what hunting meant. Its associations over centuries with military skill and virtue, with robust athleticism and manliness, with chivalry, bravery, and honor may not have differed markedly from the understandings of the value of hunting among the Kenyans whom these Imperial Hunters would encounter. However, the idea that hunting was the exclusive privilege of an upper class, that it betokened wealth and status denied to common folk, and that it conferred on the hunter exclusive membership in an aristocracy of blood and wealth would set them directly at odds with their African hunting counterparts.[34]

The experiences of early European hunters in what would become Kenya involved the transmission to an African setting of traditions and practices of hunting which were ceasing to bind together the rural communities in which they had developed. Indeed, the "invented traditions" of the new upper middle class in England found fertile soil in the translocated aristocracy of the Kenyan White Highlands. This miniature and often grotesque re-creation of aristocratic and genteel society was made possible by the virtual exclusion of immigrants from the artisanal and peasant classes of Britain and the promotion of settlement in Kenya of members of the upper and upper middle classes—"the public school educated with a patrimony, a military pension, investment income or an assurance of family support."[35] This process of cultural translocation and the re-creation of the ethos of the Hunt in Kenya during the imperial age is what we must now examine.

WHITE HUNTERS IN KENYA, 1895–1939

It can be said with some justification that every European man, and quite a few women, who came to Kenya in the late nineteenth and early twentieth centuries engaged in hunting. The prevalence of white hunters in Kenya would have a profound effect on the lives and livelihoods of African hunters. Big game hunting would profoundly alter the ways in which Africans hunted big and small game; how they thought about this age-old activity how they conceived of wildlife and the relationship between themselves and the world of nature. From a postmodern perspective, an age-old European discourse on hunting and a philosophy of man's place in nature would attempt to impose itself on Africans, interacting with equally old and established African views of the relationship between culture and nature, especially as it focused on the pursuit of animals for food and material goods as well as for "sport" and recreation, and as an assertion of manhood.[36]

The careers and initial influence of the earliest carriers of the imperial hunting tradition, who reached Kenya between the late nineteenth century and the outbreak of the First World War, will be examined here in a reading of the attitudes and hunting practices which they recorded in a range of hunting adventure accounts.[37] I will not examine in this essay two groups of extremely influential white hunters: those who came as pioneer settlers to remain as permanent residents, usually as large-scale farmers, who began to arrive in large numbers during the first decade of the twentieth century and would permanently reshape the Hunt after 1909; and the visiting safari hunters and their guides and companions, the professional white hunters. Following the safari of Teddy Roosevelt in 1909, the Imperial Hunt would become the romantic and truly East African safari, no longer an imperial sport, but the preferred outdoor recreation of the world's rich and famous. Rather, the following section will restrict itself to those early influences that laid the foundation for the emergence of the East African safari and the transformation of African and sport hunting that it would entail.

Ivory Hunters

In the first phase of Kenya's white hunting history, the hunting of big game was an activity of both those who came to Kenya exclusively to hunt for sport and profit and those who came for other purposes, but for whom hunting was an economically crucial and in many ways an ideologically fundamental activity. Explorers, colonial administrators, missionaries, scientists and technicians, and a handful of pioneer settlers including a number of titled aristocrats and gentlemen of high repute as sportsmen would set the tone of privilege and luxury which became the hallmark of the imperial hunting tradition in Kenya. Kenya would also become a happy hunting ground in those early days for what have been termed the "Ivory Hunters."[38] Often rogues and adventurers, these men of means came to Kenya in pursuit of game, trophies, and the thrill of the chase, but also for the high profits which a haul of elephant ivory or rhinoceros horn could bring on the international trophy market. In the pioneer days of imperialism, before the beginnings of hunting regulation—which included limits on the number of animals permitted to be shot and their trophies exported—this form of hunting would have an equally profound impact, both on the representations of white hunting to an international public and on the reputation and impact of white hunters among Kenya Africans who came to see or know of the work of slaughter done by these professing sportsmen.

The earliest phase of Kenya big game hunting was that in which the pursuit of trophies and of sport were contaminated by what even in their own terms could only be characterized as "butchery," the wanton killing of game animals for profit.[39] Yet, because elephants were considered noble animals, difficult and dangerous to pursue and worthy of the most expert of sportsmen, the hunting of elephants for their tusks, which might produce enormous revenue, was masked behind the facade and ritual of the Hunt.

As with other white hunters in Kenya, it is important to establish a distinction between ivory hunters who came exclusively to hunt and those who did so as an adjunct to their other activities and purposes: to explore, soldier, administer, missionize, police, and pioneer in a new land. It is among the first group of "pure" hunters with no other agenda that we will find the clearest examples of the ivory harvesting type. I want to examine briefly a few representatives of this group, from whose adventure or travel accounts we can best abstract the outlines of ivory harvesting as a hunting practice in Kenya and East Africa: W. D. M. "Karamoja" Bell, Arthur H. Neumann, A. Arkell-Hardwick, and Captain Charles H. Stigand.[40]

However, before we examine their exploits and influence, let us identify a few preconditions to the emergence of ivory hunting in the late nineteenth century. First of all, the practice of elephant hunting by trophy-seeking specialists was neither original nor unique to East Africa or Kenya. Its origins lay far to the south in the Cape Colony, the Transvaal, and the interior of southern Africa during the first half of the nineteenth century. The hunting grounds on and beyond the frontier of European expansion in "Boer" and British South Africa would see the first flourishing of this pursuit of elephants by such famous hunters as R. G. G. Cumming, William C. Harris, and later Frederick Selous.[41] By the late nineteenth century, two changes had caused a shift in the hunting frontier from southern to East and Central Africa. First was the exhaustion of the game in the south through overhunting and continual disruption of the hunting grounds in the absence of any regulation of the relatively dense resident population of white hunter-farmers. The growth of commercial hunting and poaching in South Africa and Mozambique also contributed to this shift, as did eventual efforts to develop a conservation policy in the Transvaal.[42]

Matching the decline of elephant hunting opportunities in the far south was the opening of East Africa through the surge of exploration and intensification of contacts with the coast that accompanied and constituted the Scramble for Africa in the region. Even a cursory reading of the exploration accounts of Richard Burton, and most especially Joseph Thomson, who traveled across Kenya in the 1880s, gives ample evidence of both the extensive possibilities of big game hunting which their accounts revealed to a large metropolitan readership and the fact that hunting was a key factor in enabling them to advance across the East African coastal plain to the unexplored savannahs and forest and the unnamed lakes and mountains of the interior.[43]

Among the first rank of hunter-adventurers to exploit the elephant herds of East Africa and Kenya in particular was Walter D. M. "Karamoja" Bell, who, the dust jacket of his autobiography tells us, was "the greatest elephant hunter of all times."[44] Born in 1880, too late for the precolonial heyday of hunting in East Africa, Bell nevertheless displayed an unreserved enthusiasm for killing elephants and an unapologetic belief in the naturalness of killing as many of them as he desired. His first trip to Africa when he was seventeen took him to Mombasa, but at that age he was unable to get a safari to take him up-country. After adventures in the Yukon gold rush and the Boer War he returned to East Africa where he was able to travel to Uganda in pursuit of elephant and to learn how to kill them. The indiscriminateness with which he took whatever elephant killing opportunities presented themselves was qualified only by the fact that he considered hunting the solitary bull elephant the "most interesting and exciting form of elephant hunting"—so much so that he made it the subject of the very first chapter of his book. To reinforce the notions both of his daring and of his total absorption in the pursuit of elephant with the biggest ivory, the second chapter offers the reader a detailed discussion of his preference for small-bore rifles of .275 or .256 caliber to the heavier and more powerful elephant guns of the day, as well as of his preference for the brain shot to the heart shot—a frequently debated subject among elephant killing cognoscenti at the time.[45] All in all, Bell reveals little introspection about the nature of his chosen path which he says claimed him from childhood for a life of adventure. In his defense, I would say that he does offer respectful and considered opinions on African beliefs, termed "medicine," and "its Bearing on Sport" which include a careful and accurate description of Kenya's "Wa-Boni" hunters and an illustration of an "M'sanye Bow and Poisoned Arrow" with laudatory comments on the remarkable ability of these hunters to use these rudimentary weapons.[46] As we will see, such willingness to offer respect to African hunters was by no means common.

A similar lack of introspection can be found in the writings of Abel Chapman, who hunted in East Africa, especially the western districts of Kenya, the Rift Valley, and the Athi River plains in 1904 and 1906.[47] Chapman considered that "British East Africa affords today probably the most glorious hunting-field extant, certainly the most accessible," contrasting it sharply with South Africa where the butchery of game has left "devastated herds." This he blames mostly on settler hunters and the overhunting of the region by sportsmen.[48] Although Chapman's interest in other game besides elephant, his observations on animal behavior and bird life, and his seeming lack of monetary motives may distinguish him from the group of "pure" ivory hunters, he does offer the reader an informative aside on the attitudes and motives of many of the early ivory hunters. Regarding the Empire as an escape from "sordid 'Trades Unions' and such like" debasements of the life of privilege at home, Chapman issues an exhortation to "Wake Up, England" which speaks loudly to the underlying class pressures in the construction of an imperial hunting ethos.[49]

Arthur H. Neumann ranks with "Karamoja" Bell among the most prolific killers of elephant and harvesters of their ivory. But unlike Bell he appears to have been sharply aware of the moral questions raised by such profligacy and its possible inhumanity. At least, he was introspective and concerned enough to offer an elaborate apologia and defense of sport hunting in his account of his hunting travels.[50] Neumann entered the Kenyan hunting field in 1894, earlier than Bell but as a more mature man. Using Dorobo guides in the Meru area of the Nyambeni hills and north to the Uaso Nyiro River, he seems to have been driven to kill almost continually by three factors. First, he "shot something every day to keep up a supply of beef [*sic*] for all hands, generally oryx." The need to supply porters, African auxiliaries, and camp followers with meat is a recurrent theme in the hunting practices of virtually all the early travelers in East Africa. Neumann is no exception, but rather an exemplar as he appears in his account to be almost continually shooting animals as if unwilling to miss the chance of a shot at any animal he sees pass.[51]

Second, Neumann is motivated to collect as much ivory and rhinoceros horn as he can "for paying the expenses of my rather large caravan. . . ."[52] Both by providing meat and thereby saving on the purchase of provisions and by harvesting as much ivory as he could, regardless of age, sex, or condition of the elephants he targeted, Neumann displayed the same attitude to hunting as I have described for Kamba hunters in the same era: shoot anything in case it is valuable. He knowingly pursues "a troop of cows" near the Tana River and on several occasions will shoot rhino which he has little chance of following up or hippo which he regularly loses to the river currents.[53] To the sensibilities of later generations of white hunters and poachers, these were criminal behaviors, not the deeds of either gentleman or sportsmen.[54]

Finally, Neumann seems to be willing to kill continually and without limit because he shares with his African cohort the belief that the animals he hunts are provided by nature in an infinite supply. Although he expresses contempt for certain African hunting methods such as the spear drops used by the Nyambeni,[55] this appears to be a matter of aesthetics and sportsmanship, not of reaction to the randomness, wantonness, or excesses of such methods. No expressions of the need to curtail either African or white hunting in the interests of conservation of the elephant herds, which Neumann seems to have considered virtually limitless, can be found in his 435–page account.[56]

The account of the ivory hunting and trading expeditions of the British trader, A. Arkell-Hardwick, is a curious combination of unself-conscious greed and bloodlust on the one hand, and a keen awareness of the author's place within a historical tradition of big game hunting on the other.[57] In the opening pages of the account, he offers the reader a brief historical account of his predecessors in the hunting field in Kenya, beginning with Joseph Thomson in 1885, the explorer whom we will meet below, and ending with A. H. Neumann and Dr. Kolb in 1893. This is less to establish his pedigree than to inform the reader of the basis on which the hunting field in general and the Lorian swamp area of northeast Kenya in particular was selected by him as "practically unexplored."[58] On his expedition in 1900, Arkell-Hardwick was accompanied by a mysterious English adventurer referred to only as "El-Hakim" (The Doctor),[59] who if anything was an even more profligate elephant hunter than the author. Speaking of El-Hakim, Arkell-Hardwick wrote: "He had shot elephants persistently for the previous four years in Somaliland, Galla-land and the country around Lake Rudolph, having killed over 150, on one occasion shooting twenty-one elephants in twenty-one days—a fairly good record." Commenting on the size of the tusks available in the districts north of the Waso Nyiro River, he mentioned that his largest pair weighed just over 218 lbs. and measured nine feet in length."[60] On this safari, the two hunters also shot a considerable number of rhino, including five in a two-day period—perhaps not a record—but done with such casualness as to deserve comment: "On the plains to the eastward of the camp [on the Waso Nyiro] roam vast herds of game—zebra, oryx, water-buck, and grantei [Grant's gazelle]. Rhinoceros were disgustingly frequent, El-Hakim shooting two more that had evinced an impertinent curiosity regarding his movements, when he was walking abroad one afternoon."[61]

Perhaps what makes the story of this expedition most intriguing is that its main purpose was the accumulation of elephant ivory by commercial means as well as by hunting. Although relatively little attention is paid to the commercial aspects of the expedition, the purchase of ivory from Africans with payment made in sheep and cows is not hidden or treated as a violation of the hunter's code.[62] Arkell-Hardwick's self-identification as an "ivory trader" is clearly not ironic. However, he just as clearly knows that his readership is interested in the romance of the chase. His frequent detailed descriptions of elephant and rhino hunts and his occasional paeans to the sporting life—in which he addresses his audience directly and in rapturous terms—amply testify to his appreciation of that aspect of his adventures in north Kenya:

> Game was more than plentiful, the climate glorious, and we were free as the pure air we breathed. Only those who have been placed in similar circumstances can appreciate the full value of that word "free." We did precisely what seemed good to us in our own eyes. We rose early, bathed in the warm spring, ate our breakfast, and then went shooting, or, if disinclined for that, we sat in a folding-chair in the shade of the trees and read . . . ever and anon raising our eyes to watch the herds of game walking steadily past our camp on the way down to the river to drink. . . . It was a perfect Arcadian existence, which we left with real regret . . . to resume our weary march over the sun-scorched desert country down river.[63]

Arkell-Hardwick, despite his commercial purposes and unrestrained pursuit of elephant ivory, is clearly not one to short change the aesthetic and sporting side of his own endeavors or the sheer pleasures of the outdoor life. Yet the absence of any indication of the part played by his "dark companions"—his African guides, gunbearers, porters, skinners, and cooks—except to mention their appetite for fresh meat, would indicate that he failed to recognize any such aesthetic or sporting sensibility among the Africans who accompanied him or those, like the Dorobo, who hunted in their own right but also acted as guides on occasion.[64] But his disregard for the talents and contributions of African hunters is nothing compared to the contempt and vilification expressed by my last example of the ivory harvesting hunter, Captain Charles H. Stigand.

In many ways, Stigand's three books on African hunting are unexceptional. The first, *Central African Game and Its Spoor,* was written as a "handbook . . . intended as a help or guide to the young or inexperienced sportsman who intends making a hunting trip to Central Africa."[65] As such it is replete with advice on techniques, especially regarding animal behavior and ethology, useful in tracking, shooting, and following up dangerous game. His other books contain much of this naturalist interest and this may account for their popularity as well as for the fact that he was able to persuade ex-President Theodore Roosevelt to offer a foreword to his account of *Hunting the Elephant in Africa.*[66] But most important for understanding the impact on the relations between white and black hunters is the fact that all his works contain disparaging remarks about the African character, intelligence, and capacity for hunting. Drawing on his experiences with Chewa, Yao, and Ngoni hunters, Stigand declares that "The native is a keen hunter, but we are inclined to think that his hunting instinct is derived from a love of meat and a lust of killing than from any sporting feeling." In defense of imposing restrictions on African hunting which "some people think [indicates] that the native is ill used" Stigand argues that if "the native . . . [is] allowed a free hand to shoot in his own country . . . it would only be for a few years . . . as with the weapons of precision the traders would at once bring him, . . . and the indiscriminate slaughter of the young and females he would indulge in, he would soon denude the country of game. . . ."[67]

Besides their lack of a true "sporting feeling," Stigand attacks the Africans for lacking the basic skills and physical and intellectual equipment for being hunters. He asserts that they have no skills of bushcraft once taken from their familiar environs. They are poor stalkers and trackers of game, lacking acuity of eyesight and other physical attributes. Moreover, they are incapable of judging distances properly and have no sense of time. As a group, Africans, he believes, are wanting not only in hunting skills, but also in a sense of profession and division of labor, as well as in the capacity for accurate description of the naturalist and true lover of hunting. Nonetheless, after completing a thoroughly demeaning portrait of the African hunter, Stigand assures his readers, "but I am none the less very fond of him."[68]

This derogation of African skill can be understood only as the consequence of the racist stereotype of Africans which Stigand seems to exemplify. His acceptance and propagation of an image of "the good-humoured, happy go lucky Bantu savage" is classic in its disregard for any real evidence drawn from particular individuals.[69] "The improvidence of the savage is wonderful," he tells us, and seems to be the counterpart of his "irresponsible" attitude to the preservation of game which threatens to "irreparably upset the balance of nature."[70] Improvidence, matched with modern weapons in the hands of the African, Stigand warns the reader, can lead to the denuding of the country as quickly as the threat posed by "advancing civilization"—the settlement of African wilderness by white farmers and ranchers.[71] Only the true sportsman, a trophy hunter who will shoot "a few selected heads (generally of males only) of each of certain animals," can assure the preservation of game in the interests of future generations of sportsmen.[72]

Similarly, the lack of true "sporting feeling" attributed to Africans by Stigand is simply the application to hunting of his racially stereotypical demeaning of African emotional capacities. If Africans do not love animals and their pursuit it is because they lack true human affections. "Affection between husband and wife, parent and offspring, as we know it, is practically non-existent in many of the savage peoples, whilst the mother's love for the child is much the same as that of an animal."[73] In a chapter on the subject of punitive expeditions against African peoples significantly entitled "Stalking the African," Stigand tells us that "The Native is so innocuous and appears so brainless that one is apt to grow careless with him."[74]

Despite these dehumanizing and prejudiced attitudes, it is important to recognize that when Stigand went in pursuit of the shy and elusive Bongo on the Kenyan Kinangop forests he sought and obtained the aid of Dorobo hunters as guides and trackers.[75] And his expeditions in Central and East Africa were no different from others in using Africans in a wide range of capacities from unskilled porters and carriers of loads, to trackers, skinners, cooks, and gunbearers recruited through local African, Swahili, and Arab agents, who would be the prototypes of the "dark companions" accompanying virtually every Kenyan safari of the colonial era.

Individual ivory harvesters would continue to pursue elephants for profit into the 1930s—first to the Lado Enclave in the Sudan, then to remote areas of Portuguese East Africa where the enforcement of game regulations was weak or nonexistent. But it was increasingly hard for ivory hunters to describe themselves as sportsmen, and in British East Africa nonsport hunting would meet with increasing vigilance by game departments. Although some ivory hunters would romanticize their pursuits as poaching, drawing on the images of defiance and rebelliousness of the European tradition,[76] they were clearly a dying breed. The end of the era came with the outbreak of war in Europe and the steep drop in ivory prices to below ten shillings a pound. After a last swansong of ivory hunting in Ubangi Shari, now the Central African Republic, "one by one these poachers became game wardens or white hunters," and the ivory hunters among whites in East Africa disappeared from the scene.[77]

Explorers and Administrators

The ivory hunters were not alone in the Kenya hunting field. They were joined by others, who, although they had other purposes for being in Kenya beyond stalking and shooting big game, were often prodigious hunters in their own right. I speak of the pioneer explorers, soldiers, and civil administrators who came on behalf of the Imperial British East Africa Company and the Foreign or Colonial offices. While passing through Kenya en route to some treaty-making rendezvous, or enforcing previous agreements as soldiers, or simply moving out from their stations which spread across the Kenya landscape as depots, strongpoints, or governmental outposts, these men found it central to their routines to pursue wildlife. Whether hunting for the pot or to supplement their salaries with trophies or for sport, these early colonial officials and their missionary and merchant counterparts made an important and indelible contribution to the emergence of Kenya's hunting culture.

The literature on exploration of East Africa is considerable and illuminates the careers of several figures who in pursuit of their quests of discovery and conquest managed to pursue large numbers of animals. John MacKenzie has even argued that without being successful in the pursuit of game it is unlikely that these men could have been successful in the recruitment and control of the African labor and the supply of food for their caravans necessary for meeting their political objectives. What he terms a "meat subsidy" was an essential element in successful African exploration along with the financial subsidy provided by the ivory, horn, and hides which the explorers accumulated.[78]

Kenya's most famous nineteenth-century explorer was no doubt Joseph Thomson, whose book, *Through Masailand,* drew attention to the wealth of ivory and game available in Kenya and amply demonstrates MacKenzie's thesis.[79] Thomson's Kenya experiences, repeating in many ways his own earlier experiences in Central Africa and those of V. L. Cameron in Tanganyika a decade earlier, confirm the importance of hunting as part of both the "Day's Routine" and the "Dangerous Moments" of exploration.[80] Hunting for the pot to feed the numerous porters and laborers attached to the caravan of the trader or explorer (often difficult to distinguish) formed a central part of the daily routine of African exploration and travel. Alongside confrontations with "hostile" native peoples, life-threatening encounters with wild animals lay at the heart of both the adventure of travel and the art of retailing that experience to the reading public. As early as Thomson's Kenya expedition of the mid-1880s, the dramatic scene of a charging rhino or elephant and the hastily loaded gun and "cool and collected" marksman had become the standard fare of the African travel account.[81]

By the late 1880s, the penetration of Kenya's interior by representatives of the Imperial British East Africa Company was adding to the adventure of exploration, the excitement of travel and shooting in the name of imperial conquest. Frederick Lugard's meticulously documented travels make clear the importance of hunting, especially of ivory, for the financing of Britain's imperial mission, but also the emotional importance of hunting to the sense of accomplishment and purpose of the pioneer soldier-administrator[82] However, it is the work and writings of Frederick Jackson, Lugard's contemporary and successor as both a company and colonial official, that best exemplify the centrality of hunting to early colonial administration.

Jackson came out in the 1880s to hunt, but quickly entered the employ of the British East Africa Company, and in the course of over twenty years residence in East Africa (1883–1911) eventually rose to become governor of

Uganda. He would write and reflect on these experiences in the 1920s, displaying his veritable obsession with hunting and his role as a model of the hunter-administrator which was conveyed to those who worked with and under him.[83] His first visit to Kenya, in 1885, included a hunting trip in the company of the novelist, Rider Haggard, in the country behind the Lamu area of the north coast and in the Tana River valley where he added sixteen new species to his bag while killing at the rate of three animals a day.[84] Extensive hunting followed during his company days in 1885–86 when Sir Harry Johnston was his traveling companion and where in addition to giraffe he adds impala, hartebeest, sable antelope, and greater kudu, "all new antelopes to me," to his list of newly shot species.[85] A two-month "shooting trip round about Kilimanjaro" in 1886 is followed by a three-week trip "devoted to elephants."[86] His employment by the company in 1888 barely slows his hunting pace, but it adds new purposes. In 1889, Jackson finds it necessary to feed his caravan traveling in western Kenya with the carcasses of three elephants when the cattle provided by a local chief, Mumia, as a "gift," all die. He also comments that the people of this region, Kavirondo, "were out and away the greatest gluttons for meat I have ever met with."[87] It was clearly among his purposes to sweeten his administration's reception by providing the local "gluttons" with what they wanted, and by doing so, to be able to do what he most wanted—to hunt.

Why did Jackson become such an exemplar of the hunter-administrator? It is difficult to answer this question with any assurance, but perhaps we can see Jackson as a "happy" hunter. In the forward to Jackson's reflections, Lord Cranworth, one of Kenya's leading, or perhaps ranking, aristocratic settlers, wrote: "The sportsman will find it hard to lay down the chapters that tell of the *safaris* over the plains teeming with herds of game, or through the vast elephant-haunted forests. . . . happiness was a characteristic of the man. He was happy . . . in his domestic life . . . in his work . . . in his natural history. . . . Happy in his sport. . . ."[88]

Untroubled by the destruction of animals and humans which was part of his work and his sport, Jackson would set a example to others later in the field and lower in the colonial pecking order. It was with apparent clear conscience and no hint of irony that he could relate "big game hunting yarns" about himself or stories of Count Teleki, the Hungarian hunter-traveler, who killed "35 elephants and 300 'niggers'" on his Kenya safari.[89]

Ultimately, the fact that Jackson saw himself as a hunter before everything else sets him apart as a singular figure. It was not the sheer quantity of game, but passion and devotion to sport hunting that attracted him to the African field and kept him there through his long administrative career. Perhaps this is best exemplified by the final chapter of his book, devoted to

"Hunters, Professional and Otherwise."[90] Here, as a coda to a memoir filled with hunting exploits, he finds it important to provide his readership with a series of moral admonitions and anecdotal examples of what it means to be a sportsman, gentleman, and hunter in colonial Africa.

The high regard for sport hunting which Jackson exemplified became a pervasive value in the Kenyan administrative service. Clearly this was a result of more than Jackson's example or the influence of the gentlemen-hunters who pioneered and dominated the establishment of colonial rule in Kenya. Indeed, the values and culture of the English countryside and leisure class which became proverbial among Kenya's white settlers were deeply ingrained in the Kenyan civil service. In the words of Bruce Berman: "The Colonial Office looked for modest intellectual achievement, athletic prowess, a taste for the outdoor life, and, implicitly, unquestioned acceptance of the 'aristocratic' ethos of rule and the ideals of imperialism."[91]

At the local level of Kenya administration, no one better exemplified this than S. L. Hinde, a district collector in Ukambani and Fort Hall districts at the turn of the century. Amongst his other attributes, Hinde was an amateur ethnographer and naturalist, and a prodigious hunter.[92] Having served in military capacities in West Africa and the Congo Free State, he came to Kenya in the 1890s somewhat older than most junior officers and married. Nonetheless, Hinde found ample time in addition to his official duties to hunt regularly. Like the Meru district administrators of a slightly later period, Messrs. Horne and Hopkins, Hinde was known for his regular weekly expeditions away from the collectorate offices in pursuit of his avocation. Ample evidence of his proficiency and preoccupation as a hunter is provided by the photographs which accompany his ethnographic account of Masai life, a third of which is devoted to "Field Notes on the Game of East Africa," a guide to would-be visiting hunters.[93]

Despite this evidence of proficiency as a hunter, in 1903, no less a hunting stalwart than Captain Richard Meinertzhagen would impugn Hinde's hunting prowess, accusing him of being "a humbug" for stretching the skin of a lion which he had been given, allegedly so as to later claim that he had shot the now oversized lion himself. Hinde would also stand accused in Meinertzhagen's book of "a breach of the game regulations" in stalking and shooting rhino while accompanying a traveler on a hunting expedition. This "unsportsmanlike action" was an act of "pure slaughter"—a far more serious breach of hunting etiquette and ideology than merely overshooting the limit.[94]

The question of violation of the game laws by those colonial officers empowered to enforce them would become a serious problem for gamekeepers and conservationists later in the century. Often these incidents would be both laughable and provocative. One case from Meru District deserves note as it illustrates very clearly the conflict between the hunting activities and the official responsibilities of civilian officials. J. G. Hopkins, the Meru district official in the 1930s, "was a small sharp eyed man who looked as though he was planning to sell you something." According to Charles Chenevix Trench,

> He was the only DC [District Commissioner] to educate his children entirely from the profits of elephant-hunting. He hunted both inside his own district on control and outside it on licence. The tusks of elephants shot on control, however large, had to be handed in to the government. The tusks of elephants shot on licence belonged to the shooter. As DC, Meru, Hopkins shot a huge tusker on control in his district. But it was within yards of a cairn marking the district boundary, so he wrote himself a cheque, issued himself with a licence and moved the cairn, thus starting a dispute between Meru and Isiolo districts which was still unresolved thirty years later.[95]

The boundary dispute initiated by this incident is exceptional, but the practice of colonial officers who had both local authority to issue hunting licenses and their own private hunting purposes, whether venal or sporting, frequently led to technical violations of the game regulations by colonial officials. Officials could abuse their power by issuing licenses to themselves or their friends after the fact of the shooting of game animals. No doubt they considered themselves to be sportsmen and gentlemen and therefore inherently deserving of a variance in the law or a blind eye. Only the separation of administration from game preservation and regulation would obviate this problem later in the century.

Not only the general administrative officials, but many of the technical staff of the colonial establishment proved to be hunters of great ardor and proficiency. Perhaps the best known single—or I should say double—hunting exploit to capture the imagination of the reading public, which shaped the legend of white hunting in Kenya, was that of an engineering officer attached to the construction of the Uganda Railroad by the British East Africa Company. Colonel J. H. Patterson had served in India building bridges and was called on to design and oversee the construction of a railway bridge across the channel of the flood-prone Tsavo River in 1898. The company had employed thousands of Indian "coolies" to undertake the actual construction. But shortly after Patterson's arrival at Tsavo station, a pair of marauding male lions began to persistently and mysteriously attack the workers' compound, carrying off the hapless and soon terrified workers.

Patterson, a keen hunter who had already shot lion, undertook with the aid of other local administrators to calm the fears of his workmen by shooting the rogue pair. It would take him fully nine months and over one hundred fatalities before the lions were killed and work could be resumed. The mysterious behavior of the man-eaters of Tsavo in attacking entrenched positions by day and night, and evading the many traps, snares, and baited traps set for them made this one of the most adventurous of Kenya's early hunting narratives.[96]

Police and Soldiers

In addition to the provincial and district administrative officers and technicians, Kenya's civilian police force also included among their numbers, from early in the century, hunters of ability and distinction. Although they were less likely to write about their experiences than the more upper-class officials and soldiers, there was at least one individual who would record his experiences as both policeman and big game hunter: Major W. Robert Foran. Foran was the author of no fewer than fifteen works of fiction, history, biography, travel and reminiscences. He joined the Kenya police force in 1904, and served as a senior police officer for many years; on the basis of this long experience during the last days of colonial rule, he wrote the first history of the Kenya Police.[97] But it is his writings on hunting and wildlife that make him exceptional.

The interface between police work and hunting would involve Foran in several hunting and game preservation activities early in his police career. He would be called on during his first years as a Kenya policeman to do game control work on behalf of an understaffed Game Department which included only three Europeans. This meant that Foran was called upon to kill rogue animals, field raiders, and other dangerous game, especially lions, who interfered with the work of settlement and cultivation. He would also be involved in police operations against ivory poachers as early as 1906 which would involve him in hunting as well. In his official capacity as a police officer as well as because of his reputation as a sportsmen, Foran was attached to visiting dignitaries such as Theodore and Kermit Roosevelt, and the duke and duchess of York on their highly publicized hunting safari.[98]

Foran's first major work on hunting is thinly disguised as a work of natural history, "a study of wild life in its natural environment . . . [for] those about to take up big game shooting or to others who have already done so."[99] Published in 1958, *A Breath of the Wilds* makes a spirited defense of sport hunting, blaming "confirmed game-butchers" for the modern extermination of wildlife and describing sportsmen as "the trustees of wild life for future generations."[100] However sharp the distinction, he makes an exception to his objection to "heartless profit seeking" for those sportsmen who kill elephants and sell the ivory to cover the costs of their safaris, an excusable form of shooting for profit when done by proper sportsmen. Of course, his own use of the significant proceeds of his elephant tusks for supporting the sporting life would require such an exemption.[101] After the excuses and apologies for sport hunting—given at length in spite of the fact that he says they are unnecessary[102]—the remaining chapters of his natural history are given over to the recounting of hunting exploits and nostalgic tales for the armchair hunter-naturalist of the mid-twentieth century.

Foran's account of his hunting exploits, *A Hunter's Saga,* was written near the end of his long career. By that time the genre of hunting account was well established and he deviated but little from the successful formula. He records his hunting experiences before coming to Kenya in chapters on *shikar* (pig sticking from horseback) in India and big game hunting in the western United States and Mexico. He describes his ivory harvesting phase when the proceeds of the sale of bull elephant ivory paid for his hunting safaris in Kenya.[103] He recounts exciting encounters with elephant, buffalo, and rhino.[104] He even gives a rough count of the number of elephant—four hundred—killed by him in a six-year period.[105]

Of particular interest to us, he includes a paean to his faithful and "stout-hearted gun-bearer, Hamisi, and my 'Ndorobo tracker, Juma."[106] Indeed, Hamisi's death as a result of an encounter with a rogue elephant while on safari with the author in Uganda leads to his renunciation of elephant hunting.[107] Like many long-term African hunters, he is generous in his praise of "an efficient gunbearer . . . the unsung heroes of countless shooting safaris."[108] This fulsome, if patronizing, posthumous tribute is a regular motif in the sagas of many a big game hunter. It serves not only to pay an otherwise unacknowledged debt to the "dark companions" of the safari; it also opens our eyes to the innumerable intersections of African and European hunting traditions, stories, and practices. The relationship of white hunter to African gunbearers, trackers, and skilled skinners as well as safari porters, cooks, and laborers was the active interface on which the combined European and African tradition of Kenya hunting, from the early ivory expeditions to the champagne safaris of the heyday of the white hunter, was created. I introduce this theme here because it is in the work of the policeman Robert Foran that this intimate connection is most vividly expressed.

Robert Foran is unique among Kenya's policeman-hunters for having written extensively about his experiences in and attitudes to both his roles. In the same way, Colonel Richard Meinertzhagen, whom we met earlier as a critic of the hunting practices of the administrator S. L. Hinde, will serve as our exemplar of the soldier-hunter. He was at once the colonial soldier, an officer and gentleman par excellence, and the most ruthless and proficient of Kenya's pioneer hunters. Although he was far from the first or only person in East Africa to combine the roles of hunter and soldier, his own character, marked by the extremity of his opinions, the brashness of his expression, his remorseless self-righteousness, and his contradictory and contentious attitudes to Africans, to officialdom, and most strikingly, to animals make him a thoroughly engrossing figure.

The connection between soldiering and hunting, resulting from the mutually reinforcing skills and values they inculcate, has been frequently noticed and commented upon.[109] But seldom have the two been so closely entwined as in East Africa. Most frequently, it was a matter of soldiers on active duty, like Meinertzhagen or Major R. MacDonald, who combined military duty with their passion for the pursuit of game. Occasionally, those who came to East Africa for sport, such as the soldier-adventurer Sir Claude Champion de Crespigny, a companion of Meinertzhagen in hunting and his comrade in arms on the Nandi Expedition in 1904–5, would find themselves recruited or volunteering for active duty in what has been termed a "chasse de nègres." As their accounts indicate, there was a close, functional and symbolic relationship between the two activities.[110] But it is in the controversial career and contentious character of Meinertzhagen that this amalgam of hunting and soldiering can be most clearly contrasted and deconstructed.

The first thing to note about Meinertzhagen was the sheer quantity of blood which flowed in Kenya as a result of his hunting and soldiering activities. In the preface to his published diaries, he states: "Two criticisms may occur to the reader: one, the excessive taking of human life; the other, the slaughter of game." However flimsy the justifications in terms of war and "pacification" that are offered for the former, he is unabashed in his defense of the latter. "As regards the second criticism, the fighting and hunting activities of man require an outlet; if suppressed hatred and malice occur. There are outlets which are not harmful. Active sport and competitive games are substitutes; hunting, the joy of adventure and facing danger, discovery and creation are other outlets. We should neither ignore nor regret them."[111]

There follows an account of Meinertzhagen's exploits in which he expresses no regrets for the lives of people or animals which he has taken by his own hand. If anything, he sometimes expresses regret that not enough people have been killed or their livestock confiscated to allow them to be easily controlled.[112] Paradoxically, he does express regret at the killing of elephant by Africans, which he considers to be a regrettable loss of intelligent and sensitive life. He does not himself include elephants among the list of animals killed and weighed "by me personally" presented in an appendix. But that list runs to some 448 individuals of fifty-seven species, including nineteen lion, five cheetah, sixteen rhinoceros, seven hippopotamus, six hyaena, two giraffe, five eland, three oryx, and two waterbuck. No such listing is given of the people who were killed "by me personally."[113]

There are two even more curious paradoxes. First is Meinertzhagen's awareness of the need for game preservation in the face of his own wantonness in the destruction of wildlife. In conversation with the early game warden and hunter, Blayney Percival, he recommends the creation of a large game reserve where hunting would be strictly controlled and limited. He sees (correctly) the threat to wildlife as coming principally from new arriving white settlers rather than sport hunters like himself. He believes (wrongly) that he is the first person to suggest such measures, and sees it as a precursor of the policy of game reserves, sanctuaries, and eventually national parks that became the keystone of Kenya's wildlife conservation policies.[114]

The second paradox stems from Meinertzhagen's uncanny ability to recognize the wrongs being done to Africans by conquest, colonization, and the expropriation of their property, even while he is himself engaged in the very campaigns of conquest and in the selfsame practices of expropriation of land and livestock that make colonization possible. At the conclusion of his account of the first phase of the Nandi Expedition in November 1905 during which he estimates that five hundred Nandi young men were killed and ten thousand head of cattle and seventy thousand sheep and goats confiscated, Meinertzhagen states: "The reserve they [the colonial authorities] contemplate is much too small and does not allow for expansion; I fear it is all based on requirements for white settlement and not on the welfare of the Nandi. This is a very short-sighted policy and must lead to grievances; after all, it is African land, not ours to dispose of."[115] Moreover, he recognizes that the Nandi have legitimate grievances against a corrupt administrative official and weak, ineffectual government. Yet he believes he is entirely justified in meting out brutal punishments for the hostility which the Nandi leadership and young men display towards the same colonial authorities who would dispossess and despoil them.[116]

How can we reconcile these oxymoronic attitudes and behaviors? They appear to me to be resolvable by an examination of the symbolic congruence of Meinertzhagen's dual identities as soldier and hunter. Two incidents which bear a striking structural similarlity illustrate this congruence.

On July 17, 1904, on leave near Nairobi, Meinertzhagan went for a walk with his pet dog, Baby. Uncharacteristically he went unarmed. On route to the Athi River the soldier and his "companion" chanced upon a troop of baboon.

> On approaching the Athi River . . . I saw a large mob of baboon close to the river, and Baby and I decided to stalk them and give them a good fright. We did it most successfully, and getting to within about 60 yards of them we jumped up with a shout and chased them into the river, when I called to her. Normally she would have returned, but there was a strong head wind and I suppose she did not hear me. On she went, for she feared nothing. Suddenly the whole mob turned on her. Undismayed she went straight for the nearest baboon, but the contest was hopeless from the first.[117]

The dog was quickly killed and dismembered by the baboons in clear view of her master, and his expressions of grief are immediately suffused with a desire for revenge. "I saw the disgusting creatures making off with her limbs. Never did I long so much for a rifle."

If the story ended there it would be unremarkable. The author's anthropomorphizing of his animal friends and enemies was certainly not unusual if somewhat excessive. Baby is not only a "companion," but has been one for so long that "we . . . thoroughly understood each other." The decision to attack and frighten the baboons was a mutual decision between dog and man. And the dog's behavior, her seeming disobedience, lack of fear, and absence of dismay when she is faced with a counterattack are all described as if Baby were a thoroughly human actor. Moreover, the "mob" of baboon are similarly portrayed in human terms: alternately fearful, panic-stricken, and vengeful. But they are "disgusting creatures," showing none of the loyalty of the dog nor appreciation of the good joke and sport that Meinertzhagen's sudden assault represented. But it does not end there.

> When my first horror was over I sat down and contemplated how I could teach those baboon a lesson. [Having identified a rocky cliff that served the baboons as a roost,] if I could get out to their roosting place before light, surround it with men and then open fire from the front, I should most certainly stampede them and kill the bulk of the brutes. As I returned sadly that evening I had murder in my heart.[118]

Quickly, Meinertzhagen organized a patrol of thirty men and a sergeant major which departed for the roosting place at 3 A.M. "with rifles, bayonets and 100 rounds each. . . . The baboon were to be shot down without mercy." In a military operation that went from first light to before noon, the baboons were attacked with rifle fire and driven down through a gauntlet of Meinertzhagen's troops, and then forced to retreat to the cliff face where two senior male baboons were pursued, surrounded, and killed. "We collected altogether 25 baboon, only some 15 or so escaping. We killed every full-grown male, and I was pleased," remarked the commander of this operation. As a reward, Meinertzhagen "shot an impala and a haartebeest [*sic*] for the men who had assisted . . . in avenging Baby."[119]

The close association and mutually supportive skills involved in hunting and soldiering could not have been lost on so keen a sportsman and soldier as Meinertzhagen. He is as unapologetic about the one as the other:

> The hunting of big game gave me good healthy exercise when many of my brother officers were drinking rot-gut or running about with somebody else's wife; it taught me bush craft and how to shoot straight. After all, the hunting of men—war—is but a form of hunting wild animals, and on many occasions during the First World War I thanked my God that I had learned several tricks of my trade when hunting wild and dangerous game.[120]

Nonetheless, Meinertzhagen seems to be unaware of the more deeply rooted psychological and symbolic connections between his vocation and his avocation.

His account of the death of the Nandi Laibon (properly *orkoiyot* or priest) Koitalel reads, as does the account of the revenge for Baby, like the report of a well-prepared and well-executed military operation. Moreover, the elements of provocation, vengeance, and bloodlust which they share are more than suggestive of the psychological and symbolic elements of the hunting-soldiering nexus. On October 18, 1905, Meinertzhagen entered in his diary his plans "for the undoing of the Laibon."[121] Koitalel, as the chief politico-religious leader of the Nandi people on the Mau escarpment in western Kenya, had long been a hindrance to British occupation and the distribution of land to settlers. Orders to remove the Nandi to a native reserve had fallen to Meinertzhagen to enforce. In an effort to locate Koitalel's residence so that he might attack and arrest him at night, Meinertzhagen arranged for a meeting with him on the morning of October 19. Koitalel asked that Meinertzhagen come accompanied by only five men. For his part, Meinertzhagen prepared an escort detachment of eighty men and a machine gun, as he was convinced that the interpreter and messenger who arranged the meeting was "a spy of the Laibon's, for [my agents] have heard him conspiring to destroy us when we meet tomorrow." He continued in his diary: "If he attacks me at our meeting I know how to defend myself. However, come what will, I will meet him tomorrow, and during the next 24 hours I suspect that either he or I will have said goodbye to this world; I do not really very much care which of us it is."[122]

The following day Meinertzhagen entered an account of the meeting in which he stated that as he had suspected, the Laibon attempted to ambush him as they stood facing each other after shaking hands under a flag of truce. "But we were ready, and he [Koitalel], the interpreter and several others, some 23 in all, were left dead. I had my clothes torn by a spear and arrow and one of my men was slightly scratched by a spear."[123]

How could Meinertzhagen have escaped from this face-to-face confrontation almost unscathed? "Before going down to meet the Laibon I had warned Butler [Meinertzhagen's second in command] to open fire at once he saw us being overwhelmed. He mounted the machine gun and covered the place of meeting," along with seventy-five men left in reserve. In addition, after withdrawing from the scene of the ambush, harried by the Nandi, the column of eighty men was rescued by "a welcome patrol which had been sent out on my instructions to cover our retreat."[124]

The inherent unlikelihood of such a complete reversal of the ambush and Meinertzhagen's obvious preparations for a confrontation with the Nandi leader, along with his admitted intention of seeing the Laibon "undone," would lead to questions being raised in the Kenya government about his conduct of the meeting with Koitalel. Meinertzhagen would be forced to defend himself against charges of foul play amounting to the murder of the Laibon, leveled against him by members of the civil administration. A series of three courts of inquiry was held between December 1, 1905, and January 11, 1906, at which Meinertzhagen was officially exonerated, but in the wake of the inquiry, rumors persisted that his was "an unconscionable act of treachery."[125] It was this stain on his honor as a gentleman, even more than the threat to his military career, which seems to have concerned him most deeply and indeed, would haunt him for the rest of his life.

Clearly, we will never know with certainty the exact events which transpired between the two leaders, leaving the Laibon and his retinue dead and Nandi-British relations shattered for the next several years. But elements of Meinertzhagen's mind-set do emerge clearly from the evidence of his own pen, which bear directly on his dual identity as a hunter-soldier. First, it is clear that Meinertzhagen viewed the confrontation as involving the same kind of "risk and excitement" as in the pursuit of dangerous game. Moreover, he was so convinced that his prey was wily and treacherous, and that it was in his very nature as an African and a Nandi to be so, that he was justified in setting a trap for him. In many ways we might describe the act of killing the Laibon and what Meinertzhagen believed to be his entire entourage and immediate male family members as an act of "pre-revenge" justified by the inherent lack of honor of these "disgusting creatures." Once again, Meinertzhagen had left the field of battle strewn with the corpses of all the males upon whom he sought revenge for prior wrongs. "So may all the King's enemies perish." He expresses regret only for the wives of Koitalel, who he mistakenly believes will be killed and buried with him, and some sadness at the idea that the Laibon's dynasty was ended (or so he believes, also mistakenly) on that October morning. True, he also believed that the Laibon deserved burial and "some obituary notice," an honor he would not have accorded the unlucky baboons he had chosen to frighten together with Baby and then to attack and destroy with military force.[126]

One other aspect of Meinertzhagen's soldier-hunter character deserves mention: his relationship to civilian administrative and police authorities which reveals his class bias against those he considers are not gentlemen. These attitudes clearly predate his confrontation with civil officials over the death of Koitalel and his departure from Kenya under a cloud of suspicion raised by the civil administrators in Nandi country. He attributes the hostility of civil officials to soldiers like himself to "the low class of man who is appointed to administrative appointments. Few of them have any education and many do not pretend to be members of the educated class. . . . When such men are given unlimited power over uneducated and simple-minded natives it is not extraordinary that they should abuse their powers, suffer from megalomania and regard themselves as little tin gods."[127] Nor did he confine such blatant class-biased attacks to the privacy of his diary. Even after his affray with the civil administration, he would brashly and directly criticize the administration and its police in the person of the Nandi collector, Stephen Bagge, a respected and experienced official, for the violence and the failure of the Nandi Expedition to end the threat to colonial peace necessitating the misuse of the soldiers like himself to do the work of civilian police who lacked proper training.[128]

Although Meinertzhagen expresses his hostility to civil administrators and police by alleging their inadequate education or training, it seems clear to me that beneath these criticisms lie deep-rooted prejudices against those who lack the attributes of gentlemen, attributes which diminish as one moves down the social scale and the hierarchy of official rank within the administration and civilian police. Chief among the traits of gentlemen which Meinertzhagen valued and idealized were those of the hunter and sportsman, which he assimilated to the qualities of the good soldier. It was the love of nature and the outdoors, of the chase and conquest of animal and human enemies, which led him and men like him to "the slaughter of my fellow men and wild animals" to an extent which left him "shocked at the cost," if only half a century after the fact.[129]

Meinertzhagen was not typical of the soldier-hunters I am characterizing here; he was too exceptionally ruthless as a soldier and too determined as a hunter to be typical. He is instead an exemplar of the kind of symbiotic association of hunting and soldiering, of sport and war, that makes the one a powerful metaphor for the other and allows the taking of lives, both human and animal, to be justified as part of what Meinertzhagen and others saw as a deep-rooted psychological inheritance of modern man.

Conclusions

What can be said of Meinertzhagen holds equally for the other Imperial Hunters we have surveyed here. None can be seen as representative of the imperial hunting ideal. All were too ruthless, too avaricious, and perhaps too concerned with leaving their mark, to be exemplars of the aristocratic tradition of the Hunt we described at the outset. We can fairly state that none lived up to the ideals of sportsmanship and aristocratic virtue which the ideology of hunting set as a standard for gentlemen. Indeed, neither did the aristocratic and elite hunters who would follow them into the Kenyan bush in search of game and recreation.[130] Yet, each carried with him the images and ideals, the standards and values established for the Hunt and for the proper conduct of sportsmen. These included a contempt for peasants and the hunting practices of the lower orders of society that were readily transferred to the African "tribesman." It included a sense of social superiority confirmed by hunting prowess and a sense of entitlement manifested in the belief in the hunter's right to ownership of nature and wildlife. Despite their own deviation from the code, their lapses from true sportsmanship, and their failure to behave always as gentlemen, these pioneer imperial hunters did introduce to Kenya the ideals and standards of the Hunt which they upheld but could not meet.

After 1909, increased settlement and land alienation by white farmers and the growing popularity of Kenya as a place of recreation and big game safaris among Europe and America's leisure classes would profoundly alter and enhance the image of the Imperial Hunt. Teddy Roosevelt's 1909 safari would draw international attention to Kenya's hunting fields, and in the years following, the prince of Wales and other members of the European aristocracy would add glamour and respectability to the Kenyan hunting scene. By the 1930s, Ernest Hemingway and Elspeth Huxley would add new elements of masculine bravado and literary visibility to the big game safari. But the foundation stones of the white hunting idyll were laid by the flawed men, seeking adventure, profit, and the confirmation of their status as gentlemen and sportsmen—types which had been inextricably linked in the idea of the Hunt. Even the most avaricious and ruthless of the ivory harvesters, the most callous and bloodthirsty of the imperial soldiers or police, carried with them the imprint of the ideal of the sportsman and gentleman hunter. This ideal would influence their behaviors and color the way in which they presented themselves to the European public. And it would profoundly influence their ideas about African hunters, peasants, and herdsman and their relationships to wildlife.

By the time of the outbreak of war in Europe in 1914 and the extension of hostilities to East Africa the following year, the pioneer days of big game hunting in Africa were already over. The elephant harvesters were already rapidly disappearing and the end of the period of "pacification" and the establishment of colonial rule would shortly bring to an end the era of the soldier and the administrator as dominant figures on the hunting landscape. New figures would emerge, such as the visiting safari hunters and their professional guides and African staff, who would reshape the imperial ideal in creating the big game safari of the mid-twentieth century. In the course of doing so they would have a vital impact on the culture of hunting by whites as well as on the displacement and destruction of indigenous African hunting traditions.

NOTES

1. See J. M. Mackenzie, "Chivalry, Social Darwinism, and Ritual Killing," in *Conservation in Africa,* ed. D. Anderson and R. Grove (Cambridge, 1987), pp. 41–61.

2. For the place of animals and hunting in general within the Western tradition, see Keith Thomas, *Man and the Natural World* (New York, 1983); Harriet Ritvo, *The Animal Estate* (Cambridge, 1987), and Matt Cartmill, *A View to a Death in the Morning* (Cambridge, 1993).

3. For an extreme example of the aristocratic view of hunting, see J. Ortega y Gasset, *Meditations on Hunting* (New York, 1972). In contrast, Paul Shepard's *The Tender Carnivore and the Sacred Game* (New York, 1973), offers an equally romantic view of the universality of the hunting ethos.

4. J. K. Anderson, *Hunting in the Ancient World* (Berkeley, Calif., 1985), offers the best available overall view of ancient hunting.

5. Cartmill, *A View to a Death,* pp. 60ff.

6. R. B. Manning, *Hunters and Poachers* (Oxford, 1993), 135–70.

7. Cartmill, *A View to a Death,* 61.

8. Ibid., pp. 61ff.

9. Manning, *Hunters and Poachers,* p. 4, and pp. 35–56 on medieval hunting and warfare. The connection between military prowess and hunting skill is made in an African imperial context by R. S. Baden-Powell, *Sport and War* (London, 1900), pp. 17–40.

10. Manning, *Hunters and Poachers,* pp. 5, 57–82, on the role of socially restrictive game laws.

11. Ibid., p. 58.

12. Ibid., p. 9.

13. Ibid., p. 11.

14. Ibid., p. 183.

15. Harry Hopkins, *The Long Affray: The Poaching Wars, 1760–1914* (London, 1985). Although Hopkins restricts the "Poaching Wars" to the period after 1760, I would argue that their first skirmishes and the drawing of battle lines took place in the aftermath of the English Civil Wars, almost a century earlier.

16. B. A. Hanawalt, "Men's Game, King's Deer: Poaching in Medieval England," *Journal of Medieval and Renaissance Studies* 18, no. 2 (1988): 175.

17. Ibid., p. 192.

18. Ibid., p. 193.

19. This passage draws heavily on the work of E. P. Thompson, *Whigs and Hunters* (London, 1975), on the social history of English poaching. See also Douglas Hay, "Poaching and the Game Laws on Cannock Chase," in *Albion's Fatal Tree,* ed. D. Hay, P. Linebaugh, and E. P. Thompson (London, 1975), pp. 189–253. Cf. P. B. Munsche, *Gentlemen and Poachers: The English Game Laws, 1671–1831* (Cambridge, 1981).

20. Ibid.

21. C. Chenevix Trench, *The Poacher and the Squire* (London, 1967), p. 114.

22. D. C. Itzkowitz, *Peculiar Privilege: A Social History of English Fox Hunting, 1753–1885* (Hassocks, 1977), p. 1.

23. Ibid., pp. 26–27.

24. Ibid., p. 16. In those cases in which local hunts allowed farmers or peasants to ride to hounds, they were clearly marked off from their social betters by the distinguishing hats or coats they were required to wear. Prof. J. Middleton, personal communication, April 1997.

25. Beryl Markham, *West with the Night* (Boston, 1942), and Mary Lovell, *Straight On till Morning* (London, 1987), pp. 11–13, 21–22, 344.

26. Itzkowitz, *Peculiar Privilege,* pp. 152–79; David Cannadine, *The Decline and Fall of the British Aristocracy* (New Haven, Conn., 1990), pp. 360–64.

27. Ibid., pp. 139–46.

28. Thomas, *Man and the Natural World,* pp. 181, 243–54.

29. Jean Renoir, *The Rules of the Game* (1939), with Marcel Dallo and Nora Gregor.

30. Alan Bridges, *The Shooting Party* (1985), with James Mason, Edward Fox, Dorothy Tutin, and John Gielgud. On the impact of new technologies and social classes on shooting, see Cannadine, *Decline and Fall,* pp. 364–66.

31. W. K. Storey, "Big Cats and Imperialism: Lion and Tiger Hunting in Kenya and Northern India, 1898–1930," *Journal of World History* 2, no. 2 (1991): 135–73, for the imperial implications and ritual elements of Indian as well as African big game hunting. Cf. J. M. Mackenzie, *The Empire of Nature* (Manchester, 1988), on British hunting in India.

32. M. Brander, *Hunting and Shooting* (New York, 1971), pp. 204–20 in particular on the pattern of hunting in nineteenth-century Africa. See also Stanley Trapido, "Poachers, Proletarians and Gentry in the Early Twentieth Century Transvaal" (seminar paper, Institute of Commonwealth Studies, University of London, October 1983).

33. Cf. B. Stoddart, "Sport, Cultural Imperialism, and the Colonial Response in the British Empire," *Comparative Studies in Society and History* 30, no. 4 (1988): 649–73.

34. For a full discussion of these attitudes among white hunters and their connections with European expansion and imperialism see Mackenzie, *The Empire of Nature,* pp. 7–53, and his "Chivalry, Social Darwinism, and Ritual Killings."

35. M. G. Redley, "The Politics of a Predicament: The White Community in Kenya, 1918–1932" (D.Phil. diss. Cambridge University, 1976), cited in T. O. Ranger, "The Invention of Tradition in Colonial Africa," in *The Invention of Tradition,* ed. E. J. Hobsbawm and T. O. Ranger (Cambridge, 1983), pp. 217–20. Cf. Dane Kennedy, *Islands of White* (Durham, 1987), and Cannadine, *Decline and Fall,* pp. 438–43.

36. See my conference paper, "White Hunters in Kenya, 1895–1909: The Ivory Harvesters and Pioneers" (African Studies Association Annual Meeting, Toronto, November 1994).

37. It is impossible to say if the examples selected here are in any way representative of the many hunters who left no written record of their exploits. The narrators' interest in recording their stories argues against their being representative, but suggests instead that they are the most self-

conscious exemplars of a hunting practice which they championed in word and deed.

38. Dennis Holman, *Inside Safari Hunting* (New York, 1970), pp. 26–36, presents both a definition and a description of this type of hunter in his four-part typology of white hunters.

39. Although sport hunting was frequently justified by appeals to a sporting ethos, the evidence of the concern with turning a profit or at least covering the expenses of travel, equipment, porterage, etc. belies the non-commercial principles of the ivory hunters we will be examining. For an early defense of sport hunting in these terms, see James Greenwood, *Wild Sports of the World* (London, n.d. [1862?]), p. 250.

40. The books by Bell, Neumann, Arkell-Hardwick, and Stigand as well as other related hunting accounts will be cited in full below as they are discussed in the text.

41. See Mackenzie, *The Empire of Nature,* pp. 85–117. Selous would extend his hunting range to East Africa as well, see his *A Hunter's Wanderings in Africa* (London, 1890) and *African Nature Notes and Reminiscences* (London, 1908). For Selous's influence on imperial ideas about hunting and character, see Kathryn Tidrick, *Empire and the English Character* (London, 1990), pp. 52–87.

42. Mackenzie, *The Empire of Nature,* pp. 107–16. On early conservation efforts, see Jane Carruthers, "Game Protectionism in the Transvaal, 1900–1910," *South African Historical Journal* 20 (1988): 33–56, and her "Creating a National Park, 1910–1926," *Journal of Southern African Studies* 15, no. 2 (1989): 188–216. On southern African poaching during the early twentieth century, see her "'Police Boys' and Poachers: Africans, Wildlife Protection and National Parks: The Transvaal 1902–1950," *Koedoe* 36, no. 2 (1993): 11–22; and Trapido, "Poachers, Proletarians and Gentry."

43. Richard Burton, *Zanzibar: City, Island and Coast,* 2 vols. (London, 1872); Joseph Thomson, *Across Masailand* (London, 1968 [1885; 2d ed., 1887]). Equally impressive for its descriptions of hunting prospects, but less influential was the earlier account by the pioneer missionary-explorer, J. L. Krapf. See his *Travels, Researches and Missionary Labours* (1860; London, 1968). On the general character and influence of these East African explorers, see N. Pavitt, *Kenya: The First Explorers* (New York, 1989); R. Rotberg, ed., *Africa and Its Explorers* (Cambridge, Mass., 1970); and J. Casada, "The Imperialism of Exploration" (Ph.D. diss., Vanderbilt University, 1972).

44. W. P. M. Bell, *The Wanderings of an Elephant Hunter* (London, 1923), provides an unabashed introduction to the braggadocio and bloodlust of the ivory harvester. The second edition, entitled *Bell of Africa* (London, 1960),

includes a useful biographical sketch and additional materials compiled by Colonel Townsend Whelen.

45. Ibid., pp. 1, 5–11. Cf. Denis D. Lyell, *Memories of an African Hunter* (London, 1923), pp. 98–123, 247–64.

46. Bell, *Wanderings,* pp. 16–19. In addition to the opening chapters on elephant hunting, there are additional chapters on "Buffalo," pp. 170–74, "Lions," pp. 175–78, etc., as well as one devoted to the choice of "Rifles," pp. 179–83.

47. Abel Chapman, *On Safari* (London, 1908), pp. 201–57, for his hunting in eastern Kenya in particular.

48. Ibid., preface, and pp. 5, 295–302. Cf. Lyell, *Memories,* p. 261.

49. Ibid., p. 188.

50. Arthur H. Neumann, *Elephant Hunting in East Equatorial Africa* (London, 1898), pp. viii–ix.

51. Ibid., pp. 83, 33–93 passim, for evidence of continual hunting.

52. Ibid., p. 93.

53. Ibid., pp. 150, 102–3 on pursuit of cow elephant; pp. 33–34, 148 on the loss of rhino and hippos killed or wounded. On Kamba hunters see my *Black Poachers, White Hunters* (in preparation), chap. 3.

54. Cf. J. A. Jordan, *Mongaso: Man Who Is Always Moving* (London, 1956), pp. 18–19, for an expression of these values by a self-confessed poacher.

55. Ibid., p. 33. "They are no hunters and the only way they ever kill elephants is by setting traps consisting of javelins (poisoned) in heavy shafts suspended over their paths, . . . so that when a large animal passes along it falls on its head or back after the manner of a school 'booby-trap.'" These Nyambeni we have already met as the hunters of Meru District's Nyambeni Hills.

56. Where he does note a scarcity of game, it is attributed to natural limits ("Of game there is little, as might be expected in so water-less a district") or the effects of diseases like rinderpest on the buffalo herds. Cf. ibid., pp. 141, 84.

57. A. Arkell-Hardwick, *An Ivory Trader in Northern Kenya* (London, 1903).

58. Ibid., p. 5.

59. El-Hakim is misidentified by Kenneth Cameron, *Into Africa* (London, 1990), p. 61, as a Dr. Rainsford. However, a more contemporary and reliable source named Neumann's companion as Dr. Donaldson Smith. Cf. George Fitzhugh Whitehouse, "To Lake Rudolph and Beyond," in *Hunting and Conservation,* ed. George B. Grinnell and Charles Sheldon (1925; New York, 1970), p. 261.

60. Arkell-Hardwick, *Ivory Trader,* p. 142.

61. Ibid., p. 191.

62. Ibid., pp. 123–31.

63. Ibid., pp. 191–92; see also pp. 142–47, 182–88, 303–9.

64. E.g., ibid., pp. 23–24, 201. On Dorobo guides, see p. 194.

65. Captain C. H. Stigand, *Central African Game and Its Spoor* (London, 1906), p. iii.

66. T. Roosevelt, foreword to Captain C. H. Stigand, *Hunting the Elephant in Africa* (New York, 1913), in which Roosevelt states (p. v): "More and more of late years the best type of big game hunter has tended to lay stress on the natural history of the regions into which they have penetrated, and to make his book less and less a catalogue of mere slaughter."

67. *Central African Game,* pp. 34–39.

68. Stigand, *Hunting the Elephant,* pp. 225, 209–24 passim.

69. Ibid., pp. 205, 207.

70. Captain C. H. Stigand, *The Game of British East Africa* (London, 1913), preface.

71. Ibid., pp. 224–25.

72. Ibid., pp. 225, 217–36 passim.

73. Stigand, *Hunting the Elephant,* p. 208.

74. Ibid., p. 315.

75. Ibid., pp. 325ff.

76. Cf. Jordan, *Mongaso,* p. 13.

77. Holman, *Inside Safari Hunting,* p. 36; see also pp. 26–36 for his account of the ivory hunting era.

78. *The Empire of Nature,* pp. 130–31, 148–64. Cf. his "Hunting in East and Central Africa in the Late Nineteenth Century, with Special Reference to Zimbabwe" (draft essay, July 1993).

79. J. Thomson, *Through Masailand* (London, 1968 [1885; 2d ed. 1887]).

80. Ibid., pp. 171–76, 137–41. Cf. V. L. Cameron, *Across Africa* (1877; New York, 1969), and MacKenzie, *The Empire of Nature,* pp. 150–51.

81. Thomson, *Through Masailand,* pp. 139, 131–32.

82. F. Lugard, *The Diaries of Lord Lugard,* 2 vols. (Evanston, Ill., 1959), and his *The Rise of Our East African Empire,* 2 vols. (Edinburgh, 1893).

83. F. Jackson, *Early Days in East Africa* (London, 1969).

84. Ibid., pp. 27–43, 44–59, 49.

85. Ibid., p. 103.

86. Ibid., pp. 104, 123.

87. Ibid., pp. 224, 225.

88. Foreword, ibid., p. ix.

89. Jackson, *Early Days,* pp. 225–27.

90. Ibid., pp. 368–88.

91. B. Berman, *Control and Crisis in Colonial Kenya* (London, 1990), p. 100.

92. S. L. and H. Hinde, *The Last of the Masai* (London, 1901). See also Mrs. S. L. Hinde, "Man-Hunting by Lions," *Blackwood's Magazine* 178 (1905): 192–98.

93. Hinde and Hinde, *The Last of the Masai,* pp. 113–76. For the Meru administrators, see Interviews C/7 Mate and C/8 Kibunjia.

94. R. Meinertzhagen, *Kenya Diary* (London, 1957), pp. 99, 110. My thanks to Dr. Nigel Rothfels for bringing this to my attention.

95. *The Men Who Ruled Kenya* (London, 1993), pp. 83–84.

96. J. H. Patterson, *Man-Eaters of Tsavo* (New York, 1986).

97. W. Robert Foran, *The Kenya Police, 1887–1960* (London, 1962), pp. v–viii. On the wide-ranging duties of the police in early colonial and frontier situations, see *Policing the Empire,* ed. D. M. Anderson and D. Killingray (Manchester, 1991), esp. pp. 5–13, 183–200, 106–23.

98. Ibid., pp. 32–33, 60, and his *With Roosevelt in Africa* (London, 1924).

99. W. Robert Foran, *A Breath of the Wild* (London, 1958), p. 15.

100. Ibid., p. 18. There is a certain anachronistic quality to these descriptions, with "game-butchers" very much a term of the early twentieth century and "trustees of wild life for future generations" coming straight from the language of postwar international conservationism.

101. Ibid., pp. 18–23. See his *A Hunter's Saga* (London, 1962), p. 98.

102. Foran, *A Breath of the Wild,* pp. 20–22.

103. Foran, *A Hunter's Saga,* pp. 98–99.

104. Ibid., pp. 115–24, 146–54, 169–78.

105. Ibid., p. 98.

106. Ibid., p. 125.

107. Ibid., pp. 184–88.

108. Foran, *A Hunter's Saga,* pp. 188–90.

109. The classic statement of this functional connection is found in Baden-Powell, *Sport and War.*

110. Major R. L. R. MacDonald, *Soldiering and Surveying in British East Africa 1891–1894* (1897; Folkstone, 1973); Sir C. Champion de Crespigny, *Forty Years of a Sportsman's Life* (London, 1925). In addition, many military officers serving in India would be among the visiting hunters who after 1909 helped establish Kenya's reputation as destination for sportsmen seeking good hunting.

111. Meinertzhagen, *Kenya Diary,* p. vi.

112. In March 1904, he records: "To my mind the people of Embu have not been sufficiently hammered, and I should like to go back at once and

have another go at them. During the first phase of the expedition . . . we killed about 796 niggers, and during the second phase . . . we killed about 250." Ibid., p. 152.

113. Ibid., pp. 337–46. On the insufficiency of the punishment of the Nandi, see p. 265. For a more sympathetic treatment of Meinertzhagen in Kenya, see N. West, *Happy Valley* (London, 1979), pp. 49–57.

114. Ibid., pp. 156–57. Meinertzhagen also claims that he advocated game conservation in a meeting with his uncle Sir Charles Eliot, then governor of Kenya, in 1904. However, the claim to priority in advancing the idea of game reserves must go to Sir Harry Johnson, who had earlier recommended their creation in East Africa. These inconsistencies, and the seeming ability of Meinertzhagen to foretell the future, may be the result of hindsight, as the original manuscript diary at Rhodes House, Oxford, and the edited and published *Kenya Diary* are separated by half a century. My thanks to D. Anderson, personal communication, June 1, 1997, for this point.

115. Ibid., p. 266. The estimates of Nandi dead and stock losses appear on p. 249. Cf. G. H. Mungeam, *British Rule in Kenya, 1895–1913* (Oxford, 1966), p. 156, for the official estimates, which are not vastly different except for smaller figures of stock losses.

116. Ibid., pp. 184–85, 215–17, on colonial corruption and maladministration.

117. Ibid., pp. 171–72.

118. Ibid., p. 172.

119. Ibid., p. 173.

120. Ibid., p. 179. A training manual for British forces engaged in the suppression of the Mau Mau insurgency in the 1950s makes this same point, graphically comparing the skills needed to pursue Mau Mau fighters to those required "to track down and shoot shy game." Government of Kenya, *A Handbook on Anti–Mau Mau Operations* (Nairobi, n.d. [1954?]), p. 11. My thanks to Dr. John Lonsdale of Trinity College, Cambridge, for bringing this to my attention and allowing me to read his copy of the manual.

121. Meinertzhagen, *Kenya Diary,* p. 232. The background to the situation is given from Meinertzhagen's perspective on pp. 224–32. Unfortunately, the standard work on Nandi resistance by A. T. Matson, *Nandi Resistance to British Rule, 1890–1906* (Nairobi, 1970), fails to include anything on the incident. Instead, the best critical account of the background and encounter is D. M. Anderson, "Visions of the Vanquished," in *Revealing Prophets,* ed. D. Anderson and D. Johnson (London, 1995), pp. 168–78. Cf. Mungeam, *British Rule,* pp. 145–46, 156–61, for an administrative perspective on the expedition.

122. Meinertzhagen, *Kenya Diary,* p. 233.

123. Ibid.

124. Ibid.

125. Ibid., p. 334. Pages 234–38 contain Meinertzhagen's official report of the incident, entered in his diary at the time. Rumors of his conspiring to murder the Laibon first reached him in November and he insisted on defending his honor in a military court. Ibid., pp. 249, 257, 275. Cf. J. Lord, *Duty, Honor, Empire* (New York, 1970), pp. 230–53, on both the Nandi expedition and Meinertzhagen's removal from Kenya.

126. Meinertzhagen, *Kenya Diary,* p. 234.

127. Ibid., p. 132.

128. Ibid., p. 300.

129. Ibid., pp. 333–34.

130. See my *Black Poachers, White Hunters,* chaps. 5 and 6, for a description of hunting by Kenya settlers and visitors after 1909.

6

Catching Animals

Nigel Rothfels

In his 1917 "A Report to an Academy," Franz Kafka explores themes of captivity, freedom, and art, through the voice of a "civilized ape." The story is written as an address by the ape, Red Peter, to the members of a scientific society. But while the academy had requested an account of the life he had formerly led as an ape, Red Peter insists that such a project remains impossible because, although his existence among—and as one of—the civilized has consisted of only a few short years, he has nevertheless advanced so far from his former life that his memories of that earlier time have all but disappeared. Red Peter concludes, "your life as apes, gentlemen, insofar as something of that kind lies behind you, cannot be farther removed from you than mine is from me."[1]

But if Red Peter is unable to recall substantially his life as an ape, he is nevertheless prepared to relate how he left his ape-ness behind to become human. The steps to that, for Red Peter, partly repellent accomplishment began with his capture by a "hunting expedition sent out by the firm of Hagenbeck." Red Peter's actual memories begin, however, "between the decks in the Hagenbeck steamer inside a . . . three-sided cage nailed to a locker." "The whole construction was too low for me to stand up in," he complains, "and too narrow to sit down in. So I had to squat with my knees bent and trembling all the time, and also, since probably for a time I wished to see no one, and stay in the dark, my face was turned toward the locker while the bars of the cage cut into my flesh behind."[2] Red Peter soon concluded that by becoming human he could escape his cage, even if he could never regain his freedom. The steps to his humanity were fairly straightforward: learn to shake hands, learn to spit, learn to drink, learn to smoke, learn to speak—learn, in short, to be a human on the variety stage. On a simple level, Red Peter, the artist, finds his "way out," as he calls it, through conforming to and exceeding expectations.

While admittedly it is the general theme of the relation between art and freedom that drives the narrative of this story—a theme taken up again by Kafka in such other animal works from the early 1920s as "Josephine the Singer," "The Burrow," and "Investigations of a Dog"—it is important to remember that in the "Report" Kafka is also, in fact, discussing the capture, transport, and exhibition of exotic animals. This would perhaps be a point of only passing interest were it not for Kafka's repeated references in the work to the name "Hagenbeck"—a name, which perhaps more than any other in the period in Germany, could bring forth images of the wild, the exotic, and the animal. As the dramatist Carl Zuckmayer put it at the end of the 1940s, "When a child of the South of Germany thinks of Hamburg, he paints a picture in his mind of a small town of red-brick buildings directly on the open sea, and encircled on all other sides by a huge and magic kingdom, Hagenbeck. Hagenbeck is not a proper name, but rather, like Alaska or the Wild West, the expression of a mysterious, unexplored land, where one yearns for adventure."[3] This essay describes one of the central elements of the "idea" of Hagenbeck, alluded to by Zuckmayer, which grounds Kafka's essay: that of catching wild animals in far-off places of the world.

The beginning of the Hagenbeck exotic animal business can be traced to March of 1848.[4] One day early in the month, sturgeon fishermen employed by Carl Hagenbeck's father, the fishmonger Claus Gottfried Carl Hagenbeck (1810–87),[5] had been working the lower Elbe and returned with six seals that had become caught in their nets. Carl Hagenbeck recorded in his memoirs, *Von Tieren und Menschen,* that his father "came to the lucky idea of letting people see the animals for the price of an entry fee, and to this end put them in two large wooden basins on the Spielbudenplatz in St. Pauli." According to Carl, the success of the family's first attempt at exhibiting nondomestic animals went far beyond expectations. After the Hamburg public's interest in the seals began to wane, Carl Hagenbeck Sr. was induced through business acquaintances in Berlin to take the seals to the Prussian capital to be exhibited again, and in the middle of violent domestic political upheavals, he made a neat profit showing his seals to anyone interested before selling them to another entrepreneur. As Carl Jr. noted, "that was the beginning of our animal business."[6]

After the success with the seals, Carl Hagenbeck Sr. increased his forays into the exotic animal market. In addition to regularly exhibiting and selling seals, he collected all kinds of animals for his house menagerie. Hyenas, opossums, and a variety of birds and monkeys found temporary quarters in the house; the father even bought a raccoon, which escaped on its way from Bremerhaven to Hamburg and evidently lived happily for a year and a half marauding local farms before ending up as someone's extraordinary hunting story. By the early 1850s, Carl Hagenbeck Sr. regularly bought animals that arrived in ships in the port of Hamburg. While there were already a number of exotic animal dealers in Hamburg who traded in birds, monkeys, and other small mammals, unusual reptiles, and such popular collectors' items as butterflies and beetles, Carl Hagenbeck Sr. was often willing to buy what others rejected as unsalable—especially the larger animals for which the market was relatively weak. Other than the zoological gardens in Berlin and Frankfurt—both still in their infancy—and the traveling circuses and menageries, there was little call in 1850s Germany for such animals as zebras, antelopes, ostriches, bears, and the great cats. Nevertheless, Carl Hagenbeck Sr. had a certain taste for the grand and imposing.

Despite some sensational purchases and sales, however, throughout the 1850s the Hagenbeck exotic animal business was more a hobby to Carl Hagenbeck Sr. than a business. His primary enterprise remained the buying, selling, and processing of fish, and with the increasing popularity of sturgeon caviar, that business was doing well. In contrast, the riskier animal business, according to the younger Carl, presented recurring losses. Quite simply, through inexperienced handling many animals died before they could be sold or profitably displayed. Despite the risk, Carl—having at fifteen years of age completed his formal education which, rather typically, had been more practical than scholarly—had decided to devote himself entirely to the animal business.[7] His younger siblings followed in his steps: his sisters Luise and Christiane took over the care of the birds, for example, and his brother Wilhelm helped with the other animals and experimented with new training methods. The deals that the younger Carl began to close were becoming larger, and soon he began to cut away at the limited markets of his competitors, including those of his main rivals, the English firm of Charles and William Jamrach which, along with a number of smaller English companies, had controlled the greatest share of the market for large exotic animals in Europe.[8] Attending the annual fall animal auction at the gardens in Antwerp in 1862, for example, the eighteen-year-old Hagenbeck, using his contacts with Dr. Bodinus, director of the zoological gardens in Cologne, and Director Schöpf of the gardens in Dresden, quickly sealed contracts with Geoffroy St. Hilaire of the Paris gardens, and Drs. Westermann of the Rotterdam and Martin of the Amsterdam zoological gardens. Shortly thereafter, the young Hagenbeck purchased Christian Renz's entire menagerie, including an evidently magnificent North African lion, an arctic wolf, a jaguar, and several panthers. He then sold most of the animals immediately on the spot at the fair in Krefeld bringing the family business a profit of some two thousand talers.[9]

In 1863, Carl's father responded to the growth of the family animal business by moving to a larger property at Spielbudenplatz Nr. 19—a residence that had previously housed the menagerie of Jacob Gerhard Gotthold Jamrach (the father of Charles Jamrach), who had moved the focus of his family's business to London and Liverpool.[10] The house and its rear court were illustrated and described by Heinrich Leutemann in 1866 for the popular magazine *Daheim* (figure 1).[11] On the street side of the house were two shops, one rented by a shoemaker and the other used by the Hagenbecks as a bird shop, a branch of the business which was taken over completely by Carl's sister Christiane in 1870. As with all of the properties on Spielbudenplatz, behind the house was a large court, where most of the menagerie was housed. Leutemann's illustration presents an assortment of presumably fairly common sights at the Hagenbecks. A sea captain has just brought a small animal and Carl unpacks it from its crate. Near the front rests the perennial wooden basin used for displaying the seals. Two men to the left struggle with an apparently stubborn reindeer as a polar bear looks out from its cage located in the right rear. A monkey rides on the back of one of the unpackers, a peccary with free rein stands in the foreground, and a dog and a peafowl look on while other assorted larger birds carry on in cages in the rear. At the front of the scene more birds look for the food thrown by one of Carl's sisters who is seen holding a small basket. The image, a composite of the many activities taking place at the Hagenbeck animal business, presents well what one might have expected to see upon visiting the small business and has become one of the central icons of the family history.

By the mid-1860s, Hagenbeck's business began slowly to change its character. On one hand, the company continued to purchase the haphazardly arriving animals brought by the ships docking at the Hamburg port. In 1863, for example, a quartermaster of the HAPAG line brought the five large bears that had belonged to Grizzly Adams—the one-time trapper and later impresario—to Hamburg after having purchased them in the Adams estate sale.[12] The two grizzly bears, one black bear, and two brown bears were too much for any dealer in Hamburg to manage other than the Hagenbecks, so Carl and his father purchased the animals for an unusually low price. But while incoming ships continued to be met by the Hagenbecks and their representatives, Carl also attempted to secure more regular and dependable lines of supply by building contacts in England, France, Holland, and Belgium. Still, at this point, most of the business consisted of buying and reselling animals that had already been brought to Europe and

1. "Hagenbecks Tierhandlung." Reproduced with permission, Hagenbeck Archive.

were being sold by menageries, circuses, zoological gardens, or other dealers. Recalling that in 1864 he made his first trip to England to purchase a South American giant anteater, Carl noted that his "dependence on the London animal market ceased only later after the founding of the German Empire and the growth in German overseas relations."[13]

We will have occasion to return to the connections between the development of Hagenbeck's business and the growth of colonial ambitions and imperial sensibilities, but for now, while it is clear, as Hagenbeck maintains, that the unification of Germany and the drive toward and eventual acquisition of formal colonies did give a boost to the Hagenbeck animal business,[14] it is important to recognize that already in 1864, Carl Hagenbeck made his first clear step toward the elimination of middlemen and the acquisition of his own sources of supply from what would eventually amount to outposts all over the world. Two years before, in 1862, an Italian named Lorenzo Casanova, who had found his way to Africa after losing his dog and monkey show in St. Petersburg to fire in 1859, had brought to Europe a large collection of animals that he had obtained from the Egyptian Sudan. Hagenbeck recalled that the transport had "consisted of six giraffes, an elephant, and many other rare animals." Casanova had encountered some difficulty getting rid of the animals, however, and had sold them cheaply to a traveling menagerie. Receiving a telegram in the summer of 1864 that Casanova was returning with more animals, Carl immediately set out to meet the man, to purchase whatever was available, and, most importantly, to sign an exclusive contract with the catcher. The 1864 shipment was small, consisting only of "two young lions, three striped hyenas, a collection of very beautiful large monkeys, [and] a number of birds," but Hagenbeck bought the lot and Casanova signed the contract.[15]

A year later, in July 1865, Casanova returned with the first contracted shipment of animals consisting of "three beautiful African elephants, several young lions, a number of hyenas and leopards, young antelopes, gazelles, and ostriches."[16] Similar contracts were extended during the following years to other adventurers, and by the late 1860s, it seems, Carl Hagenbeck could sell almost any animal that arrived at his business. By the 1870s, with the completion of the Suez Canal and the increasing prominence of German trade and Hamburg businesses in Africa and the Far East, Carl Hagenbeck's business was rapidly developing into the leading firm in the international trade of exotic animals. With almost every year the shipments of animals grew in frequency and size. In the spring of 1870, for example, Hagenbeck received telegrams from Africa indicating that two of his agents there, Casanova and Migoletti, were underway to Suez and Alexandria respectively. Casanova's report included the request that Hagenbeck meet him in Suez because he had fallen fatally ill. Carl left for Suez with his eighteen-year-old brother, Dietrich, who himself died three years later in Zanzibar during a catching expedition. Arriving at the train station in Suez, Hagenbeck found a number of animals already loaded onto wagons, but it was the scene at Casanova's hotel that really caught his attention. Carl noted in his memoirs:

> I will never forget the unique scene with which I was greeted when we entered the court of the hotel. Had a painter seen this sight, he would perhaps have immortalized it under the title, "chained wilderness." Elephants and giraffes, antelopes and buffalo were bound to the palm trees. In the background sixteen full-grown ostriches ran about freely, and in sixty crates there were lions, leopards, cheetahs, thirty spotted hyenas, jackals, lynx, serval cats, monkeys, marabou storks, rhinoceroses, birds, and a large number of birds of prey.[17]

With some difficulty, Hagenbeck and his brother took over Casanova's caravan and managed to ship the animals by train to Alexandria where they were joined by the other transport. The entire shipment of animals, which boarded a ship bound for Trieste, represented, according to Carl, "the largest animal transport that to that point had been brought to Europe."[18] Among the animals were

> a rhinoceros, five elephants, two warthogs, four aardvarks, fourteen giraffes, twelve antelopes and gazelles, four wild Nubian buffalo, sixty large and small beasts of prey, including thirty spotted and striped hyenas, seven young lions, eight leopards and cheetahs as well as several other wild cats, etc. Beyond all of those, came twenty-six African ostriches, among which were sixteen full-grown birds. . . . The transport was completed by twenty large crates with monkeys and birds, as well as seventy-two Nubian milk-goats, a wandering dairy which provided milk for our young animals.[19]

This impressive shipment more or less sealed Hagenbeck's reputation as the leading figure in the international trade in exotic animals. En route to Hamburg, Carl dropped off animals in Vienna, Dresden, and Berlin, the delivery at the last being the largest during the first year of Heinrich Bodinus's tenure as director of the zoo—a time during which the number of species at the zoo almost doubled and the number of specimens tripled.[20]

If many of the elements of Hagenbeck's reputation were more or less in place by the early 1870s, the image of animal catching which rested behind

Kafka's story of Red Peter still lay decades ahead. To begin to get a sense for the background of that image, we must turn away from Hagenbeck for a moment and direct our attention to the animal catchers themselves, and how those hunters and their apologists presented animal catching to the German public.

First of all, it is important to recognize that the Hagenbeck animal catchers were part of a notably diverse group. Among the participants were professional commercial hunters, such as the elephant hunter Hans Hermann Schomburgk and the walrus and whale hunter Ole Hansen,[21] who took up sideline interests in catching the young of the animals they harvested. There were naturalists and explorers, such as the photographer Carl Schillings,[22] who during travels occasionally found themselves collecting live specimens; soldiers of the colonial armies, such as Hans Dominik;[23] and settlers in German and other colonial outposts, such as Carl Hagenbeck's half-brother John,[24] who found themselves well situated to catch or purchase animals for Hagenbeck and other dealers. There were also a group of more or less independent entrepreneurs, such as Casanova, Bernard Cohn, and Josef Menges,[25] who found their peculiar niche in the animal trade fairly early on, and often by accident, but who were particularly adept at working with indigenous peoples to obtain animals, and who often had long-term relationships with Hagenbeck. Finally, there were those professional catchers, including Christoph and Elisabeth Schulz, Jürgen Johannsen, and Wihelm Grieger, whom Hagenbeck himself, at least initially, outfitted.[26] The backgrounds of these people were, of course, also quite varied. Hans Schomburgk, for example, was the nephew of Sir Robert Schomburgk, the explorer of Guayana and later English consul general in Siam. A man of education and some, if limited, means,[27] Schomburgk took up commercial elephant hunting to finance his interests in natural history, but after success in catching a young elephant which he later sold to Hagenbeck, Schomburgk gave up elephant hunting to become an animal catcher and later a photographer. Josef Menges, on the other hand, though clearly a first-rate explorer and adventurer, had little interest in hunting and little time for studies in natural history. After arriving in Africa in the 1870s and participating in Charles Gordon's explorations of the White Nile, Menges stayed on in Africa to work for Hagenbeck from 1876 to 1890 in the Sudan and Somalia, with additional trips to Sri Lanka, before struggling on more or less independently, apparently often at the point of bankruptcy, through to around 1909.[28] In short, all sorts of people ended up catching animals for all sorts of reasons. Beside elite hunter-naturalists like Carl Schillings, a man much admired by Theodore Roosevelt,[29] there were adventurers of lesser means such as Christoph Schulz who initially went to Africa through a desire for adventure but who soon found substantial support from Hagenbeck's firm.

Just as a survey of the people working for Hagenbeck between 1860 and 1914 presents a diversity of backgrounds and approaches to catching animals, it is also clear that during this period the practice of catching animals in colonial lands evolved dramatically. What had really only begun to take off at the middle of the nineteenth century in response to the proliferation of zoological gardens, circuses, and hunting parks, was, by the beginning of World War I, a relatively mature industry, the evolution of which can be summarized in three basic points: (1) While the trade had begun with almost complete dependence on both indigenous animal catchers and indigenous middlemen, with European traders essentially purchasing only those animals which reached coastal and later inland trading stations, by the end of the century Europeans had taken over the management, step by step, of each stage of the process. Thus, for example, while Hagenbeck's first "travelers," such as Casanova and Menges, bypassed coastal middlemen by setting up their own trading stations in the interior where they dealt directly with indigenous catchers, later Hagenbeck catchers, such as John Hagenbeck and Christoph Schulz, took over the catching process as well, albeit relying, of course, on the substantial labor of local peoples. (2) Although throughout this period, a number of remarkable catches (for examples, the first pygmy hippopotamus, the first German colonial elephant, the first walruses) and indeed full collections, were organized by a varied group of people whose primary interest lay elsewhere than in catching animals—naturalists, colonial administrators, and various kinds of hunters—the catch soon became dominated by people whose primary occupation was collecting animals for zoos and circuses. Thus, although the rhinoceros collected by Carl Schillings in German East Africa in 1903 created something of a sensation, the large corralled collection of zebras at the Hagenbeck-Schulz farm in German East Africa, and Wilhelm Grieger's heavily financed expeditions beginning in 1901 to capture specimens of the Przewalskii horse in Mongolia, represent the more significant part of the Hagenbeck business.[30] (3) Whereas, during most of the nineteenth century, portrayals of animal catching in popular articles and books often display an enthusiasm for describing the particularly bloody and destructive methods employed in most catches, by the end of this period, apologists began to assiduously avoid discussions of the "losses" both during and after the catch. Clearly, by 1914, a code of catching had begun to develop—partly in response to popular criticism and partly as a result of the professionalization of the practice—which argued both that catchers embodied a deep love of nature and animals and that their activities as catchers represented an integral part of larger projects of conservation and education. In short, the business was whitened and professionalized in much the same way as other colonial industries had been, and this evolution becomes more clear if we briefly look a little more closely at a few of Hagenbeck's catchers and some of the stories of catching animals from this period.

A good place to begin is with one of the very early accounts of Hagenbeck animal catchers, the 1870s souvenir pamphlet *Carl Hagenbeck's Thier-Karawane aus Nubien.*[31] According to the pamphlet, "for years the most important animal dealer, Mr. Carl Hagenbeck in Hamburg, sent his travelers every year to Nubia and Sudan," in order to "take over and ship to Europe" animals caught by the "Nubian hunters." Hagenbeck's agents, we are told—and here we should count such people as Casanova, Cohn, and Menges—are chosen for their familiarity with the land and its "half-savage inhabitants," and their fluency in Arabic. The text continues, describing the animal station:

> The camp of the animal trader is regularly situated close to a river, is always very extensive, and forms a small village around itself. It holds the straw huts for the Europeans and their black servants, areas for the captured animals which are fettered in the shade of thick trees, the stable for the horses, and the kitchen in which one or more slaves prepare food for the group. The whole thing is surrounded by a high, thick thorn hedge made from felled trees. A thornbush set up nightly closes the only gate to this camp. Through the nights fires burn to keep the wandering carnivores away, and rarely does a dark night pass in which a lion or leopard does not visit the lonely camp, roaring for hours at the inhabitants until they scare the cat away with thrown torches or a few shots into the air. During the day the station is governed by bustling activity, familiar natives arrive in a constant flow, often out of sheer curiosity, but mostly to offer their service as hunters and to obtain hunting horses.[32]

Once the native hunters have been supplied with horses, ropes, and the other necessary materials, they leave the zareba—their thorn-and brush-enclosed camp—to return weeks later with the animals. In general, according to the pamphlet, the herd animals are simply chased until the young fall behind, traps are set for leopards and other cats, and young elephants and rhinos are captured after their mothers have been killed. After their capture, the young are gradually transported to the station where negotiations begin. The pamphlet notes that transactions could often last "for several days before the trader comes into possession of the animal through a sum of money and small gifts such as tobacco, sugar, soap, and (especially coveted) empty bottles. The captured animals are always quite exhausted, and a third to half soon die from the stress of the hunt."[33]

After a stay of often several months, Hagenbeck's early "animal catchers" would begin the transport from the interior to the coast. Again, the example of the "Nubian caravan" described in the pamphlet gives us some idea of the difficulty of getting the animals to Europe. First, the size of these caravans is often startling. In addition to smaller animals, the pamphlet's caravan included three rhinoceroses, fourteen elephants, nine giraffes, fifteen antelopes, and seventeen ostriches. Moreover, Hagenbeck's agents contracted two to four men to walk with each elephant, three for each giraffe, and two for each antelope and ostrich. Those animals too small or dangerous to be walked to the coast—the young lions, leopards, birds, and the like—were packed in cages and carried by camels. Finally, a large herd of goats was brought along to provide milk for the young and meat for the carnivores. According to the pamphlet, the "Nubian caravan" took fifty-four days to reach the Red Sea from the interior, marching during the cool hours around sunset and dawn. Barely half of the animals obtained in the interior survived to the port of Suakin. From there, they were transported by steamship to Suez, then by rail to Alexandria, by steamship to Trieste, and finally by rail to northern Europe. The pamphlet concludes by noting that "from all of this one can appreciate the trouble and expense that such a transport requires, and it is hoped therefore that the public might compensate Mr. Hagenbeck for his great cost and sacrifice through friendly attention to the animal transport and its black Nubian attendants and hunters."[34]

Of course, as the trade evolved in this early phase, a certain regularity set into the business. Soon it became less and less necessary for Casanova, Menges, Cohn, and the other traders from predominantly Germany, Greece, and Italy who were active in northeast Africa to travel deep into the interior to contract with catchers. As Josef Menges noted in an 1876 article on the animal trade in northeast Africa:

> The animal trade, as it is now handled by the Germans, has become through years of practice strongly regulated and operates in general as smoothly as any other business. In October and November when the unhealthy period following the rainy season comes to an end, the traders travel from Cairo or Alexandria to Suez, where they catch the Egyptian steamship which travels twice a month to Suakin and Massawa. The trip to Suakin lasts between four and six days. Suakin is the beginning point of the great caravan routes which lead through Kassala and Gedaref to Khartoum and through Gallabat to Amhara. The

> trading caravans travel for sixteen days to get from Suakin to Kassala; the path is fairly well traveled, but it is naturally only passable with camels, water being available every two to three days. In Kassala the trader rents a home where he will be sought by the nomadic Arab peoples of the interior. They will bring him those animals they have already captured which they wish to sell and will enter into further binding agreements for other animals. The only Arabs who systematically engage in the capture of animals are the . . . Hamran who have their hunting grounds on the Setit (a lower fork of the Takkaze which flows from Abyssinia). The land between the Setit, Gasch, and Atbara is extraordinarily rich in wildlife and therefore there is no shortage of the young and catchable animals.[35]

With the arrival of the trader, Menges notes, the hunt would be stepped up with the catchers giving special attention to those animals especially coveted by the trader. The animals would then be delivered to Kassala where they would be purchased by the trader, along with those animals which had fallen by chance into the hands of other peoples. Menges notes that while this more systematic character of the trade had made the European's task somewhat easier, if an especially rare animal was sought, the trader would usually have to go into the field to contract directly with indigenous catchers. If, on the other hand, the trader was interested in only the standard assortment of giraffes, elephants, buffalos, hyenas, lions, leopards, antelopes, ostriches, and other birds and small animals, he could reasonably expect to wait in the relative comfort of Kassala for the animals to be delivered.[36]

It is not simply a coincidence, of course, that the procedures described in the "Nubian caravan" pamphlet and in Menges's article echo the histories of other early colonial industries; indeed, the two works capture fairly well that moment in the consolidation of power in colonial lands when Europeans increasingly bypassed coastal traders and chiefs to move through the interior collecting their trade goods. Nor is it surprising, therefore, that the methods employed by the animal catchers paralleled the practices in other more firmly established industries. Still, as Endre Stiansen has emphasized in his study of the ivory and gum Arabic trades in the Sudan, clear generalizations about colonial trades are difficult.[37] While in the case of gum Arabic, European traders tended to travel throughout the region and purchase the gum wherever it was available, and in the case of ivory, trade tended to occur at more permanent stations, different practices were employed in different areas and local traditions and the trader's abilities and expertise all played a part in how and where a given transaction took place. It seems fairly likely, though, that the trading stations used by the Hagenbeck "catchers" were at the same time being used for other trades, and that similar techniques were employed in all these businesses. Traders, for example, often utilized variations of the trust system suggested by the "Nubian caravan" pamphlet, whereby supplies and goods would be given in advance to the indigenous people in return for the promised future delivery of goods. Largely stemming from practical trade considerations in areas without established banking systems, the incurred debt in this system would drive the market as debtors trading ivory, gum, oils or, in this case, live animals sought both to maintain their valuable contacts with the traders and to extricate themselves from debt. In any case, general rules for the animal trade, as for other colonial trades, are difficult to establish.[38]

Nevertheless, it remains clear that the kinds of trading systems used by Hagenbeck's catchers carried significant speculative risk and required substantial capital support to cover the transportation and maintenance of the catchers themselves, the establishment and provisioning of trading stations, possible prepayments to catchers, and the transport and insurance of the animals. In the cases of the semi-independent traders such as Casanova, Menges, and Cohn, most of that cost was born by the traders themselves. It is clear, however, that Hagenbeck increasingly provided up-front money to his catchers, although it is difficult to establish exactly when and to what degree. This was, of course, the case with Dietrich Hagenbeck's expedition to East Africa in 1873, described below, but it is also clear that more independent traders such as Menges were actively recruiting needed capital for their ventures.[39] By the turn of the century, Hagenbeck seems to have been entirely funding a majority of the animal catching expeditions supplying his establishment. The reason, of course, for his desire to have complete financial control becomes clear when one considers the potential profits of successful catching expeditions. According to figures provided by Josef Menges for the purchase price of elephants, giraffes, rhinoceroses, antelopes, and lions in Kassala and their sale price in Europe in 1876, an elephant, for example, could be purchased in Kassala for 80–400 marks and sold in Europe for 3,000–6,000 marks; a giraffe purchased for 80–200 marks could be sold for 2,000–3,000 marks; and a rhinoceros obtained for 160–400 marks could be sold for 6,000–12,000 marks.[40] To be sure, these huge markups—in the case of the rhinoceros some 3,000 percent—have to be considered in light, Menges argues, of the often extraordinary costs of getting the animals to Europe. The costs of such trips can be imagined, he notes, when we consider that to move the animals to Suakin on the coast one must obtain fifty to one hundred camels at 24 marks each for the trip, fifty to sixty people to walk with the animals, and supplies of water and food for the animals and people. Shipping the animals then from Suakin to Suez could cost as much as 150 marks for each elephant or giraffe and 130 marks for each smaller animal, with animals in cages costing 34 marks per cubic meter. Each railroad wagonload of animals from Suez to Alexandria would cost 330 marks; and tariffs were as much as 8 percent of the animals' value. Add into all of this the costs of loading and unloading the animals at every stage of the trip, of housing and feeding the animals along the way, of the required tips to move the whole process along, and of having a third, half, or even two-thirds of the animals die along the way, and, Menges argues, the markups do not seem so extreme.[41]

That said, for the purposes of getting a general idea of the potential profits involved in this trade, Menges's 1876 figures lead to some pretty remarkable, but not wholly unexpected, conclusions. Taking an easily imagined shipment of, for example, one rhinoceros, five elephants, fourteen giraffes, twelve antelopes, and seven young lions,[42] the animals, according to Menges's figures, could be obtained in Kassala for as little as 1,976 marks or as much as 7,200 marks. Taking the larger figure, and adding twice that amount to cover all the incidental costs of shipping the animals to Europe (my estimates suggest the actual cost to be under 10,000 marks for these associated expenses), these animals will have cost Hagenbeck a maximum of 21,600 marks—an admittedly impressive sum of money. Nevertheless, for this shipment of thirty-nine animals, Hagenbeck, again using Menges's figures, could expect to obtain at least 61,000 marks and as much as 136,800 marks. Even halving those numbers to allow for a 50 percent mortality rate yields impressive profit margins. When we consider the sheer volume of animals moved by the firm from the 1860s to 1880s, it becomes clear how Hagenbeck rose to such wealth and fame. By the mid-1880s, for example, Heinrich Leutemann noted that, by Hagenbeck's estimate, in the first twenty years of the company,

> He had sold at least a thousand lions, three to four hundred tigers, six to seven hundred leopards, a thousand bears of different varieties (at one time he had forty-two at the same time), and around eight hundred hyenas. Some three hundred elephants had passed through his hands, of which sixty-three Indian and four African elephants had come in 1884 alone. He had sold seventeen rhinoceroses of the three Indian species and nine of the African, while 150 giraffes, of which in 1874 alone he sold thirty-five, and six hundred antelopes of diverse species including the rarest, largest, and most beautiful, had been traded by the company.

Leutemann's account continues by noting that in the same period the company sold 180 Bactrian camels, 120 dromedaries, and 150 reindeer, and that in one remarkable shipment of animals, some 600 snakes had been sent of which 374 arrived alive in Hamburg.[43]

It is precisely within the context of Hagenbeck's remarkable success in delivering animals to Europe that we must understand his decision in 1873 to send his twenty-year-old youngest brother Dietrich to Africa specifically to catch young hippopotamuses for the company. This decision marks an important moment in the history of the firm. Although by 1873 Hagenbeck's animal business had become the most important source for animals from Africa primarily through its stations in the Sudan, the deliveries were still generally based on whatever Hagenbeck's agents had managed to get their hands on, mostly through trade and purchase. By sending Dietrich into the field to catch a specific animal, and, in this case, one of the most valuable animals on the market, the firm attempted to gain more control over the types of animals delivered.[44] At the same time, however, Dietrich's appearance as a new kind of catcher highlights a moment in the development of colonial trade. Whereas Casanova and Menges had bypassed the traditional middlemen to trade directly with the peoples of the interior, albeit on similar terms, thereby diminishing the power of the coastal elites and consolidating the control of colonial administrations, Dietrich, and those who followed, took the next logical step and assumed control over the catching process itself while using local peoples merely as labor.

In the end, however, Dietrich's attempts to bring home even a single hippopotamus were unsuccessful despite months of effort. In a letter of March 14, 1873, the young hunter noted that after four weeks of searching, having seen many hippos including as many as twenty animals in a group, and having shot four adults, he had been forced to retire to Zanzibar to recover from fever.[45] A letter from the end of April from Dar es Salaam reports another month of unsuccessful effort, increased frustration with the "laziness" of the "natives," and unrelentingly bad weather during a three-day pursuit of a female hippo and her baby.[46] Finally, in a long letter of June 25, Dietrich reports that he managed to capture a young hippo, but that the animal had apparently been bitten by a crocodile sometime earlier, and after six days had died from the earlier wound.[47] Dietrich himself died shortly thereafter from a virulent form of malaria.[48] As the bourgeois magazine *Die Gartenlaube* noted, he represented "unfortunately yet one more victim in the long list of those who have not returned from Africa."[49]

While Hagenbeck's success at taking increasing control over all stages in an animal's delivery to Europe or the United States—from the animal's capture to transport and eventual sale to a zoological garden or other collector—helped the company become the preeminent conduit for the larger exotic animals into the world market, more modest collections of animals put together by naturalists, colonial administrators, "great white hunters," and commercial hunters continued to be purchased by Hagenbeck and other traders. How these hunters and their hunts were portrayed varied, of course, according to the interests of the collector himself, but a closer look at most of the stories about them will show, often in sharp contrast, the differences between these amateur catches and the type of professional catching which would become the hallmark of the Hagenbeck business by the turn of the century. Hans Dominik's elephant adventure, "Catching Elephants among the Mwelle," from his 1901 *Kamerun: Sechs Kriegs- und Friedensjahre in deutschen Tropen,* presents a rather striking example. Prompted by a conversation in the mid-1890s with Ludwig Heck, the director of the Berlin Zoological Gardens, Dominik, a captain in the German colonial army in Cameroon, had decided to catch a baby elephant. Heck had told Dominik, who had already brought a variety of animals from the colony, that the day when the first elephant arrived in Berlin from German Africa would be a day of celebration.[50] Dominik concluded that his best chance of success would be if he captured an entire herd and then obtained the young by killing off the adults. Though bloody, Dominik's plan was practical, and his ability to execute the task was repeatedly noted with admiration by other writers.[51]

After spreading the word that the first people to lead him to a herd of elephants would collect "a large reward," Dominik finally received information in early October 1898 that Mwelle hunters had discovered a herd and were awaiting his instructions. After his arrival, Dominik observed the herd with his guides Zampa and Amba:

> There was little activity among the animals. The calls of the working humans which carried clearly through the quiet forest hardly appeared to bother them. One bull stood apart, preoccupied with tearing twigs off branches with his trunk and consuming the leaves. Closest to us stood a cow using her trunk to lovingly caress her baby, which was barely larger than a pig and stood between her legs. A few animals ate—sweeping together and ripping up low-growing grasses and using their trunks like sickles—most of them appeared to be sleeping. . . . We seemed so small, so insignificant when compared to the mighty animals in the mighty wild.[52]

A fence was constructed during the night to thwart any attempt by the elephants to escape, and the hunt began the next morning. In his account,

Dominik begins by describing the scene when he and his guides came across two adults:

> One after the other, a head turning to the left to pull up something green from here and there, the elephants came slowly toward us. The safeties were released. "You, the second," I whispered to Zampa. Now we had the animals ready. I fired at the right ear of the foremost animal. At the sharp crack, the elephant threw its trunk into the air and trumpeted loudly. The short tail stretched out far, he turned upon himself like a top. In this moment, Zampa had also fired. Close before me the second animal buckled at his knees, but quickly stood up again and followed the incessantly bellowing and bleeding lead bull which pushed up the hill.

Firing after the animals and then tracking them over a small rise, Dominik found the two elephants beside each other:

> There lay one of the animals; apparently the spine had been hit because the elephant had only collapsed in the rear and was in a sitting position. Like columns, the forelegs projected from out of the ground, the head and trunk swung left and right: a muffled moan sounded, thick clumps of blood flowed at the side, a sign that the lungs were also wounded. The other stood next to him, motionless except for his trunk. He blew frequently, and with his trunk threw soil on himself. Our approach didn't seem to bother the animals. We crept around them. I had the eye of the sitting giant exactly in the rifle sight, when beside me Zampa fired. The standing elephant trumpeted loudly. Now I fired, and the animal collapsed onto its side. The other elephant was still standing; finally with the first shot from my second chamber he collapsed. Close beside one another lay the two giants in a massive pool of blood. Amba and Balla were already there; with their sharp machetes they cut through the trunks, half the thickness of a man, of the still breathing animals. As if from a fountain, the red blood sprayed up from the thick arteries onto our clothes as we stood beside the animals examining our guns and discussed how we should proceed with the hunt.[53]

The fascination with grisly detail which permeates this narrative continues as the hunt progresses. Soon Dominik encountered a female with a young calf; after several shots, also graphically described, the female was dispatched with a shot in the left eye. The calf was roped and tied to a tree where it "churned up the soil with its small tusks, bellowed and moaned, charged backwards, stood on its head, and foamed at the mouth in rage as bloodshot eyes protruded from its head."[54]

Three remaining calves were soon captured as well, one dying of suffocation after having its trunk pulled between its forelegs and tied to its rear legs so that it "breathed with difficulty and lay on the ground like a large gray sack."[55] Another calf died during the night of wounds sustained in the capture, but Dominik had still managed to secure two calves from the herd and soon added three more to his collection. Two died a month later, but the remaining three apparently thrived in their new environment and one found its way through Hagenbeck to the Berlin Zoo where it was seen by literally thousands of Berliners who lined up to view the newest acquisition from the colonies.[56]

In his 1887 biography of Hagenbeck, Heinrich Leutemann was frank about the destructiveness of animal catching. "For the animal trader," Leutemann noted, "the method of capture is, from a business point of view, a trivial issue. For him the most important thing is that the animals arrive healthy, that they are, therefore, well packed." As Leutemann made clear, the animal catcher's primary concern was not obeying the cultured rules of the hunt, but doing his business as quickly and efficiently as possible. Thus, animal catching rarely concerned itself with the ritualized confrontation of man and animal on a field of honor; rather, for quite practical reasons such as care and transport, the catcher generally ridded himself of the adults as efficiently as he could, and concentrated on obtaining the young with as little impact on the animals as possible. As Leutemann put it:

> Without exception, lions are captured as cubs after the mother has been killed; the same happens with tigers, because these animals, when caught as adults in such things as traps and pits, are too powerful and untamable, and usually die while resisting. . . . The larger anthropoid apes can, in addition, also only be captured—taking into account occasional exceptions—quite young beside the killed mother. The same is the case with almost all animals; in the process, for example, giraffes and antelopes, when hunted, simply abandon their young which have fallen behind, while in contrast the mother elephant more often defends her calf and therefore must be killed, as is the case with hippopotamuses. . . . Also in the case of the rhinoceros, the young are captured from the adults, which are usually killed as a result.[57]

The point is, that at least as far as its general approach to the efficient collecting of elephants, Dominik's tale seems generally predictable; we should not be shocked that this type of hunt was a relatively bloody affair. Or should we? In the end, what is absolutely basic about Dominik's tale, what marks it as a particular kind of hunting story contrasting in sharp relief to the stories of Menges and the works of Hagenbeck's professional catchers described below, is Dominik's decision to present, and indeed revel in, all the gruesome details of his encounter with the elephants. His description, for example, of standing essentially in a blood fountain examining his gun after dispatching the first two elephants—so particularly preoccupied with and fascinated by blood and death—suggests that this story is only peripherally about catching elephants. Centrally, in fact, Dominik's story is about killing and the conspicuous display of power.

In this respect, Dominik's narrative is echoed by a whole range of "catching stories" from this period. A typical item in this genre, for example, is Hans Hermann Schomburgk's 1910 *Wild und Wilde im Herzen Afrikas,* in which Schomburgk describes his life as a professional ivory hunter and later animal catcher for Hagenbeck. Having arrived in southern Africa in 1898 ostensibly to start a farm, and after several unimpressive expeditions to the interior and a stint in the Natal Mounted Police, Schomburgk traveled north and by the fall of 1906 had become a professional ivory hunter, he tells us, for largely pecuniary reasons. Fascinated by Africa and passionate about being out in the bush, the young explorer nevertheless needed a livelihood, and hunting elephants provided a sufficient income to fund all manner of enthusiasms for exploration, sport, and natural history. In the introduction to his *Wild und Wilde,* however, Schomburgk is sensitive to possible criticism about the large number of elephants he in fact killed and the many more he was reputed to have slain. In response, he notes succinctly, "In my travels I have received financial support from no one, and if I should be accused of having killed many elephants, I must acknowledge myself as guilty, but at the same time console myself that I always hunted according to the rules and with very few exceptions—where there was mortal danger—didn't bag any cow elephants."[58] In fact, the first two elephants Schomburgk shot—in a situation that was far from mortally dangerous, at least for him—were females, but the author glosses over this unfortunate detail, writing plainly, "It is impossible for me to describe just how proud I was to sit myself down on the first elephants I had slain by myself."[59] In any case, after hunting ivory for a year and a half, Schomburgk took a lead from Carl Schillings who had tried several times to catch and keep alive a baby elephant in German East Africa. After some failed efforts, the hunter eventually found himself favorably situated to catch the perfect calf. On April 21, 1908, Schomburgk sighted a small herd and discovered among the cows a suitably young specimen. After separating the cow and her calf from the rest of the group, Schomburgk tracked the animals for three days before, quite unexpectedly, he came across the pair in high grass. Immediately the mother cow—"a huge tuskless animal" according to Schomburgk—charged the hunter, who "nevertheless luckily succeeded in landing a shot at the base of the animal's trunk."[60] Schomburgk continues: "She collapsed in the rear and gave me the opportunity to jump quickly sideways and bring to bear a deadly shot, after which she immediately died. Obeying the laws of nature, the young animal remained standing beside its mother. . . . Until my men arrived, I observed how the pitiful little baby continuously ran about its mother while hitting her with his trunk as if he wanted to wake her and make their escape."[61]

In his reminiscences, Schomburgk included a photo of "Jumbo" standing in front of his dead mother with the caption, "obeying the laws of nature, the young animal remained standing beside its mother" (figure 2).[62] Killing the mother was, of course, only the beginning of Schomburgk's task in catching Jumbo, but soon the animal was under control and before long it had formed an attachment to the hunter and his men and was eating well, eventually taking milk from a bottle. In the middle of August, some four

2. The elephant Jumbo beside his mother. Reproduced with permission, © Hans Schomburgk/Jutta Niemann Archive.

months after the elephant's capture, Schomburgk began the trip to the coast. The walk was slow because the young animal's feet often became sore on the march, necessitating that the caravan stay, for example, for eleven days in Tringa. By mid-September, Schomburgk's party reached Morogoro where both he and Jumbo were a sensation. From there the party took a train to Dar es Salaam where Jumbo became famous as a sandwich-sign carrier announcing upcoming concerts and other newsworthy events. Despite Ludwig Heck's suggestions to Schillings that the day the first German East African elephant came to Berlin would be a day of celebration, Heck, according to Schomburgk, did not respond to cables offering the animal for sale. Instead, one of Hagenbeck's collectors—Christoph Schulz—saw the animal, bought it, and immediately sent it to Stellingen.[63] Schomburgk reported that "on April 5, 1910, I was able to greet Jumbo back in Germany, and he now lives in Carl Hagenbeck's Animal Park where I visit him often." "That is how Europe," Schomburgk incorrectly declared, "came into possession of its first east African elephant."[64]

The modest but critical point to the stories of Dominik and Schomburgk is that these hunting tales are primarily about killing, about the ostentatious affectation of "courage," about sitting on top of dead elephants and standing in pools of their blood. In describing his career, for example, Schomburgk writes: "No matter what people believe, it nevertheless takes courage to do battle with an outraged bull elephant in the thick brush or high grass. Three times I've been under an elephant; once I was thrown in the air by an elephant; an elephant trampled and ripped apart one of my carriers; and if I have managed to come out of all of these situations with a relatively intact skin, I have been lucky."[65] Fond of writing about the lack of sentimentality in nature and about how, in nature, the death of an individual is meaningless and only the survival of the herd significant,[66] Schomburgk found repeated occasions in his many books to praise the unappreciated courage of the great hunters who risked their lives in pursuit of the most dangerous quarry.[67] It is precisely this type of writing that would become a sheer impossibility in later accounts of animal catching.[68] Indeed, the general sense that "unnecessary" killing must be avoided in descriptions of animal catching became more and more clear to everyone in the business. During negotiations to procure a young rhino for the Bronx Zoo in 1902, for example, then director William Hornaday asked Hagenbeck to be circumspect in talking about how the animals were obtained. Hornaday writes:

> I have been greatly interested in the fact that your letter gives me regarding the capture of the rhinoceroses; but we must keep very silent

> about forty large Indian rhinoceroses being killed in capturing the four young ones. If that should get into the newspapers, either here or in London, there would be things published in condemnation of the whole business of capturing wild animals for exhibition. There are now a good many cranks who are so terribly sentimental that they affect to believe that it is wrong to capture wild creatures and exhibit them—even for the benefit of millions of people. For my part, I think that while the loss of the large Indian rhinoceroses is greatly to be deplored, yet, in my opinion, the three young ones that survive will be of more benefit to the world at large than would the forty rhinoceroses running wild in the jungles of Nepal, and seen only at rare intervals by a few ignorant natives.[69]

While it is clear, however, that Hornaday's letter to Hagenbeck suggests a certain sensitivity to discussing the unfortunate slaughter of animals in the process of catching them that contrasts to the tales of Dominik and Schomburgk, the two approaches to discussing animal catching have much in common.

Most importantly, both approaches essentially conflate the animals, people, and larger environments of territorial possessions into generic objects valued specifically for their importance to the colonial powers. Thus, Hornaday, in a justification for animal catching that is common enough even today, argues that the captured rhinoceroses will far more "benefit the world" than forty adults "running wild in the jungles . . . seen only at rare intervals by a few ignorant natives." Sentiments such as these helped drive the often brutal nature of this business and, moreover, the often almost despotic behavior of the catchers themselves. Referring to the peoples of Cameroon which he was charged with subduing, Dominik, for example, declared, "they must know that I am their master and that I am the stronger, and as long as they do not believe that, they must be made to feel it, and I mean severely and pitilessly, so that rebellion will pass them by for all time."[70] Typically, in signing peace treaties with Dominik, "freed" villages would be required to pay tribute in goods and men, with the men being taken away chained for carrying and plantation work. Dominik's deliveries of hundreds of forced laborers to plantations and factories over the course of his years in Cameroon seem rather typical in this regard, as does his use of hostages to obtain carriers and his whipping and shooting of disobedient workers, even if his habit of having the heads of his dead enemies presented to him in a sack seems somewhat eccentric.[71] In this environment, we must remember, beating and chaining workers was not considered especially exceptional or barbaric behavior, but appropriate and necessary in order to get the "natives" to work. As one Cameroon planter succinctly responded to critics at home in Germany in 1904, "Our people must not be allowed to forget that the politics of colonialism is the politics of conquest, that we are invaders in order that the land be brought to culture and developed for the German nation. We take it from the natives to exploit it for our purposes. We force the natives to work for the cultural development of the colony. This work must be demanded from them otherwise the colony will remain at the status quo at which we took it over."[72]

What we must recognize is that however romantic or adventuresome the exploits of the animal catchers of this period may have been represented as being, in many—indeed probably in most—cases, this business had a ruthless side in its dealings with animals, people, and its environment in general, which was almost a necessary component of the trade. The animal trade was a labor-intensive, extractive enterprise and, like most colonial industries—including tobacco, coffee, cocoa, saltpeter, rubber, ivory, tropical woods, palm nuts, and palm oil—was often based on forced labor and was highly destructive in its use of land. Dominik is not, in this regard, somehow unusual. To be sure, with readers at home increasingly uncomfortable with the use of force against their colonized peoples,[73] few authors in their published adventures of catching animals dwelt on their dealings with indigenous people beyond emphasizing their own abilities to work closely with "lazy," "uncooperative," "savages."[74] Nevertheless, even descriptions of actual beatings of carriers and other laborers are present in these accounts. In his story of catching pygmy hippopotamuses for Carl Hagenbeck in 1911–12—printed in full in the *New York Times,* as well as in the *Bulletin* of the New York Zoological Society—Schomburgk, for example, refers repeatedly to beating and whipping his workers and even holding a "chief" at gunpoint to secure additional carriers for his precious cargo. Early on in the story, for instance, he writes, "I thought the time had come to teach my carriers a lesson. We were too far from civilisation already to fear desertion. When I called the boys in the morning to start, nobody came, so I called up my headman and asked him very quietly if the boys were packing up. 'No,' was the reply, 'they do not want to start yet.' Without saying another word I took up my Browning and put 7 shots through the roof of the boys' hut.—And they came quickly! From this minute I took up the reins and after I had picked out the biggest and laziest of the motley crowd and had given him personally a good hiding, I had no further troubles."[75] "Troubles" followed, however, in his second expedition into the interior after pygmy hippos, during which at one point, Schomburgk claimed, he faced "open rebellion" to which he responded as follows: "I slipped the Browning into my pocket, took my hunting crop and went among them. Clash, crack went the whip on the naked body. A few straight hits from the shoulder on the jaws of those who did not move and quicker than I can tell it I drove the mutinous crowd before me like a herd of sheep! The result of the rebellion for the boys was that I stopped their rations for three days and the allowance of gin for a month."

Nevertheless, just as in this period the often callous aspects of colonial rule began to draw criticism at home, so, too, did the often wasteful and "inhumane" methods of the animal catchers working within colonial regimes. On one hand, those criticisms took form in the "sentimental" condemnations issuing from Hornaday's "cranks," the same type of sentimentalization of animals driving such diverse cultural phenomena as the Steiff stuffed animal business and the "nature faker" controversy surrounding nature writing at the turn of the century.[76] Other critics soon came forward, however, from often surprising quarters. Richard Meinertzhagen, for example, a particularly destructive if also complicated hunter, naturalist, and member of the colonial army in Kenya, could not countenance what he considered a lack of sportsmanship in catching animals for zoological gardens. Arguing that the commercialization of hunting was responsible for the devastating declines in animal populations in Africa, Meinertzhagen believed that perhaps the most horrible of those commercial enterprises was the trade in living exotic animals. "I wholeheartedly disapprove of the catching up of wild animals for zoological gardens and condemning them to solitary confinement for life amid squalid surroundings," Meinertzhagen wrote. "I have seen a good deal of the 'catching-up' stage and can vouch for the high rate of mortality, for the shooting of mothers to obtain the young, for deaths in transit and for the cold-blooded nature of the catchers-up."[77] Even figures within the world of zoos began to take a stand against what was seen as the pernicious effects of the commercialization of animal catching. Friedrich Knauer, for example, the director of the zoological gardens in Vienna, admonished his colleagues shortly before World War I:

> It is clear that the high prices which are paid for rarities have spurred the professional animal catchers and dealers to extreme efforts. Were the demand from zoological gardens for living musk oxen . . . not so brisk, it would hardly have occurred to the Norwegian fishermen to search the east coast of Greenland and slaughter whole herds of this, until recently, rarely endangered form of cattle. Twenty-eight musk oxen had to die so that the Danish expedition sent to Greenland in 1900 . . . could get possession of a young live musk ox for the Copenhagen Zoological Garden. . . . And so it is with many other rare

> species—we need to think only of the white-tailed gnu, of the varied representatives of the ancient Australian fauna, of the giant tortoises of the Galapagos Islands, of the birds of paradise.[78]

Partly in response to the growing criticism of their business and partly as a consequence of the careful refinement of how Hagenbeck wished his company to be perceived, heroic stories such as Dominik's and Schomburgk's were gradually displaced by narratives highlighting the more "civilized" hunting practices of Hagenbeck's quite new professional catchers such as Christoph Schulz.

Of the many tales written by or about the Hagenbeck animal catchers, perhaps that which marks best the first decades of the twentieth century is Christoph Schulz's *Auf Großtierfang für Hagenbeck: Selbsterlebtes aus afrikanischer Wildnis.* Based on his experiences in German East Africa in the decade preceding the First World War, *Auf Großtierfang für Hagenbeck* went into several editions and was also abridged into a more popular form.[79] In the book's introduction, Eduard Elven, who met Schulz in Africa before the war and later found himself a fellow prisoner of war with the animal catcher in Malta, suggests the book's main appeal. Comparing it to the broader literature about hunting big game in east Africa, Elven concluded that the book is unique because although the chase is an important element, keeping animals alive is Schulz's main task. Consequently, Elven noted, the book focused on "the sufferings and joys of an animal catcher, his life on the steppes and during the transport of the animals, as well as the fashion in which he is able, in the middle of the wilderness, to catch the animals which we admire as the ornaments of our zoological gardens."[80]

For those familiar with hunting literature from the period, the book presents few surprises; it is, essentially, an edited field journal. The book has no plot; rather, the author has adopted a typical approach for the genre and organized his field notes into a series of adventures divided by animal type—the hunt for buffalo, for giraffe, for zebra, for lion. The detail of geographical information is extensive, while the observations of animal life are told in an unstudied and matter-of-fact style which demonstrates throughout a primary focus upon keeping the captured animals alive. By and large, *Auf Großtierfang,* which shared many features with highly popular works of travel literature in the period, transported readers to a land both strange and familiar—filled with "exotic" animals and people, where ordinary Germans could build an almost ordinary life—and, consequently, the book was generally more concerned with detail and problem solving than with drama.[81]

Thus, unlike the heroic hunter-catchers such as Hans Schomburgk who somehow always managed to bring the deadly shot to bear on the charging animal in the final moment, and unlike also the tales of Casanova and the "Nubian caravan" which posited animal collecting stories as a series of exciting and deadly adventures with savage peoples and arduous tasks, Schulz is simply not particularly interested in describing his life as filled with danger and death. Whereas Schomburgk, the professional hunter, illustrated his famous catch of a young East African elephant with a picture of "Jumbo" beside his dead mother, Schulz illustrates his books with pictures of his wife tending animals at the Hagenbeck-Schulz farm on the foothills of Mount Meru in German East Africa. Indeed, more than a farm: Schulz and Hagenbeck, as the latter made clear in a letter to William Hornaday about Hornaday's newly published *Our Vanishing Wildlife,* began to think of this property in Africa as a reservation. Hagenbeck writes, "You are quite right about the destroyment there is going on with all kinds of animals and birds. And not alone that I fight against that—I get more and more friends which assist me, and I have brought it so far for our German East-African districts that we have worked out very severe restrictions. It will interest you that I have got from the Government more than 100,000 acres of land as a wild animal reservation."[82] Schulz corrals adult zebras and lassos young giraffes. On his farm, he and his wife settle down to rear animals, care for them, and ship them to Europe. While he is out surveying, his wife, an altogether new presence in a work such as this, feeds and tends the tame animals at home. Schulz writes that he found "great assistance in the care of the animals in my wife who, during my absence, faithfully cared for our charges and never shied from inconvenience or work. She is in large measure to be thanked for the great success I had in the rearing of the animals, and for the fact that I was able to bring back alive and healthy the first giraffes, oryxes, elands, aardvarks, and many other representatives of the East African fauna to Germany."[83] When Schulz sets out after a baby rhinoceros he attempts to do so without killing the mother by separating the young from the adult with the use of trained dogs. In catching hippopotamuses, he does not shoot the adults, but builds instead special pits designed to catch only the young. Not surprisingly, Schulz has little affection for the thrills and joys of more traditional hunting. "For me," he writes, "there is no greater pleasure than to observe wildlife in the wilderness, to obtain young animals alive and unhurt, and to take care of them, so that they can be kept as long as possible in the zoological gardens where they will serve the advancement of science and the popular education."[84]

The general impression that one gains in reading works like that of Christoph Schulz is that while animal catching may, at one time, have been a wasteful, perhaps even cruel profession in which Europeans sought little else than the maximization of profit, enlightened catchers and dealers now control the trade. These new catchers, we are told, truly understand the needs of animals, indeed these new men, and women too, cherish animals, and catch them largely because the public back home needs to be shown just how wonderful the creatures of the world are. They do not hunt for mere sport, but because a captive live animal can show the world the need to treasure and protect all animals. Of course, by the time that Schulz went to Africa to work for Carl Hagenbeck in 1908, the company had already come a long way in defining itself as an organization primarily motivated by a love of animals. Indeed, Hagenbeck's Animal Park in Stellingen—which opened in 1907 and which over the course of the twentieth century has become the central public feature of the company—gained something of an echo in the Hagenbeck-Schulz African "farm" or "animal reservation": they were both quickly cast as places where animals could live in safety, protected from the predations of man and removed from the fight for survival by the kindly Hagenbeck.[85]

Schulz and Hagenbeck sought to reconstruct themselves and their professions as true friends of animals, utilizing throughout references to a beneficent and ameliorative process of progressive civilization, as well as metaphors of and literal references to "families" of animals and people. Not surprisingly, in his memorial in the Animal Park, then, Hagenbeck is portrayed standing with a calm hand resting on his favorite lioness, "Triest," who, the story goes, threw herself into battle against a suddenly dangerous group of cats in order to protect her keeper during one of his customary visits to the lions' den. As Ludwig Zukowsky, one of Hagenbeck's assistants, noted, "It was touching when [Hagenbeck] spoke of his lioness 'Triest' who was devoted to him like a dog. . . . Never did Hagenbeck appear greater among his animals as when the royal lioness lay overcome at his feet, not as a slave, but as a friend beside a friend."[86] At Hagenbeck's grave, "Triest" is seen again lying across her master's tomb; a gigantic representation of fidelity and love.

The basic problem with this new image put forward by Hagenbeck, however, is that it remained consistently difficult for the company to portray itself as a protector of animals when its business rested so firmly on the capture and selling of whatever came into its hands. This point is made in a cartoon of 1893 showing a wagon jolting its way through a jungle as all manner of animals flee before it screaming "Hagenbeck Is Coming!" (figure 3) and two ostriches stick their heads in the ground. The cartoon was meant primarily to be humorous and, with the caricatured animals, most people today would probably agree that the illustration is still funny. At the same

Hagenbeck kommt!

Zeichnung von Hans Oberländer in den „Fliegenden Blättern" (1893).

3. "Hagenbeck Is Coming!" *Die fliegenden Blätter,* 1893. Reproduced in Carl Hagenbeck, *Von Tieren und Menschen* (Leipzig, 1908).

time, the satiric undertone of the picture—its playful reversing of the Noah's Ark motif—would have escaped few, and it was to this type of critique that Hagenbeck and his catchers began to respond. Indeed, throughout the world of Hagenbeck in the 1890s and first decades of the twentieth century, we see the company wrestling with a public image complicated by the very nature of its business. This struggle is apparent, for example, in the popular images of Hagenbeck and his walruses (figure 4)—images which have been brought forth repeatedly over the last eighty years to demonstrate Hagenbeck's love for his animals. The photographs present an elderly Hagenbeck surrounded by his charges who mob him for special treats as a crowd of onlookers observe with fascination and delight. As Zukowsky recalled, "Carl Hagenbeck could occupy hours with these enchanting walrus babies and the Park's visitors have often been able to see with their own eyes what a deep love of animals lay hidden in the Hanseatic businessman. Over and again he stroked the rough hide of the little ones with a loving hand, praised them, and gave them treats, while the thick bodies affectionately snuggled up close to him."[87]

In his own memoirs, Hagenbeck briefly explained the method used to capture the young animals: "In order to catch the young, it is usually necessary to kill the mother. One of the animals to be found at Stellingen was seized this way: the killed mother was pulled up close to the boat and all maintained a complete silence, until the young one came and climbed upon its dead mother's back. At that point it was, naturally, not difficult to overcome the unassisted young animal." In the end, according to Hagenbeck, "with the capture of the five young ones recently brought to Stellingen, sixty-eight other walruses were killed."[88] Even in this case where the captive walruses were essentially a by-product of the commercial hunt of the animals—the sixty-eight walruses were clearly not killed in order to catch the young—it remained tenaciously difficult for the company to dissociate itself from the commerce, death, and more general imperial arrogance at the heart of its business.

All of these points are driven home in one of the more remarkable works about a Hagenbeck animal catcher, Wilhelm Munnecke's 1931 *Mit Hagenbeck im Dschungel.* At its most basic, Munnecke's work poses as an account of the life and adventures of Carl Hagenbeck's half-brother, John, who lived a large part of his life in what was then known as Ceylon. In 1884, at about the age of eighteen, John Hagenbeck set out on his first trip to the East to organize an exhibition of people for his brother. More trips followed and by the end of the 1880s, John had settled in Colombo where, by the beginning of World War I, he had established himself as a successful entrepreneur

Fütterung der Walrosse.

4. "Feeding the Walruses." Reproduced with permission, Hagenbeck Archive.

whose wealth stemmed from coconut, rubber, and cocoa plantations, partnership in a number of short-lived companies, and also largely from catching animals in India, Sumatra, Java, and Ceylon.[89] John Hagenbeck published several volumes of his travels and adventures, and more were issued after his death.[90] Munnecke's account, however, published in 1931 but focusing on the years before World War I, though presenting itself as an account of the famous animal hunter, is at least partly fictional. Barely suggesting a time line (Munnecke brings forth a handful of dates which are usually vague and often inaccurate)[91] and depicting main characters and events which seem to be merely modeled after real people and occurrences, the book resembles more a historical novel than a memoir or record of someone's experiences in the wild—the staple form of the hunting chronicle.

Munnecke begins his work in the suffocating stillness of a midday jungle heat in which all life rests in the shade. "Not a blade of the thick, man-high grass," he writes, "shows a sign of life. Long, slender stands of bamboo rise menacingly like lances into the heights, protecting in an endless phalanx the hulking muddy primeval forest."[92] In this stillness, a bull elephant is banished from his herd for contesting the dominance of the lead male. Now, as he crashes through stands of thick bamboo, he becomes the "rogue" Munnecke describes as the greatest fear of the jungle. The elephant crashes through villages where the "huts of the natives rest peacefully on stilts under protecting palms" and where "children play naked with shaggy dogs" and "nothing disturbs the peace."[93] Children and huts are left crushed in the wake of the terror. In the end, the only force which can stop the mad elephant is a European—John Hagenbeck—and his gun. Munnecke describes the eventual confrontation:

> With an unholy, threateningly raised trunk, the Rogue rages out of the primeval forest and stands still in the middle of the path. . . .
>
> Bang! resounds a shot.
>
> The hollow echo is still making its rounds when the giant sinks forward on his knees. Breathing laboriously and hoarsely, he thrusts his powerful tusks into the ground, steadies for a few seconds and then pushes himself up again with his hind legs and tusks.
>
> Then a second shot cracks into the massive skull slightly behind the ear, and silently the colossus falls on its side.[94]

This "dethronement" of the "tyrant of the jungle" foregrounds recurring themes of Munnecke's work. Told through a disembodied voice able to follow animals into the jungle and describe their world through their eyes, the narrative focuses on the dramas in the lives of jungle animals and the place of the European hunter in those dramas. In the stories of Munnecke's *Mit Hagenbeck,* there is a pervading sense that things are rarely what they seem, that violence and death are always close at hand, and that the European is somehow both the deliverer and undoing of the jungle.

After the opening story of the elephant, Munnecke introduces us to John Hagenbeck as simply "the big European" who has followed his trackers to the den of a tigress. "'Don't shoot, whatever happens don't shoot!'" are the first words from Hagenbeck as the party observes the mother tigress leaving her lair. "'Without reason,'" he warns, "'one does not kill.'"[95] When the party is assured that the mother tigress is no longer in the vicinity, they approach the den and Hagenbeck sends his tracker in after the cubs: "Quietly he creeps deeper into the gradually widening passage. A lost, dull light breaks from above into the dark hole and plays on the floor in small moving specks. Quickly he sets to work, rushes outside with his booty, turns it over to the others, and returns back just as quickly. Yet another time he scurries through the dark passage, and yet one more time he grabs the whimpering, squirming golden flecks. Then the nest is empty."[96] With the cubs tucked into their clothes, the hunting party hurries back to their camp. Hagenbeck requests milk from a nearby village for his new charges, the hunters eat happily after their day's labor, and all retire for the night. Munnecke's text, however, returns to the tigress: "There, above in the cave, however, a mother moaned complainingly, a mother whose three children had been abducted by the cruel hands of humans. By humans against whom she had never done anything, whose path she had stepped aside from as soon as her nose caught wind of them. By humans, who, with malicious, refined, and ingenious weapons, attack, torment, jail, abduct, and in cowardly fashion kill the animal from behind."[97] After realizing what has happened to her "children," the mother tigress sets off in search of vengeance. Eventually, after a night spent watching the fires of humans, she begins a killing rampage against the natives. Then, still moved by her wrath, she attacks and kills the stabled horse of a Dutch surveyor, who is erecting a telephone wire through the jungle. Again alluding to the power of "civilization" intruding into the natural world, Munnecke describes how this surveyor not only "pulls the long thin wire which at every moment connects him out of the depths of the jungle with civilization,"[98] but also shoots the "the mother" the next night when she returns to feed on the horse. The following day, the tigress's corpse is brought to a village suspended on bamboo poles to the joy and relief of all. Meanwhile, Hagenbeck rears the young tigers to become performers for circuses.[99]

The unease with which Munnecke describes the death of the "mother tigress" is picked up repeatedly in *Mit Hagenbeck* as we watch the anguish and confusion of animals caught to be brought to Europe and, indeed, as we watch the slow and painful death of the jungle itself. Indeed, though the book is ostensibly intended to tell the heroic story of the animal catcher John Hagenbeck, it seems to be more about the death and suffering of animals. If the tone of the first half of the book, for example, is set by the story of an elephant gone mad who must be killed by Hagenbeck, the second half of the work—which focuses on Hagenbeck's adventures in Ceylon—begins with the story of another elephant, this one an ancient member of a herd dying a prolonged and agonizing death outside the view of man.

For weeks the ancient elephant had been trailing its herd, its withered trunk hanging to the ground while its five-foot-long tusks fatigued the huge head. Occasionally a sign of life could be seen in the ancient animal's trunk, but then it would soon fall lifeless once again, and though the old bull tries desperately to grasp at leaves with his mouth, he becomes weaker and hungrier as the days pass. "Hunger gradually overcomes the Old One," the narrator explains, "He does not get angry, he does not complain. All he wants is shade and cool. Protection from the scorching, draining sun." Not even a small pool can ease his pain, however, for he is unable to use his all but dead trunk to drink. "Alone he carries himself along the jungle path. Abandoned by his herd, abandoned by his huge strength—the fate of the old."

One evening, however, a cool breeze from a lake reaches the elephant who is standing quietly in the woods, and he breathes deeply. Knowing that a lake is near, the elephant begins to walk slowly to the water. "Through the thick hedge by the water," the narrator relates, "he forces his massive body, wades ponderously—his wide feet sink in the sludge at the bottom—up to his mouth in the water: water! Finally water!—And as if in a mating prayer, he closes his eyes reverently and drinks—drinks long and heavily." After taking his fill, however, the old elephant does not leave the water, but stands quietly, listening to "sweet, loving, tempting" voices. The old animal's life slowly ends:

> Life becomes easy and dreamlike for him. Herds, vast elephant herds roll by him in waves. He runs with them, falls in battle, for one more time hot blood courses through his trunk and he swings it out of the water—and once more a surge of wind streams through it.
>
> With the fleeing of his life, he trumpets out in a roar his last breath. Then he sinks into the water.[100]

Munnecke's stories are deeply imbued with a romantic foreboding, a sense that the essence of the jungle is embraced by tragedy and that civilization brought to the primitive forest carries an almost unavoidable destructiveness. Thus, while the Indian jungle is in some ways liberated by the power of John Hagenbeck (when he kills the rogue, for example), that jungle, as told in the story of the old, dying elephant, is itself fading away. *Mit Hagenbeck im Dschungel* is filled with animals who die in the wake of civilization, but whose deaths are never seen as heroic or somehow just. In an adventure to catch orangutans, for example, two adults are caught in a frenzy of fear and distress, but their "baby lay crushed under a branch" and the two survivors soon die of sorrow.[101] In a sense, Munnecke's hero represents the last stage in the development of Hagenbeck's animal catchers. From Casanova and Menges through Schomburgk and Schulz to the fictionalized "John Hagenbeck," we can trace the evolution of the modern animal catcher. By the time we reach Munnecke's "hero," the animal catcher's task has become increasingly precarious and the focus of accounts has moved away from the catcher to the animals themselves, and those animals are caught as much by the needs of civilization, science, education, and recreation as by some heroic catcher.

And Red Peter? Not fully remembering his capture, he explains:

> I belong to the Gold Coast. For the story of my capture I must depend on the evidence of others. A hunting expedition sent out by the firm of Hagenbeck—by the way, I have drunk many a bottle of good red wine with the leader of that expedition—had taken up its position in the bushes by the shore when I came down for a drink at evening among a troop of apes. They shot at us; I was the only one that was hit; I was hit in two places.

Red Peter was hit, he tells us, once in the face leaving the red scar which led to his name, and once "below the hip." Although, he acknowledges, he has been criticized in the press for doing so, Red Peter reports that he enjoys showing his public where the second shot hit him, insisting that nothing untoward is to be seen but "the scar made by a wanton shot."[102] Red Peter's desire to show his audience where he was hit by the second shot alerts us to a crucial point in his analysis of his capture: as brutal as the actual event may have been, and as brutal as his "middle passage" between the decks of the Hagenbeck steamer may have been, it is, in the end, his exhibition which makes sense of his capture. Thus, Red Peter's critique of his capture turns necessarily on his critique of his life as an exhibited object, as an artist, among the civilized. If, however, the expression "Hagenbeck Is Coming!" meant one thing in the far reaches of the globe where Hagenbeck's animal catchers plied their trade, the expression meant something altogether different in the places where Red Peter would eventually perform.

Notes

1. Franz Kafka, "A Report to an Academy," trans. Willa and Edwin Muir, in *Franz Kafka: The Complete Stories,* ed. Nahum N. Glatzer (New York, 1971), p. 250. While the debate over the evolution of humans from apes does play a role in the "Report," that debate, along with the issue of the lives of performing apes, provides only a general framework for Kafka to discuss his larger concern about the artist and his unappreciative and misinterpreting public. Thus, like Kafka's "The Hunger Artist," although the "Report" is based on the facts of a certain kind of exhibit—the trained ape that drinks, smokes, shakes hands, spits, etc.—of which there were many examples in the day, this story is not, in its most important sense, about an ape.

2. Kafka, "Report," p. 252.

3. Carl Zuckmayer, "Hamburg, Hafen, Hagenbeck," *Arche Noah* 1 (1949): 3. As quoted in Herman Reichenbach, "Carl Hagenbeck's Tierpark and Modern Zoological Gardens," *Journal of the Society for the Bibliography of Natural History* 9 (1980): 573, and Günter H. W. Niemeyer, *Hagenbeck: Geschichte und Geschichten* (Hamburg, 1972), p. 7.

4. The idea that the seal exhibition in 1848 represents the beginning of *Hagenbecks Tierhandel* has been the official story of and by the company since at least the 1880s. Whether March 1848 is an entirely accurate or useful date to mark the beginning of the company—the seals were shown as mermaids in much the same way as the elder Hagenbeck exhibited all manner of domestic "freaks" during the period—we should recognize that the date has a certain additional appeal for the company.

5. Heinrich Leutemann, who was acquainted with the elder Hagenbeck, gave the name of Carl Hagenbeck's father as "Claus Carl Gottfried" in his *Lebensbeschreibung des Thierhändlers Carl Hagenbeck* (Hamburg, 1887); Niemeyer in his *Hagenbeck: Geschichte und Geschichten* refers to the father as "Carl Gottfried Claus." Meanwhile, Erna Mohr in her article, "Das Geschlecht Hagenbeck," *Der zoologische Garten* 21, no. 1 (1954): 2–9, refers to him as Gottfried Clas Carl Hagenbeck. I will follow Mohr's order and Herman Reichenbach's good advice, and refer to the elder Hagenbeck, the father of Carl Gottfried Heinrich Hagenbeck, as Carl Hagenbeck Sr. (1810–87) and to the son as Carl Hagenbeck Jr. (1844–1913), or, more simply, Carl Hagenbeck.

6. Carl Hagenbeck, *Von Tieren und Menschen: Erlebnisse und Erfahrungen* (Leipzig, 1908), p. 31.

7. Although Carl had taken over most of the control of the business by 1860, he became the sole proprietor of Carl Hagenbeck's Handelsmenagerie in 1866.

8. Among the other English dealers, we should count Charles Rice, William Cross, and, later, J. D. Hamlyn. Other German traders were competing in this market as well, including August Fockelmann in Hamburg and the families of Reiche and Ruhe in Alfeld, the latter companies originating in the explosion in the canary trade by midcentury. For a brief treatment of the various dealers in Germany including Fockelmann, Reiche, Ruhe, Menges, Christiane Hagenbeck, and Heinrich Möller, see Ludwig Heck, *Heiter-ernste Lebensbeichte: Erinnerungen eines alten Tiergärtners* (Berlin, 1938), pp. 243–57; see also Richard Müller, "100 Jahre Tiergroßhandlung L. Ruhe/Alfeld," *Der Zoologische Garten,* n.s. 28, no. 2–3 (1963): 71–79. As much as Carl Hagenbeck may have been in competition with these other companies, especially the British ones, it is important to recognize that all the companies had occasion to work closely with one another. The London dealer Charles Rice, in fact, was married to one of Carl Hagenbeck's sisters, and letters between Charles Jamrach and the Hagenbecks—for example those from Dietrich to "Charles" from Suez on March 3, 1872, and from Zanzibar on March 14, 1873, and that from Carl on November 8, 1877—suggest decidedly friendly and supportive relations. Letters among the miscellaneous papers, Hagenbeck Archive, Stellingen.

9. Hagenbeck, *Von Tieren und Menschen,* pp. 58–59.

10. My thanks to Herman Reichenbach for alerting me to the Jamrach connection to the property on Spielbudenplatz discovered by Annelore Rieke-Müller and Lothan Dittrich.

11. Leutemann, *Lebensbeschreibung,* notes that the illustration and accompanying article resulted in a steep climb in the number of visitors coming to Hagenbeck's business (p. 19).

12. For more about Adams and his bears, see Joanne Carol Joys, *The Wild Animal Trainer in America* (Boulder, Colo., 1983), pp. 10–11.

13. Hagenbeck, *Von Tieren und Menschen,* p. 60.

14. Perhaps the most notable case of the Hagenbecks becoming involved in colonial issues was the German government's contract for two thousand dromedaries—with saddles—to be delivered to German Southwest Africa to help in the war against the Herero. Hagenbeck, *Von Tieren und Menschen,* pp. 377–86.

15. Hagenbeck, *Von Tieren und Menschen,* p. 64.

16. Ibid.

17. Ibid., p. 66. According to the bill, handwritten on a small piece of paper and dated June 19, 1870, Casanova's shipment was sold to Hagenbeck for £2,750 and consisted of some 128 animals—excluding small birds and milk goats and including nine giraffes, five elephants, two buffaloes, five lions, four leopards, two cheetahs, twenty-five hyenas, fifteen monkeys, twelve cranes, etc. Miscellaneous papers, Hagenbeck Archive, Stellingen.

18. Hagenbeck, *Von Tieren und Menschen,* p. 74.

19. Ibid., p. 77.

20. In a letter to me of November 4, 1998, Herman Reichenbach states that Lothar Schlawe has corrected an earlier statement by Reichenbach in "Carl Hagenbeck's Tierpark," p. 574, and by Heinz-Georg Klös in *Von der Menagerie zum Tierparadies—125 Jahre Zoo Berlin* (Berlin, 1969), p. 65, that Hagenbeck's single shipment doubled the species and tripled the specimens.

21. Hans Schomburgk, an elephant hunter who later became an animal catcher for Hagenbeck and, later still, a wildlife photographer, published his adventures in many books, including *Wild und Wilde im Herzen Afrikas* (Berlin, 1910); *Bwakukama: Fahrten und Forschungen mit Büchse und Film im unbekannten Afrika* (Berlin, 1922); *Von Mensch und Tier und etwas von mir* (Berlin, 1947); *Der Kleine Jumbo, das Schicksal eines afrikanischen Elefantenbabys* (Hanover, 1952); and *Zelte in Afrika: Fahrten—Forschungen—Abenteuer in sechs Jahrzehnten* (Berlin, 1960). Ole Hansen was a Norwegian walrus hunter who had been active in the Arctic since 1886 and who became Hagenbeck's main source for walruses beginning around 1906.

22. For further information on Schillings see Ludwig Heck, "Die Schillingsche Sendung Deutsch-Ostafrikanischer Tiere," *Die Gartenlaube* (1900): 562–64; C. G. Schillings, *Flashlights in the Jungle,* trans. Frederic Whyte (New York, 1905); *Der Zauber des Elelescho* (Leipzig, 1906); and *In Wildest Africa,* trans. Frederic Whyte (New York, 1907).

23. The works of Hans Dominik include *Kamerun: Sechs Kriegs- und Friedensjahre in deutschen Tropen* (Berlin, 1901); and *Vom Atlantik zur Tschadsee: Kriegs- und Forschungsfahrten in Kamerun* (Berlin, 1908).

24. See John Hagenbeck, *John Hagenbecks abenteuerliche Flucht aus Ceylon: Meine Ausweisung aus Ceylon und Flucht nach Europa* (Dresden, 1917); *Fünfundzwanzig Jahre Ceylon: Erlebnisse und Abenteuer im Tropenparadies,* ed. Victor Ottmann (Dresden, 1922); *Kreuz und quer durch die Indische Welt: Erlebnisse und Abenteuer in Vorder- und Hinterindien, Sumatra, Java, und auf den Andamanen* (Dresden, 1922); *Südasiatische Fahrten und Abenteuer: Erlebnisse in Britisch- und Holländisch-Indien, im Himalaya und in Siam* (Dresden, 1924); the pamphlet *Gebr.-Hagenbeck's Indische Carawane* (Hamburg, n.d.), Jacobsen-Nachlaß, Hamburgisches Museum für Völkerkunde; and Wilhelm Munnecke, *Mit Hagenbeck im Dschungel* (Berlin, 1931).

25. For more about Casanova, see, in addition to Carl Hagenbeck, *Von Tieren und Menschen,* and the works of Leutemann and Niemeyer, H. Dorner, "Casanova und Hagenbeck," *Die Gartenlaube* (1869): 42–47, and Wilhelm Fischer, *Aus dem Leben und Wirken eines interessanten Mannes* (Hamburg, 1896). Beyond the biographies of Hagenbeck and the histories of the company little can be found about Bernard Cohn (a.k.a. Bernhard Kohn). See, in addition, however, "Afrikanische Thier-Expedition," *Die Gartenlaube* (1876): 844, and *Carl Hagenbeck's Thier-Karawane aus Nubien* (Hamburg, n.d.), Jacobsen-Nachlaß, Hamburgisches Museum für Völkerkunde. A number of letters can be found pertaining to Josef Menges in the Hagenbeck Archive, the Jacobsen papers of the Hamburgisches Museum für Völkerkunde, and in various zoo archives, such as the archives of the Bronx Zoo. Menges also made regular contributions between 1881 and 1891 to *Petermanns Mittheilungen* about the geography of northeastern Africa. See also *J. Menges' Ost-Afrikanischen Karawane aus dem Somalilande* (Hamburg, n.d.), Jacobsen-Nachlaß, Hamburgisches Museum für Völkerkunde.

26. The works of Christoph Schulz include *Auf Großtierfang für Hagenbeck: Selbsterlebtes aus afrikanischer Wildnis* (Dresden, 1922), and *Jagd- und Filmabenteuer in Afrika: Streifzüge in das Innere des dunkeln Erdteils* (Dresden, 1922). His wife, Elisabeth Schulz, published a first volume of "fiction" about their adventures in Africa entitled *Afrikanische Nächte* (Hamburg, 1926), but the projected second volume never appeared. For further information on Johannsen and Grieger, see Hagenbeck, *Von Tieren und Menschen,* and Niemeyer, *Hagenbeck: Geschichte und Geschichten.*

27. Schomburgk arrived in South Africa with one thousand marks to start a farm, but spent the money immediately, he tells us, to outfit an expedition to the interior.

28. The relatively high capital outlays involved in this business, combined with the apparently conservative purchasing patterns of the directors of the European and American zoological gardens (these were people who liked to deal with "reputable" companies), help explain why Menges, despite years of effort, did not succeed in offering a competitive alternative to Hagenbeck. Although by the 1890s he was sending out printed price lists for animals he had brought back from North and East Africa, he found it difficult to sell his stock more cheaply than Hagenbeck and consistently complained in correspondence that although he could bring many more animals back, he found it difficult to find buyers for them. In a letter to Johan Jacobsen of July 27, 1894, for example, Menges writes that although he has just returned from three months in Somalia, having brought back a great many animals, "unfortunately the sale of the animals has gone poorly, even worse than last year. The gardens have no money and do not want to

buy anything. The last lot of animals are dying miserable deaths instead of being sold, the Berlin and Dresden gardens, for example, haven't spent a penny." Referring to the year before in which he had lost a substantial amount of money at the World's Fair in Chicago where he was trying to sell ethnographic artifacts, Menges laments, "If this goes on, this year will be even worse than the last in Chicago, in which I was completely devastated" (Jacobsen-Nachlaß). Given the strength and "market share" of the firm of Carl Hagenbeck, it is simply hard to imagine that Menges could compete in the large animal business in the 1890s in Germany. Only a company with a clearly established presence in the national and international market such as that of Ludwig Ruhe in Alfeld could hope to take significant business away from Hagenbeck in the large animal market.

29. In his introduction to C. G. Schillings's popular version of his *Mit Blitzlicht und Büchse im Zauber des Eleléscho* (Leipzig, 1910), Theodore Roosevelt described Schillings as "a mighty hunter and explorer, a trained field naturalist, an exceedingly interesting writer, and the most noted of nature-photographers." "His book," Roosevelt concluded, "should be circulated everywhere and translated into every language spoken by those men who love the wilderness, and the beasts of the wilderness, and the wild, hardy life of the big game hunter."

30. For a description of the Schillings hunt, see C. G. Schillings, *Mit Blitzlicht und Büchse,* p. 110. For descriptions of the Hagenbeck expeditions to Mongolia, see, among the many accounts, Hagenbeck, *Von Tieren und Menschen,* pp. 212–28.

31. Based on internal evidence, the pamphlet *Carl Hagenbeck's Thier-Karawane aus Nubien* was most likely published in 1876, and referred to a shipment organized by Bernard Cohn. This pamphlet, along with others, including Josef Menges's *Ostafrikanische Karawane aus dem Somalilande* (Hamburg, n.d.), and some private correspondence were used to complete sections in Hagenbeck's *Von Tieren und Menschen.* The whole story in Hagenbeck's memoir of how the Hamran hunters captured elephants with swords, for example, is taken from *Carl Hagenbeck's Thier-Karawane aus Nubien.* Copies of these brochures can be found in the Jacobsen Nachlaß at the Hamburgisches Museum für Völkerkunde.

32. *Carl Hagenbeck's Ther-Karawane,* p. 11.

33. Ibid., p. 12.

34. Ibid., p. 14. The "Nubian caravan" organized by Bernhard Cohn and brought back by Hagenbeck in 1876 was exhibited in Hamburg, Düsseldorf, Breslau, and Paris, attracting over thirty thousand visitors in a single day in Breslau; the "Nubian caravan" of 1877–78, again organized by Cohn, traveled to Hamburg, Berlin, Frankfurt, Dresden, and London. Josef Menges

also organized a "Nubian" show in 1878. For more information about the Hagenbeck Völkerschauen, see Hilke Thode-Arora, *Für fünfzig Pfennig um die Welt: Die Hagenbeckschen Völkerschauen* (Frankfurt, 1989).

35. Josef Menges, "Bemerkungen über den deutschen Thierhandel von Nord-Ost-Afrika," *Der Zoologische Garten* 17 (1876): 231–32.

36. It should be noted that Menges had an interest in making the trade seem easier than it probably was. With Hagenbeck controlling more and more aspects of the business, and Menges apparently hoping to become independent of the Hagenbeck operation, it was important for Menges to present the procuring of animals as a relatively easy affair so that he could convince the generally risk-averse zoological gardens to bankroll his future plans.

37. Thanks are due to the helpful suggestions of Endre Stiansen regarding potential comparisons between the animal trade and the gum Arabic and ivory trades. See Endre Stiansen, *Prelude to Imperialism* (Bergen, 1993).

38. For a more complete discussion of the credit systems in West Africa, see Colin Newbury, "Credit in Early Nineteenth Century West African Trade," *Journal of African History* 13, no. 1 (1972): 81–95. Again thanks to Endre Stiansen for pointing out the more subtle aspects of this issue and referring me to Newbury.

39. Menges's "Bemerkungen," for example, is at least partly an attempt to encourage potential investors, including zoological gardens, to fund a venture to focus on importing northeast African bird life.

40. Menges lists the prices of animals in Kassala in Maria Theresia talers (1 taler = 4 marks), and provides the following table:

	in Kasala:	in Europe:
1 Elephant	20–100 = 80–400 marks	3,000–6,000 marks
1 Giraffe	20–50 = 80–200	2,000–3,000
1 Rhinoceros	40–100 = 160–400	6,000–12,000
1 Antelope	5–30 = 20–120	400–3,000
1 lion, young	2–20 = 8–80	600–2,400

See Menges, "Bemerkungen," p. 232.

41. Figures given ibid., p. 233.

42. I have not simply imagined this shipment. Hagenbeck's 1870 Casanova/Migoletti importation included, in fact, one rhinoceros, five elephants, fourteen giraffes, twelve antelopes, and seven young lions, and further two warthogs, four aardvarks, four wild Nubian buffalo, fifty-three large and small beasts of prey (excluding the seven lions), twenty-six African ostriches (among which were sixteen full-grown birds), and twenty large crates with

monkeys and birds. The "Nubian caravan" described above represented an even more valuable shipment, consisting of three rhinoceroses, fourteen elephants, nine giraffes, fifteen antelopes, and seventeen ostriches.

43. Leutemann, *Lebensbeschreibung,* pp. 33–34.

44. It is, of course, likely that Casanova and the other early traders occasionally caught animals themselves. All the information about the early hunters suggests, however, that they rarely ventured after animals and relied on indigenous hunters who, naturally, would have had an idea about which animals were of most interest to the company—at this time, elephants, giraffes, and ostriches. Dietrich Hagenbeck's attempt to catch hippos is the earliest instance of the company sending out a hunter after a particular animal which I have been able to uncover. To give a sense of the relative value of hippopotamuses in the period, an 1883 price list from Hagenbeck shows a six-year-old female hippo as the most expensive animal in Hagenbeck's collection selling for eighteen thousand marks, followed by a seven-year-old male African rhinoceros for twelve thousand marks.

45. See Diederich Hagenbeck, letter to Charles [Jamrach], March 14, 1873, Hagenbeck Archive, Stellingen.

46. Dietrich Hagenbeck, letter to Carl Hagenbeck, April 29, 1873, published as correspondence under the title, "Ein junger Nilpferdjäger: Dietrich Hagenbeck," *Die Gartenlaube* (1873): 754.

47. Dietrich Hagenbeck, letter to Carl Hagenbeck, June 25, 1873, ibid.

48. Thanks to Edward Steinhart for identifying "blackwater fever."

49. *Die Gartenlaube* (1873): 754.

50. For discussion of Dominik's earlier animal collections, see *Vom Atlantik zur Tschadsee: Kriegs- und Forschungsfahrten in Kamerun* (Berlin, 1908). Dominik relates the Heck dialogue in *Kamerun,* p. 240.

51. Both Hans Schomburgk and C. G. Schillings noted Dominik's accomplishment with admiration. Schillings, also at the instigation of Heck, tried repeatedly to catch and then keep alive a baby elephant from German East Africa. In each case where he succeeded in catching a calf, it soon died, however. See C. G. Schillings, *Mit Blitzlicht und Büchse,* pp. 55–56.

52. Dominik, *Kamerun,* p. 242.

53. Ibid., p. 244.

54. Ibid., p. 248.

55. Ibid., p. 252.

56. It is unclear how many of the young elephants survived over the longer term. Dominik writes that three survived, but Schillings notes that only one animal survived—the one which eventually arrived in Berlin. By the end of this episode in his life, Dominik had become convinced that Europeans could look forward to a future in which tamed elephants could

be used in Africa as they were in India. He concluded: "One hears time and again in discussions of the thousands of tons of ivory which are shipped yearly from Africa and more specifically from the Congo, the anxious question, when will the last elephant in Africa be shot? Even if it will be a good while until then, it would be greeted with great satisfaction if the last wild elephant were to be carried to his grave by his domesticated brothers" (ibid., p. 253).

57. Leutemann, *Lebensbeschreibung,* p. 29.

58. Schomburgk, *Wild,* p. xiii. An exception that he notes was the necessity of killing a mother elephant in his eventually successful attempt to procure a baby elephant for the Zoological Gardens in Berlin.

59. Ibid., p. 121.

60. Ibid., p. 77.

61. Ibid., p. 278.

62. Ibid. The picture has a montage quality to it, as if Schomburgk had pasted a photo of Jumbo in front of a photo of a dead cow elephant. Without examining the original plate, however, we should avoid speculation.

63. See Schulz, *Großtierfang.* Once the connection to the Hagenbecks was made, Schomburgk mounted special expeditions for the company, the most important being his successful first capture and transport to Europe of the pygmy hippopotamus from Liberia in 1912. This adventure is described in detail in Hans Hermann Schomburgk, *Pulsschlag der Wildnis* (Berlin, n.d.). See also "Hagenbecks Zwergflußpferd," *Zoologischer Beobachter* 53 (1912): 132–34, 208–10, 338–39, and discussion below.

64. Schomburgk, *Wild,* p. 319. Schomburgk's interest in Jumbo continued long after their reunion in Stellingen. Some forty years after the appearance of his *Wild und Wilde,* Schomburgk published a rather poor but, in this context, interesting children's story centering on the adventures of the young Jumbo. Partly lifted from his earlier memoir, partly imagined, and partly combined with material unrelated to Jumbo or his capture, *Der kleine Jumbo* is more a curiosity than a work of historical interest or literary quality. This is not the place to go into depth about this light book, but it is important to note that, as with most of the stories in this chapter, the "fate" of "little Jumbo" as it was painted in Schomburgk's stories, was a mixture of pain over the past, and joy and excitement about a new adventure in a different land. But even in a children's story the central fact of the killing of Jumbo's mother was, however unpleasant, simply inescapable. The best that the hunter could do in his memoirs and memory, it seems, was to describe the confrontation with and death of the mother—despite his intentions from the beginning—as being somehow completely unavoidable, as almost an

accident, and to show that Jumbo's new life was certainly as good as, if not better than, his old one.

65. Schomburgk, *Von Mensch und Tier,* p. 39.

66. Schomburgk writes, for example, "Nature doesn't recognize sentimentality. If an animal is driven out, it is meant to be. In most cases, a sick animal will leave its herd of its own will. Animals understand how to die without making a fuss. The inner natural instinct teaches them that their individual lives are meaningless, but that their death can be useful to the community, that their death serves the species. This is so because sick or weak animals can bring the downfall of the entire herd. As in a chain, which is also only as strong as its weakest link, the weakest member of an animal community must always be at least strong enough to meet fully all the demands of nature" (ibid., p. 110).

67. On the death of Major Philip Pretorius, who had killed some 555 elephants during his career, Schomburgk referred to "English statistics" (figures he habitually brought up in his books, which suggested that an elephant hunter could expect to live only two years before being trampled by an elephant, murdered by the natives, fatally bitten by a snake, or perishing from malaria or some other tropical disease. Returning to Pretorius, Schomburgk writes: "555 elephants! Who can possibly understand what that means! Hunters will marvel at this Nimrod, 'only' friends of animals will damn him. I say 'only' intentionally, because every true hunter is also a friend of animals, and by the 'only' I refer to those enthusiastic and sensitive friends of animals who are not hunters and who, in their admirable eagerness, carelessly overshoot their target" (ibid., p. 121).

68. Elizabeth Hanson, in her 1996 University of Pennsylvania dissertation on the American animal trade and American zoological gardens, "Nature Civilized: A Cultural History of American Zoos," insightfully describes Frank Buck, one of the most famous American animal catchers of the twentieth century, not as a heroic animal catcher per se—especially since he himself rarely participated in the actual capture of the animals—but rather as a type of heroic businessman who struggles, under the most adverse conditions, to get his animals to market.

69. William T. Hornaday to Carl Hagenbeck, June 11, 1902, Outgoing Correspondence, Director's Office, New York Zoological Park, The Wildlife Conservation Society.

70. See Renate Hücking and Ekkehard Launer, *Aus Menschen Neger machen: Wie sich das Handelshaus Woermann an Afrika entwickelt hat* (Hamburg, 1986), p. 87. In this regard, Dominik seems to be a good match to Cameroon's governor at the time, Jesko von Puttkamer—a man sympathetic to the needs of German merchants for increased control in the colony who

described the Duala in his memoirs as "the most lazy, false, and base rabble the sun ever shone upon, and it would certainly have been better if, during the conquest of the land in 1884, they had at least been kicked out of the country if not exterminated" (ibid.).

71. Helmuth Stoecker, in his *Kamerun unter deutscher Kolonialherrschaft,* vol. 1 (Berlin, 1906), provides a fairly gruesome catalogue of atrocities visited upon the indigenous peoples of Cameroon (pp. 196–221).

72. Hücking and Launer, *Aus Menschen Neger machen,* pp. 89–90.

73. Hücking and Launer excerpt, for example, a 1907 article in the newspaper *Kolonie und Heimat*: "The honest citizen is overcome by a slight horror when he reads about corporal or chaining punishments in the colonies. He is inclined to regard them as relics of a sinister barbarism and the perpetrators who are forced to use these forms of punishment as inhuman" (ibid., p. 88). The article, however, goes on to explain that while these types of punishment are naturally abhorred by civilized people both at home and in the colonies, they remain the only truly efficacious methods to discipline indigenous peoples.

74. Josef Menges, for example, describes the Somalis as "unbathed, shameless, brazen, and presumptuous savages" in his letter to Johan Jacobsen of March 30, 1891. Jacobsen-Nachlaß, Hamburgisches Museum.

75. This and the next quotation are taken from the signed unpaginated typescript sent by Hans Schomburgk to William Hornaday and housed with the Director's Incoming Correspondence at the New York Zoological Society archives. The published versions of the story received only minor orthographic and grammatical alterations.

76. My thanks to Harold James for bringing to my attention the history of the Steiff stuffed animal business. The success of the various manufacturers of stuffed animals in the late nineteenth century, of which the German company Steiff is only the most prominent, is, of course, closely connected to the growth of zoological gardens in the period. John Berger's observation that "the manufacture of realistic animal toys coincides, more or less, with the establishment of public zoos" can be extended, however, to include a number of other concurrent sentimental developments including the literary/scientific controversy known as the "nature faker" debate that raged in North America at the turn of the century. Focusing on highly anthropomorphized natural historical writing, the debate featured John Burroughs and Theodore Roosevelt on one side and William Long and Ernest Thompson Seton on the other in an often furious discussion about the oversentimentalization of animals in works of natural history. See John Berger, "Why Look at Animals?" *About Looking* (London, 1980), pp. 19–20, and Ralph H. Lutts, *The Nature Fakers: Wildlife, Science, and Sentiment* (Goldon, Colo., 1990).

77. Colonel Richard Meinertzhagen, *Kenya Diary, 1902–1906* (Edinburgh, 1957), p. 178.

78. Friedrich Knauer, *Der zoologische Garten: Entwicklungsgang, Anlage und Betrieb unserer Tiergärten* (Leipzig, n.d. [?1914]), pp. 117–18.

79. See Christoph Schulz, *Aus Hagenbecks Jagdgründen: Abenteuer eines Tierfängers in den Steppen und Urwäldern Afrikas* (Dresden, 1922). Schulz's *Jagd- und Filmabenteuer in Afrika,* separately published in 1922, focuses on one adventure during the 1910–13 period described in *Großtierfang* when he shot some nine thousand meters of film in Africa to be shown in the theater at the Stellingen Animal Park.

80. Schulz, *Großtierfang,* p. 9.

81. Schulz's work does, of course, pause for moments of more poetic description; in general, though, the work has few literary aspirations. In a description of a particularly striking morning, for example, Schulz's narrative quickly develops into the predictable: "On the next morning I was astonished at the beginning of dawn by a spectacular natural display. All was still covered in darkness, then I saw in the far distance the two snow-covered summits of Kilimanjaro, the Kibo and Mawenzi, light up in the first rays of the climbing sun. As if spellbound I stood and enjoyed the splendid scene. Gradually it became lighter and the outline of the huge and imposing mountain became more and more clearly visible. Over a wide area the two summits, which jutted high above a layer of clouds, glimmered. For the first time I saw the giant of the African mountains and my heart rejoiced in bliss over the overpowering and magical splendor which this view offered to me. Over there, then, lay German territory, and with pleasure I greeted the king of the German East African mountain world, which the preceding day had been covered in mist" (*Großtierfang,* p. 78–79).

82. Carl Hagenbeck to William T. Hornaday, February 3, 1913, Incoming Correspondence, Director's Office, New York Zoological Park, The Wildlife Conservation Society. Hagenbeck typically corresponded with his English-speaking clients in English—even if his language was sometimes awkward.

83. Schulz, *Großtierfang,* p. 90.

84. Ibid., p. 92.

85. I am not able to write at length here about Hagenbeck's Animal Park, which is discussed in my book *Savages and Beasts: The Birth of the Modern Zoo* (Baltimore: 2002). I will note here, though, that Ludwig Zukowsky in his biography of Hagenbeck, *Carl Hagenbecks Reich: Ein deutsches Tierparadies* (Berlin, 1929), writes that "Hagenbeck called his park an animal paradise. . . . One could more justifiably compare it with that paradise of which we are told in Genesis where animals lived beside one another in

peace and where the struggle for existence was nonexistent" (p. 58). In the early years of the Animal Park, Hagenbeck and his promoters tended to conflate several images of the gardens. The image which first came to mind—drawn from Hagenbeck's striking panoramas of diverse animals, including large carnivores and herbivores, living beside each other—was of the Paradise where the animals lived in peace. This image, and my thanks here to Gavin Lewis for pointing this out, was quickly complicated, however, by the image of the lion laying down with the lamb, thus referring to the Kingdom of God here transposed as the Kingdom of Carl Hagenbeck (*Hagenbecks Reich*). In fact, in Hagenbeck's panoramas, lions actually appeared to rest comfortably beside zebras, antelopes, and water fowl. Finally, the Paradise and Kingdom motifs were soon dominated by a third idea of the Park as an Ark—a step made more readily by the image of Hagenbeck, himself: a sort of white-haired, congenial-looking Noah figure.

86. Zukowsky, *Carl Hagenbecks Reich,* p. 62.

87. Ibid., p. 234. It is important, however, to note that although Zukowsky argues that the Hagenbeck family rules insist that in the effort to catch animals, killing other animals should at all costs be avoided, Heinrich and Lorenz Hagenbeck had stepped up the importation of the walruses, receiving ten in 1928; and Zukowsky himself notes that the method of catching the animals was virtually unchanged since the nineteenth century. He writes: "After a mother with a catchable young is sighted, the old one is either driven onto the land or harpooned from the ship and then killed with a shot to its head; the young one is then caught with a lasso. Touching scenes take place between the dying mother walrus and her baby. Until the last moment the old lady attempts to bring her baby to safety. Because the young walrus itself does not leave its dead mother, it is easy to pull it by its flippers on board or catch it with a lasso from the boat. Young walruses become accustomed relatively easily to fish for food, so it is not especially difficult to keep the valuable animals alive" (ibid., pp. 148–49).

88. Hagenbeck, *Von Tieren und Menschen,* p. 414.

89. John Hagenbeck, like his brother, was always ready, it seems, to put his not inconsiderable talents to diverse enterprises. According to his own account, he first went to Ceylon in 1884 to organize a *Völkerausstellung* for his brother, although there remains some confusion about which if any show he helped to organize (see, for example, Thode-Arora, *Für fünfzig Pfennig um die Welt,* p. 46). After various activities for the company, including assisting his brother-in-law Heinrich Mehrmann with the Hagenbeck circus and helping with various *Völkerausstellungen,* John settled in Colombo. His first business interests centered around the pearl fishery, but there he met with little success (*Kreuz und quer durch die Indische Welt,* p. 10). Frustrated in the

pearl business, John Hagenbeck took a position with a ship chandlery and before long was turning his hand to a wide variety of activities. Of primary interest throughout remained the catching and shipping of exotic animals to Europe and America, but he remained consistently involved in organizing *Völkerausstellungen* as well. With his brother Gustav, moreover, he founded an exporting company concerned chiefly with tea. He fled Ceylon with the outbreak of World War I, leaving his wealth and possessions behind, and worked beside his nephew Lorenz in Stellingen for a short while before becoming involved in a movie house in Berlin. He first returned to Ceylon in 1920, traveling there more regularly beginning in 1924 to organize *Völkerschauen.* According to Thode-Arora, he returned with his family to Ceylon in 1927, regaining a position in the ship chandlery and reestablishing himself as a planter. He died in British internment during World War II on December 16, 1940.

90. See above, note 24.

91. Munnecke notes, for example, that John Hagenbeck went to India in 1885 to catch animals, it seems, and that in the 1920s he did not return to the Indian subcontinent before May 1929; *Mit Hagenbeck im Dschungel,* pp. 73–74, 201.

92. Ibid., p. 9.

93. Ibid., p. 15.

94. Ibid., pp. 58–59.

95. Ibid., p. 17.

96. Ibid., p. 18.

97. Ibid., pp. 19–20.

98. Ibid., p. 23.

99. Munnecke tells a longer story of the most famous of the three, Radja. Despite being a completely trustworthy performer, and indeed a sensation since he was never shown behind bars, Radja one day accidentally killed a monkey. Having tasted blood and having thereby heard the atavistic "call of the wild," Radja, the Indian prince, could no longer be trusted by the circus owners and was forced to live out his days behind bars. The motif here of the Indian prince who can only partly be civilized is taken up elsewhere in the work (ibid., pp. 129–32).

100. Ibid., p. 93.

101. Ibid., pp. 78–79.

102. Kafka, "Report," p. 251.

Part Three

ANIMALS EXHIBITED: THE ZOO

7

Constructing the Zoo

Science, Society, and Animal Nature at the Paris Menagerie, 1794–1838

Richard W. Burkhardt Jr.

The first public zoo of modern times was the menagerie of the Muséum National d'Histoire Naturelle in Paris.[1] Founded at the height of the French Revolution, it was championed as having not only great scientific potential but great social, political, and moral potential as well. Naturalists and political leaders alike projected it as a site where citizens and savants would be able to learn from the natural and extend the range of the possible.

Theory, however, tends to be simpler than practice. The human actors whose interests intersected at the menagerie were a diverse lot. The animals of the menagerie were more heterogeneous still. And neither the humans with their own particular stakes in the menagerie nor the animals confined to their respective pens and cages proved entirely malleable when others sought to script their behavior. The Paris Muséum was the material and intellectual center of the greatest ordering enterprise in the natural sciences in the first half of the nineteenth century. The menagerie in the meantime was an arena where the potential for disorder always lurked close beneath the outward and visible signs of domination and control.

Our aim here is to explore the role the menagerie played in the ongoing life of the Paris Muséum. Doing so brings us face to face with the problematic place that *live* animals occupied in the study of *life* in the late eighteenth and early nineteenth centuries. Historians and philosophers of science have had more than a quarter of a century to grapple with Michel Foucault's provocative suggestion that the concept of life did not exist before the nineteenth century,[2] but the relation of the general concept of life to the daily practices of the naturalists who developed it bears further investigation. When Georges Cuvier and J.-B. Lamarck sought to comprehend life's manifestations across the whole sweep of the animal kingdom, they almost invariably did so using specimens that were dead. And yet the era of Cuvier and Lamarck was the same period that saw at the Muséum the development of a novel institution—the menagerie. The menagerie literally brought life to the Muséum in a way that had not been experienced at the old Jardin du Roi (also known as the Jardin des Plantes), the Muséum's predecessor. It is commonly acknowledged that the menagerie was the part of the Muséum that proved most attractive to the general public. But what sort of difference the menagerie made to the zoologists of the Muséum is another question.

We will begin by addressing the different functions and interests that the menagerie was conceived to be capable of serving. We will then consider the circumstances under which these early expectations were realized or restructured in the course of practice. Then we will look specifically at how Frédéric Cuvier (the younger brother of the famous comparative anatomist) sought in the newly created role of *garde de la ménagerie* to found a science of animal actions based on a study of the animals in his charge. The younger Cuvier's experience underscores the precarious role that living animals played in French zoology—and that the menagerie played in the life of the Muséum—during the golden age of French zoology. Furthermore, the study of the menagerie's first half-century allows one to see how the animals of the menagerie, symbolic as they may have been of France's imperial reach in particular and of the dominance of humans over the animal world in general, could still offer resistance to the plans that humans conceived for them.

The menagerie came into being, some five months after the Muséum itself was established by decree of the Convention of June 10, 1793, as the result of contingent events that were neither anticipated nor necessarily desired by the new professors of the new institution. On November 3, 1793, the Conseil général of the Commune de Paris decided that the lions and leopards and other live animals that were being exhibited on the streets of Paris were both a public danger and, in their cages and chains, an affront to republican sensibilities. The police were instructed to confiscate these animals and transport them to the Jardin des Plantes (i.e., the new Muséum). It came as a complete surprise to the professors of the Muséum the following day when the police appeared at the gates of the Jardin des Plantes with a polar bear, a panther, a civet cat, and a monkey, accompanied by the former owner of the animals, who was demanding to be compensated for them and requesting in addition that he be hired to continue to take care of them. Two days later, the professors found themselves facing eleven more animals and two more former owners, who were retained to care for the animals. The animals were kept in their cages, which were placed on carts and sheltered in an empty carriage house.[3]

The issue of whether there should be a menagerie attached to the Jardin des Plantes was not new. Though Buffon in 1771 had stated that most animals confined in menageries displayed behavior that was "altered, constrained, and scarcely worthy of the attention of the philosopher," he himself in the following decade attempted to have the animals of the menagerie at Versailles transferred to the Jardin des Plantes.[4] Buffon was unsuccessful in this effort, but with the coming of the Revolution, the existence of the king's menagerie at Versailles reemerged as an issue. It was this that led Bernardin de Saint-Pierre in 1792 to compose his "Memoir on the Necessity of Adding a Menagerie to the National Botanical Garden of Paris."[5]

Saint-Pierre identified a variety of reasons—scientific, political, artistic, economic, expedient, and even personal—for establishing a menagerie at the Jardin des Plantes. On the personal side, he believed that attaching a menagerie to the Jardin des Plantes was the single greatest contribution that he, as a nonscientist, could make as Buffon's successor at the institution. On the scientific side, he felt that a menagerie would allow natural history to escape its domination by comparative anatomy. The latter, he maintained, was not in itself sufficient for the understanding of animals, and particularly when it came to shedding light on *human* needs and existence, it was not nearly as important as "the study of their tastes, their instincts, and their passions." These were the topics, he claimed, that had made the writings of Buffon so interesting, "not just to savants, but to everyone."[6]

On the political side, Saint-Pierre believed, a menagerie was necessary "for the dignity of the nation": that is to say, France needed a menagerie in order to be able to accept properly the live animals sent as gifts by non-European rulers. "What will we say to the powers of Africa and Asia, which, from time immemorial, have made a custom of presenting us gifts of animals? Are we to kill the animals, and make skeletons from them? That would be an insult. Are we to refuse them, saying that we do not have the means to lodge and feed them? Our political relations require the existence of a menagerie."[7] At home, moreover, a menagerie would be a means of fostering political and social stability in Paris. Saint-Pierre's argument was that the Jardin des Plantes was located in the faubourg Saint-Marceau, the poorest district of Paris, and social unrest was most likely when very poor districts existed side by side with very rich districts. The placement of a menagerie in the faubourg Saint-Marceau, he claimed, would benefit the district economically, reduce economic inequities across Paris as a whole, and thus encourage social and political harmony.[8]

Among Saint-Pierre's additional arguments were the advantages a menagerie would provide on the one hand for the liberal arts and on the other hand for *l'économie rurale.* It would give artists, poets, and writers the opportunity to base their works on living animals. Agriculture in the meantime would benefit from the opportunities the menagerie would provide for naturalizing animals, ameliorating their natures, and crossing them with domestic races.

Saint-Pierre's memoir became the subject of a report presented to the Société d'histoire naturelle de Paris by a committee of that society in December 1792.[9] The three members of the committee, A. L. Millin, P. Pinel, and Alexandre Brongniart, strongly endorsed the establishment of a menagerie, while taking care to distinguish the kind of institution they had in mind from the costly and useless menageries of kings. They argued that natural history, physiology, and agriculture would all benefit from the establishment of "une ménagerie sans luxe" in Paris. Their report reinforced many of Saint-Pierre's claims and added several new points as well.

Millin, Pinel, and Brongniart asserted that "a menagerie would advance natural history from two points of view: the observation of the distinctive characters of animals, and the study of their behavior [*mœurs*]." Such an institution would in fact bring a new dimension to the continuing competition between field naturalists and cabinet naturalists. Field naturalists would no longer be able to claim to be the only ones to have viewed living nature. Though a menagerie might not suffice to make known all the characteristics of an animal in its free-living state, a living animal would still be more instructive in this regard than an animal skin. In addition, a menagerie would allow for certain kinds of experiments to be conducted. It would be possible to study the extent to which domestication altered the character of animals, and to conduct physiological experiments. The reason physiology had not progressed as fast as it might have, the committee suggested, was the relative paucity of different species on which physiologists had been able to experiment. Among the subjects they suggested for experimentation was generation. If it was the state of captivity that kept animals from mating in menageries, they said, then Spallanzani's recent demonstration of artificial insemination introduced the possibility that this new technology might be used to raise species from other lands, cross different races, and produce new hybrids.[10]

Such were some of the early suggestions of what might be accomplished with a menagerie. The reality of the matter was that between November 4 and 10, 1793, the professors of the Muséum unexpectedly found themselves in charge of a dozen or so live animals, several of which were dangerous. This collection grew, particularly the following spring. In April the four animals that remained at Versailles, consisting of a lion, a dog that was the lion's companion, a quagga, and a bubal (a kind of antelope), were transferred to Paris. The bubal died almost immediately from injuries sustained on the trip. The fraternal relations of the lion and dog, however, served as a happy moral lesson of how a beast as ferocious as a lion could, under the proper circumstances, come to live in harmony with an animal of another species.[11] In May some thirty-six additional animals, including sheep, goats, and deer, were transferred to Paris from Raincy, the former estate of the duc d'Orléans. On May 16, 1794, the menagerie was officially authorized by a decree of the Convention. By this time the menagerie had fifty-eight animals (thirty-two mammals and twenty-six birds). Confined to cramped pens and cages, most of the animals were in a sorry state. By a decision taken by the professors on May 1, 1794, two signs were placed above the animals' quarters, one bearing the inscription "Provisional Menagerie," and the other instructing the public "Do not throw anything to the animals."[12]

If there was anyone who could look beyond the unsuitable provisional quarters of these animals to envision an ideal zoological garden, it was surely the Comte de Lacépède, a composer of music, follower of Buffon, and provider of grandiloquent disquisitions on Nature. Three times between 1795 and 1801 Lacépède published pieces in which he repeated many familiar claims about the utility of menageries. He also championed them in terms of the moral power of images of nature, as well as maintaining that, properly constituted, they could elevate the morality of those who frequented them, and reflected the state of civilization of the peoples who created and maintained them.[13]

Beginning with a letter to the *Décade philosophique* in 1795, Lacépède sketched out what an ideal menagerie might look like and expressed not only his concern over the sorry facilities of the provisional menagerie but also his sense of the moral consequences of the kinds of spectacles that were presented to the public. Thus he urged "that the images of constraint and the appearances of slavery be removed as much as possible from the eyes of a free people." He suggested also that resources be devoted primarily to those animals that, though less rare than the others, were better suited to inspiring the mind of the observer and satisfying the soul of the citizen. It would be good for the people, he explained, to see "the lion, so proud and superb, but so useless and so dangerous, cede its place to the cashmere sheep, so gentle, so humble, so peaceful, and so useful." [14]

Among the intellectual projects that Lacépède began in the 1790s was the construction of a conjectural history of human social evolution.[15] In 1801, he proceeded to identify four types of menageries corresponding to four successive stages in the development of civilization. The first type of menagerie was "the product of *need,*" created when the chiefs of primitive tribes gathered in large enclosures the animals they used for hunting and warfare. The second type of menagerie was a function of *pride.* It came into being when powerful rulers decided to symbolize their might by showing how even the most formidable of animals, such as lions and tigers, were subservient to their power. After the military successes of a people led to the conditions of independence, prosperity, and leisure time necessary for science to flourish, a third type of menagerie, inspired by *curiosity,* came into existence. In this case, rare species were assembled alongside the animals of the two earlier types. The fourth and most advanced type of menagerie was a function of "the devotion of the citizen." The continued progress of civilization, Lacépède explained, awakened new needs with respect to food, clothing, agriculture, and transportation. These in turn led man to work toward the perfection of the races of animals that were useful to him, and this goal was the defining feature of the fourth type of menagerie. "These menageries of the fourth kind," he stressed, "belong only to a very advanced civilization." So it was, he said, "that a general menagerie finds itself, as it were, successively established by the need of the hunter-warrior, the pride of the dominator, the curiosity of the scientist, and the devotion of the citizen."[16]

In 1801 a new, more splendid and permanent menagerie was being designed at the Jardin des Plantes to replace the early, provisional one. Lacépède anticipated that the new structure would represent the work of human politics "carried to the highest degree of civilization to which the present state of civilization can give rise." The menagerie would be like "a varied and cheerful countryside," where different animal species would "enjoy all the liberty it is possible to allow them without endangering the numerous and sometimes imprudent spectators." So far as possible, the animals would be surrounded by plants from or similar to those of their native lands. "Feeling neither their exile nor their loss of independence," Lacépède said, the animals would "present to the eye of the observer the faithful picture of the productions of nature living in the most remarkable lands of the globe."[17]

A detailed history of the menagerie of Paris will need to examine at length how the ideals expressed by Saint-Pierre, the commission of the Société d'histoire naturelle, and Lacépède corresponded to actual practice. The following comments may serve to illustrate a number of the ways in which the functioning of the menagerie proved more problematical than was hoped.

When one considers the sources of the animals that were brought to the menagerie, one finds that some of the finest of them were gifts from rulers in other countries, just as Saint-Pierre had anticipated. This was true, for example, of the three lions given to the Republic by the bey of Algeria in 1798, or a famous giraffe that was sent as a gift to the king of France by the pasha of Egypt in 1827. It is worth noting at least in passing, however, that the menagerie of the Muséum did not end up receiving all the animals that foreign rulers designated for the rulers of France. Joséphine, the wife of Napoleon, established a menagerie of her own at Malmaison along with her roses, exotic plants, and works of art. In the early 1800s there was often a question of whether a new prize animal would go to Her Majesty the Impératrice or to the Muséum. Needless to say, the professors always took care to thank Joséphine profusely when she turned a special animal over to them. Conversely, they had to hide their disappointment when she asked them in 1804 to send her the two giant kangaroos that the menagerie had acquired from England two years earlier.[18]

In addition to gifts and to the appropriation of the animals from Versailles, Raincy, and the animal shows of Paris, the menagerie also gained new animals through purchases, trades, and naturalist ventures like the French "voyage of discovery to Australia" (1800–1804), commanded by Captain Nicolas Baudin.[19] Nevertheless, through the first decade of its existence the menagerie's single most important source of new animals was military conquest. In Holland, French armies confiscated the menagerie of Stathouder William V, which resulted in forty-nine animals, including two elephants, being added to the menagerie in Paris.[20] In Switzerland, Napoleon's armies appropriated the bears that were the very symbol of the city of Berne. As *La Décade philosophique* reported in 1798: "The bears of Berne have arrived at the menagerie of the Muséum d'Histoire Naturelle. Few nations have had the privilege, as we have, of populating their cabinets of natural history with a succession of victories each more glorious than its predecessor. In peacetime, a soldier of the country will be unable to walk about Paris without noticing on all sides the fruits of his efforts and bravery."[21] Though the lions that Félix Cassal brought back from Africa in 1798 were acquired as gifts or purchases rather than by way of military conquest, the French government arranged to have Cassal's wagons full of African animals meet up with the wagons bringing the art treasures of Italy to Paris so that they all could enter the city in a triumphal procession on July 14, the anniversary of the Revolution.[22] Though Lacépède in 1801 wanted the Paris menagerie to reflect "the curiosity of the scientist and the commitment of the citizen," the menagerie's inhabitants in 1801 still represented in large measure the motives of an earlier stage of menagerie building—"the pride of the dominator."

The distance between Lacépède's lofty aspirations for the menagerie and what transpired there in practice is illustrated again when one considers his idea of populating the institution with peaceful and useful species and thereby elevating the behavior and moral vision of the public. Félix Cassal, the first keeper of *les animaux féroces,* was certainly not interested in the sort of menagerie Lacépède projected. A zoo guide for young people in 1798 represents Cassal as saying: "I would give the little I possess to see this menagerie populated with all kinds of ferocious animals. When the public comes to visit, I am ashamed to have only ordinary animals to show them. I would like to be able to say to you, 'here is the tiger, here is the panther, here is the leopard, here is the jackal, here is the hyena, here is the lion.'"[23] As for the moral elevation of the public, manuscript accounts from the period indicate that the public was not content to watch herbivores resting or grazing peacefully. Visitors to the menagerie threw stones at the animals, or sent their dogs after them, to make them move about. Meanwhile, the animals' behavior was frequently no more ennobling than the public's. The herbivores in their pens and stables took to locking horns with one another. And then there was the problem of the monkeys. As one 1804 guide to the Jardin des Plantes noted with respect to the two monkeys of the menagerie: "they should not be exhibited to women and children, especially in certain seasons, because the baboon then seems to make a game of showing off his lewdness."[24]

With regard to the living conditions of the animals, these remained very poor in the late 1790s as the Muséum waited for funds and architectural plans for new buildings. With many of the animals dying and most of the others in very unhealthy quarters, the professors were confronted with the possibility that unless the conditions at the menagerie were quickly improved, the menagerie was more likely to hurt than help the reputation of the Muséum as a whole. Under such circumstances, the opportunities to study animal behavior, let alone experiment upon it, were limited. There were some genuine triumphs, such as the birth in 1800 of three lion cubs. Nonetheless, the scientific promise of the menagerie at the turn of the century remained precarious at best.

The popular appeal, extended symbolism, and scientific limitations of the menagerie were all evidenced in one of the most remarkable events of its early history, a concert that was held for a pair of elephants on May 29, 1798. The elephants in question were a male and a female, named Hanz and Marguerite, which had been the crowning ornaments of the stathouder's menagerie in Holland. The other quadrupeds and birds of the stathouder's collection had been transported to Paris with little difficulty, arriving there at the end of August 1796, but the transportation of the elephants proved more troublesome. It was not until the evening of March 24, 1798, that the elephants arrived in Paris by way of the Seine. Preparations had been made at the Muséum to accommodate not only the elephants but also the large crowds of people who would want to see them. The populace was allowed to view the elephants on the even days of each *décade* from eleven to one and from three to five, but so many people came that it was difficult to get a good look at the great beasts. The influx of visitors to the Jardin des Plantes was accompanied by a flood of merchants seeking to sell food and merchandise.[25]

On May 29, some two months after the elephants arrived in Paris, the elephants were the subject of an experiment, when a concert was performed for them by fourteen musicians from the Conservatoire de Musique. An account of this event, written by Georges Toscan, the head librarian of the Muséum, was published at the time, and has formed the basis of all subsequent published accounts of this concert to date.[26] In 1992 the present writer discovered in the Archives Nationales in Paris a second document of the event, a previously uncited manuscript which turns out to be nothing less than field observations made at the time by members of the musical troupe. The manuscript lists in parallel columns the pieces performed by the musicians and the elephants' responses to each piece. The musicians had different questions in mind than did the librarian, and their reports differ accordingly. Where the musicians were interested to see how the elephants responded to different instruments, different rhythms, and different keys, Toscan was particularly interested in the idea that noninvasive, painless, and indeed even pleasure-giving experiments could be conducted on living creatures.[27]

The musicians initially set themselves up out of sight of the elephants, while the keeper kept the animals occupied eating so that they would not be aware of any strange goings-on in their vicinity. Then the musicians began to play. Their first piece began with a "trio de petit air" in B major for two violins and bass. When the music started, the elephants stopped eating and manifested surprise. As the music moved to the more vigorous dance air in B minor from Gluck's *Iphigenia in Tauris,* the elephants became increasingly agitated and seemed to move in time to the music. They pushed at the barriers of their enclosure, and emitted cries and whistling sounds. They were calmed by a bassoon solo in C minor based on the air, "Oh ma tendre musette," but were then stirred to "une agitation prodigieuse" by the full orchestra playing the revolutionary song "Ça ira." This was performed in the key of D and featured the piercing sound of a piccolo. The female in particular seemed especially excited by the music. The postures she adopted and the caresses she gave the male seemed to onlookers to be evidence that the music had aroused her romantic ardor. The animals were then calmed by a choral duet in E flat from Sacchini's opera *Dardanus.* The musicians' manuscript account of the concert indicates that as the duet was sung, "The male appeared to reflect profoundly."[28]

A second rendering of "Ça ira" by the orchestra, this time in the key of F, gave observers the opportunity to conclude that animals experience different effects not only according to the different character of music but also according to changes in key, "since they have been infinitely less agitated by this air in F than the first time, when it was in D." A flute duet followed, which produced no excitement on the elephants' part. This was in turn followed by a solo of the song "Charmante Gabrièle" in C major. According to the observations recorded by the musicians at the time, "This piece so moved them, that the whole gallery noticed that the female shed tears."

A full account of the concert will have to be reserved for another occasion. For the present, the following may be noted.

1. In all, the musicians presented fourteen musical arrangements for the purpose of observing how different musical pieces might affect the elephants. Among their offerings, they played the revolutionary song "Ça ira" a total of four times. The elephants seemed most excited during the third playing of the piece, when it was once again performed by the whole orchestra in the key of D, but this time with vocal accompaniment. Perhaps because the musicians wanted to do follow-up experiments with the elephants, they did not at this point draw any special conclusions from their observations except to note that the elephants seemed to respond to the character and key of the music being played.

2. Toscan for his part stressed that both elephants, in particular the female, appeared to be sexually aroused by the music. This behavior was not, for Toscan, a trivial matter. Elephants were notorious for their unwillingness to breed in captivity, and getting them to do so was the primary obstacle to domesticating them. They were said to be so modest that no naturalist had ever seen them copulate in nature, let alone in captivity. Toscan's report gave the encouraging news that several nights after the concert the elephant keeper heard a noise in their enclosure and surprised them in what the reporter supposed (incorrectly) to be the elephant's true mating position, with the female on her back and the male astride her. Toscan allowed that under the proper circumstances—his suggestions included a moonlit night, with soft music in the background—the two elephants might be induced to mate in the Paris menagerie. In the meantime a French artist, J.-P. Houel, spent many hours at the menagerie drawing the elephants, eventually publishing a book with numerous pictures of them, including a speculative depiction of two elephants mating in nature.[29]

3. The experiment was conceived by the musicians, not the professors of the Muséum. What is more, the primary published account of the concert was produced by Toscan, the Muséum's librarian, and not by a professor. Indeed, though Etienne Geoffroy Saint-Hilaire, the professor of mammals and birds at the Muséum, later claimed that the menagerie had been his responsibility from its first beginnings in 1793 up until 1837, the fact is that two librarians of the Muséum, Georges Toscan and Mordant Delaunay, took as much interest in the menagerie from 1796 to 1801 as did any of the professors. Delaunay was in effect the menagerie's director for this period, while Toscan was the one who reported its news to the public.[30]

If the scientific prospects of the menagerie looked problematical in 1798, they appeared considerably better in 1804, as the plans for the permanent facility began to be executed. The first stone of the *rotonde,* initially intended to house *les animaux féroces,* was laid that year. Nonetheless, problems remained. On the same day that Frédéric Cuvier was appointed *garde de la ménagerie*—a new position created late in 1803—Félix Cassal, the *garde des animaux féroces,* was dismissed for having tried to arrange the murder of another member of the staff. Valuable animals, such as a splendid lion named Marc, continued to die. As Frédéric Cuvier indicated early in his tenure, it was difficult to know just what the animals needed:

> Who can one consult in constructing the most suitable shelters for wild and often dangerous animals, which for the most part come from climates different from our own? What nourishment do most of them require? How must they be treated to maintain them in good health and to manage them? To what diseases are they subject? What remedies can one administer to them? . . .
>
> There is practically no way of responding to these questions and a hundred other similar ones. Nothing has been written, almost nothing has been seen, and everything remains to be done. There is no help other than some general knowledge and some analogies which might well help in the search for the truths needed, but which do not teach them.[31]

Let us turn to the question of what Frédéric Cuvier managed to accomplish in the more than three decades during which he occupied the position of *garde de la ménagerie.*[32] At the time the position was created, its duties were spelled out in a set of sixteen articles, drawn up by Geoffroy. The *garde* was placed under the authority of the professor of mammals and birds and was to communicate with the administration through him (article 2). The *garde* in turn was to have charge of the various animal keepers, the menagerie's day workers, and the commissioner of provisions (article 4) for the dual purposes of the good health of the animals and the maintenance of proper order when the public was present (article 5). Among the additional articles relating to the duties of the *garde* we may note in particular article 10, instructing him to take every opportunity to acclimatize and naturalize foreign species that might be economically useful; article 11, stating that he was to conduct a program of experiments devised by the professors; and article 12, indicating that the study of animal behavior was also part of his province and that he was to ascertain with care the onset and conclusion of the periods of rutting, gestating, laying, molting, and so forth that animals experienced annually.[33]

The position of *garde de la ménagerie* having been created, Geoffroy recommended to the assembly of professors that the position be filled by Frédéric Cuvier, the younger brother of Geoffroy's illustrious colleague and friend, Georges Cuvier. Geoffroy admitted it might not be apparent that Frédéric Cuvier was particularly qualified for the job, since he was best known for his work in chemistry, but Geoffroy insisted: "I know him to be well versed in comparative anatomy, natural history, and general medicine; he is the one who composed the largest number of zoology articles for the *Nouveau dictionnaire d'histoire naturelle*. . . ." Geoffroy also testified to the quality of the younger Cuvier's character. Following Geoffroy's recommendation, the professors of the Muséum on December 28, 1803, elected Frédéric Cuvier unanimously to be *garde de la Ménagerie.* Frédéric was then thirty years old.[34]

Over the next three and a half decades, the younger Cuvier proceeded to develop a more comprehensive and broadly informed assessment of mammalian behavior than anyone had offered previously. Not surprisingly, his successes and his frustrations were intimately related to the particular setting in which he found himself. The material conditions of the menagerie, the kinds of research that could be conducted on captive animals, the menagerie's functions within the aims of the Muséum as a whole, and the dependent status of the *garde* in the formal institutional hierarchy all influenced what he was able to accomplish.

It is not rare in the history of science to find an investigator making a special virtue out of conditions of practice regarded by other investigators as confining. That is precisely what J.-B. Lamarck did after the reorganization of the Jardin des Plantes moved him from the role of "Botanist Affiliated with the King's Cabinet" to that of "Professor of Insects, Worms, and Microscopic Animals." Lamarck initially worried that "there was much more to be gained and more interest to be excited in the demonstration of the characters, manners, and habits of the lion than of the earthworm."[35] He soon was reassuring himself and his students, however, that the study of the invertebrates had major advantages over the study of the vertebrates. He proceeded to maintain that the simpler animals displayed more clearly than did the more complex animals precisely what the essence of life actually was. He argued further that the simpler animals revealed more clearly than did the more complex animals the course (*la marche*) nature had followed in bringing all the different forms of life successively into existence.[36]

In the case of Frédéric Cuvier, here was a young naturalist, best known as the younger brother of Georges Cuvier, also known slightly for some work in chemistry, who was entrusted with the care and study of animals under conditions that many naturalists regarded as so artificial that nothing of value could be learned about the lives of the animals in question. Frédéric Cuvier confronted this challenge and proceeded to construct a rationale for menagerie studies that justified, exploited, and even extolled what one could learn from animals in captivity. In his very first article on animal behavior, "Du rut," the one-time chemist wrote: "Menageries can be for those who occupy themselves with zoology what the laboratory of the chemist is for those who devote themselves to the study of brute matter. One does not see there what takes place in nature, but what could take place there."[37]

Frédéric Cuvier pushed this analogy further in 1809 in a new edition of a guide to the menagerie, remarking that Galileo would never have discovered the law of falling bodies had he not had the opportunity to manipulate such bodies experimentally. If one wished to discover the laws of animal behavior, Cuvier maintained, one would have to engage in comparable sorts of intervention. The reason contemporary ideas of animal intelligence were so inadequate, he insisted, was because they were based on isolated facts involving free-living animals on which it was impossible to experiment.[38] He expressed this attitude most forcefully in his 1825 article on the domestication of animals. There he observed that the study of free-living animals in nature could reveal their role on the earth, but it was only with the aid of captive animals that one could uncover the general causes of their actions and the nature of their intellectual faculties.[39] He indeed went so far as to suggest that zoo studies could render field studies unnecessary. He allowed that once one learned the general faculties and dispositions of a species from studying it in captivity, one would then be able to determine "even in advance" how it would act in its native surroundings. It would no longer be necessary to track animals down and follow them through the details of their existence, so long as one was familiar with the details of the environments from which they came. Compared to the study of the behavior of individual animals, Cuvier indicated, this knowledge was "much easier [to obtain] and much less subject to error."[40]

These happy formulations represented Cuvier's thinking at its most optimistic—or at least at its most combative in the face of skepticism about what one could learn from the study of animals in a menagerie.[41] Frédéric Cuvier faced, however, a variety of difficulties in his work. He was not a professor; indeed his brother Georges convinced him that it would be inappropriate for him to aspire to a professorship at the Muséum as long as Georges himself was a professor there.[42] As a result, Frédéric did not have the opportunity to develop and present a course of instruction on his subject. He remained officially subordinate to Geoffroy until almost the end of his career. Over this period, furthermore, his relations with Geoffroy deteriorated seriously.

The animals as well as the professors posed genuine problems for Frédéric Cuvier. The conditions under which the animals were kept were frequently so confining that they had little opportunity to display any of their natural habits or abilities. Even when they displayed interesting behavior, it was often difficult to keep them alive long enough to study the behavior thoroughly. Of the animal subjects of Cuvier's early writings, the zebra, the orangutan, and the seals all died before he could make all the observations he wanted to make on them.[43] Significantly, while the death of an animal in the menagerie was a setback for him, it was generally not a problem for his colleagues. To the contrary, their practice essentially began when they were presented with a specimen that was no longer living. Dead menagerie animals were a major source of specimens for the cabinets of both comparative anatomy and natural history.[44] When the ménagerie's male elephant died of pneumonia on January 3, 1802, that was when work on him by the comparative anatomists commenced. Hanz, the magnificent creature who less than four years earlier had been stirred by the strains of "Ça ira," was of more service to science dead than alive. Georges Cuvier and his prosector and other assistants spent more than forty days anatomizing Hanz's cadaver.[45] Hanz and another Indian elephant who followed him became a means for Georges Cuvier to strengthen one of his most important claims, namely, that the species of quadrupeds of the present, elephants included, were distinct from the species of the past.[46]

Frédéric Cuvier himself, of course, could not devote all his time to observations and experiments on animals when he also had to be worrying about all the other dimensions of running the menagerie, including making sure the animals had fresh water, seeing to the fixing of the heaters when they were not working properly, suggesting how to arrange the buildings for the animals, questioning the recommendations of the architect when these did not seem to fit the needs of either the animals or the public, and so forth.

Frédéric's tenure as *garde de la ménagerie* corresponded with the most active building period of the menagerie's history. The *rotonde* for the elephants (initially intended to house the large carnivores), was begun in 1804 and completed in 1812. The bear pits were dug in 1805. The gallery of *animaux féroces* was started in 1818 and finished in 1821. The aviary for the birds of prey was completed in 1825; the pheasantry in 1827; and the monkey house in 1836.[47] Beyond his responsibilities for the menagerie, Cuvier was appointed *inspecteur de l'Académie de Paris* in 1810 and *inspecteur général de l'Université* in 1831, and he devoted a significant part of his time to these additional activities.

In his role as *garde de la ménagerie,* Cuvier proceeded to address the kinds of questions that the professors in setting up that position had wanted the *garde* to address, beginning with his article "Du rut" of 1807, which included a number of breeding experiments, and continuing with his publication in 1825 of two important articles, one on animal sociality and the other on animal domestication.[48] Limitations of space do not permit here a thorough review of all the animals Cuvier studied or the major observational and theoretical points he established, but we may mention some of the conclusions in which he took particular pride.

In his work with a young female orangutan, and then later with other animals, he concluded that the intelligence of animals develops most dramatically when they are young, and then grows weaker as their physical strength grows.[49] Again in his work with the orangutan, but more especially with his 1811 study of seals, he demonstrated, contrary to popular opinion, that intellectual perfection is not directly correlated with the perfection of the sense organs.[50] Also contrary to popular opinion was his observation that herbivores are not by nature more gentle than carnivores. Another of his major claims, the result of comparing those species of animals that had been successfully domesticated with those that had not, was that domestication was only possible in species that already possessed a preexisting, instinctive disposition to sociability.[51]

In his experiments with beavers, he proved that the constructions of beavers are not the result of projects they undertake intelligently and communally, when not bothered by man (as Buffon had supposed), but instead are only "the fruits of any industry that is wholly mechanical, the results of a new need that is purely instinctive." He determined this by observing young beavers that were reared in isolation and thus never had the opportunity to learn from conspecifics. He found that these individuals still displayed an instinctive tendency to build.[52]

In his study of the intellectual faculties of animals Cuvier was necessarily led to consider the similarities and differences between the faculties of animals and humans. In his 1810 article on the orangutan, he stressed this animal's circumspection, which he said was made necessary by the fact that "nature has given the orangutan rather little means of defense." He also credited the animal with being able to generalize. He observed that if the orangutan did not want to be brought down from a tree, it would shake the tree branches if people started to climb up after it. It did so, Cuvier believed, because it had reasoned from its own experience "that the fear of falling would prevent them from climbing higher." Likewise it moved a chair in order to be able to reach and open a lock, and it took care to cover itself at night, thereby exhibiting in addition to the ability to generalize "the awareness of a future need."[53]

It is not recorded whether Cuvier's article of 1810 led him to be criticized for crediting the orangutan with such faculties. Nor is it recorded whether anyone objected the following year when his observations on seals led him to endorse Gall's opinion "which gives the brain the primary influence on ideas."[54] This was, however, at just the time that Lamarck was speculating on how an orangutan or similar creature, in taking up new habits, might have developed the faculty of speaking.[55] In any case, in January of 1812, when Cuvier presented to the Société philomathique his "observations on the intellectual faculties of animals," he took special care to identify "the principal differences existing between man and the animals of the highest classes with respect to intelligence." He located these in "the faculty of meditating and reflecting," which he claimed was unique to humans. He argued "that the animals, like us, have the faculties of being attentive, of feeling, of judging, of remembering, etc., and that all our actions, in which we use only these faculties, are absolutely similar to theirs; but we distinguish ourselves from the animals when we apply to these faculties that of meditating."[56]

In his later writings Cuvier continually oscillated between underscoring the fundamental differences between humans and animals and insisting that the study of animal behavior could enhance the understanding of human behavior. In his 1822 article on instinct he insisted that "all animals without exception lack the intimate sense of the perception of self and the faculty of reflection."[57] In his 1825 article on animal sociality he expressed his belief that "human sociability is originally instinctive." Insofar as this instinct was modified in man by reason, the better means of understanding instinct and its direct consequences was to study it in the simpler state in which animals displayed it.[58]

Two further points of historical interest are that Cuvier took a keen interest in the species question and at the same time strongly endorsed the idea of the inheritance of acquired characters. He wrote: "Experience proves incontestably that intellectual faculties developed accidentally are passed down over time by generation and become hereditary."[59] He regarded this as the primary cause of the different races of mammals. He did not suppose, however, that such changes could extend so far as to produce new species. The whole subject of the origin of species seemed to him to be beyond the reach of the modern scientist. As he put it in 1819: "The force that gives them birth has ceased to act, or hides itself in depths where our eye cannot see them." He thus dodged the issue of transformation at the species level but embraced it at the level of races or varieties. At this lower level of operation, furthermore, what he promoted was not merely intellectual understanding but practical power. In his words: "The causes of varieties . . . are under our control."[60]

Cuvier's thoughts on the inheritance of acquired characters had an important bearing on the general project of animal domestication. His experience at the menagerie had shown him that at least some animals could have their wildness tamed and their intelligence enhanced through careful treatment and training. This experience, combined with his belief that habits become hereditary if exercised over a sufficient number of generations, gave him hope that additional animal forms could be domesticated. He represented domestication not as the degeneration of an animal's original nature, but as an improvement of it. Denying the common equation of domestication with slavery (though he had used this language himself earlier), he came to speak of domestication not in terms of domination but of education.[61]

Cuvier did not, however, regard animals as pliable beings that could be readily improved for human purposes. As he saw it, only those animals that already possessed a strong instinct of sociability were candidates for domestication. Other animals had natures that were permanently resistant to this enterprise, the hereditary transmission of acquired physical and mental modifications notwithstanding. Yet it was not just animal nature that stood in the way of certain domestication projects. Cuvier recognized that humans, too, displayed resistances in this regard. The agricultural class of French society, as it appeared to him, was disposed to resist almost any innovation. French farmers seemed to carry "natural human prudence" to such an extreme that it virtually became "a simple instinctive movement analogous to that which causes an animal to distrust and flee everything unknown to it."[62]

In the early 1820s Cuvier expressed his intention of publishing eventually a general work on the origin of animal actions.[63] He never succeeded in doing so. He did not, for that matter, publish any articles specifically on animal behavior after 1825, though he continued to publish his large work on mammals, which ultimately included the descriptions of some five hundred species. Had he become a professor at the Muséum earlier in his career than he did, things might have been different. His professorship, however, came too late. Upon his brother's death in 1832, Frédéric asked the professors of the Muséum to request of the minister of public instruction the establishment of a new chair there, dedicated to the study of living animals and their education. The professors declined to act upon his proposal.[64] Late in 1836, when the minister gave the Muséum the opportunity to create two new chairs, Cuvier sought to claim one of these. He again submitted to the assembly of professors his proposal for "a course on the nature and on the education of animals." He argued that a zoology that dealt only with the physical characteristics of animals and not their faculties was "a truncated science." He insisted further that if the study of human anatomy had benefited as much as it had from the study of animal anatomy, the same would certainly be the case for human psychology, when a whole range of psychological phenomena from the whole animal series was brought to bear on it—particularly since these phenomena exhibited a simplicity that was rarely found in humans. By studying what animals have in common with us, he said, one would be able to offer "a *natural* analysis of human intelligence."[65]

As justification for his proposal, Cuvier explained that in his more than three decades as warden of the menagerie, "my researches and my work have been directed principally toward the knowledge of living animals, towards the means of observing them and of submitting them, without physical suffering, to planned experiments, to understand the laws of their existence, as well as the nature and extent of the modifications to which these laws are subject." He had continued his researches, he said, sustained "by the hope that one day I would be able to disseminate them with more success by teaching them, and by giving a more utilitarian direction to the menagerie."[66]

The consideration of the proposal for new chairs at the Muséum led to a general review of the instructional needs of the museum. A committee consisting of Joseph Louis Gay Lussac, Pierre Flourens, and Alexandre Brongniart was instructed to conduct the review and report back to the assembly. On the particular question of the study of living animals, the committee was unable to come up with a recommendation, saying that insofar as the facts of animal behavior were yet to be united as a body of science, it would be difficult to judge how they could collectively serve as a basis for a course of public instruction. The committee turned the issue back to the assembly as a whole, which on February 14, 1837, decided by a vote of seven to five not to create a new chair along the lines that Frédéric Cuvier had suggested.[67]

The following two weeks witnessed a bitter dispute between Geoffroy and Cuvier over the direction of the menagerie.[68] In May Cuvier requested and received a two-month leave in his capacity as *inspecteur général de l'Université.* Geoffroy's son, Isidore, was named to take Cuvier's place in Cuvier's absence.[69]

By December 1837, the assembly found itself considering a new proposal from the minister "to create a new chair having as its special subject the animals of the ménagerie." On December 12, by a vote of ten to one, the professors agreed to create a new course and chair of comparative physiology. On December 19, a decision was reached on the activities and competence of the chair: among other things, the holder of the chair was given the direction of the menagerie.[70] On December 24, Frédéric Cuvier was officially named professor of comparative physiology. This enraged Geoffroy, who regarded himself as the founder of the menagerie, believed the ultimate responsibility for it belonged to him, and felt that it should be his prerogative to pass the responsibility for it on to his own son, Isidore.[71]

As it was, Cuvier's opportunity to see his life's work at the menagerie and Muséum come to fruition was cut short. On April 17, 1838, Cuvier informed the assembly of professors that he was going to leave Paris at once to fulfill his administrative responsibilities as university inspector. That was the last assembly that he attended. He died in Strasbourg on July 24, 1838, in the course of administrative duties.

Though the chair of comparative physiology was continued, the professors decided to modify the chair's responsibilities and prerogatives. They returned the direction of the menagerie to the professor in charge of mammals and birds. They chose not to reestablish the position of *garde de la ménagerie,* however, giving the supervision of this part of the Muséum instead to the professor's *aide-naturaliste.*

The study of animal behavior at the Paris menagerie would never again have the aspirations or successes that it had under Frédéric Cuvier. As zoos were set up in other capitals and major cities throughout Europe, particularly in the second third of the century, the menagerie ceased to be the finest of such institutions in the world. However, these other institutions were not necessarily more successful in furthering the study of living animals than that in Paris had been. Given that the Paris menagerie came into being during the Revolution with the idea that the animals from the street shows of the capital and the king's menagerie at Versailles should be liberated from their chains and cages, it is ironic to find the London Zoo at the beginning of the twentieth century characterized, because of "the most unmercifully severe confinement" of the animals there, as "the bastille of the beasts."[72]

In the twentieth century, zoos (and those of Berlin and Amsterdam in particular) played a significant role in the establishment of a new scientific discipline devoted to the study of animal behavior, the discipline of ethology. In the early nineteenth century, the only place that came close to being an important center for the study of animal behavior was the menagerie of the Muséum d'Histoire Naturelle in Paris. Though it did not become all that its early proponents hoped it might, it did end up serving multiple functions. It was a magnet that attracted the public to the Jardin des Plantes. It was a matter of national pride, a symbol of France's power over nature and over other nations. It was a source of specimens for the cabinets of natural history and comparative anatomy. Under Frédéric Cuvier, it was the site of the most important studies of comparative psychology of the period. If ultimately it remained marginal to the nineteenth-century zoologist's quest to understand life, it nonetheless provided citizen and scientist alike a special setting in which to ponder the complex relations, and possibly to create new relations, between *Homo sapiens* and the rest of the animate world.

NOTES

1. On the history of the Paris menagerie, see especially E.-T. Hamy, "Les anciennes ménageries royales et la ménagerie nationale fondée le 14 brumaire an II (4 novembre 1793)," *Nouvelles Archives du Muséum,* 4th ser., 5 (1893): 1–22; Gustave Loisel, "Histoire de la ménagerie du Muséum," *Revue scientifique* (1911): 262–77, 301–4; Yves Laissus and Jean-Jacques Petter, *Les animaux du Muséum, 1793–1993* (Paris, 1993); and Richard W. Burkhardt, Jr., "La ménagerie et la vie du Muséum," in *Le Muséum au premier siècle de son histoire,* ed. C. Blanckaert et al. (Paris, 1997), pp. 481–508, of which the present paper is a revised version.

2. Michel Foucault, *The Order of Things: An Archeology of the Human Sciences* (New York, 1971), pp. 127–28.

3. The published accounts of precisely when and which animals were conveyed to the Jardin des Plantes are not consistent. The list provided in the text above is based on the unpublished "Mémoire de la dépense occasionnée pour la nourriture des animaux composant la Ménagerie du muséum d'histoire naturelle du 14 brumaire jusque compris le 30 fructidor l'année 2e de la République française une et indivisible," in Archives Nationales (hereafter AN), Minutes des procès-verbaux des assemblées des professeurs, AJ/15/577; and on the *procès-verbaux* of the *séance* of 9 frimaire an II (AN, AJ/15/96), which gives the expenses claimed by Marchini, Louzardy, and

Henry (the owners of the animals). A manuscript by A.-L. de Jussieu entitled "Notes relatives à l'établissement d'une ménagerie dans le museum d'hist. naturelle," undated but written several months after the event (and before the animals from the Versailles menagerie arrived in April 1794), indicates that the owners claimed reimbursement for thirteen animals. See AN, AJ/15/844.

4. Buffon, "L'outarde," *Histoire naturelle des oiseaux,* 9 vols. (Paris, 1770–83), 2: 202. On Buffon's ideas on animal behavior, see Richard W. Burkhardt, Jr., "Le comportement animal et l'idéologie de domestication chez Buffon et chez les éthologues modernes," in *Buffon 88: Actes du Colloque international Paris—Montbard—Dijon,* ed. Jean Gayon (Paris, 1992), pp. 569–82. On Buffon's attempt to bring the animals of Versailles to Paris, see Hamy, "Les anciennes ménageries royales," p. 9.

5. Jacques-Bernardin-Henri de Saint-Pierre, *Mémoire sur la nécessité de joindre une ménagerie au Jardin national des plantes de Paris* (Paris, 1792). On Bernardin de Saint-Pierre, see Roselyne Rey, "L'animalité dans l'oeuvre de Bernardin de Saint-Pierre: Convenance, consonance, et contraste," *Revue de synthèse* 113 (1992): 311–31.

6. Saint-Pierre, *Mémoire,* p. 4.

7. Ibid., pp. 12–13.

8. Ibid., p. 59.

9. A. L. Millin, P. Pinel, and Alex. Brongniart, *Rapport fait à la Société d'histoire naturelle de Paris, sur la nécessité d'établir une ménagerie* (Paris, 1792).

10. Ibid., pp. 2–3. Bernardin de Saint-Pierre was not in favor of physiological experiments on animals. See Rey, "L'animalité dans l'oeuvre de Bernardin de Saint-Pierre," p. 319.

11. G. Toscan, "Description de la ménagerie du Muséum d'histoire naturelle: Histoire du lion et du chien," *La Décade philosophique, littéraire et politique* 3 (an III [1794]): pp. 129–38, 193–99.

12. The animals in the menagerie as of May 15, 1794, are listed in Étienne Geoffroy Saint-Hilaire, "Etat des animaux, vivans, existans à la ménagerie provisoire du Museum national d'histoire naturelle, ce jourd'hui 26. floréal an 2eme . . . ," AN, AJ/15/844. These animals included those from Raincy. Gustave Loisel's identification of May 27, 1794, as the date when the animals were transferred from Raincy to Paris is incorrect (see Loisel, "Histoire de la ménagerie du Muséum," p. 264). For the placing of the signs above the animals' quarters see Minutes des procès-verbaux des assemblées des professeurs, Séance du 12 floréal an 2, AN, AJ/15/96 (p. 111).

13. Lacépède, "Lettre relative aux établissemens publics destinees à renfermer des animaux vivans, et connus sous le nom de MÉNAGERIES,"

Décade philosophique 7 (an IV [1795]): 449–62; "Sur les établissemens publics destinés à renfermer des animaux vivans, et connus sous le nom de ménageries," *Séances des Ecoles normales* 8 (1800): 303–18 (this lecture was virtually identical to Lacépède's 1795 paper); and "Introduction," in *La Ménagerie du Muséum national d'histoire naturelle, ou description et histoire des animaux qui y vivent ou qui y ont vécu; par les citoyens Lacépède et Cuvier, avec des figures peintes d'après nature, par le citoyen Maréchal, peintre du Muséum* (Paris, 1801), pp. 1–9.

14. Lacépède, "Lettre," pp. 454, 456.

15. On this topic see Claude Blanckaert, "Une anthropologie de transition: Lacépède et l'histoire naturelle de l'homme (1795–1830)," *Annales Benjamin Constant,* no. 13 (1992): 95–111.

16. Lacépède, "Introduction," pp. 1–2.

17. Ibid., p. 6. Here Lacépède voiced again his concerns about the way that images of constraint could affect adversely "la morale de la multitude."

18. The sources of the animals that existed at the menagerie in 1804 were described in a book published anonymously by Frédéric Cuvier, *Notice des animaux vivans de la Ménagerie, leur origine et leur histoire dans cet Établissement* (Paris, an XII [1804]). For Frédéric Cuvier's feeling that the giant kangaroos were essentially lost to science when they were sent to Malmaison, see F. Cuvier, "Le Kanguroo Géant," in Etienne Geoffroy Saint-Hilaire and Frédéric Cuvier, *Histoire naturelle des mammifères, avec des figures originales enluminées, dessinées d'après des animaux vivants,* 7 vols. (Paris, 1818–42), vol. 1.

19. On the Baudin expedition, see Richard W. Burkhardt, Jr., "Unpacking Baudin: Models of Scientific Practice in the Age of Lamarck," in *Jean-Baptiste Lamarck, 1744–1829,* ed. Goulven Laurent (Paris, 1997), pp. 497–514; and Jacqueline Bonnemains, Elliott Forsyth, and Bernard Smith, eds., *Baudin in Australian Waters: The Artwork of the French Voyage of Discovery to the Southern Lands 1800–1804* (Melbourne, 1988).

20. See Gustave Loisel, *Histoire des ménageries de l'antiquité à nos jours,* vol. 2 (Paris, 1912), pp. 36–49, 353–55; also Ferdinand Boyer, "Le transfert à Paris des collections du Stathouder (1795)," *Annales historiques de la révolution français* 43 (1971): 389–404.

21. *Décade philosophique* 17 (an VI [1798]): 240–41.

22. See Minutes des procès-verbaux des assemblées des professeurs, séance du 14 ventôse an 6 (letter from Thouin dated 29 pluviôse) and séance du 14 messidor (letter from the minister of the interior dated 11 messidor), AN, AJ/15/581.

23. L. F. Jauffret, *Voyage au Jardin des Plantes* (Paris, an VI [1798]), pp. 126–27. Though the words attributed to Cassal here were perhaps not a

direct quotation, the sentiment attributed to him fits what is known of his activities and personality.

24. J.-B. Pujoulx, *Promenades au Jardin des Plantes, à la ménagerie et dans les galeries du Muséum d'histoire naturelle,* 2 vols. (Paris, an XII [1804]), 1:95–96.

25. *Décade philosophique* 17 (an VI [1798]): 44–46, 180. Although the flood of merchants to the Muséum may have constituted a realization of Bernardin de Saint-Pierre's hope that a menagerie at the Jardin des Plantes would benefit the neighborhood economically, it cannot be supposed that everyone benefited equally. The *procès-verbaux* of the assemblies of professors record the case of one woman who complained that the increased competition was making it impossible for her to continue to earn a living by selling fruits in a doorway of the Abbaye St. Victor. She solicited the assembly of professors to allow her to sell her fruit, *limonade,* and *tisane* in the "Cour de Régie" of the Jardin des Plantes. See Minutes des procès verbaux des assemblées des professeurs, séance du 24 germinal an 6, AN, AJ/15/581.

26. G. Toscan, "Du pouvoir de la musique sur les animaux, et du concert donné aux éléphans," *Décade philosophique* 18 (an VI [1798]): 257–64, 321–29, 498–99. Accounts based on Toscan's original publications are Georges Toscan, *L'Ami de la nature, ou choix d'observations sur divers objets de la nature et de l'art; suivi d'un catalogue de tous les animaux qui se trouvent actuellement dans la ménagerie* (Paris, an VIII [1800]); J. P. L. L. Houel, *Histoire naturelle des deux éléphans, mâle et femelle, du Muséum de Paris, venus de Hollande en France en l'an VI* (Paris, 1803); Yves Laissus and Jean-Jacques Petter, *Les animaux du Muséum, 1793–1993* (Paris, 1993); Michel Lemire, "La France et les collections du stathouder Guillaume V d'Orange," in *Een vorstelijke dierentuin: De menagerie van Willem V; Le zoo du prince: La ménagerie du stathouder Guillaume V,* ed. B. C. Sliggers (n. p. [Walburg Instituut], 1994), pp. 87–114; James H. Johnson, *Listening in Paris: A Cultural History* (Berkeley, Calif., 1995).

27. The musicians' manuscript is dated 10 prairial an VI (May 29, 1798). See AN, AJ/15/581, Séance extraordinaire du 27 prairial an 6.

28. The account provided here is based on the musicians' unpublished account and Toscan's *Décade philosophique* article. The latter was reproduced in Toscan, *L'Ami de la nature.* The *procès-verbal* of the professors' assembly of 27 prairial an VI indicates that the musicians expressed the desire to conduct further tests on the effect of music on the elephants. The secretary of the assembly was charged with thanking the musicians for what they had done already and with inviting them to postpone their next experiment until the park for the elephants was completed and the male had recovered from a slight infirmity. Houel, *Histoire naturelle des deux éléphans,* describes in detail the story of the two elephants and states (p. 101n) that a trumpet

concert was given for the elephants sixteen days after the first concert. There is, however, no mention of this concert in the *procès-verbaux* of the professors' assemblies.

29. Toscan, "Du pouvoir de la musique sur les animaux." Houel, *Histoire naturelle des deux éléphans.*

30. One should note that for part of this period Geoffroy was away in Egypt. Note also that Georges Cuvier wrote most of the articles for the volume *La Ménagerie du Muséum national d'histoire naturelle.* On Toscan, his sensibilities, and his contributions to *La Décade philosophique,* see Jean-Marc Drouin, "L'histoire naturelle à travers un périodique de l'époque révolutionnaire: *La Décade philosophique,*" in *La nature en révolution 1750–1800,* ed. Andrée Carvol (Paris, 1993), pp. 175–81.

31. Frédéric Cuvier, *Notice des animaux vivans de la Ménagerie, leur origine et leur histoire dans cet Établissement* (Paris, an XII [1804]), pp. ix–x. The book was published anonymously, but Cuvier's authorship is certain, for he presented it to the assembly of professors on 20 thermidor an XII (August 8, 1804). See the Minutes des procès-verbaux des assemblées des professeurs for that date, AN, AJ/15/592.

32. For additional biographical information on Frédéric Cuvier see Pierre Flourens, "Eloge historique de F. Cuvier," *Mémoires de l'Académie des Sciences* 18 (1842): i–xxviii; Franck Bourdier, "Frédéric Cuvier," *Dictionary of Scientific Biography,* 3:520–21; and Robert J. Richards, *Darwin and the Emergence of Evolutionary Theories of Mind and Behavior* (Chicago, 1987), pp. 65–69.

33. Minutes des procès-verbaux des assemblées des professeurs, séance du 29 frimaire an 12 (December 21, 1803), AN, AJ/15/106.

34. Minutes des procès-verbaux des assemblées des professeurs, sessions of 29 frimaire an 12 (December 21, 1803) and 6 nivôse an 12 (December 28, 1803), AN, AJ/15/106.

35. Lamarck, "Discours d'ouverture pour le cours de 1816," in *Inédits de Lamarck,* ed. Max Vachon, Georges Rousseau, and Yves Laissus (Paris, 1972), p. 28.

36. See Richard W. Burkhardt, Jr., *The Spirit of System: Lamarck and Evolutionary Biology* (Cambridge, Mass., 1977, 1995).

37. Frédéric Cuvier, "Du rut," *Annales du Muséum d'Histoire Naturelle* 9 (1807): 119.

38. *Notice des Animaux vivans de la Ménagerie du Muséum d'histoire naturelle* (Paris, 1809), p. xii. Published anonymously, the book was written by Frédéric Cuvier. The earlier edition, as indicated above, was entitled *Notice des animaux vivans de la Ménagerie, leur origine et leur histoire dans cet Établissement.*

39. F. Cuvier, "Essai sur la domesticité des mammifères, précédé de considérations sur les divers états des animaux, dans lesquels il nous est possi-

ble d'étudier leurs actions," *Mémoires du Muséum d'Histoire Naturelle* 13 (1825): 420–21.

40. Ibid., p. 422.

41. As Cuvier explained in a letter of October 29 or 30, 1826, to P. Flourens, his "Essai sur la domesticité des mammifères" consisted of two parts, the first of which was devoted to responding "à ce qu'on avait dit, à propos de mes recherches sur les animaux vivants, que des animaux captifs ne pouvaient offrir aucune observation utile." This letter is reproduced and dated by Georgette Legée, *Pierre Flourens, 1794–1867: Physiologiste et historien des sciences,* vol. 1 (Abbeville, 1992), pp. 136–38.

42. See ibid., p. 139.

43. Frédéric Cuvier, "Note sur l'accouplement d'un zèbre et d'un cheval," *Annales du Muséum d'Histoire Naturelle* 11 (1808): 237–40; "Description d'un orang-outang, et observations sur ses facultés intellectuelles," ibid., 16 (1810): 46–65; "Observations zoologiques sur les facultés physiques et intellectuelles du Phoque commun (*Phoca vitulina,* L.)," ibid., 17 (1811): 377–97.

44. Of the 1,008 mammalian specimens added to the natural history cabinet over the fifteen-year period from June 1793 to the end of December 1808, 175 came from the menagerie. This was the third largest source of mammalian specimens during this period. Only the numbers of specimens acquired through Geoffroy's personal correspondence (193) and appropriated from the stathouder's collection at the Hague (176) were larger. These figures are taken from Etienne Geoffroy Saint-Hilaire, "Sur l'accroissement des mammifères et les oiseaux du Muséum d'Histoire naturelle, depuis le 10 juin 1793, epoque où j'ai été chargé de leur administration jusqu'au 1 janvier 1809," Muséum d'Histoire Naturelle, ms. 2290.

45. G. Cuvier, "L'Éléphant des Indes," in Lacépède et Cuvier, *La Ménagerie du Muséum national d'histoire naturelle.* The prosector was the *aide-naturaliste* Rousseau.

46. G. Cuvier, "Sur les éléphans vivants et fossiles," *Annales du Muséum National d'Histoire Naturelle* 8 (1806): 1–58, 93–155, 249–69.

47. The dates of construction for the different parts of the menagerie are taken from A. Milne Edwards, "Rapport sur l'état du ménagerie du Muséum," n. d., AN, AJ/15/844. See also Loisel, "Histoire de la ménagerie du Muséum," pp. 271–72.

48. Among Frédéric Cuvier's publications, in addition to those cited above, notes 37, 39, and 43, were "Observations sur le chien des habitans de la Nouvelle-Hollande, précédés de quelques réflexions sur les facultés morales des animaux," *Annales du Muséum d'Histoire naturelle* 11 (1808): 458–76; "Observations sur l'accouplement d'un cigne chanteur mâle et

d'une oie domestique femelle: Description du mulet qui en est provenu," ibid., 12 (1808): 119–25; "Essais sur les facultés intellectuelles des brutes," *Société philomathique, nouvelle bulletin* 3 (1813): 217–18; "Instinct," *Dictionnaire des sciences naturelles* 23 (1822): 528–44; "Examen de quelques observations de M. Dugald-Stewart, qui tendent à détruire l'analogie des phénomènes de l'Instinct avec ceux de l'habitude," *Mémoires du Muséum d'Histoire naturelle* 10 (1823): 241–60; "De la sociabilité des animaux," ibid., 13 (1825): 1–27.

49. Cuvier, "Description d'un orang-outang."

50. Cuvier, "Observations zoologiques sur les facultés physiques et intellectuelles du Phoque commun."

51. Cuvier, "Essai sur la domesticité des mammifères."

52. Cuvier, "Essai sur la domesticité des mammifères," p. 419; "Castor," *Dictionnaire des sciences naturelles* 7 (1817): 242–52. Cuvier reported one instance of a beaver that did not display any tendency to build, which he interpreted as the result of this beaver having been tamed while very young and never having had its instincts developed by any real needs. See Cuvier, "Le Castor du Canada," in [Etienne] Geoffroy Saint-Hilaire and Frédéric Cuvier, *Histoire naturelle des mammifères, avec des figures originales enluminées, dessinées d'après des animaux vivants,* 7 vols. (Paris, 1818–42), vol. 1.

53. Cuvier, "Description d'un orang-outang," pp. 56, 58, 62, 64.

54. Cuvier, "Observations zoologiques sur les facultés physiques et intellectuelles du Phoque commun," p. 392.

55. Lamarck, *Philosophie zoologique,* 2 vols. (Paris, 1809), 1:349–57.

56. Cuvier, "Essais sur les facultés intellectuelles des brutes," pp. 217–18 ("que les animaux ont, comme nous, la faculté d'être attentifs, de sentir, de juger, de se ressouvenir, etc., et que toutes nos actions, dans lesquelles nous ne faisons entrer que ces facultés, sont absolument semblables aux leurs; mais que nous commençons à nous distinguer des brutes lorsque nous appliquons à ces facultés celle de méditer").

57. Cuvier, "Instinct," p. 537 ("tous les animaux sans exception sont dépourvus du sens intime de la perception du moi et de la faculté de réfléchir; c'est-à-dire, de considérer intellectuellement, par un retour sur eux-mêmes, leurs propres modifications").

58. Cuvier, "De la sociabilité des animaux," p. 8.

59. Cuvier, "Observations zoologiques sur les facultés physiques et intellectuelles du Phoque commun," p. 393.

60. Cuvier, "Le Bouc de la Haute-Egypte," in Saint-Hilaire and Cuvier, *Histoire naturelle des mammifères,* vol. 1.

61. See Cuvier, "Essai sur la domesticité des mammifères."

62. See Cuvier, "Le Couagga," in Saint-Hilaire and Cuvier, *Histoire naturelle des mammifères,* vol. 3.

63. Cuvier referred to his plans to write a general work in his "Examen de quelques observations de M. Dugald-Stewart," pp. 241n, 258.

64. Cuvier's letter to the professors and his "Programme d'un cours sur la nature des animaux et leur éducation" are reproduced in Legée, *Pierre Flourens,* pp. 138–43.

65. Cuvier, "Considérations sur l'étude des actions des animaux, suivies du programme d'un cours sur la nature des animaux et sur leur éducation," *Annales des Sciences Naturelles; Zoologie,* 2d ser., 12 (1839): 141 (emphasis added).

66. Ibid., p. 142.

67. Minutes des procès-verbaux des assemblées des professeurs, January 31, 1837, AN, AJ/15/134; Minutes des procès-verbaux des assemblées des professeurs, February 14, 1837, AN, AJ/15/653.

68. See Geoffroy's "Rapport sur la Ménagerie. Lu le 21 février à l'assem. des professeurs," in Muséum d'Histoire Naturelle, ms. 2290; and the copy of the letter of February 22, 1837, from Geoffroy to Cuvier in Muséum d'Histoire Naturelle, ms. 2295.

69. Minutes des procès-verbaux des assemblées des professeurs, May 9, 1837, AN, AJ/15/653.

70. See Minutes des procès-verbaux des assemblées des professeurs, December 12 and 19, 1837, AN, AJ/15/653.

71. See Etienne Geoffroy Saint-Hilaire, *Fragments biographiques, précédés d'études sur la vie, les ouvrages et les doctrines de Buffon* (Paris, 1838), pp. 143–57.

72. Anonymous, "The Old Zoo and the New: I. The Bastille of the Beasts," *Saturday Review* 91 (1901): 330–32. The author of this article was Edmund Selous, an important pioneer of field studies of animal behavior in the early twentieth century.

8

Violent Health and the Moving Image
The London Zoo and Monkey Hill

Jonathan Burt

The Trouble with Zoos

In this essay I am going to look at aspects of London Zoo history during the secretaryship of Peter Chalmers Mitchell, which ran from 1903 to 1935. The essay begins with discussion of some general problems that relate to the theorizing of zoo history; changing ideas concerning animals in captivity as regards their health and representation during this period are then described in some detail; and finally, the essay proceeds to a detailed study of an exhibit known as "Monkey Hill," a community of Hamadryas baboons on an artificial rockscape opened in 1925, which serves to bring together the major themes of my argument.

The period of Mitchell's secretaryship was a time of considerable change and development at the London Zoo. It began with a systematic emphasis, for the first time in the zoo's history, on the health of captive animals, and culminated with the main achievement of Mitchell's secretaryship, the founding of Whipsnade, a park for animals in the English countryside thirty miles north of London that was opened to the public in 1931. This country park fulfilled many of Mitchell's ambitions for wild animals to be kept in the open air in "natural" surroundings. At the same time the original zoo in London was never more popular and saw increasingly large crowds, particularly after World War I, when attendance figures rose steadily from 1.5 million visitors a year to a record 2 million in 1924. Attendances around that figure continued for the rest of the 1920s. Bank Holidays also saw huge crowds of sixty to seventy thousand people pouring into the cramped thirty-four-acre site on the edge of Regent's Park, in the northwest of central London. In fact, on Easter Monday 1928, the figure reached 89,218.[1] The zoo also received considerable coverage from the press and, significantly, the growing cinematographic industry.

Whilst the zoo could always be said to have negotiated, with greater and lesser degrees of success, the differing needs of science and spectacle, this was a period in which these two principles were most intensively figured. Far from being opposed, they served each other. Healthier animals made better exhibits and cut down on costs. Furthermore, the kinds of health reform envisaged, providing the animals with fresh air and ultraviolet light treatments, for instance, were particularly appropriate to visual spectacle.

This period of London Zoo history reveals much about what zoos, and the practice of keeping animal menageries, tell us about human-animal relations in Western culture. Zoos have many functions. They are sites of fantasies about the exotic or the bizarre; they can be social meeting places or grand ornaments for the prestige of nobility, kings, and states; they can even be sites of experiment and eccentric architectural follies.[2] It would be tempting to generalize that they represent the stranger side of human-animal relations, although it is never clear in such relations what one might count as normal. In Western Europe we accept as "normal" the slaughter of millions of animals a year, both wild and domestic: but this level of killing can also be seen as in some sense pathological. Bearing in mind this tension between the strange, or uncanny, and the normal, I am going to pursue the idea that, even though zoos and their inhabitants may be the bearers of all sorts of cultural values, they are also the sites where these values actually break down.

Zoos are often places out of joint, as manifested by the well-recorded mixed reactions and sentiments of visitors; the discordant relations of animals to their surroundings; and, particularly of note with reference to the London Zoo, the internal divisions within the zoo hierarchy.[3] The zoo is not a microcosm of human-animal relations even though it involves a variety of activities that include acquisition, trade, and hunting; the large-scale politics and economics of the management of fauna; and science, breeding, education, and exhibition. It sits on the cusp of all these practices, but awkwardly so.

There are two related reasons why the quirkiness and individuality of the zoo as an institution have been obscured. The first lies in the fact that writers have underestimated the ways in which the animal can be a resistant object and fails to fit some of the grander anthropomorphically oriented cultural paradigms, such as those of empire, class, and capitalism.[4] Animals are also often treated as marginal, or liminal, figures. According to such schemes, the reasons why an animal may hold an observer in a kind of spell, and hence the basis of our fascination with animals, are lost in the idea that the animal constantly stands for something else. It is either the "raw material" for human representations of order, or the boundary marker between the human and "nature." A good example of the exorbitant symbolic weight that animals are presumed to carry can be found in R. W. Jones's recent study of the London Zoo in the nineteenth century: "animals were to be viewed as metonyms for imperial triumph, civic pride, the beneficence of God or scientific discovery."[5] What strikes one about this remark is less the extent of its truth, given that animals did stand for these things in many respects, than the fact that the animal appears to bear every aspect of Victorian society so unproblematically on its back. Indeed, attitudes to animals could be highly ambivalent.

Within the London Zoo for as many animals that were seen as reflections of imperial glory, there were others that elicited nothing but disinterest or even disgust. Murie's description of a *Macacus cyclopus* in estrus whose buttocks "acquired such hideous proportions, that it was necessary, for decency's sake, to remove her from the public gaze," is a case in point.[6] Moreover, animals that were traditionally seen as noble beasts could also be described by others as exactly the opposite. In Bennett's 1829 description of the Tower Menagerie in London he describes the lion, long the symbol of imperial and civic pride, as cowardly, bloodthirsty, and lazy.[7]

The location of cultural registers in projections onto animals is inevitable, but such projections are the outcome of a deeply ambivalent process. Indeed, the fact that anthropomorphism, which is both an intellectual and an emotional need, is a form of transformation should alert us to the instabilities of what is, symbolically, a very labile process. Running through all human-animal relations are various forms of fluctuation between, on the one hand, veneration or fascination and, on the other hand, repulsion. One manifestation of this can be seen in James Serpell's example of the way we admire the dog, which is of marginal economic use, as contrasted to the way we look down on the pig, which provides us with so much materially.[8] However, at another level these conflicting attitudes can come together around a single animal when we remember that the faithful dog is also a creature that "eats shit, sniffs genitals and bites people."[9] Some writers go so far as to suggest that the capacity for ambivalence and paradox actually marks the difference between human and animal.[10] According to this view, it is the structure of our thinking about animals that also marks what makes us different from them.

Another version of the anthropomorphic paradigm is where the zoo itself is compared to other kinds of enclosed human institutions such as prisons and mental hospitals. Certainly psychiatric patients and captive animals can show similar symptoms such as forms of depression and lack of will, or stereotypical movement.[11] However, types of animal will often vary in their susceptibility to these traits, whilst the variability and eccentricity of zoo architecture makes zoos very different, even unique, kinds of physical spaces.[12] To get further towards theorizing the idea of the zoo we have to look more closely at what marks it out from other kinds of institution or cultural practice. This also means less emphasis on what the zoo represents, and more on the particular types of experience of human-animal relations within the zoo.

One of the few texts that attempt to get to grips with the idea of the interlocking of human and animal within the space of the zoo, particularly with reference to the mutual gaze, is John Berger's essay "Why Look At Animals?" In his view the boundary between captive animals and the human crowd delineates two alienated groups, both marginalized from their worlds by the processes of capitalism. In place of an encounter with the animal is the sharing of a space that frames the impossibility, or disappearance, of such an encounter. For many people, according to Berger, the animal that appears in the zoo is frequently seen as "less than I believed," or an "image out of focus," rather than anything else.[13] I would argue, however, that the animal is very much in focus and starkly present before one's eyes. The mutual isolation of human and animal in the zoo highlights the strangeness of the zoo and confronts us with creatures without the frameworks, implied by various forms of romanticism, that might make the animal in some sense plenitudinous. Certainly the zoo exploits myths. Indeed, it frequently relies on them to attract spectators. But it also exhibits the problematic animal body in both its beauty and its physical vulnerability. If we have to look away from it to bring it into focus, then clearly there are psychological, not physical, barriers to our seeing the animal.

The first problem of zoos, therefore, lies in the peculiar disjunctions between the physical presence of the animal and the contradictory sentiments, fantasies, or anthropomorphisms we have about them. Second, even when animals do stand as visible emblems of human social relations, they embody the anxieties and violence of those relations. Thus, for all the pleasure and interest zoos can provide, they are also a particular site of cultural anxiety. In one sense, their combination of strangeness and banality arises from a mismatch of, on the one hand, those features of the animal that mark it out for display such as exoticism, ferocity, or beauty, and, on the other hand, its physical vulnerability and evident proneness to mortality. In the example I am going to describe it is worth noting that when the zoo attempts to bridge some of these internal contradictions the disjunctions become even more marked.

THE DISCIPLINES OF ANIMAL HEALTH

Although the London Zoo drew its ideas of health from contemporary preoccupations with the human body, it still set out to define itself as a strictly delineated community adhering to the regulations of the Zoological Society. Its preoccupations with public access to the Zoological Gardens and the privileges of its members, as well as its complicated ticket systems, all indicate this. It was, in many ways, a little fiefdom on the edge of Regent's Park. As Solly Zuckerman put it in his memoirs, "Peter Chalmers Mitchell ruled his domain with a rod of iron."[14] But although Mitchell sought to establish a functional and disciplined order in contrast to the regime that had preceded his own, he was not always successful. This was partly due to the divided nature of the Zoological Society, but also, as we shall see, to the vagaries of the animal body. To begin with it is important to set the context for the changes Mitchell and his regime introduced and then focus on the impact of this for human-animal relations.

In 1902 a committee was established by the Zoological Society of London to examine the state of its zoo.[15] In the copy of the committee's report in the London Zoo library there are annotations in the hand of one of the members of the council of the society, Henry Woodward. Through the following passage he has drawn a series of lines to indicate that it is too strong for the consumption of the fellows of the society, even if true. The passage runs: "one Superintendent generally crippled by gout, one Assistant Superintendent liable to epileptic fits, one Head Gardener who is unused to Menagerie work, one or two boy helpers and two Watchmen with a rooted objection to the system of tell tales are a wholly inadequate staff to rely upon to deal with an outbreak of fire or the accidental escape of an elephant or hippopotamus. It should be noted that when a fire did occur it was not the Society's Watchmen who gave the alarm, but the Watchmen on Primrose Hill."[16] The list of complaints was extensive and included the following: the boundary fences were insecure; the slaughterhouse was next to the place where animals were waiting to die, causing unnecessary distress; the housing of animals was in poor condition; some buildings, such as the Monkey and Antelope houses, were rife with tuberculosis; drunkenness had been reported amongst the keepers; and the death rates for animals were alarmingly high. "In some places the bleeding skins of freshly slaughtered animals were actually suspended over the heads of living goats. . . . The Bear Pit is a relic of the middle ages."[17] There are two strands to the report which are of relevance here. The first concerns the question of animal health and the second concerns the social class of those working at the zoo. As we shall see, the two strands are related.

The senior officers in charge at the zoo were skeptical with regard to the possibility of veterinary medicine curing sick animals. Frank Beddard, then prosector and vice secretary of the Zoo, felt that veterinary science was not advanced enough and only of use for dogs and horses.[18] The extent to which it was felt that little could be done is summarized by the secretary, Philip

Sclater: "I do not think myself there is much to be done, even in the case of a simple factor. It is very difficult to treat a sick animal."[19] Most disturbing of all was the fact that no one in the Zoological Gardens would be able to recognize a disease like anthrax.[20] Whilst there was genuine concern amongst members of the inquiry committee with regard to animal health it was also bound up with the issue of the social status of those presenting the public face of the zoo: ideally what was wanted was someone who was both a man of science and a gentleman. In fact, this had already been a consideration prior to the 1902 inquiry. During the questioning of Frank Beddard it was remarked that his appointment in 1898 had been due to the fact that the zoo needed a "man of science" to greet visitors of similar standing who "would naturally be disappointed on coming to the first Zoological Society's Gardens in the world if they met only a rough but practically sound good man, but who in no way represented men of science."[21]

This preoccupation with class not only affected the more senior management of the zoo but ran right through the zoo hierarchy, including keepers and laborers. Keepers had tended to be ex–farm laborers and the committee wished to recommend a more systematic method of recruitment to obtain "a better class of helpers, because the keepers as a rule are selected from amongst the helpers."[22] The types of men preferred were the sons of gamekeepers or countrymen. As a number of staff mention, in the years prior to the report many of the workers were city men or "ex–gentlemen's servants." Worse still, whereas in previous years they had tended to be uneducated, increasing literacy and foreign travel had raised the expectations of such workingmen particularly in relation to pay: "nowadays . . . many of these keepers buy books and things, and sit and read and all that sort of thing . . . (Question) You mean they waste their time reading books?"[23]

It is interesting that the changing social status of the keeper should have been of concern. The zoo wished to change the conditions of its community and its face to the world without changing the social relations within it. A more educated keeper should really have been a plus. In fact, the hierarchy of the zoo's organization became more sharply defined in the years coming up to World War I. In April 1909 the garden staff were reorganized into a more strictly regulated body with a greater emphasis on closer inspections and reports on feed, the conditions of the buildings, and sanitation.[24] Reforms included ensuring that keepers wore a visible designation of their rank, whether senior keeper, junior keeper, or helper, and a number on their uniforms. Furthermore, helpers new to the job were to be instructed specifically in "ventilation, cleaning, and disinfection." Thus, health and social reforms went hand in hand as the zoo sought to internalize a more scientific and functional order.

One way in which this new order was to be manifest was by an increasing emphasis on fresh air and light. The idea was, quite literally, to create a new atmosphere in the middle of London that would keep the animals healthy, and also, by implication, those around them. The emphasis on fresh air and light had been partly based on studies to improve the conditions for workers in factories so its public dimension was well in place. But, more significantly for the Zoo as a business based on spectacle, this was also a form of health that was both visible and "natural."

Carl Hagenbeck, the leading animal dealer in the world by the 1870s, famously gave the idea of fresh air as being necessary to animal health a considerable boost with his spectacular displays at his zoo at Stellingen, just outside Hamburg, which he opened to the public in 1907. However, a number of writers had already commented on temperature and atmosphere in regard to animals.[25] Studies of animal health in the zoo did appear in the *Proceedings of the Zoological Society of London* prior to 1903 although occasionally rather than systematically. In 1830 and 1832 J. C. Cox noted how animals endured great extremes of temperature but recommended that temperatures should be kept constant in the animal houses and should, as far as possible, imitate the climate from which the animals concerned came. However, he did remark that ventilation was crucial to prevent problems arising through the recirculation of damp air: "This is doubly necessary where the air is vitiated, not only by the animals themselves, but by numerous visitors."[26] A number of writers noted that, despite the long tradition of keeping animals in captivity, there was little in the way of either systematic investigation, or even a literature on the subjects of health and the treatment of disease.[27] Edwards Crisp remarked in 1860 that captive animals would make good experimental subjects given the advantages of knowing their exact diet and the fact that their "deviations from the natural state can be readily ascertained."[28] In an extensive study of diseases in the Zoological Gardens begun in association with the Pathological Society of London in 1881, J. Bland Sutton, a lecturer in comparative anatomy at the Middlesex Hospital, conceded that there was little to be learnt medically regarding the animals themselves, although there was a great deal to be gained regarding knowledge of human pathology.[29]

The risks of captivity to animal health were considerable. The high death rates were partly accounted for by the large numbers dying within the first few months of arrival at the zoo, due to either shock or the conditions of the journey. Monkeys, particularly, made up a high proportion of the deaths and were especially prone to lung diseases, such as tuberculosis, and stomach infections. For this reason it was felt that, from a health point of view, monkeys needed to be protected from people rather than the other way round. Other problems which arose as a direct consequence of the conditions of captivity included rickets and various forms of parasite.

From 1906 onwards, a regular annual report appeared in the *PZS* summarizing the mortality statistics in the Zoological Gardens. This was part of a systematic program to analyze the patterns of disease in captive animals and suggest ways of improving the situation. Clearly matters were serious. For example, C. G. Seligman noted that for the year 1905, apart from birds, primates were the highest casualties, with 148 deaths of which 11 were due to rickets, 53 to tuberculosis, 21 to pneumonia, and 29 to enteritis.[30] In his report for the following year Seligman noted a decline in tuberculosis attributed to a thorough cleaning of the Monkey House, although figures for other lung diseases, such as pneumonia, were unchanged. In 1910 Mitchell put into action a scheme to have all the viscera of dead animals from the zoo examined by a pathologist while further samples were forwarded to William Nicholl, at the Lister Institute, for parasitological work.[31] H. G. Plimmer, who wrote the mortality reports between 1909 and 1918, pointed to two areas that required particular attention. First, there was the fact that half of all animals died within their first six months at the zoo with most of these casualties occurring in the first three, which indicated a need for correct forms of quarantine. Second, the large numbers that died from inflammation of the alimentary tract would require an examination of diet by "experts."[32]

The problems faced by the zoo matched contemporary anxieties about the fitness of the population and the prevalence among humans of diseases such as tuberculosis, the principal cause of death amongst adults at the turn of the century, and rickets.[33] In keeping with issues being explored in society at large, the zoo looked for solutions in the areas of diet, fresh air, light, and cleanliness. In retrospective remarks celebrating the 1929 centenary of the founding of the zoo, Mitchell notes that the utility of animals to science begins when they are dead. However, he then says, "we have returned to one of our great scientific functions, and we have tried to study the relations of animals to their environment, especially as regards fresh air, heat, light, and food. We may claim that some of our most recent installations, notably the Baboons' Hill, [and] the new Monkey House . . . are pioneering experiments based on close investigation of the animal physiology."[34] These experiments were not really part of an investigative science in the strict sense of the term, but rather an intensification of natural processes (more air, more light) the purity of which was intended to provide for collective health. Indeed, for the zoo, one of the advantages of heliotherapy, which was used in the treatment of tuberculosis and certain forms of war wounds, was that it avoided the need for surgery and cut down the possibility of bodily mutilation.[35] As a preventive therapy it did not seek to cure particular diseases for individual animals. In fact, this lack of a sense of agency in the application of medical benefits seems apt for the display of the bodies of a community to be seen and appreciated by the community. A hygienic environment is, in effect, a constant and all-encompassing application.

These applications also had the advantage of economy. In fact, Mitchell recommended that housing for animals should be cheap and easily destructible so that changes could be made in response to scientific progress.[36] This had a parallel in public medicine. Improvements in ventilation, or, as Deborah Dwork notes, putting the onus of hygienic practice onto the woman in the home, were a lot less expensive than the wholesale reform of housing conditions or wages.[37] In the case of the zoo, added advantages would accrue if healthier animals lived longer and bred more efficiently. Mitchell noted, in a massive survey of the mortality of 20,000 animals in the Zoo between 1870 and 1902, that the main purpose behind the institution was not to collect new or rare species for the "glory of a Zoological Society," but to enable animals to live for the duration of their natural longevity.[38]

As we shall see with the Monkey Hill display, the showpiece manifestations of this thinking were bare environments which, by implication, avoided rich surroundings cluttered with physical objects. Instead, we find that they exemplified the process of, and gave clear passage to, the elements deemed important for health. In the Experimental Monkey House, a temporary building intended to exploit the advantages of fresh air and ultraviolet light, all the surfaces with which the animals came into contact were made of some form of impermeable material, whether glazed bricks, concrete, or asphalt.[39] A good example of the emphasis on bare function can also be found in a description of Lubetkin's Gorilla House, built in 1932, as "a scientific mechanical contrivance adapted to specific purposes, and the architects have tried rather to emphasize this purpose than to conceal it by architectural 'features.'"[40]

There were strong ideological agendas behind the desire for this basic environment, which were linked to changing attitudes to empire, the domestic interior, and the efficiency of the laboring body. Leonard Hill, who had published a number of studies concerning the necessary relation of fresh air and ventilation to health, labor, and a decent working environment, had been a colleague of Mitchell's at the London Hospital Medical School, and Mitchell claimed a direct influence from Hill's ideas.[41] Hill noted, in a series of books, that fresh air and sunlight promoted health and bodily vigor whereas confinement "in still, warm atmospheres and lack of open-air exercise dull the fire of life and, together with lack of sunlight, produce the deleterious effects of city life on the physical development."[42] Hill had humans in mind for his work but he frequently mentioned the relevance it had for animals and in one interesting passage conflated the two categories: "We want to aim at the perfect neuro-muscular condition of the wild animal."[43] The language of hardiness and severity runs throughout Hill's texts: a fit population will work hard and have, at the same time, a greater resistance to disease.[44] With reference to the question of empire it was also argued that tropical environments promote neither work nor health and indeed explain why the most vigorous, and therefore successful, races live in cold regions.[45]

Despite the severity and emphasis on discipline of much of this thinking, it is important to remember that the attitude of what could be described as "unsentimental caring" was the result of a far from simple combination of detachment from animals and concern for their welfare.[46] Mitchell notes in his autobiography, commenting on wild animal shows, "All my life I have tried to avoid these emotional judgements which interpret the behaviour of animals in terms of human feelings."[47] He also reveals a marked aversion to making pets of wild animals.[48] At the same time, he and the Zoological Society constantly sought to draw attention, or bring pressure to bear, on a variety of issues that related to the problem of cruelty. In fact, Mitchell saw little use for zoos unless they went hand in hand with efforts to preserve animals in their native habitats.[49]

Many of these sentiments also find expression in the pronouncements of the Zoological Society. As far as animal welfare was concerned, the position of the Zoological Society, in relation to both state authority and its own sense of its internal powers and interests, meant that it tended to pursue causes that directly related to its activities. The areas in which it was prepared to act, or not, were sharply and consistently defined. A brief survey of its administrative pronouncements during the 1920s and early 1930s reflects this well. Thus, it had a problem responding to a request from the University College of Science, Calcutta, with regard to proposed cuts in funding for the Zoological Survey of India. Although the Zoological Society agreed that this would be a great loss, "it was pointed out that there were difficulties in the form of any protest as the British authorities in India had been pressing for reduction in expenditure."[50] The society also turned down a request from the Royal Society for the Protection of Birds to exhibit seabirds killed by oil on the grounds that it was not suitable, although they did take an interest in the Oil Pollution Conference held in Washington in June 1926.[51] On the other hand, the society took a definite stand on vivisection. In answer to a letter from Stephen Coleridge concerning the appointment of Geoffrey Vevers as zoo superintendent in 1923, the society's council noted that Vevers had given up his vivisection license on entering his new post, and stated that it was "the practice of the Society not to permit experiments requiring licence under the Vivisection Act to be conducted on animals belonging to the Society or in their charge."[52] The society also set up a committee in August 1927 to examine the state of transportation of wild animals; attempted to ensure that no animals that it sold went to animal trainers; and in 1931 protested at the possibility of animal cruelty in the making of films.[53]

The fact that the zoo was keen to define its parameters of action carefully was paralleled by its control over imagery, most notably in the field of photography. Furthermore, the arrival of the rapidly growing cinematographic industry meant changing possibilities in the representation of animals. Animals were no longer to be seen as static images but as figures in events that would provide the narratives and stories that the public loved. Mitchell saw the implications of this very quickly and was enthusiastic about the possibility of representing the animal as a vital object, interacting with its environment.[54] Indeed, given that the early animal films were also part of a business that attempted to combine science and entertainment it is not surprising that there were close connections between the zoo and certain aspects of the film industry. In fact, exhibits like the Monkey Hill, where the attraction of the spectacle was rooted in the fact that the monkeys were constantly doing things, reflect a parallel between zoo exhibits and film displays, particularly during the 1920s.[55] The zoo established it as society policy from 1914 onwards that it should run its own filmmaking. This was reiterated after World War I: "the Society should exercise, through the Secretary, a right of veto, on the titling of films for exhibition."[56]

The four most notable filmmakers connected with the zoo before and after the war, Charles Urban, F. Martin Duncan, Percy Smith, and Mary Field, were all well suited to the agenda of combining science and entertainment. There was an added advantage to being associated with Charles Urban in that he had gained considerable respectability in a business that had aroused concern in many people about its relation to public morals.[57] Urban, who had pioneered as a mainstream commercial filmmaker much of the early filming of animals and natural history, assiduously invited dignitaries and foreign government officials to his shows and became something of a favorite with a Royal Family that had been initially hostile to the idea of motion pictures.[58] When he exhibited his newly developed color process in New York his main example was film from the London Zoo.[59] F. Martin Duncan, who worked with Urban, developed techniques for microscopic photography, and also exhibited to the society at the zoo. His work combined technological novelty with a concern for public health, particularly in relation to the study and identification of bacteria for the dairy and food industry.[60] From 1922 the Zoological Society allowed British Instructional Films to film in the zoo, and as a result, a series of films was made by Mary Field, who also worked with Percy Smith, called by the Daily Mail the "Darwin of the films."[61]

The impact of early cinema on the culture of animal exhibition, and vice versa, is yet to be fully explored. Most of the films of the London Zoo that survive in the period up to 1930 tend to show close-ups of animals in short episodes, even when the technology for longer continuous reels of film was already available. Although there are shots of spectators, there are few sequences where the camera draws back and takes in the zoo itself. Animals are seen to be doing things, although in the context of the zoo their activities are somewhat limited.[62] However, the importance of film is to propose to the viewer that even within the confines of the zoo the life of an animal is eventful and part of an ongoing story. The idea of nature as a visual narrative runs through every level of natural history films, from Percy Smith's speeded-up images of the growing flower, or of mold growing on cheese, to the larger-scale visions of Charles Urban. In 1921 Urban produced *The Four Seasons,* which followed the life cycle of animals during a single year; and then *Evolution,* in 1926, which began with the birth of the planets and ended with scenarios of animal and human development.

At the height of the zoo's popularity in the mid-1920s, the *Times* was prompted to comment on the extent of the public naming of animals: "Some of them have become both public characters and personal friends."[63] But many of these characters, in order to maintain interest, needed to produce outstanding events. One good example of these different registers coming together can be found in the treatment of the impending death of Sam the polar bear, a popular favorite, in August 1925. The zoo had announced that since the death of Sam's mate, his condition had rapidly deteriorated and he had become somewhat distressing as an exhibit. The press followed his last days with considerable interest, and a film was made entitled *Tragedy Of Sam* (Topical Film Company). The title boards run: "Tragedy of Sam / Zoo's idol for 24 years / Shot at dawn to end his infirmities / Mr. Brown Bear is brokenhearted / Mr. Grizzly grieves / Everywhere children are crying for . . . / Poor Old Sam!" Sam remained a star even in death. However, the idea of his decaying body alongside the inevitably burlesque cinematic treatment of his demise implies that he was the object of a peculiar combination of sympathy and detachment. It is difficult to determine historically the ways in which moving film may have changed, or merely intensified, emotional and visual attitudes to animals in the early twentieth century. But such films do reflect the fact that the humanizing of the animal is integral to the process of displaying the animal's alterity.

There is a parallel theme to this whereby the technology of film itself marks the alterity of the animal. In other words, film cannot help but draw attention to itself and highlight its own "spectacular" nature through its filming of animals. As natural history films developed through the twentieth century, they would always integrate the idea of a spectacular subject matter with a self-conscious technical virtuosity. This is the best example of the fusion of science and spectacle in the image of the animal. One way in which this was manifested in the case of the zoo was in films of animals listening to the gramophone or even watching films of themselves. The tradition of playing music to animals for quasi-experimental purposes is, in fact, a long one.[64] In 1910, Urban's shows for the Zoological Society included film of animals listening to the gramophone, whilst in 1932 Martin and Osa Johnson's film *Congorilla* was projected onto the back wall of the cage of Mok and Moina, the newly arrived and much-prized gorillas.[65] The disjointed relation of animals to their means of exhibition was never more clearly marked.[66]

Thus, at one level, as we have seen, the zoo came to express in varying degrees the naturalism of a healthy environment, and the realism of its depiction or spectacle. At the same time the animals became widely known via the press and the intensity of public interest. The zoo stressed, as far as was practicable, its scientific emphasis on the needs of the animal body whilst also making this emphasis an element of novelty in its attractions. The advantage of air and light was their transparency, which intensified the visibility of the animal body, rather like in film. Yet in the 1920s the zoo had also become increasingly claustrophobic with its large crowds. The two factors are related: as society becomes more dense, spaces become more necessary. However, there was a cost to this, as we shall see with Monkey Hill. The open-air community of monkeys would exemplify all the themes discussed in this section: the importance of public health and the idea of the animal exhibit as a spectacular event. Indeed, it was the one exhibit in the whole zoo that could be said to reflect most perfectly the fusion of science and spectacle. But, although it did bear on issues like community health and social relations, its strangeness showed that there was nothing quite like Monkey Hill. In this most anthropomorphic of exhibits we find the animals at their most distinct.

MONKEY HILL AND THE SACRED BABOONS

Many of the developments at the zoo that took place in the 1920s had, in fact, already been proposed in some form or other prior to World War I.

The constraints of space and the particular geography of the zoo meant that it could not develop anything like the grand vistas of Hagenbeck's displays. In Captain Swinton's "Report of Proposed Improvements in the Zoological Gardens," dated 1914, the lack of space and the need for a uniform design are stressed. Among his many suggestions he proposed to replace the old Elephant House, situated beside the Regent's Canal, which runs in a cutting with sloping sides on the zoo's northern edge. He writes, "This would be an ideal place for almost the whole Monkey and Baboon tribe, where they could have not only houses and cages but an open space or hill over which the Common Monkeys could play."[67] Part of the problem the Zoo faced with regard to development was financial, and in the end the various new exhibits there, such as the aquarium and Monkey Hill, and eventually also Whipsnade, had to be developed project by project as money allowed. It was, thus, never possible to institute a uniform design for the zoo. However, the ideas about health and environment could be said to have made up for that, which is one reason why they were so heavily emphasized. At least one could pursue a uniform principle, even if there could be no uniform design. It is also worth noting that Mitchell constantly complained in various memoranda about the resistance of many members to his progressive plans.

The Monkey Hill was part of a general project to reorganize the rather scattered displays and inadequate housing of monkeys in the mid-1920s. Monkeys were, and still are, an important attraction in any zoo and were most suitable for exhibits that maximized both the health ideal and the public interest. In the "Secretary's Memorandum on Garden Improvements," dated June 1924, proposals for the Experimental Monkey House and Monkey Hill were put forward together. In fact, the two displays were going to be placed near each other. The Hill itself was an artificial rock formation in the center of an oval measuring one hundred by sixty feet, with a surrounding ditch. Inside the hill there were heated quarters containing shelves and perches and also quartz bulbs for ultraviolet irradiation. Some of the ledges on the outside of the hill were also heated and contained quartz bulbs.[68]

In October 1924 Geoffrey Vevers, the zoo superintendent, went to Stellingen to have a look at the monkey hill display erected by the Hagenbecks. In this exhibit a variety of baboons were shown including the Arabian, Chacma, and Anubis species. Vevers was advised that only males should be shown "as they make a better display and do not fight so much as when they are in the company of females."[69] Although the London Zoo, famously, did not stick to this principle, the reasons were to some extent accidental, given that it had every intention of making the exhibit a single-gender one. During the trapping of the baboons in Abyssinia it had been hard to sep-

1. View of Monkey Hill from the nearby Mappin Terraces bear compound, 1931.

arate some of the females from the males and it appears as if it was less trouble to keep a handful of the females than to separate or kill them.[70]

The numbers of the baboon population on Monkey Hill vary very slightly in the different records and accounts.[71] This may seem a minor point but it exemplifies the fact that the display was seen in a discrepant way by both the zoo, the press, and the public. Basically, ninety-seven hamadryas baboons were placed on the Hill in May 1925 including six females. In June 1927, thirty females were added along with five immature males. The introduction of the second batch of females was intended "to make the conditions more natural" even though it exacerbated violence within the community for a while.[72] Whilst the initial appearance of the females was somewhat fortuitous, the increase in their numbers has to be seen as a mistake, although it meant that Monkey Hill became a more dramatic exhibit as the newcomers became a source of violent conflict in the ensuing months.

There were periods of time when little violence took place on the Hill. For instance, from May to December 1928 there were four male deaths, of which three were due to disease, and two females were killed. From 1928 the colony settled down to some extent and the annual figures for deaths until 1931, according to Zuckerman's charts, are five, two, three, and one from disease, and nine, three, four, and zero from killings. However, in figures that appear to cover the period from 1925 to 1930, when the last surviving females were taken off the Hill, Zuckerman notes that fifty-three males died from disease, and eight in fights; but female deaths included thirty in fights and three from disease.[73] Although many of the males died of various forms of lung disease and enteritis, none showed evidence of tuberculosis. The females died as "prizes fought for by the males. The injuries inflicted were of all degrees of severity. Limb-bones, ribs, and even the skull, had been fractured. Wounds have sometimes penetrated the chest or abdomen, and many animals showed extensive lacerations in the ano-genital region. . . . two died shortly after miscarriages apparently precipitated by fighting."[74] Most of the deaths that occurred on Monkey Hill were clustered around two periods: the initial arrival in 1925, and the addition of the extra females in 1927. In 1927, of the thirty new females placed on the Hill, fifteen were killed within two months, including eleven in July alone. Of the fifteen baboons born on the Hill, only one, a female, survived. Some died of strangulation due to be being held too tightly during fights.

In Mitchell's account of the Hill the more unpleasant aspects of the exhibit are played down. He wrote that the colony had survived extremely well. "As was inevitable in so large a number . . . there have been some deaths, but none that could be attributed to exposure. . . . On the whole,

2. Unpacking baboons on Monkey Hill, 1928.

these baboons lead a harmonious communal life." If anything his main concern was over the question of whether their tendency to copulate in public had a "demoralising" effect on visitors.[75]

The press reports provide an interesting slant on the exhibit emphasizing both the strangeness of the creatures and the conflicts arising from the disparity of the sexes. The strength and beauty of the males was the source of all the drama and excitement, and the press reveled in the contrasts between them and the "miserable" females. "DRUDGES. You know the women baboons at once. They are meek, subdued, tractable creatures. No beautiful cape for them. They live on the leavings and take life sadly."[76] Descriptions of the strong male leader with as many wives as he could get, and what was called either the bachelor, aide-de-camp, or lodger, gave the impression of admirable, well-defined groups. But for all the press rhetoric, it was obvious from these accounts that the colony could be highly vicious.

That the baboons seemed to have a sense of death was the main source of the idea that they were somehow uncanny. Writers connected them, in the light of the recent discovery of the grave of Tutankhamen, with the Egyptian symbolism of the baboon. Baboon mummies had been unearthed in quantities in ancient cities and it was also noted that they were dedicated to Thoth, the god of ancient words, who would take the form of a baboon when announcing to Osiris whether the soul about to pass into the underworld had led a blameless life. The mythic aspect of the baboons fed stories of their so-called "death rites." After one particularly disturbing fight the baboons were described as suddenly going quiet when they realized that a female was dead. They lined up in two rows, while a few baboons rolled the body between the rows down into the water. "Twenty or thirty of them watched the body disappear into the pond and then one by one they stole away without another murmur and went inside their dens."[77] Wherever these stories came from, they would certainly have been influenced by the fact that the baboons were very loath to give up dead bodies to the keepers. Furthermore, it was frequently the case that a mother would carry a dead infant body around for some days and the corpse would become increasingly putrescent.[78]

In Zuckerman's notebooks on Monkey Hill one can get some sense of what this exhibit was like to an observer.[79] Most of his observations are sketchy and routine, and mainly note visible signs of the reproductive cycles of particular individuals. Longer passages can be very striking indeed. Although they are not frequent, they remind us that when events did happen on the Hill they could be disturbing and grotesque. "Babe with a Redface bach[elor]. Who walks over to another bach[elor] no sound. Stands over him and has his erect penis gripped. He has had an erection playing

3. A group photo taken in 1925.

with babe."[80] Spotty's baby dies: "The body seems a bag of decomposition";[81] "Spotty carries her babe in her left hand by the remains of its head or its hand. She puts it down frequently tastes smells."[82] "Scraggy's babe dead. Green and decomposed. . . . The body lies limp over her arm. She feeds with the thing between her knees. . . . [Scraggy still carrying the remains of her infant] Said to have been biting at its head."[83] "Motherbabe looks on very interestedly while Cripple masturbates and fellates himself."[84] It was also the case that weak, or sickly, baboons would be picked on and attacked. The following notes concerned a baboon who was evidently sick, and also conveys the way that a particular behavior, such as fighting, could spread very quickly through the community. The keepers had reported a bachelor collapsed and surrounded by a hostile crowd: "Head in shoulders. Pale face. Laboured shallow breathing. . . . Others all interested and threatening barking, but kept back by keeper's threats. . . . Attention of whole Hill conc[entrated] on him." The decision is taken to kill him: "Shot in left shoulder and both lungs penetrated. Pleural cavities filled with blood. Defecated on point of death. Dragged off by grappling hook."[85]

Monkey Hill was of primary importance to Zuckerman's thesis on the centrality of dominance and sexual receptivity in social relations. Whilst he did not avoid the issue of violence he saw the artificiality of the environment as intensifying aspects of social behavior, rather than changing them substantially. As noted above, the air and light therapy implied a principle of noninterference with the community of animals. The environment was created in order that they might live in a healthy and "natural" way, in some sense. Zuckerman believed that this principle enabled the animals to behave as they would in the wild. The only difference was that, since their food was provided for them, the foraging energies of the baboons were released into other channels. Zuckerman wrote that "there is no reason for supposing that [sexual behavior] is artificially perverted. . . . from the point of view of the observer, confinement concentrates a normal response both temporally and spatially."[86] However, he did admit that the exhibit might have worked better if the females had outnumbered the males, and that "the social group of which they formed a part was in some way unnatural."[87] Zuckerman saw the colony very much as a laboratory in that certain features of behavior were isolated and could be studied undisturbed. Confinement in itself did not qualitatively alter the social behavior of the baboon.

As we can see from Zuckerman's notebooks, Monkey Hill is above all a visually accessible scientific object. The phenomena that Zuckerman singles out, the cycles of estrus and sexual behavior, are those that are most strikingly visible. His analyses of social behavior are not simply formulated by attention to modes of contact, or conflict, between the baboons but in the main by a specific focus on the varying states of the highly marked female genitalia. This was also something that was measurable and could be backed up by pathological examinations to determine the relation between what was going on on the surface of the genitalia and what was going on within the animal. Furthermore, Zuckerman saw his work as avoiding the anthropomorphism that he felt surrounded all previous observations of baboons. In a letter to V. Stefansson he wrote, "I was brought up in South Africa, and, long before I had ever seen a Baboon, I had heard innumerable tales about their human qualities, about their leaders, sentries, military organisation, skill, and the beautiful way they behaved towards each other. All this influenced my first observation of these animals, but closer study soon showed me that most of the tradition was mere fantasy."[88]

More recent studies of the hamadryas baboon conducted in the wild show that Zuckerman's description of its social structure is not wrong in its basic outline. As Hans Kummer notes in his extensive studies of both captive and wild hamadryas baboons, the basic social unit seen in the zoo is the same as that found in the wild: the dominant male with his group of females. Indeed, the structure of this unit and the strictness with which it is maintained is peculiar to the hamadryas and is not found in other savanna baboons nor most other Old World primates.[89] In observations of a group kept at Zurich Zoo Kummer and Fred Kurt noted, like Zuckerman, that forms of sexual behavior took up a great deal of time as compared to the wild. They concluded that although actual copulation was no more frequent in the zoo than in the wild, mounting, presenting and other activities definitely were.[90] It is also worth noting, with regard to the question of violence, that in fights observed in the wild killed or severely wounded animals were never found.[91] The implication is that the violence found in the wild does not result in fatalities in the way it did on Monkey Hill.

The accounts of Monkey Hill, whether they are scientific, journalistic, or part of the zoo's expression of its own achievement, all play on a variety of registers that include sexuality, gender difference, the culture of violence, and politics. Indeed, one could make Monkey Hill something of an exercise in cultural overdetermination and extend the cultural resonances to include the maimed bodies of World War I veterans, the Surrealist and modernist interest in the primitive, or the contemporary vogue for things Egyptian.[92] However, for all the apparent clarity of hamadryas social organization, the baboons also cut an indeterminate figure between health and perversity, and "naturalness" and artificiality. It is this that enabled the possibility of their symbolic overdetermination. In other words, it was their indeterminacy that allowed the wealth of anthropomorphic projections that the exhibit could generate, whilst at the same time making the anthropomorphism

4. Onlookers from the perspective of Monkey Hill, late 1920s.

5. Popular awareness of Monkey Hill extended to political cartoons. (*Evening Standard,* November 6, 1930.)

symbolically unstable. If these monkeys were taken as visible emblems of human social relations, they also represented the anxieties and disorder of such relations.

The importance of sexual difference, which is one of the dominating features of the exhibit, for capturing public attention, at least as it was presented by the press, was highlighted by the fact that when the females were removed from the Hill in 1930, the newspapers ceased to cover the exhibit. "Men are a peaceable race, once you eliminate the women. It has proved a happy thought, this clearing out of the female baboons from Monkey Hill, where the bad girls from Abyssinia used to raise whoopee."[93] In the light of the violence of Monkey Hill this seems in poor taste. In fact, it reminds us of another dimension of anthropomorphization which concerns not simply the projection of human values onto animals, but a parodying of both human and animal.

CONCLUSION

The conclusion that I wish to draw from this general account of the London Zoo in the opening decades of the century, and Monkey Hill in particular, concerns the way in which we might analyze the zoo as a unique space in which certain specific forms of human experience with animals are enacted. This is a different perspective on zoo history from that which sees the zoo as an exercise in the cultural categorization of nature and the symbolic reinforcement of human power over animals and, in some contexts, colonial subjects. Such perspectives play down the ambivalences and paradoxes that constitute human projections onto the animal world, as well as avoiding the more immediate issue of the tension between animal alterity and anthropomorphism that characterizes the zoo experience. Furthermore, the fact that the London Zoo, in this particular example, saw itself as negotiating the differing requirements of a science, concerned with health and environment, and of public spectacle and entertainment, required a certain flexibility of policy which changed with time and still does to this day. In other words, zoos themselves can be quite changeable entities.

In 1927 the zoo launched, for the summer months, the Chimpanzees' Tea Party, described by the *Daily Mail* as "Zoo Tea Room for Apes—Lessons in Table Manners."[94] The animals' orderly behavior, passing round mugs and plates of food, clearing the table, and so forth, drew large crowds. It seems significant that this exhibit coincided with Monkey Hill. The baboons represented a society of hierarchy and ritual, but they had no decorum. Taken together, the two exhibits point to the fragmentary nature of anthropomorphism. If humans are equally capable of violence and decorum, then animals, inevitably, are only ever partial representatives of our behavior. In a sense it could be said that the mirroring that anthropomorphism implies means that certain features we project onto animals return to us transformed. In such instances the zoo becomes a space in which human and animal identities become mutable. In fact, one of the sources of the attraction of zoos may therefore be a combination of a sense of comfort in the different and curious nature of the animal body, and a sense of anxiety about our own curious practice of keeping them there so close to us.

NOTES

I could not have written this piece without my fellow zoo historians, Robert Bieder, Rory Browne, and John Edwards; and my fellow anthropomorphizer, Nicolette Zeeman. I am extremely grateful to them. I would also like to thank the curators and assistants who provided such considerable help and information: Ann Sylph and Michael Palmer at the Library of the Zoological Society of London; Deirdre Sharpe and Michael St. John at the University of East Anglia Archive, U.K.; Simon Baker at the National Film and Television Archive, London; Robert Sharp at the Science Museum Library, London; and also Professors Mary Henninger-Voss and William Jordan at the Shelby Cullom Davis Center, Princeton, where a version of this paper was first given in October 1996; David Feldman, and Ian Patterson.

1. *Reports of the Council and Auditors of the Zoological Society of London,* 1928 (hereafter *RCZS*). This area was at the time just over a third the size of the Hagenbeck Tierpark at Stellingen; the New York Zoological Park in the Bronx, which was one of Mitchell's inspirations for the animal park at Whipsnade after a visit in 1907, was 270 acres. See his memorandum entitled "Policy of Council Regarding Possible Dissolution of the Botanic Society," November 11, 1907, in which he lays out ideas for purchasing land outside London. Chalmers Mitchell Historical Papers File, 1908–1927, Library of the Zoological Society of London, Regent's Park (hereafter HPF).

2. The main source of information is still Gustave Loisel, *Histoire des ménageries de l'antiquité à nos jours,* 3 vols. (Paris, 1912).

3. This has been a particularly marked feature of the Zoological Society of London from its inception, one reason for which is the heterogeneous composition of its membership. See Adrian Desmond, "The Making of Institutional Zoology in London: 1822–1836," *History of Science* 23 (1985): 153–85, 223–50, esp. p. 228.

4. Well-known examples that do not take enough account of the countercurrents to anthropomorphism include Donna Haraway, *Primate Visions: Gender, Race, and Nature in the World of Modern Science* (London, 1992); and Harriet Ritvo, *The Animal Estate: The English and Other Creatures in the Victorian Age* (Cambridge, Mass., 1987). See also Kay Anderson, "Culture and Nature at the Adelaide Zoo: At the Frontiers of Human Geography," *Transactions of the Institute of British Geographers,* n.s. 20 (1995): 275–94.

5. Robert W. Jones, "'The Sight of Creatures Strange to Our Clime': London Zoo and the Consumption of the Exotic," *Journal of Victorian Culture* 2 (1997): 5.

6. James Murie, "Observations on the Macaques: III. The Formosan or Round-faced Monkey," *Proceedings of the Zoological Society of London* (hereafter *PZS*) (1872): 772. I accept that this concerns a different kind of cultural register relating to the issue of public sexuality, decency, and the female body. But, in Murie's case, there is also a sense that the extremity of this example sets it apart from anything he has seen before.

7. E. T. Bennett, *The Tower Menagerie: Comprising the Natural History of the Animals Contained in that Establishment; with Anecdotes of their Character and History* (London, 1829), p. 19. See also Bennett's remarks on tigers, ibid, pp. 29–30; for examples of the civic value of lions in European history, see Loisel, *Histoire,* 1:152–57. On the internal contradictions of imperialism with regard to the environment and animal conservation, see Richard Grove, *Green Imperialism: Colonial Expansion, Tropical Island Edens and the Origins of Environmentalism 1600–1860* (Cambridge, 1995).

8. James Serpell, *In the Company of Animals: A Study of Human-Animal Relationships* (Cambridge, 1996), pp. 5–20.

9. James Serpell, "From Paragon to Pariah: Some Reflections on Human Attitudes to Dogs," in *The Domestic Dog: Its Evolution, Behaviour and Interactions with People,* ed. James Serpell (Cambridge, 1995), p. 254.

10. Jeffrey Masson, *Dogs Never Lie about Love: Reflections on the Emotional World of Dogs* (London, 1997), p. 26; Marjorie Garber, *Dog Love* (London, 1997), p. 137.

11. H. F. Ellenberger, "Zoological Garden and Mental Hospital," in *Psychopathology in Animals: Research and Clinical Implications,* ed. J. D. Keehn (London, 1979), pp.17–30.

12. For a very interesting discussion of this see Robert Bieder, "The Constructed Gaze: The National Zoological Park and the Representation of Animals" (unpublished ms., 1996). My thanks to Robert Bieder for allowing me to cite this article.

13. John Berger, "Why Look at Animals?" in his *About Looking* (London, 1980), p. 21.

14. S. Zuckerman, *From Apes to Warlords: An Autobiography* (London, 1988), p. 37.

15. The conflict behind this report and the subsequent election of Peter Chalmers Mitchell as secretary of the London Zoo in 1903 has been excellently described by Rory Browne, "Between Menagerie and Zoopark: The Movement to Reform the Zoo at the Turn of the Century" (unpublished ms., 1996). My thanks to Rory Browne for permitting me to cite this article. The committee was chaired by the Duke of Bedford and the proceedings were printed verbatim as *Report of a Committee Appointed to Enquire into the Organisation of the Society's Offices and Gardens* (1902) (hereafter cited as *Report*).

16. *Report,* p. 2. Primrose Hill is a small park immediately to the north of Regent's Park.

17. Ibid., p. 1.

18. Ibid., p. 9.

19. Ibid, p. 76. See Browne, "Between Menagerie and Zoopark," pp. 18–22.

20. *Report,* p. 41. Ironically, there was an outbreak of anthrax amongst the elephants in December 1926 and the zoo was fortunate not to have any human casualties in addition to the animals that died.

21. Ibid., p. 14. This is reiterated by Sclater who obviously feels that the current superintendent, Clarence Bartlett, is not educated enough (ibid., p. 76). What is significant about these exchanges is the idea that "scientific" gentlemanly knowledge needs to replace, or supplement, the hands-on practical knowledge of the common man, and, by implication, the keepers. In many ways this would serve to reinforce various kinds of conflict within the zoo. Thus, Zuckerman in a letter of January 9, 1932, to D. Morris, one of his laboratory assistants at the Zoo between 1931 and 1932, writes, "Your remarks re. Shelley, Bud and Harwood [the monkey keepers at the time] are interesting, but I find the attitude of the three somewhat difficult to understand. . . . I fear that they consider any attention, other than their own, paid the monkeys to be uncalled for whether or not it be scientific, and I am quite certain they do not appreciate the nature of scientific work." Solly Zuckerman Papers (hereafter SZ), file no. SC/1, University of East Anglia, Norwich, U.K.

22. *Report,* p. 4.

23. Ibid., p. 29.

24. Minutes of Council (hereafter CM), April 7, 1909, Library of the Zoological Society of London, Regent's Park. The Council met once a month and coordinated the committees of the society, which included the Finance Committee, the Prosectorial Committee, and the Gardens Committee. I am

deeply indebted to Rory Browne for pointing this passage out to me and for emphasizing the inward-looking aspect of the zoo.

25. See Loisel, *Histoire,* 2: 324–25, on mid-eighteenth-century experiments of leaving sheep out in the cold; for Carl Hagenbeck see his *Beasts and Man* (London, 1909); also H. Reichenbach, "Carl Hagenbeck's Tierpark and Modern Zoological Gardens," *Journal of the Society of Bibliography in Natural History* 9 (1980): 573–85; Nigel Rothfels, "Bring 'em back alive: Carl Hagenbeck and Exotic Animal and People Trades in Germany 1848–1914" (Ph.D. thesis, Harvard University, 1994).

26. J. C. Cox, "On the Preservation of a Proper Temperature for Exotic Animals," *PZS* (1830): 19; also his "On Atmospheric Causes as Influencing the Health of Exotic Animals Kept in Confinement in England," ibid. (1832): 33–38. In 1844, a Professor Percy wrote on the conditions for the keeping of monkeys, and he also noted the importance of fresh air and the risk to health of keeping animals at a uniform temperature: "The object in the construction of a menagerie for monkeys should be to enable these animals as much as possible to keep themselves warm." Percy, "On the Management of Various Species of Monkeys in Confinement," ibid. (1844): 82. David Armstrong notes that the first open-air TB sanatorium was built in 1840, whilst the first open-air school was opened by the London County Council in 1907; D. Armstrong, *Political Anatomy of the Body: Medical Knowledge in Britain in the Twentieth Century* (Cambridge, 1983), p. 16.

27. For a perspective outside London see M. Schmidt, "On the Duration of Life of the Animals in the Zoological Garden at Frankfort-on-the-Main," *PZS* (1880): 299–319.

28. E. Crisp, "On the Causes of Death of the Animals in the Society's Gardens, from 1851 to the Present Time," ibid. (1860): 176. Crisp also had a penchant for cooking and eating the animals he dissected. See E. Crisp, "On the Probable Cause of Death of a Wart-Hog," ibid. (1862): 212; and his "On Some Points Connected with the Anatomy of the Hippopotamus (*Hippopotamus amphibius*)," ibid. (1867): 601.

29. See, for instance, J. Bland Sutton, "On the Diseases of Monkeys in the Society's Gardens," ibid. (1883): 581–86; his "On the Diseases of the Carnivorous Mammals in the Society's Gardens," ibid. (1884): 177–87; and his "On Some Specimens of Disease from Mammals in the Society's Gardens," ibid. (1886): 206–17.

30. C. G. Seligman, "Notes on Deaths Occurring in the Society's Gardens," ibid. (1906): 234–36.

31. W. Nicholl, "Statistical Note on the Worm Parasites Collected from the Animals Dying in the Zoological Gardens from December 1910 till April 1912," ibid. (1912): 856–58.

32. H. G. Plimmer, "Report on the Deaths Which Occurred in the Zoological Gardens during 1913, Together with a List of the Blood-Parasites Found during the Year," ibid. (1914): 186.

33. See D. Dwork, *War Is Good for Babies and Other Young Children: A History of the Infant and Child Welfare Movement in England 1898–1918* (London, 1987), p. 61.

34. *RCAZS* (1929): 86–192. However, Mitchell was well aware of the restricted amount of practical knowledge that his research was likely to provide, given the limitations of both animals and the zoo environment; see P. C. Mitchell, *Centenary History of the Zoological Society of London* (London, 1929), p. 200. It is worth noting that there was a constant international background to these changes. See Mitchell's "Report on a Visit to Brussels, Berlin, and Dresden in December 1913," January 1, 1914 (HPF), where he observes the great change that has taken place with regard to exhibiting animals outdoors in recent years in Berlin and Dresden. He adds that London Zoo has much improved since his first official visit to Germany and that "we have now surpassed the German Gardens in cleanliness, absence of smell, and general care of the animals."

35. See R. A. Hobday, "Sunlight Therapy and Solar Architecture," *Medical History* 41 (1997): 464. Hobday also notes examples from the first three decades of the century whereby architectural and town planning proposals were made which incorporated ideas about the therapeutic benefits of sunlight.

36. P. C. Mitchell, *My Fill of Days* (London, 1937), p. 244.

37. Dwork, *War is Good,* p.19.

38. P. C. Mitchell, "On the Longevity and Relative Viability in Mammals and Birds; With a Note on the Theory of Longevity," *PZS* (1911): 545. Mitchell also introduced a card system for documenting the life and circumstances of every creature that came into the zoo. Some of his thinking about longevity was inspired by the work of Ilya Metchnikoff, two of whose books Mitchell translated: *The Nature of Man* (London, 1903), and *The Prolongation of Life* (London, 1907). For Metchnikoff's interest in diet, see L. M. Barnett, "'Every Man His Own Physician': Dietetic Fads, 1890–1914," in *The Science and Culture of Nutrition,* ed. Harmke Kamminga and Andrew Cunningham (Amsterdam, 1995), pp. 166–67.

39. P. C. Mitchell, *Centenary History,* p. 204. Experiments involving irradiating animals with ultraviolet light, often termed "artificial sunshine," had been performed in a variety of ways using either special bulbs, or a glass known as Vitaglass which did not filter out such rays from sunlight. A number of press commentators pointed out the incongruity of providing monkeys with this treatment when it was not made available to children

and adults living in the poorer parts of London. See *Daily Chronicle,* September 26, 1926; and *Referee,* June 3, 1928.

40. *RCAZS* (1932): 40. Lubetkin was well aware of the question of health and, amongst other ideas, drew on designs he was also working on in 1932 for a chest clinic in East Ham, London, when planning the Gorilla House. In John Allan's monograph on Lubetkin he notes that Mitchell had the idea of providing a removable screen which gave summer and winter permutations to the design making it more adaptable to changes in climate through the year: J. Allan, *Berthold Lubetkin: Architecture and the Tradition of Progress* (London, 1992), pp. 114–17, 202. See the celebration of air and sunlight in Lubetkin's Finsbury Health Centre, built 1935–38; Peter Coe and Malcolm Reading, *Lubetkin and Tecton: Architecture and Social Commitment: A Critical Study* (London, 1981), pp. 140–44.

41. Mitchell, *Centenary History,* p. 189.

42. L. Hill, *Sunshine and Open Air: Their Influence on Health with Special Reference to the Alpine Climate,* 2d ed. (London, 1925), p. 107.

43. Ibid., p. 102: see also p. 99.

44. L. Hill and Argyll Campbell, *Health and Environment* (London, 1925), p. v; L. Hill and Mark Clement, *Common Colds: Causes and Preventive Measures* (London, 1929), p. 68.

45. On heat and the empire, see Hill and Campbell, *Health,* pp. 96–97. For the implications of this "politics of the cold" in ecology movements of the period see Anna Bramwell, "Back to the Northland," in her *Ecology in the Twentieth Century: A History* (New Haven, Conn., 1989).

46. For a comparable attitude see Winifred Felce's memoir of working with apes at Munich Zoo between 1931 and 1939. There is the same combination of a distancing from sentimentality, a celebration of fitness, and recommendations for the corporal punishment of chimpanzees. This is directly contrasted with the cruelties perpetrated on them by African natives; she remarks that the chimpanzees tended to throw excrement at chimney sweeps and coalmen, mistaking them for Africans; Winifred Felce, *Apes: An Account of Personal Experiences in a Zoological Garden* (London, 1948), p. 48.

47. Mitchell, *My Fill of Days,* p. 122.

48. Ibid., p. 216.

49. Mitchell was president of the Society for the Protection of the Wild Fauna of the Empire from 1923 to 1927, and vice chairman from 1927 until his death in 1945. See F. Hindle, "Peter Chalmers Mitchell 1864–1945," *Obituary Notices of Fellows of the Royal Society* 5 (1947): 370. See Mitchell's worries about the fur trade, *PZS* (1918): 195–96; and for his

preference for the camera over the gun see Mitchell's remarks in his introduction to Hagenbeck *Beast and Man,* pp. vii–viii.

50. CM, November 18, 1931.

51. CM, June 5, 1925, and August 18,1926.

52. CM, February 21, 1923. Vevers had required a vivisection license for his research, which necessitated feeding live parasites to animals, prior to joining the zoo. It is worth noting that in answer to another letter from Stephen Coleridge in May 1933 with regard to the society's honorary parasitologist, Professor Leiper, the council noted that his vivisection license did not apply to the Zoological Gardens but to his laboratory work elsewhere, so that they saw no need to dismiss him; CM, May 17, 1933. On Stephen Coleridge (who was in fact a member of the Zoological Society), and his protests concerning the zoo and vivisection, see Coral Lansbury, *The Old Brown Dog: Women, Workers, and Vivisection in Edwardian England* (Madison, Wis., 1985), pp. 9–12. In a protest meeting against keeping animals and birds in confinement Coleridge had said that the London Zoological Gardens "were a source of sadness . . . and [were] made more poignant by the dull insensibility of the staring crowds," *Daily Telegraph,* March 6, 1926.

53. CM, March 19, 1924, and March 18, 1931. See also the *RCAZS,* (1934): 13–14, where this disquiet about animals in the cinema is repeated.

54. His lectures on film do not appear to have survived, but see his remarks as recorded in *Times,* March 17, 1922. Mitchell was highly impressed with Charles Urban, who was one of the first makers of a considerable number of natural history films, and fully realized the benefits of film for the zoo at a number of levels: "I fully recognise that your exhibitions are of value to this Society, directly as advertising in the Gardens and indirectly as causing an interest in Zoology." Letter to Charles Urban, February 1, 1907, in file no. URB1, The Charles Urban Collection (hereafter URB), Science Museum Library, South Kensington, London. In fact, Mitchell is quoted on the subject of the benefits of cinema for zoology in an advertising pamphlet for a chain of cinemas, c. 1909. See Clegg, *A Revolution in the Worlds of Amusement* (London, n.d.), p. 14, in file no. URB 2.

55. Zoo animals appeared in the first films shown by the Lumière Brothers in London in February 1896. See J. Hunningher, "Première on Regent Street," in *Cinema: The Beginnings and the Future,* ed. Christopher Williams (London, 1996), pp. 41–54. On listings of the earliest animal films see also Simon Popple, "The Diffuse Beam: Cinema and Change," ibid., p. 105; John Barnes, *The Beginnings of Cinema in England* (Newton Abbot, 1976); Barnes, *The Beginnings of Cinema in England 1894–1901,* vol. 2: *1897* (Exeter, 1996); Stephen Herbert, in association with Colin Harding and Simon Popple, *Victorian Film Catalogues Reprinted from the W. D. Slade Archive* (London, 1996); Colin Harding

and Simon Popple, *In the Kingdom of Shadows: A Companion to Early Cinema* (London, 1996), p. 40. Certainly a number of early films of the London Zoo still survive in the National Film and Television Archive, London, such as *Pelicans at the Zoo* (British Mutoscope and Biograph Company, 1898); an English film with German titles, *Zoologischer Garten in London* (Warwick Trading Company, 1905); *Visite au Jardin zoologique de Londres* (Pathé Frères, 1909); and *Natural History Studies at the Zoo* (British and Colonial Kinematograph Company, 1911). On the impact of moving film on the study of natural history, see Greg Mitman, "Cinematic Nature: Hollywood Technology, Popular Culture, and the American Museum of Natural History," *Isis* 84 (1993): 637–61.

56. CM, April 19, 1922. There was some resentment over this exclusivity. The *Daily Express* noted that what it saw as the highly elitist Zoological Society barred the ordinary film photographer "lest he should convey to the world at large the interesting features of the Zoo for which the public pay." "Zoo Films for Greybeards," March 17, 1922.

57. Moral and physical health was also an issue of concern both in the debate on the possible effects of cinema on eyesight, and because cinemas were thought of as places in which diseases could be easily caught. In 1917 the National Council of Public Morals published *The Cinema: Its Present Position and Future Possibilities* (London). The inquiry combined the issues of light, health, and the conduct of audiences (the composition of which was usually seen as working-class), and focused, amongst other things, on the problem of "The Moral Danger of Darkness" (p. xxv).

58. See the *London Standard & Evening Standard,* September 14, 1909, for his relations with royalty.

59. For a description of this, see "Kinemacolor: First American Exhibition of Moving Pictures in Natural Colors, Madison Square Garden Concert Hall, New York, December 11th, 1909, Special Criticism of the Process," *Moving Picture World,* December 18, 1909, pp. 873–74. He showed these films to the Zoological Society in 1910. See *PZS* (1910): 252.

60. Urban also made films concerning health, including ones on malaria, hygiene, and the house fly, and on the benefits of physical fitness and being outdoors. For an example see the list and description of these in "List No.1 Kineto Review—The Living Book of Knowledge" (c. 1921), file no. URB 3.

61. December 2, 1931. Percy Smith was an ingenious filmmaker who had worked for Charles Urban from 1908 and made a series of films at the zoo in 1911. Among other things, he was renowned for making stop-motion studies of the growth of plants and his film *Birth of a Flower* (1911) was a great success. Most of his work was done on a shoestring budget and in 1925 he moved to British Instructional Films. See the papers in URB 8: the 1911 films are listed in a notebook entitled "Some of My Film Work

and Some of Its Associations during the Year 1911." See also Mary Field and Percy Smith, "Zoo Films," in their *The Secrets Of Nature* (London,1934).

62. There were other reasons for these constraints, such as the practice of showing nature films as part of cinema programs made up of more attractive films such as comedies and melodramas, or more exciting documentaries such as war films. Keeping the zoo films, and other natural history subjects, short avoided the perceived risk of boring the audience.

63. July 26, 1926.

64. On the concert given to the elephants in Paris at the end of the eighteenth century, see Loisel, *Histoire,* 2:47; and Richard Burkhardt, "Constructing the Zoo: Science, Society, and Animal Nature at the Paris Menagerie, 1794–1848" (appearing in this volume). On playing music to animals to observe their reaction, see E. G. Boulenger's account in the *Daily Telegraph,* March 30, 1927; C. J. Cornish, *Life at the Zoo: Notes and Traditions of the Regent's Park Gardens* (London, 1895), pp. 115–38; and for an example from the Washington Zoo, see Charles Musser and Carol Nelson, *High Class Moving Pictures: Lyman H. Howe and the Forgotten Era of the Traveling Exhibition 1880–1920* (Princeton, N.J., 1991), p. 302.

65. Geoffrey Vevers remarked that "Mok obviously recognised the close-ups. His beating of his chest was a response to what he saw on the screen, and the cry he made was one I had never heard him make before." *Daily Mail,* September 23, 1932. In an undated listing of Urban's films, under the title "Animal Drolleries," there is a description of a film in which bears smash up a record and a gramophone; see file no. URB 3.

66. It was also felt that cinema might be of indirect benefit to animals. Thus, in an advertising pamphlet by Charles Urban, *The Cinematograph in Science, Education, and Matters of State* (1907), he extols the advantages of film for education in medical science, for example, as a means of recording operations. He also notes that film records of operations on animals "would decrease vivisection by at least 90% of their present number" (p. 38); in file no. URB 3. On film and vivisection see Christopher Lawrence, "Cinema Verité? The Image of William Harvey's Experiments in 1928," in *Vivisection in Historical Perspective,* ed. Nicolaas A. Rupke (London, 1987), pp. 295–313; and more generally Lisa Cartwright, *Screening the Body: Tracing Medicine's Visual Culture* (Minneapolis, 1995).

67. P. 14; HPF.

68. Mitchell, *Centenary History,* pp. 201–2.

69. G. Vevers, "Report on a Visit to Carl Hagenbeck's Animal Park at Stellingen," p. 2; HPF.

70. See "How I Caught the Baboons for the Zoo: Mr. Frank Hayter's Own Story," *Evening News,* May 28, 1925.

71. According to the entry in the "Daily Occurrences" (daily records of births, arrivals, and deaths at the zoo) one hundred baboons arrived on May 11, 1925. Mitchell noted that eighty baboons were put on the Hill in 1925, whereas Zuckerman recorded ninety-seven, although three may well have died on arrival at the zoo. He also noted that some female deaths had been attributed to males for that year. Zuckerman also records thirty-five new arrivals for June 27, 1927, whereas the zoo's "Daily Occurrences" show thirty-seven. Zuckerman's monthly chart for Monkey Hill, recording deaths and killings there, can be found in SZ, file no. SC/2.

72. Mitchell, *Centenary History,* p. 202.

73. S. Zuckerman, *The Social Life of Monkeys and Apes* (London, 1932), pp. 219–23.

74. Ibid., p. 220.

75. Mitchell, *Centenary History,* pp. 201–3.

76. *Daily Mail,* June 4, 1925.

77. *Daily News,* January 25, 1927. One might compare this with Zuckerman's entry in his notebook marked "Hill 3," August 31, 1929, describing the death of a female: "[The keepers] drove the others out, and then with bars after half an hour managed to clear RF-LT [Redface-Longtip] out. He tried to pull Mother's carcass up to the level of the floor but failed. Keepers then found that the baboons were gathered either round door of den or above at fanlight. They waited half an hour and then in spite of general opposition got her off. RF-LT posted himself then on top of the small hill and did an hour's wailing."

78. For an example of this in the wild, see Jeanne Altmann, *Baboon Mothers and Infants* (Cambridge., Mass., 1980), pp. 129–30.

79. There are three of these notebooks ("Hill 1," "Hill 2," and "Hill 3") covering a period from April 5, 1929, until June 13, 1930. Zuckerman draws up a specific list of names of selected baboons, numbering about twenty-two, on whom he mainly concentrates; see SZ, file no. SC/2.

80. "Hill 2," June 22, 1929.

81. "Hill 3," December 19, 1929.

82. "Hill 3," December 24, 1929.

83. "Hill 3," November 8–9, 1929.

84. "Hill 3," January 6, 1930.

85. "Hill 3," December 31, 1929.

86. Zuckerman, *Social Life,* p. 217.

87. Ibid., p. 220. Zuckerman was well aware that there were other methods of managing hamadryas colonies. On his visit to the Munich Zoo in June 1928 he saw an exhibit which contained a larger number of baboons in a slightly greater space. In a set of short notes he describes the display

which, at the time, contained about 150 monkeys with approximately 25 adult males and 25 adult females. He notes, "Must give them plenty of room. To breed isolate a male with 5 or so females. (Successful in all species in Berlin). To introduce his new hamadryas (40 males) in the spring, he [i.e. the director of the exhibit] will isolate five of these with one of his single females and leave them 8 days." (It would be interesting to know whether the violence in this community was lessened by this environment.) Handwritten notes inserted in "Hill 3" notebook; see also Zuckerman, *Social Life,* p. 211.

88. August 12, 1930, SZ, file no. SC/4. One might also note his remarks in an application for a London University chair in social biology, January 22, 1930: "I soon recognised that the structure of baboon colonies is fundamentally a reflection in behaviour of the baboon reproductive process, and I accordingly commenced research upon the physiology of the primate menstrual cycle." SZ, file no. OU. The link between the contemplation of the hamadryas baboon's genitalia and a science of reproduction reminds us how important it is to consider the ordinary experience of the animal body.

89. H. Kummer, *Social Organisation of Hamadryas Baboons* (Chicago, 1968), pp. 1–3.

90. H. Kummer and F. Kurt, "A Comparison of Social Behaviour in Captive and Wild Hamadryas Baboons," in *The Baboon in Medical Research: Proceedings of the First International Symposium of the Baboon and Its Use as an Experimental Animal,* ed. H. Vagtborg (Austin, Tex., 1965), p. 74. Kummer takes a different view from Zuckerman as regards the role of sex in the formation of social units. Indeed, he sees relations developing out of the young male's taking of infants, or the adopting of juveniles of both sexes, as formative. When the young adult male adopts a juvenile female Kummer terms this an "initial unit." "The male and the juvenile female in the initial unit still stimulate the behaviour of mother and child. We therefore hypothesized that the male-female bond in the hamadryas baboon may originate from a transferred mother-child relationship" (p. 82); see also Kummer, *In Quest of the Sacred Baboon* (Princeton, N.J., 1995), pp. 216–19.

91. Kummer, *Social Organisation,* p. 49; *Quest,* p. 150.

92. On figurations of the masculine body during and after World War I, see Joanna Bourke, *Dismembering the Male: Men's Bodies, Britain and the Great War* (London, 1996).

93. *Daily Express,* June 9, 1930.

94. April 6, 1927.

Part Four

GOOD BREEDING: INSIDE THE DOMESTIC SPHERE

9

The Sincerest Form of Flattery

Harriet Ritvo

By the middle of the nineteenth century, most British naturalists considered the classificatory principles of Linnaeus to be quaint and artificial. But one major component of his work survived with its authority apparently little diminished. A committee of the British Association for the Advancement of Science reported in 1842 with regard to "the *binomial system of nomenclature,* or that which indicates species by means of two Latin words, the one generic, the other specific," that it was "invaluable."[1] Many of those with reservations about the Linnaean system exempted his nomenclature from their general critique. Near the close of the eighteenth century, for example, Thomas Pennant emphatically rejected Linnaeus's arrangement of quadrupeds, primarily on grounds of inappropriate associations—he objected as much because "the half-reasoning *Elephant* is made to associate with the most discordant and stupid . . . *Sloths, Ant-eaters,* and *Armadillos*" as because "my vanity will not suffer me to rank mankind with *Apes, Monkies, Maucaucos,* and *Bats.*" But Pennant went on to mitigate his harsh assessment, acknowledging that Linnaeus's classification of fishes, insects, and shelled animals merited praise, and that, especially, "he hath, in all classes, given philosophy a new language."[2]

More than a generation later, even the nationalistic John Fleming paused in the midst of denouncing the "greatly overrated" Swede to make a grudging nomenclatural concession: "Linnaeus was not, it is true, without much merit, in rendering trivial names [that is, technical species names] popular."[3] Fleming used the word "popular" in a rather restricted sense, referring only to the acceptance of "trivial names" within the community of naturalists. Indeed, wider acceptance might have made such names less popular among Fleming's own cohort. For at the same time that it provided a newly systematic means of referring to animals and plants, Linnaean terminology also offered a newly definitive means of discriminating between the zoological knowledge of specialists and the implicitly less significant and reliable information about animals that was broadly available to ordinary Britons. It redefined a reservoir of traditional lore as the turf of experts, thus inspiring predictable resistance, most frequently, although not exclusively, outside the community of specialists.

Not everyone with a potentially competing stake in the material of natural history resented the increasing hegemony and prestige of scientific nomenclature, however. One group of lay specialists in animal matters, the breeders of pedigreed domesticated animals, admired this arcane technical terminology in more or less the terms intended by the naturalists who deployed it. Although disinclined to express their appreciation theoretically, these hands-on experts demonstrated it persuasively through their own nomenclatural practice. They borrowed the terminology devised for systematic natural history and adapted it to their own rather different ends.

Such appropriation constituted a striking exception to the ordinary relationship between the two groups. Although the subject matter of eighteenth- and nineteenth-century zoology inevitably overlapped substantially with that of domestic animal husbandry, the literatures of those subjects intersected much less predictably. Occasional individuals participated in both communities and contributed to both discourses. For example, Joseph Banks, who served as president of the Royal Society, also wrote widely on agricultural matters, promoting such pet projects as the naturalization in Britain of Spanish merino sheep.[4] But, on the whole, the two voluminous discourses evolved in parallel rather than interactively. Naturalists gave domestic animals more than their share of ink in the zoological surveys and overviews that they industriously produced, but seldom incorporated the practical insights of stockbreeders and pet fanciers. (Charles Darwin was a famous exception to this generalization in his eagerness to profit from the accumulated wisdom of working-class pigeon breeders, as well as of agriculturalists.)[5]

Reciprocally, animal breeders tended to prefer homegrown expertise, loyally adhering to time-hallowed attitudes and practices, even where these conflicted with the most solidly grounded and widely held zoological consensus. They firmly resisted the occasional attempts of scientifically minded colleagues to bring them up to date—like Everett Millais's *Theory and Practice of Rational Breeding,* which urged breeders to stop "trusting to luck" and instead base their decisions on "Darwin's theory . . . the stupendous work of a stupendous mind."[6] As James Cossar Ewart, then Regius Professor of Natural History at the University of Edinburgh and, like Banks, a straddler of the boundary between zoology and animal husbandry, suggested in 1899, the roots of this stubborn imperviousness might stretch deeper than mere indifference: "in England, where there has been . . . in many quarters a dread of scientific methods akin to that of the supernatural, breeders have failed to distinguish between facts and beliefs."[7] Punch proffered an alternative explanation when it imagined the ordinary agriculturalist's reception of Darwin's *The Variation of Animals and Plants under Domestication.* This work was widely appreciated in the stockbreeding and pet fancying press, although perhaps more for the simple recognition it awarded their enterprise than for its weight of evidence and argument. Because of the resulting publicity, "Old John Stockwell . . . as good a judge of a beast as any man in Midlandshire . . . borrows the two stout volumes from his clergyman. . . . he makes a conscientious effort to master the polysyllabic difficulties, but fails. . . . He tries again. . . . [Finally] he throws Darwin down, is quite laborious with his pipe, empties his tumbler, takes off his boots, and . . . goes up-stairs to bed."[8]

Whatever general reservations John Stockwell and his ilk may have harbored about Darwin's vocabulary, they showed themselves favorably inclined toward at least one of its polysyllabic components. The latinate binomials and trinomials of zoological nomenclature surfaced frequently in the literature of animal husbandry. Such terms were not, however, generally employed to designate the species of primary concern to breeders. In print, as in everyday conversation, the zoologists' *Bos taurus, Equus caballus, Equus asinus, Ovis aries, Capra hircus, Sus domesticus, Canis familiaris, Felis catus,* and *Oryctolagus cuniculus* were routinely referred to by stockbreeders and pet fanciers as the ox, the horse, the ass, the sheep, the goat, the pig, the dog, the cat, and the rabbit. Instead, animal husbandry experts were apt to reserve Linnaean nomenclature for more subtly differentiated and controversial taxa, where its connotation of scientific authority enhanced not only their prestige but their credibility. A latinate designation could certify the reality of a "breed" or "strain" when more concrete evidence for its existence seemed thin on the ground.

THE CASE OF THE WILD WHITE CATTLE

The debate that swirled for over a century around what were known as "wild white cattle" or "white park cattle" clearly illustrated the argumentative value of latinate terminology, although it was a rather anomalous debate in terms of the predominant preoccupations of animal husbandry. These striking and famous animals—they were celebrated in verse by Walter Scott and commemorated in oils by Edwin Landseer[9]—roamed a few great estates in Scotland and northern England untroubled by the restraints that burdened ordinary domestic beasts. They were left largely to their own devices

1. A wild white English bull, as portrayed by Thomas Bewick.

as far as food and mating were concerned, and they were shot rather than slaughtered when their herds were culled. Like the emparked deer that lived under similar conditions throughout the British countryside, the white cattle were treated in this way because the magnates who owned them considered them to be wild animals. This attributed wildness transfigured them into creatures inherently different from cattle destined for the butcher or the dairy. If the white cattle left their rugged native haunts, it was to grace the display cases of natural history museums or the enclosures of zoos.[10]

The behavior of the white cattle was often adduced as evidence of their untamed nature. They formed circles to menace intruders before charging; cows hid their newborn calves in high grass, returning to nurse them only occasionally; when a member of the herd was seriously wounded or ill, the others turned on it and gored it to death. But their most telling claim to wildness was historical or genealogical. They were described by admirers as genetically distinct from other British bovines—the ignoble descendants of domesticated animals imported by the various human conquerors of the island—because their ancestors had been the never-tamed aboriginal wild cattle, or aurochs, that had ruled the primeval forests of Europe. (The last Continental aurochs had allegedly been killed in Poland in 1627.)[11] At the end of the eighteenth century, Thomas Bewick, relying heavily on the account offered by George Culley in *Observations of Livestock* (1786), characterized the herds emparked at Chillingham (Northumberland), Chartley (Staffordshire), and a few other ancient estates as survivors of "a very singular species of wild cattle in this country, which is now nearly extinct."[12] Almost a century later, in his introduction to *The Wild White Cattle of Great Britain,* John Storer asserted that "there . . . have existed in this country from the earliest historic times, herds of White Cattle, perfectly distinct, and of a different breed from its ordinary domestic races. Some of these herds seem to have been always wild."[13] His contemporary James Harting claimed that "the few scattered herds of so-called Wild White Cattle which still exist in parks in England and Scotland may be said to form a connecting link . . . between the wild animals which have become extinct in this country within historic times, and those which may still be classed amongst our *ferae naturae.*"[14] When the Earl Ferrers offered the Chartley cattle for sale in 1904, the advertising circular described them as "the lineal descendants of the original British Wild Ox."[15]

The distinguished, often aristocratic, landowners over whose estates the cattle wandered cherished these claims to wildness, aboriginality, and pure descent; they often regarded the animals as family mascots or totems. They were seconded in these feelings by many humbler compatriots, to whom the white cattle appeared as "ancient Britons," representatives of the historical dignity, nobility, and ethnic isolation of their country.[16] To perform this symbolic function, the emparked herds had to be as distinct as possible from "the quiet, servile steer [which] is probably as unlike the original wild cattle of this country, as the English gentleman of the present day is unlike the rude baron of the age of King John."[17] But there was a large element of myth-making or wishful thinking in these assertions of ancient heritage and contemporary wildness, and not everyone was equally transported by nationalistic passion. Less sentimental observers claimed that the vaunted trappings of uncontaminated wildness reflected human meddling rather than the untrammeled workings of nature.

It was asserted, for example, that even the white color of the herds—an obvious symbol of purity and also, because it was unusual among British domestic breeds, of reproductive isolation—was maintained by killing the off-color (usually black) calves that were frequently produced. The apparent wildness of the white cattle could analogously be explained as the consequence of their style of life; it was claimed that similar behavior would emerge in any group of domestic bovines left to their own devices for a generation or two. Skeptics further pointed out that what looked like wildness to eyes accustomed to the disciplined countryside of the eighteenth and nineteenth centuries had been the standard condition of medieval domestic cattle.[18] In this view, the emparked herds were merely the froward, if interesting, remnants of ordinary cattle which had been liberated at some period in the unfathomable past: possibly feral (that is, having reverted from an earlier state of domestication), but certainly not aboriginally wild.

In the heat of combat, advocates of both positions enlisted the formal authority of zoological nomenclature to buttress their arguments. The advocates of ancient isolation and uncompromised wildness were apt to project the distinction between the emparked herds and other cattle into the past, and to assign the remote ancestors of the white park cattle to a different species than the forebears of their domesticated relatives. Numerous bovines, domesticated and otherwise, had inhabited northwestern Europe in prehistoric and early historic times. In the nineteenth century their skeletal remains were widely available for examination, and many experts assigned them to two or three separate species, based on the size and shape of their skulls.[19] The largest bones, which belonged, by implication, to the wildest and most powerful animals, were assigned to the aurochs and called *Bos primigenius.* Since the process of domestication frequently resulted in reduced size, at least in its earlier stages, remains of smaller animals were designated *Bos longifrons* and *Bos frontosus.*

At the end of the nineteenth century Jacob Wilson boldly, if somewhat nostalgically asserted that "our British cattle . . . are derived from two species: *Bos Longifrons* and *Bos Urus* or *Primigenius*" and "that the wild cattle of Chillingham are closely allied to the Urus is beyond dispute."[20] But not everyone who sympathized with this position embraced it with such confidence. Although James Harting included a chapter on the "wild white cattle" in a book entitled *British Animals Extinct within Historic Times,* he hedged about "whether . . . [they were] descended . . . from the aboriginal wild breed of the British forests—the Urus of Caesar (*Bos primigenius*)," taking refuge behind "the weight of scientific opinion . . . [which] seems to favour the view that these wild white cattle were descended from the Urus, either by direct descent through wild animals from the wild bull, or less directly through domesticated cattle deriving their blood principally from him." He did, however, firmly distinguish between the urus, for which there was no evidence of early domestication in Britain, and "*Bos longifrons* . . . [which] was everywhere subjugated and used by man."[21] A compromise position, which endorsed the distinctive descent of the white cattle but not their persistent wildness, could assign them different forebears than other cattle, but not the ferocious aurochs or urus: "Our . . . Chillingham cattle may . . . be descended from *Bos longifrons,* . . . Our indigenous kyloes and runts . . . we may assign, perhaps, to *B. frontosus.*"[22]

Some believers urged an ancestry still more separated from that of ordinary domestic *Bos taurus.* The author of a mid-nineteenth-century agricultural manual suggested that the aboriginal British wild cattle "still preserved in their native purity . . . in Chillingham park" were descended from a distinct genus (*Bubalus*), that of the "buffaloes" of Asia, Africa, and southern Europe.[23] Thomas Pennant characterized the white cattle as the "*Bisontes jubati* of *Scotland,*" thus connecting them to the European bison or wisent, by then nearly extinct except in protected tracts of the East European forests.[24] More radical isolationists proposed a genus that would include only the aurochs and its progeny, typically denominating the white herds as *Urus scoticus.*[25] On the other hand, the observation that the emparked cattle interbred enthusiastically and successfully with ordinary British breeds whenever they got a chance sometimes moderated the taxonomic construction of difference, prompting their placement in a separate species of genus Bos, rather than an independent genus. For example, William Swainson distinguished "Bos Scoticus," which he characterized as "fierce" and "untameable," from the "pre-eminently typical" "Bos Taurus" within his quinary scheme; because of the temperament of the white cattle, he viewed their position in the bovine circle as analogous to that of the carnivores in the circle of mammals.[26] And authors on the skeptical end of the continuum could reinforce their position by assigning the white cattle to the same species as shorthorns and Herefords, perhaps acknowledging by terms like "Bos taurus (Linn.) var. Scoticus" or "*Bos Taurus. Restricted Variety,*" a difference roughly equivalent to that between ordinary domestic breeds.[27] Or they could emphasize their conflation of the white cattle with more humdrum herds by refusing them any latinate nomenclature at all, simply declaring, for example, that "it is . . . highly probable that these animals are the remains of a breed which was formerly kept tame in the farms in many parts of England."[28]

CREATIVE ACTS

The constitutive effect of naming and the added prestige of latinate designation did not depend solely on wildness. With reference to more conventional livestock, both binomial nomenclature and the abstract language of classification (that is, terms like "genus," "species," and "variety") were routinely deployed to define, enhance, and celebrate the act of domestication. Eighteenth- and nineteenth-century agriculturalists prided themselves on the extent to which they had "improved" the motley local strains which

2. Robert Bakewell's improved Leicestershire or Disley sheep offered an early example of the benefits of the new breeding and marketing techniques. Engraving by Thomas Bewick.

they had found in British yards and pastures, but the categories in which they consolidated their advances could seem as much the product of imagination as of the breeding pen. Pedigreed breeds and sub-breeds were therefore reified in the language of subspecies or varieties. After all, not only were they, inevitably, of recent origin, but their characteristic physical qualities could be difficult to demonstrate. Breeders invoked the prestige and authority of zoological rhetoric to emphasize the often subtle distinctions that separated one breed from another, and well-bred from ill-bred animals—that is, to emphasize and reinforce the refined discriminations on which their success or failure depended.

The stakes involved in this enterprise, economic as well as intellectual or technical, were extremely high. The institutional trappings of breed recognition and consolidation—show classes, breed societies, studbooks—were often accompanied by striking and gratifying corollaries. As the editors of the *Sussex Herd Book* complacently noted, within six years of the appearance of their first volume, the prediction there expressed "that Sussex Cattle would come more into favour, has been fully borne out" in increased demand, and consequently larger profits.[29] The Kennel Club was similarly founded to promote the interests of dog breeders by holding shows and field trials, and by ensuring "the proper registration of all exhibited dogs, with their pedigrees."[30] The lack of such trappings could impede the recognition of a breed, and consequently the willingness of prospective customers to pay premium prices for pedigreed animals.

The categorical weakness or lack of focus that breed associations aimed to correct reflected human failings as well as refractory animal material. All breeders, whatever the species or variety to which they dedicated their efforts, strove to produce animals of high and predictable quality. But the apparent simplicity of this common goal was somewhat compromised by its vagueness and abstraction. It assumed as given something that, in many cases, had to be energetically established: the nature of the breed to be improved and propagated. There was great disagreement among breeders of every domestic species about which physical attributes characterized an excellent animal. And even if such standards could be established, there was further disagreement about which animals should be subjected to them. Breeders concurred in principle that each domestic species should be further divided into subcategories, each with its own aesthetic and functional desiderata; but defining the boundaries and standards for each subcategory inevitably invited further contention. Establishing the existence of a previously unrecognized subcategory—or the validity of a previously disrespected one—could be still more more problematic. Since an impressive and authoritative label constituted an implicit argument for the reality and value of the thing labeled, breed partisans frequently attempted to buttress their claims by borrowing the latinate language and the hierarchical categories of zoological taxonomy.

Insubstantial as it might seem, therefore, the argument from nomenclature could prove more powerful than other arguments; and it was certainly more readily available.[31] After all, though pedigree was essentially a genealogical guarantee of purity and excellence, even the best-established breeds rested on uncertain historical foundations. For example, the shorthorn reigned incontestably preeminent among early nineteenth-century cattle. It was among the first cattle breeds to be judged separately at the annual Smithfield Show, the showcase of British livestock, and in 1822 it was the first breed of livestock animals (as opposed to sporting animals like the thoroughbred horse or the greyhound) to achieve the dignity of its own studbook, or compendium of individual genealogies.[32] Yet the breed had only coalesced around 1780, with the emergence of the superstar bull Hubback.[33] Although he figured prominently in subsequent pedigrees, Hubback's own origins were obscure. He had begun life as "a calf belonging to a poor man who grazed his cow on the sides of a highway"; it was pure good

3. The Godolphin Arabian, whose progeny were much better documented than his forebears. Courtesy of the National Horseracing Museum, Newmarket (UK).

fortune that a prominent stockbreeder had passed by and perceived his great potential.[34] The roots of thoroughbred horses, the best-bred animals of eighteenth- and nineteenth-century Britain, were not much deeper. All traced their descent from three imported stallions of the late seventeenth and early eighteenth centuries, animals who were, though renowned, semi-anonymous, identified only by their English ownership and their somewhat imprecisely supposed country of origin: the Darley Arabian, the Godolphin Barb, and the Byerley Turk.[35]

The logic that attracted breeders to the language of formal classification might be extended still more boldly and imaginatively. In addition to underwriting or reifying differentiations already incarnated in barn and showyard, zoological terminology might be used to create distinctions that

4. Shorthorn cattle, like the bull here portrayed by George Garrard, belonged to the first livestock breed to have its pedigrees documented in a herd book.

were desired but not yet achieved. For example, James Long acknowledged in 1886 that, as a result of the difficulty of distinguishing one kind of swine from another, "it is questionable whether agricultural stock of any description has been subjected to such extraordinary classification as the pig"; he nevertheless hopefully asserted that "the Small White may . . . be termed the 'fancy' breed of the pig genus."[36] An equivalent process of projection or wishful thinking could be applied to human beings, who were widely recognized to share many physical and behavioral tendencies with other domesticated animals. For example, Robert Knox asserted that "men are of various Races; call them Species, if you will; call them permanent Varieties; it matters not. . . . in human history race is everything."[37] Reversing the analogy, Judith Neville Lytton separated horses "into two distinct types, the Northern and the Southern, which roughly follow the human types of North and South"—labeling them "*E.* [*Equus*] *Frigidus*" and "*E. Ardens.*"[38]

A further attraction of taxonomic jargon was that the term "breed" itself had become unstable. Originally what defined a breed was shared provenance and shared function, along with some degree of physical resemblance. By the middle of the eighteenth century, however, at least among the most enlightened stockbreeders, the term was generally applied in a narrower sense. As one agriculturalist dedicated to improvement put it, certain "varieties have been usually distinguished among farmers by the appellation of different *breeds*; as they have supposed that their distinguishing qualities are, at least, in a certain degree, transmissible to their descendants."[39] But retrograde usages persisted, even within the community of elite breeders, especially as a way of disparaging unimproved animals. In 1794, for example, a surveyor commissioned by the Board of Agriculture referred to the "original black breed" of Carmarthenshire cattle as "ill shaped and unprofitable to the pail," and to the local "breed of swine—a narrow, short, prick-eared kind."[40]

These general definitional problems were exacerbated by the lack of consensus about the basis for subdividing particular species. The dog was the most emotionally versatile and physically plastic of common domesticates, "more tractable than any other animal, conform[ing] himself to the . . . habits of life of his master."[41] When dog fanciers attempted to make sense of the dizzying variety of canine types, they often found it necessary to identify some intermediate category between the species and the breed. The principles according to which these categories were defined were both numerous and variable. Perhaps the most commonly cited scheme was based on the work of John Caius, whose *On Englishe Dogges,* originally published in 1576, remained influential for several centuries. It divided dog breeds into three unequal categories: "the most generous kinds," further subdivided into "dogs of chace" (hounds and terriers), "fowlers" (spaniels and setters), and "lapdogs" (small spaniels); "farm dogs" (shepherd's dogs or collies and mastiffs); and "mongrels."[42] According to a still simpler principle of analysis, "we may divide our domestic dogs into three great classes:—1. Whole-coloured dogs . . . 2. Parti-coloured dogs . . . 3. Pure white varieties."[43]

Of the fanciers who sought more sophisticated categories, some paradoxically separated dog breeds into the "wild" (which usually included collies, huskies, and others with some perceptible resemblance to wolves) and the "domestic."[44] The resulting groups overlapped significantly with those generated by the more anatomically oriented separation of breeds with "erect pointed ears" from those with "pendant ears."[45] Some Victorian breeders acknowledged Cuvier as their authority for a tripartite division based on the conformation of the parietal bones of the skull (they could be convergent, parallel, or divergent), which yielded categories represented by the greyhound, the setter or spaniel, and the mastiff.[46] And fanciers often grouped dog breeds on the basis of function, a criterion which could itself give rise to further disagreement. Thus the strictly utilitarian categories proposed by Peter Lund Simmonds—"farm dogs . . . hunting dogs . . . and shooting dogs"—excluded the majority of pets, while the more relaxed categories of the Kennel Club—"Sporting Dogs" and "Non-Sporting Dogs"—left no one out.[47]

The incompatibility of these competing vernacular differentiations inevitably threatened to undermine the authority of those who made them. To shore up the eroding foundations of their expertise, nineteenth-century fanciers frequently dignified their systems of breed classification with terminological trappings borrowed from the rich zoological literature on domestic dogs, the attractiveness of which was undiminished by the fact that naturalists were as polyphonic about the interrelationships of dog breeds as were fanciers. Charles Hamilton Smith attached a latinate binomial to each domestic breed in the *Dogs* volume of the Naturalist's Library; Edward Griffith repeated Linnaeus's own labels for domestic dog breeds in his survey of the carnivora.[48] Surveying the animals of Britain, John Fleming more modestly labeled each of the "principal races of dogs" with a trinomial or subspecific tag. The collie was *Canis familiaris pastoralis,* the Newfoundland *C. familiaris amphibius,* the bulldog *C. familiaris taurinus,* the mastiff *C. familiaris mastivus,* and even the cur *C. familiaris villaticus.*[49] In its early years, the Regent's Park Zoo displayed caged examples of "many interesting varieties of the *domestic dog*" sporting such labels as "*Canis familiaris,* var. *borealis*" (eskimo dog) and "*Canis familiaris,* var. *Australasiae*" (dingo).[50] Fanciers could thus pick from a rich array of possibilities. For example, in the introduction to his compendious *Illustrated Book of the Dog,* Vero Shaw surveyed several centuries of dog breed classification, carefully reproducing every one of the long-discarded latinate binomials listed by Linnaeus, even those which, like "Canis Hybridus, or Bastard Pug Dog" and "Canis Vertigus, Turnspit," referred to animals that his Victorian contemporaries would have found very difficult to recognize, or even to imagine.[51]

Cattle breeds were fewer and, on the whole, slightly more venerable than dog breeds; in addition, their characteristics were not quite so volatile, due as much to the smaller number and greater homogeneity of breeders, as to the lesser variability of the animals themselves.[52] Nevertheless, in order to reconfirm the claims to distinctiveness and merit of their favorite breeds, aficionados of shorthorns, Herefords, and their ilk similarly analyzed and subdivided domestic cattle. As with dogs, some of the resulting groupings were organized on completely vernacular principles. Such principles might be geographical, as in an eighteenth-century agricultural manual that recommended "your *Angleseys* and *Welsh*" as "a good hardy sort for fatting on barren or midling [*sic*] sort of land."[53] Or they might be based on superficial physical characteristics. For example, some experts claimed that "the cattle of Great Britain . . . may be divided . . . into the following primary sections: Long Horns; . . . Middle Horns; . . . Polled Cattle; . . . Short Horns." Not only were such categories grounded exclusively in practical experience, but their formulators often explicitly denied any relationship with the authority of science. In this way, after propounding his horn-based categorization, W. C. L. Martin stipulated that "the . . . sections do not . . . derive their nomenclature from points of zoological importance."[54]

He may have been protesting too much, however; his use of the word "nomenclature" itself suggested at least a tacit parallelism. And, like dog fanciers, many of his fellow breeders attempted to buttress their favorite subdivisions by explicitly enlisting the categories of zoology. An early nineteenth-century agricultural pamphleteer noted that "the different breeds or varieties of our domestic animals are, in general, so fixed and incapable of speedy alteration, that they may be deemed comparatively permanent."[55] Relatively forceful and lucid stockbreeders could make direct assertions of analogy, claiming that "the distinction between some [breeds] . . . has scarcely been less than the distinction between that variety and the whole species" or, more sweepingly, that "as most wild species consist of several varieties, so must domestic breeds consist of several strains."[56]

It was therefore a mark of distinction among cattle breeds to be further broken down into named and recognized subdivisions, each smaller category simultaneously reifying itself and validating the one above. The introduction to a late Victorian herd book explained that "TRIBE is used in referring to the whole of the descendants . . . of each originally selected Cow. . . . Distinguished collateral Branches of a Tribe . . . are spoken of as Distinct FAMILIES. . . . The representatives of such TRIBES and FAMILIES, now existing . . . are classed as forming one GROUP."[57] In the case of shorthorns, an extra layer of differentiation commemorated the competition between the eponymous "sub-breeds" established by master breeders Thomas Booth and Thomas Bates.[58] Most tribes and families of cattle, however, were named for a bovine mother rather than a human father, to the annoyance of some macho breeders, who feared that it would lead their colleagues "to overestimate the value of the female side of the pedigree."[59] In the late nineteenth century, the "Leading Families" of polled Aberdeen Angus cattle included the Queen tribe, the Ericas, the Princesses, the Baronesses, and the Emilys; an advertisement for shorthorns from the Prince of Wales's estate at Sandringham offered the Cold Cream, Seedling, and Primula families of the Booth strain as well as the Bates families of the Blanches, the Wild Eyes, the Honeys, and the Sweethearts; and Amos Cruikshank, a latter-day challenger for the mantle of Bates and Booth, had developed his own widely recognized set of strains, including the Venus family, the Lovely family, and the Spicy family.[60]

Such subdivisions—and such names—evoked the taxa of natural history without literally borrowing them. "Families" were part of the canon of zoological categories (most commonly, in order of diminishing inclusiveness, kingdom, phylum, class, order, family, genus, and species), and "Tribes," while less formally established, had long hovered on the periphery as an option for naturalists who felt constrained by the seven-step hierarchy. Both categories occupied similar stations on the taxonomical scale; indeed, "tribe" could function, like "kind," as a vernacular alternative for "family." A zoological tribe, like a zoological family, was composed of several related genera; thus William Swainson included "the genera of the elephant, rhinoceros, megatherium, and hippopotamus" within "the thick-skinned or pachydermatous tribe."[61] In confirmation of the taxonomic importance that naturalists accorded them, most species of common domesticated animals gave their latinate names to both the genera and the families of which they were part. For example, the genus *Bos,* which included domestic cattle and their near relations, belonged to the family Bovidae, along with bison and buffalo, and the genus *Canis,* which included domestic dogs and wolves, belonged to the family Canidae, along with foxes, dholes, and the so-called wild dogs of Africa.

VARIATIONS ON THE THEME

When breeders of domesticated animals applied the taxa and nomenclature developed by naturalists for wild species to their pedigreed strains, the dynamic of appropriation seemed straightforward enough. Assuming that the process of zoological classification had relatively firm intellectual moorings and that its objects had relatively real physical existence, they implicitly claimed these qualities for their own shakier enterprise. But breeders also conjured with the prestige of science in more complex, assertive, and even playful ways. Their adaptation of terms like "family" and "tribe" to designate groups of pedigreed livestock, for example, suggested that they could transform zoological categories, as well as imitate them. Despite their shared appellations, the families and tribes of stockbreeders were not at all like those of naturalists. On the contrary, they designated categories that were off the scale in the opposite direction, smaller not only than the genus and the species, but even than the shadowy "variety." Everett Millais illustrated their relations as a set of embedded circles, with the "species" on the circumference, descending through "varieties," "strains," and "families," to the "individual" at the core.[62] Thus, while mirroring the hierarchy of scientific classification, these agriculturalists had also inverted it, retaining the formal authority of its structure but dispensing with that of its content. And if the titles of these transmogrified categories nevertheless echoed the language of natural history, the names of the groups of animals they classified proclaimed that they had been transplanted to a very different, more homely and practical sphere. Naturalists might refer to the Bovidae, or they might, if their intended audience was relatively broad or miscellaneous, refer to the ox tribe, but they never referred to families or tribes with names like Booth, Bates, or Cruikshank, let alone Emily, Wild Eyes, Lovely, or Cold Cream.

By means of such provocative adaptations, stockbreeders demonstrated that they were strong readers of scientific nomenclature, not only reinterpreting its message to suit their own purposes, but confidently emphasizing that they had done so. After all, the goal of their enterprise was very different from that of natural history. The formal differentiation embodied in the hierarchy of breeds and sub-breeds was important to agriculturalists as a sign of their ability to manipulate nature directly, by reshaping the bodies of animals, rather than metaphorically, through understanding and analysis. The improved breeds of the eighteenth and nineteenth centuries constituted an unprecedented technological accomplishment; and so they were represented by their creators and boosters. For this reason, the relationship posited by stockbreeders between their breeds, tribes, and families and the genera, species, and varieties of natural history had to remain on the level of analogy.

As Darwin pointed out in *The Variation of Animals and Plants under Domestication,* breeders were as likely as naturalists to reject the notion that "natural species" and "domestic races" were identical, although for different reasons.[63] While naturalists emphasized the greater differentiation of wild species, breeders disparaged any suggestion that their strains were natural, as opposed to artificially fabricated. Even though the term "improvement" carried the connotation of inherent qualities, gradually teased out, eighteenth- and nineteenth-century stockbreeders understandably tended to stress the disjunction of their highly finished products with the raw zoological materials that nature had provided. A contrary emphasis on continuity would have inevitably undermined their claims about the value—and perhaps the very existence—of what they had to sell. Thus breeders walked a kind of figurative tightrope, implying that artificial breeds resembled natural species or varieties in their distinctiveness or individuation, but that, at the same time, their development represented a violation of natural processes rather than an extension of them.

Despite their apparent high-handedness, these taxonomic transformations were, at least in intention, as respectful of zoological expertise as the more literalistic appropriation of latinate nomenclature. In a sense the very boldness of the changes rung by stockbreeders on the practices of naturalists may have signaled a more thoroughgoing acknowledgement of scientific authority. By erecting a more precarious structure of assertions upon the foundation provided by zoological classification, they implied a proportionately higher estimate of the solidity of that base. Technical nomenclature, and the structure that it expressed, were thus assumed as givens, part of the realm of established fact—secure enough to admit even the license of humor. When aficionados of domesticated animals indulged in binomial witticism, their targets were not ordinarily the labels themselves, as was the case with populist critics of zoological terms, but the creatures that might prove unworthy of their ascribed dignity. Thus "*semi-homo canis*" satirized pets "who live with us nearly upon the footing of our fellow-man," while "Bos Composita" gently ridiculed the genealogical pretensions of shorthorn cattle.[64]

By assuming such stability, however, breeders also called it into question; the accolade contained in their imitation might be too high. Their deference to zoological authority was not itself entirely neutral, and the flow of influence it represented was therefore not entirely unidirectional. Neither the spirited transformations of cattle breeders nor the more slavish borrowings of dog fanciers precisely reproduced the nomenclature or the classificatory system of naturalists; in general, sound was preferred to sense. And this imperfect echo was then further distorted in its dissemination to a public that was very large and significantly different from the audience for technical natural history. If one consequence of this transfer—the analogy between kinds created by nature and kinds created by human effort—was to enhance the prestige and plausibility of domestic breeds, another, reciprocal consequence was to undermine the prestige and plausibility of zoological nomenclature, by impugning its absoluteness. After all, its adaptation to such a different purpose constituted a significant modification, perhaps the earnest of others. And the wilfully eccentric version of the language of classification offered by animal breeders constituted a commentary or interpretation, as well as a reflection—a parody, unintended but nonetheless telling, as well as a transformation.

The undermining effect of these admiring appropriations may have been exacerbated because the level of taxonomic discrimination most relevant to animal breeding was the level most troublesome to zoology. Questions about how to define a species, whether the species or the genus was the fundamental unit of classification, whether varieties existed in the same sense that species did, and so forth, had bedevilled naturalists long before they helped lead Darwin to the formulation of his theory of evolution through natural selection. Despite the formal assurance projected by the vocabulary of classification, the species itself was a very uncertain category. Although comparing breeds to species might help to validate breeds in the eyes of fanciers unfamiliar with scientific controversy, the same juxtaposition further problematized the species in the eyes of zoologists. And, paradoxically, if breeders explored the causes of this instability, they might find their confidence in the authority of binomial nomenclature undermined, at the same time that their comparison of breed to species was better justified. The issues of purity and descent that troubled breeders were also problematical in the realm of nature, no matter what they were called.

NOTES

1. Quoted in [William Jardine], "Proposed Reform of Zoological Nomenclature," *Edinburgh New Philosophical Journal,* n. s. 18 (1863): 266.

2. Thomas Pennant, *History of Quadrupeds,* vol. 1 (London, 1793), pp. iii–v.

3. [John Fleming], "On Systems and Methods in Natural History. By J. E. Bicheno," *Quarterly Review* 41 (1829): 304.

4. Banks's crusade has been described in detail in Harold Burnell Carter, *His Majesty's Spanish Flock: Sir Joseph Banks and the Merinos of George III of England* (Sydney, 1964).

5. See James A. Secord, "Darwin and the Breeders: A Social History," in *The Darwinian Heritage,* ed. David Kohn (Princeton, N.J., 1985), pp. 519–42, and, generally, Charles Darwin, *The Variation of Animals and Plants under Domestication,* 2 vols. (New York, 1868).

6. Everett Millais, *The Theory and Practice of Rational Breeding* (London, 1869), pp. vii–ix.

7. J. C. Ewart, *The Pennycuik Experiments* (London, 1899), p. 63.

8. *Punch,* February 29, 1868, p. 91.

9. Scott dedicated several stanzas to the cattle in "Cadyow Castle," published in *Minstrelsy of the Scottish Border* (1802); Landseer exhibited *Death of the Wild Bull* at the Royal Academy show of 1836 and *The Wild Cattle of Chillingham* at the Royal Academy show of 1867.

10. For a more extensive discussion of the cattle in relation to eighteenth- and nineteenth-century British culture, see Harriet Ritvo, "Race, Breed, and Myths of Origin: Chillingham Cattle as Ancient Britons," *Representations* 39 (summer 1992): 1–22.

11. Juliet Clutton-Brock, *A Natural History of Domesticated Mammals* (London, 1987), p. 63.

12. Thomas Bewick, *A General History of Quadrupeds* (Newcastle, 1824), p. 38.

13. John Storer, *The Wild White Cattle of Great Britain: An Account of Their Origin, History, and Present State,* ed. John Storer, Jr. (London, 1879), p. xv.

14. James Edmund Harting, *British Animals Extinct within Historic Times, with Some Account of British Wild White Cattle* (London, 1880), p. 213.

15. Jan Tavinor, "A Chapter in the History of the 'Chartleys,'" *Ark* 18 (1991): 379.

16. A. G. Bradley, *The Romance of Northumberland* (1908; reprint ed., London, n. d.), p. 89.

17. Robert Chambers, *Vestiges of the Natural History of Creation* (New York, 1846), p. 236.

18. See, for example, Boyd Dawkins, "The Chartley White Cattle," *North Staffordshire Field Club Annual Report and Transactions* 33 (1898–99): 49.

19. Modern zoological consensus derives all European and Asian domestic cattle from the aurochs. The other species identified by Victorian paleontologists were early domestic cattle. Clutton-Brock, *A Natural History of Domesticated Mammals*, pp. 63–65.

20. Jacob Wilson, "The Chillingham Wild Cattle," *Land Magazine* (January 1899): 19–20.

21. Harting, *British Animals Extinct within Historic Times*, pp. 213–14, 216–17.

22. H. T. Meunell and V. R. Perkins, "Wild Cattle of Chillingham," *Transactions of the Tyneside Naturalists Field Club* 6 (1863–64): 142–43.

23. George Henry Andrews, *Modern Husbandry: A Practical and Scientific Treatise on Agriculture . . .* (London, 1853), p. 150.

24. Pennant, *History of Quadrupeds*, 1:17.

25. For example, James Aikman, *A Natural History of Beasts, Birds, and Fishes: Or Stories of Animated Nature* (Edinburgh, 1846), p. 37.

26. William Swainson, *On the Natural History and Classification of Quadrupeds* (London, 1835), p. 235; disparagingly discussed in George Vasey, *A Monograph of the Genus Bos* (London, 1857), pp. 178–79.

27. Thomas Bell assisted by Robert F. Tomes, *A History of British Quadrupeds, Including the Cetacea* (London, 1874), p. 368; Vasey, *Monograph of the Genus Bos*, p. 140.

28. "Wild Cattle of Great Britain," *Penny Magazine* 7 (1838): 443.

29. *Sussex Herd Book, Containing the Names of the Breeders, the Age, and the Pedigrees of the Sussex Cattle*, vol. 2 (Tunbridge Wells, 1885), "Preface" (unpaginated).

30. F. A. Manning, "The History and Growth of the Kennel Club," *Dog Owners' Annual for 1890* (London, 1890), p. 138.

31. For a discussion of the construction of domestic animal breeds by Robert Bakewell and others, see Harriet Ritvo, "Possessing Mother Nature: Genetic Capital in Eighteenth-Century Britain," in *Early Modern Conceptions of Property*, ed. John Brewer and Susan Staves (London, 1995), pp. 413–26. John Borneman has offered a synchronic account of a similar process in "Race, Ethnicity, Species, Breed: Totemism and Horse-Breed Classification in America," *Comparative Studies in Society and History* 30 (1988): 25–51.

32. Leonard Bull, *History of the Smithfield Club from 1798 to 1925* (London, 1926), p. 261; George Coates, *The General Short-Horned Herd-Book:*

Containing the Pedigrees of Short-Horned Bulls, Cows, &c. of the Imported Durham Breed (Otley, 1822).

33. Stephen J. G. Hall and Juliet Clutton-Brock, *Two Hundred Years of British Farm Livestock* (London, 1989), p. 49.

34. David Low, *On the Domesticated Animals of the British Islands* (London, 1845), pp. 381–82.

35. C. M. Prior, *The History of the Racing Calendar and the Stud-Book, from Their Inception in the Eighteenth Century, with Observations on Some of the Occurrences Noted Therein* (London, 1926), p. 15. See Nicholas Russell, *Like Engend'ring Like: Heredity and Animal Breeding in Early Modern England* (Cambridge, 1986), chap. 5, for a fuller account of eighteenth-century horse-breeding.

36. James Long, *The Book of the Pig: Its Selection, Breeding, Feeding, and Management* (London, 1880), pp. 316, 124.

37. Robert Knox, *The Races of Men: A Philosophical Enquiry into the Influence of Race over the Destinies of Nations* (London, 1862), p. 2.

38. Judith Neville Lytton, *Thoroughbred Racing Stock and Its Ancestors: The Authentic Origin of Pure Blood . . .* (London, 1938), pp. 64–65.

39. James Anderson, *Essays Relating to Agriculture and Rural Affairs,* vol. 2 (Edinburgh, 1777), pp. 138–39.

40. Charles Hassall, *General View of the Agriculture of the County of Carmarthen* (London, 1794), pp. 35, 37.

41. William Bingley, *Animal Biography; Or, Authentic Anecdotes of the Lives, Manners and Economy of the Animal Creation,* vol. 1 (London, 1804), p. 202.

42. This iteration occurred in Thomas Pennant, *British Zoology,* vol. 1 (London, 1768), p. 50. A Victorian facsimile of *On Englishe Dogges* was published in 1880.

43. Everett Millais, "A Colouring Tonic for Dogs," *Dog Owners' Annual for 1892* (London, 1892), p. 86.

44. S. O. Beeton, *Beeton's Book of Home Pets: Showing How to Rear and Manage, in Sickness and in Health, Birds, Poultry, Pigeons, Rabbits, Guinea Pigs, Dogs, Cats, Squirrels, Fancy Mice, Tortoises, Bees, Silkworms, Ponies, Donkeys, Goats, Inhabitants of the Aquarium, Etc.* (London, n. d. [late 19th century]), pp. 551ff.

45. B. Waterhouse Hawkins, *The Artistic Anatomy of the Dog and the Deer* (London, 1876), p. 9.

46. See, for example, H. D. Richardson, *The Dog: Its Origin, Natural History, and Varieties* (London, 1853), p. 40.

47. Peter Lund Simmonds, *Animal Products: Their Preparation, Commercial Uses, and Value* (New York, 1877), p. 237; Edward William Jaquet, *The Kennel Club: A History and Record of its Work* (London, 1905), pp. 19–20.

48. Charles Hamilton Smith, *Dogs . . . Part II: (Domestic Dogs, Foxes, and Hyaenas)* (Edinburgh, 1840); Edward Griffith, *General and Particular De-*

scriptions of the Vertebrated Animals Arranged Conformably to Modern Discoveries and Improvements in Zoology: Order Carnivora (London, 1821), pp. 204–18.

49. John Fleming, *A History of British Animals, Exhibiting the Descriptive Characters and Systematical Arrangement of the Genera and Species of Quadrupeds* (Edinburgh, 1828), p. 11.

50. *Guide to the Gardens of the Zoological Society, March 1829* (London, 1829), p. 5.

51. Vero Shaw, *The Illustrated Book of the Dog* (London, 1881), pp. 2–5.

52. On the social dimension of nineteenth-century British livestock breeding and pet fancying, see Harriet Ritvo, *The Animal Estate: The English and Other Creatures in the Victorian Age* (Cambridge, Mass., 1987), chaps. 2 and 3.

53. J. Mortimer, *The Whole Art of Husbandry: Or, The Way of Managing and Improving of Land,* vol. 1 (London, 1721), p. 227.

54. W. C. L. Martin, "The Ox," in *The Farmer's Library: Animal Economy,* vol. 1 (London, 1849), p. 56.

55. A. Coventry, *Remarks on Live Stock and Relative Subjects* (Edinburgh, 1806), p. 39.

56. John Wilkinson, *Remarks on the Improvement of Cattle, &c. in a Letter to Sir John Saunders Sebright, Bart.* (Nottingham, 1820), pp. 4–5; J. Cossar Ewart, "The Principles of Breeding and the Origin of Domesticated Breeds of Animals," in *Twenty-seventh Annual Report of the Bureau of Animal Industry* [Department of Agriculture] *for the Year 1910* (Washington, D.C., 1912), p. 152.

57. Henry F. Euren, *The Norfolk and Suffolk Red Polled Herd Book,* vol. 1 (Ipswich, 1874), p. 7.

58. James Sinclair, ed., *History of Shorthorn Cattle* (London, 1907), p. 97.

59. Robert Bruce, *Fifty Years among Shorthorns, with Over 300 Pen Pictures of Notable Sires* (London, 1907), p. 195.

60. James MacDonald and James Sinclair, *History of Polled Aberdeen or Angus Cattle, Giving an Account of the Origin, Improvement, and Characteristics of the Breed* (Edinburgh, 1882), chap. 17; "Breeder's Directory," appended to Robert William Ashburner, *The Shorthorn Herds of England, 1885–6–7* (Warwick, 1888); Bruce, *Fifty Years among Shorthorns,* pp. 134–52.

61. William Swainson, *A Preliminary Discourse on the Study of Natural History* (London, 1834), p. 232.

62. Everett Millais, *Two Problems of Reproduction* (Manchester, 1895), p. 4.

63. Darwin, *The Variation of Animals and Plants under Domestication,* 2:297–98.

64. [Whitwell Elwin], "Blaze's *History of the Dog,*" *Quarterly Review* 72 (1843): 488; W. Holt Beever, *An Alphabetical Arrangement of the Leading Shorthorn Tribes with Notes for the Use of Breeders* (London, n. d. [c. 1885]), p. v.

10

"The Eden of Home"

Changing Understandings of Cruelty and Kindness to Animals in Middle-Class American Households, 1820–1900

Katherine C. Grier

In the 1820s and 1830s, the decades when the northeastern United States saw the flowering of a self-conscious and energetic bourgeois culture, middle-class Americans began to rethink their relationships to the nonhuman animals living in and around their households. Articles in magazines and newspapers, epistolic advice literature for parents, stories and poems directed to juvenile readers, and even inexpensive engravings and lithographs articulated a popular, domestic ethic of kindness that soon became an important aspect of middle-class self-definition. This ethic redefined some common behaviors toward animals as cruel and found broad social implications in that cruelty. It had two principal arguments. First, children, although born with natural feelings of kindness, had to be socialized into self-conscious tenderness to other sentient beings because thoughtless youthful cruelty to animals, especially on the part of boys, foreshadowed more serious antisocial behavior in adulthood. A second argument appeared more frequently as the decades passed: animals were themselves worthy of special care because they had not only the capacity to feel pain but expressed recognizable feelings—joy, sorrow, and parental love among others—and acted on an innate (innately middle-class) moral sense like that of their human stewards. By locating the origins of middle-class feelings and morals in nature, Victorian Americans naturalized their own culture, but they also opened themselves to more intense, and less embarrassed, emotional attachments to selected animals. One important consequence of this was that the practice of pet keeping, which had been viewed as a slightly eccentric personal indulgence, was transformed into an act that resonated with emotion and high moral purpose.

Although it laid the groundwork for popular support of organized anticruelty activity in America following the Civil War, the domestic ethic of kindness focused most of its commentary on the treatment of animals within the context of family life. It conceptualized the relations between humans and nonhuman animals in light of changes in the relations between middle-class parents and children and in terms of antebellum America's expanding definition of dependency. Kindness to animals became an integral part of the middle-class cultural ideal that one minister labeled the "Eden of Home."[1] This was a formulation that, to some evangelical authors, explicitly included idyllic relations with animals; Charlotte E. B. Tonna, an English evangelical whose writings seem to have been widely available in the United States, argued that animals had also fallen when Adam was driven from the Garden of Eden. The restoration of kind relations between humans and animals was a step toward the reconstitution of Paradise on earth.[2]

The domestic ethic of kindness was, in fact, overdetermined by three sets of ideas shaping bourgeois (or "Victorian") culture in America: evangelical Protestantism, gentility, and domesticity.[3] Evangelical Protestantism, presented as an alternative to orthodox Calvinism, was attractive to tens of thousands of converts in the Second Great Awakening of the 1820s and 1830s with its arguments for the importance of emotion as the wellspring of belief, the efficacy of the individual will, and the real possibilities of individual and social progress toward perfection despite the burden of Original Sin. Both individual and social progress required self-control, and we will see that cruelty to animals was diagnosed as a failure of self-control brought about by "passions" that would, in time, overwhelm the person.[4] The cultural ideal of gentility, the secular standard of personal excellence that Anglo-American elites had fostered during the eighteenth century, also assumed the efficacy of the will in its taxonomy of experience. Gentility's ideal of flawless self-control was understood as the result of personal "cultivation," a term which implied both the possibility of growth and the necessity of constant self-care. Advocates of kindness argued that improved treatment of animals was evidence of benevolence, the disinterested desire to promote the happiness of others that was the product of the undamaged, properly cultivated human "moral sense." Benevolent feelings could be an outgrowth of religious belief, but they were also a characteristic of gentility.[5] Inexpensive manners books, published in large numbers throughout the nineteenth century, continued to offer an eighteenth-century formula for the "amiable character," connecting benevolence, the "general impulse to do good," with "true delicacy," the sensitivity that allowed a genteel person to enter into the feelings of others. The anonymous author of *A Manual of Politeness* (1837) explained that the best sensibility had "a quick sense of what may give pleasure or pain, and teaches us to pursue the one, and avoid the other: and a refined understanding points out the surest means of doing this in different circumstances."[6]

In concert with these ideas, the powerful nineteenth-century cultural construct "domesticity"made the individual household simultaneously the primary medium for creating self-disciplined adults, a refuge from the masculine world of economic competition, and a model of the world as it should be, where the threat of naked power was supplanted by moral influence and feminine love.[7] Along with the greater responsibility for child rearing experienced by middle-class women in the first half of the century, the conventions of domesticity intimated that, as each woman presided over her "state in miniature," her conventional feminine attributes—a more feeling nature, gentleness, and the capacity for deep maternal devotion—shaped the particular kind of power she wielded. Her tender feelings and influence in the "softening, sanctifying environments of home" were meant to guide family members gently toward the good.[8] This understanding of the indirect, yet powerful, influence of mothers on the growth of feeling in the young also dovetailed nicely with gentility's understanding of the best methods for cultivating the human moral sense.

Thus bourgeois culture in the United States made kindness to other sentient beings a manifestation of the softened feelings that respectable people should value on multiple counts; it became a marker of class identity. The underlying premises of arguments on behalf of animals for the rest of the nineteenth century were still based upon those wielded by antebellum commentators, who included some of the era's best-selling authors—Sarah Josepha Hale, the influential editor of *Godey's Lady's Book,* and the advice and fiction writers Lydia Sigourney, Lydia Maria Child, T. S. Arthur, and Harriet Beecher Stowe—along with a cohort of now obscure figures. These antebellum arguments for kindness had, in turn, a genealogy of their own in the eighteenth century and before, although popular commentators, the focus of interest here, probably had only a fragmentary awareness of their long history. They cobbled together fragments of biblical verse, snippets of Romantic poetry, and tidbits of moral philosophy with their own particular social and cultural concerns; from this unlikely mixture came the commonsense standards of kind and cruel treatment that, I would argue, many middle-class people abide by today.[9]

While advocates of kindness did believe that young minds had to be influenced away from what was interpreted as "thoughtless" cruelty, they still assumed that infants had a natural moral sense that made it easy to direct them gently toward the good. This notion reflected significant changes in theories of child rearing among progressive parents, who had begun to adopt the ideas first outlined by John Locke in *Some Thoughts Concerning*

Education (1693) by the latter part of the eighteenth century.[10] During the first half of the nineteenth, some even argued that infants were superior to adults in their "holy innocence," their spirits "fresh and unsullied by contact with the breathing world."[11] However, even evangelical authors of advice to parents such as Horace Bushnell, who rejected what he termed the "radical goodness of human nature," argued that parents should not presume "that the child is to grow up in sin, to be converted after he comes to a mature age; but that he is to open on the world as one that is spiritually renewed . . . seeming rather to have loved what is good from his earliest years."[12] One of the interesting outcomes of this re-visioning of children was its impact on the domestic ethic of kindness. Later we will see that middle-class perceptions about animals were shaped by the increasing importance of a conventional metaphor that compared young children and some animals, especially household pets, favorably. Of course, children could be raised into civilization further, but household pets and some other animals also benefited from this particularly Victorian take on "domestication," as commentators pointed out that animals and children were both naturally innocent, kind and good.

The notion that animals and children were similar beings, however, was only one of three conventional metaphors employed by the domestic ethic of kindness and shaping perceptions of living animals in and around Victorian households: the animal as servant, the animal as child, and animal families as human families. All three tropes appeared in texts that promoted kindness—a veritable flood of cautionary stories, poems, and epistles—along with the material culture of Victorian households, where imagery operated as a silent messenger of changing sensibility. Publishers of inexpensive lithographs, for example, seem to have found a ready market for prints that suggested the expanded role of animals in the middle-class household and made manifest an enlarged, if selective, understanding of animals as models of social order, particularly its most ideal form, the middle-class family. Often the three metaphors were mingled in popular media, so that it became difficult to differentiate among the kinds of love and duty they represented—of a conscientious master, of a doting parent, and of a neighbor engaged in the shared enterprise of raising offspring.

All three metaphors not only framed animals in terms of human roles in and around the middle-class household but they were thoroughly congruent with the conventions of domesticity, which conceptualized the world into two domains opposite and parallel to one another. The first was rational and hard, commercial, public, and masculine, leaving no place for ties of the heart. The other was bounded by the pleasures and responsibilities of home; it was beyond commerce, private, and decidedly feminine. A psychological as well as physical sanctuary, its occupants were connected first by their love for one another. The domestic realm was also a nexus linking heavenly order and earthly social order. An essay titled "Home Interests" in *Youth's Penny Gazette,* a weekly paper published by the American Sunday-School Union, insisted that each household was a "little world," where the principles of good social order were "as operative . . . as in the town, state, or kingdom." Domesticity did not reject divinely ordained hierarchical relations within families; first there was the authority "God himself" conferred on parents. However, power in families was softened with a rhetoric of service and duty, using the concept of "dependence." "There is dependence—the weak upon the strong, and the ignorant upon the wise. There is a little theatre on which all the charities and graces of social life may be beautifully and harmoniously displayed."[13] The influential advice author Catharine Beecher agreed. The "special discipline of the family state" was care of the dependent. She urged "daily self-devotion of the stronger and wiser to elevate and support the weaker members." Children's "disposition to social usefulness" was expanded when their parents exercised a new law, the "law of human kindness," there.[14]

While it has been a useful justification of social inequality, the concept of divinely ordained hierarchy has also been central to Western interpretation of the place of humans in relation to the natural world. Both classical and biblical texts supported the notion that every plant and animal had been created for the purpose, directly or indirectly, of meeting human needs. Two divergent habits of thought existed within this general tradition of anthropocentrism, however. These may be called, to be concise, dominion and stewardship. The first more frankly pressed unqualified human rights to exploit animals; the second, minority position emphasized what Keith Thomas has called "a distinctive doctrine of stewardship and responsibility for God's creatures" while still acknowledging the special position of humans.[15] If the patriarchal order of traditional family life can be seen as analogous to dominion, the doctrine justifying the rule of man over all of nature, the more liberal view of family relations adopted the gentler notion of stewardship and applied it to all dependents, including children and animals. It blurred the authority of patriarchy, and the realities of economic and legal power within families, by emphasizing the emotional and spiritual obligations of husbands and wives to one another, and their obligations to their children.

The domestic ethic of kindness used the same strategy for prescribing good relations between people and animals that the general prescriptions of domesticity applied to parents and children, softening coercive power over other sentient beings with a similar rhetoric of duty and service. Authors generally agreed with Francis Wayland, who stated in his standard text on moral philosophy that "the revealed will of God" gave humans the right to use animals "to promote our comfort."[16] However, just as power in families was softened by self-discipline and love, so advocates of kindness to animals tried to soften the implications of divine hierarchy. Having agreed that humans rightfully ruled animals, the author of *A Natural History of Animals* (1828) also argued that abuse of animals "given to us for the purpose of promoting our convenience and comfort" was "foul ingratitude to the all-bountiful Donor," a stance that Wayland also supported.[17] Understanding this also allowed humans to contemplate their own relation to the divine. A. L. Carroll, author of an 1867 essay titled "An Apology for Dogs," argued that full appreciation of the implications of stewardship led to "higher philosophical considerations" of the relation of man to God, since humans were as God to "domesticated brutes." Observing with honesty the "imperfections of our management" of dogs, despite their willingness to serve, could lead to contemplation of God's justice to man "and to feel from the comparison a salutary humility." Ownership of dogs should, therefore, encourage people "to imitate the divine character in the exercise of the authority delegated to us," and teach, by analogy, "from ruling inferiors, to submit to Superior rule."[18]

This "Christianization" of relationships of power between humans and animals was a move strikingly like earlier attempts at Christianizing slavery through an argument for moral stewardship and restraint on the part of masters to the benefit of their less civilized chattel—and it suffered from similar limits as an argument, since both slaves and animals were in truth defined fundamentally by their legal status as property rather than their place in the human heart.[19] It seemed an insolvable problem, but some authors tried to get around it without proposing the unimaginable, a world of animals without human masters. They did this by euphemizing ownership, perhaps finding its implications of sentient property too close to slavery. Thus Harriet Beecher Stowe called animals "a sacred trust from our Heavenly Father."[20] Charlotte E. B. Tonna went so far as to inform her readers that animals were never really owned by people, which was "one of the things that we are apt to forget when we have a beast, or a bird, or a fish, or an insect, in our power. We are too ready to say to ourselves, 'This is mine, and I may do what I like to it.' Not so, it is a creature of God's, not of ours; and if we do any thing that he does not approve of, he will surely recon [*sic*] with us for it."[21] This argument was probably based on Psalm 50, verses 10 and 11: "For every beast of the forest is mine, and the cattle upon a thousand hills. / I know all the fowls of the mountains; and the wild beasts of the field are mine," one of a handful of biblical verses that advocates for kindness had mined in earlier times, and that appeared repeatedly in nineteenth-century anticruelty literature.

A central tenet of middle-class child-rearing practice in the first half of the nineteenth century was the gradual growth of infants, who were instinctively kind yet lacked experience and discipline, into independent moral agency. But how were young children, confined to such a small sphere of action in the world, to learn the law of human kindness that led to general benevolence? Inside the "Eden of home," the animals already present as guardians and workers or companions now had a special role to play. Lydia Sigourney's popular advice manual *Letters to Mothers* (1838) insisted that kindness to animals was one of the three fundamental lessons of the "moral code of infancy" that laid the "foundation for a future superstructure of virtue." After the lesson of obedience to parents, she suggested, the next lesson for children was "kindness to all around."

> The rudiments are best taught by the treatment of animals. If [a child] seizes a kitten by the back, or pulls its hair, show immediately by your own example, how it may be held properly, and soothed into confidence. Draw back the little hand, lifted to strike the dog. Perhaps it may not understand that it thus inflicts pain. But be strenuous in confirming the opposite habit. Do not permit it to kill flies, or to trouble harmless insects. Check the first buddings of those Domitian tastes. Instruct it that the gift of life, to the poor beetle, or the crawling worm, is from the Great Father above, and not lightly to be trodden out.[22]

Parents also had to be newly self-conscious of their own everyday relations with animals just as they had to exercise self-discipline in parenting. In *Views of Christian Nurture* (1847), Horace Bushnell enjoined parents to exercise "a constant and careful discipline of themselves" in setting examples for their children. Parental authority must never be a "reign of brute force" since the "ductile nature" of children was twisted away from the good by every bad example seen or experienced firsthand.[23] Discussing his approach to incorporating moral lessons into young lives without "formal precepts and didactic instruction," Jacob Abbott, author of the enormously popular "Rollo" stories for children, chose for his single example parental treatment of songbirds, one of the iconic animals in discussions of kindness:

> The development of the moral sentiments in the human heart, early in life, . . . is determined in a far greater degree by sympathy, and by the influence of example. . . . If a boy hears his father speaking kindly to

> a robin in the spring,—welcoming its coming and offering it food,—there arises at once in his own mind, a feeling of kindness toward the bird, and toward all the animal creation, which is produced by a sort of sympathetic action, a power somewhat similar to what in physical philosophy is called *induction.* On the other hand, if the father, instead of feeding the bird, goes eagerly for a gun, in order that he may shoot it, the boy will sympathize with that desire, and . . . there will be gradually formed within him . . . a disposition to kill and destroy all helpless beings that come within his power. There is no need of any formal instruction in either case.[24]

Assuming that their own behavior toward animals was exemplary, however, child-rearing literature still insisted that parents be alert to childhood behavior that caused pain to other sentient creatures. One of the axioms repeated by the proponents of kindness was that the practice of cruelty, like other sins, was incremental—the boy who tormented animals would inevitably become the man who tormented other people. This presumption, which sociologists have dubbed the "violence graduation hypothesis," became one of the fundamental principles of the humane education movement of the late nineteenth century, and it still informs humane education activity.[25] However, the Victorian argument that the cruel child begat an even worse adult was not new. Keith Thomas has discussed the scriptural underpinnings of arguments against cruelty and subsequent early commentary such as Thomas Aquinas's *Summa contra Gentiles,* iii, 113. Aquinas, no animal lover, noted that scriptural injunctions against being "cruel to brute animals" seemed to rest on the reasoning "lest through being cruel to animals one becomes cruel to human beings or because injury to an animal leads to the temporal hurt of man."[26] The formula was revived during the second half of the eighteenth century in discussions of what was believed to be an innate human moral sense. The ideal of this "cult of sensibility" was the "man of feeling," a being whose innate goodness had not been thwarted or twisted in youth and whose "tender-hearted susceptibility to the torments of others was the mark of his deeply virtuous nature."[27] The inverse, the man who did not feel at all, was a creature to be feared. The most powerful eighteenth-century expression of this narrative of progressive cruelty was William Hogarth's famous series of engravings *The Four Stages of Cruelty* (1750–51).[28] The first print depicts gangs of boys, including the boy Tom Nero whose path is being plotted in the narrative, hanging cats, cockfighting, gouging out the eye of a small bird with a wire, and sexually molesting a dog, among other outrages; the second includes Tom Nero beating a carriage horse who has been overburdened by a group of thoughtless, obese lawyers, and the death at the hands of a drover of a sheep too weak to walk. (In the third, Tom Nero murders his lover; in the fourth his hanged corpse is dissected by doctors, while a dog that looks suspiciously like the one he once abused eats his discarded heart.)[29]

By the 1780s, this formula had moved to the new realm of children's literature. Reverend C. G. Salzmann's *Elements of Morality for the Use of Children* (1782), translated into English by Mary Wollstonecraft and in circulation in America by 1800, used a variation on the same formula when a young boy is scolded by his father for proposing to punish a mouse as a "little thief" by cutting off its ears and tail (parroting punishments for human criminals). Mr. Jones shames him as a "cruel boy" and warns, "*He who can torment a harmless animal* . . . accustoms himself by degrees to cruelty, and at last he will find a savage joy in it: and after tormenting animals, will not fail to torment men."[30] Eighty-five years later, "Cruelty and the Gallows," a particularly hair-raising piece of admonition in the *Youth's Companion,* still used Hogarth's narrative structure when it informed its readers that a famous murderer, recently gone to the gallows at age nineteen, had admitted that as a child, he had liked killing animals "better than anything else": "Youthful reader! whenever you find yourself delighting in cruelty to any of the creatures which God has made, think of Prescott—think of the *gallows*!"[31]

Based on the premise that cruelty was the private act of an underdeveloped or thwarted moral sense, antebellum children's literature on the theme was intended as an act of intervention. Children were supposed to read these horror stories and, through sympathetic identification with the sufferers, escape the firsthand experience of cruel behavior. The authors shared their predecessors' concerns about the hardening effects of childhood cruelty, and they took particular pains to embrace all animals, no matter how lowly, in their stories. Take, for example, their critique of what the Reverend A. B. Muzzy called the "wanton sport of torturing poor insects," an example that turns up repeatedly in the genre.[32] In an essay for boys, Lydia Sigourney quoted the poet William Cowper, a favored source of exhortations to kindness, to the effect that "I would not have for my friend . . . one who carelessly sets foot upon a worm," and warned, "If I see even a young child, pull off the wings of an insect, or take pains to set his foot upon a worm, I know that he has not been well-instructed, or else that there is something wrong and wicked in his heart."[33]

It was precisely because cruel treatment of insects and other "lower animals" (authors also singled out frogs, minnows, toads, turtles, and snakes as undeserving victims) seemed so trifling that it was so important. It could easily become a "habit of cruelty." The notion that so thoughtless and small a childish act led to a more dangerous habit originated in the assumptions about human nature and the passions that evangelical Protestantism used in its interpretation of the mechanics of sinful behavior. In a series of sermons reprinted as *Lectures to Young Men* (1844), Henry Ward Beecher, one of the most influential preachers of the mid–nineteenth century, explained the physio-psychological process that led to the hardening of the heart and the habit of cruelty:

> The body is not stored with a fixed amount of strength, nor the mind with a uniform measure of excitement; but both are capable, by stimulation, of the expansion of strength or feeling, almost without limit. Experience shows, that within certain bounds, excitement is healthful and necessary, but beyond this limit, exhausting and destructive. Men are allowed to choose between moderate but long-lived excitement, and intense but short-lived excitement. Too generally they prefer the latter. To gain this intense thrill, a thousand methods are tried. . . . When once this higher flavor of stimulus has been tasted, all that is less becomes vapid and disgustful . . . for these very disgusts, and the perpetual struggle, strike a deeper chord of excitement than is possible to the gentler touches of home, friendship and love . . . and they must be continued, or the mind reacts into the lethargy of fatigue and *ennui.* It is upon this principle that men love *pain*; suffering is painful to a spectator; but in tragedies, at public executions, at pugilistic combats, at cock-fightings, horse-races, bear-baitings, bull-fights, gladiatorial shows, it excites a jaded mind as nothing else can.[34]

This ambivalent understanding of the role of violence in the physiology and psychology of excitement suggested simultaneously that the excitement of violence was actually addictive, hence virtually impossible to resist, and that falling into habits of cruelty was a failure of individual self-control with broad implications for the nature of public life. Beginning in the 1820s, a call for the abolition of public executions in the United States was successful because it drew upon this common, if confused, understanding. Part of a reform movement against public and private corporal punishment that met with mixed success, critics of public execution argued that the repeated sight of such suffering hardened and debased both the actors and onlookers who, unable to intervene, experienced a weakening of the natural sympathy they were born with. Taking pleasure in violence was a latent human passion, the obverse and twin of the gentle sensation of sympathy.[35]

Reverend William Alcott probably worried conscientious young mothers with his admonition that "all the larger cruelties of mankind have their origins in the cruelties of infancy and youth."[36] Yet everyone knew children were not naturally cruel; in fact, they were blessed with " natural feelings of kindness."[37] So, how could the abuse even infants apparently heaped upon the animals in their paths be explained? Through inability to face the possibility that causing pain inspired infantile pleasure, the actions of young masters were interpreted as "thoughtless"—at least the first time something bad happened. Stories of thoughtless cruelty to animals, guilt at the realization of pain suffered, repentance, and the "lesson of mercy" learned filled children's periodicals. They were intended as acts of literary intervention, written with the purpose of preventing cruelty and its moral consequences by encouraging preemptive sympathy. In this sense the function of these tales was similar to (and in fact some seem to have predated) the published narratives of slavery that expanded the concept of individual rights and the definition of cruelty in the 1830s and 1840s.[38] A few selections from the early decades of the *Youth's Companion,* the popular and long-lived children's weekly paper founded in 1827 by the publisher of the Boston Congregational newspaper the *Recorder,* offer a good sample of both the contents and tone of these tales.[39] Sometimes the injury done to an animal was pretty mild stuff. The protagonist of a "*true* story" titled "The Puppy" is perpetually mortified by the fear and mistrust forever felt by a dog he had thrown in a pond as a joke.[40] However, often repentance came only after the death of an innocent creature. In "The Wounded Bird," Henry and Alfred, "two thoughtless boys," go gunning for songbirds. When Henry returns home with three dead songbirds and a wounded oriole and presents them to his mother with pride, he is stunned by her response: " 'Poor bird!,' said she, taking from Henry the wounded oriole, and handling it with great tenderness. 'Can it be possible that my son has done this?—that his hand has committed so cruel a deed?' and the tears dimmed her eyes." Henry's heart is "touched in an instant" as he has "a glimpse of the truth that it was wrong to sport with the life of any creature," and he dissolves into repentant tears. Not only does he pledge to never "hurt any animal in sport," but he convinces the gun-toting Alfred to sell his prize and purchase "a handsome little juvenile library of good books."[41] Forcing a woodchuck to swim until he drowned, the protagonists of "The Woodchuck" remember how "he had approached them as though he considered them friends, but they proved to be his murderers." They are haunted by the image of the dying animal, "motionless on the bank, now and then looking at them with an expression which seemed to denote nothing but forgiveness."[42] Stories like this would have been horribly depressing to young readers, except that the young protagonists always seemed to learn the lesson of kindness so quickly and to understand evangelical Protestantism's religion of the heart intuitively when called to account—and they received forgiveness readily. The hero of *Master Henry's Rabbit,* a small paper-cover children's book published in the United States around 1840, exemplifies this formula. Punished for neglecting the comfort of his pet rabbit by being deprived of his own supper, Henry, a boy of seven or eight years, immediately "humbled himself before God for this sin, and prayed earnestly for a better heart."[43]

These fictional examples feature little boys, and, indeed, relatively few antebellum tales in this genre seem to have had girls as their principal protagonists. This had not been true in the earliest discussions of children's moral development. In *Thoughts on the Education of Daughters* (1787) Mary Wollstonecraft, perhaps drawing on Salzmann, included in her chapter "Benevolence" an anecdote about a girl who killed ants "for sport"; Wollstonecraft stopped the child by "adapting Mr. Addison's account of them to her understanding. Ever after she was careful not to tread on them, lest she should distress the whole community."[44] By the nineteenth century, little girls were almost always the untutored voices of sympathy and conscience, rehearsing the role they would play as mothers.[45] On the rare occasions when they too were the thoughtless perpetrators of cruelty, they, like their brothers, reformed immediately. In one 1833 story, "The Bird's Nest," a naughty nine-year-old girl smashes a meadowlark's nest and is found by her father praying at four in the morning for God to forgive her sin.[46]

The problem with this part of the discourse of middle-class kindness was that it stigmatized common forms of play by boys of all social classes in nineteenth-century America. Young boys spent more time outdoors than their modern counterparts, and hunting and trapping small wild animals provided outlets for youthful energies—and imitated adult practice—in both town and country. William Lyon Phelps (1865–1943), the son of a Connecticut minister, recalled that he "delighted in shooting and killing birds, any bird, edible or otherwise," especially after he received his first double-barreled shotgun at the age of twelve, although "I was not cruel by nature, and could not bear to see any animal ill-treated or in pain. Yet, when I was too small to own a gun, I would get up at dawn, armed with David's implement, and try to kill robins and bluebirds. On the rare occasions when I succeeded, I felt thrills of joy, unshaded by regret."[47] An expansive and dramatic critique of bird-nesting and songbird killing, about which more below, also ran counter to parallel advice that boys become amateur naturalists. Part of the "rational amusement" of natural history, collecting specimens and the creation of amateur "cabinets" almost always involved taking life and awkward attempts at taxidermy.[48] Boys also trapped muskrats to earn pocket money from their pelts; they trapped barn rats or used them for target practice. Even Henry Ward Beecher acknowledged that the sight of a woodchuck in his youth had caused attacks of "venatorial perturbation," and he winked at his own sons' efforts to catch and kill them.[49] Killing small wild or nuisance animals was just something boys did, a rehearsal for manhood, even the manhood of white-collar work.[50]

Yet the domestic ethic of kindness made killing for sport, the kind of leisure in which middle-class men engaged since their families no longer required game for subsistence, suspect because it dealt in blood for the purposes of amusement. This particular element of the discourse peaked in the 1850s; it waned somewhat in the 1860s—when, one presumes, the whole subject of killing grew too complicated for frank discussion owing to the moral positions staked out in the Civil War. Dr. William A. Alcott admonished women neither to smile on "these juvenile murderers" nor to "eat the fruit of their doings."[51] "Fishing and hunting for mere sport" were among the "exciting and pernicious amusements" of the senses targeted by advocates of kindness.[52] It was not that humans weren't entitled to kill animals. The difficulty came when they exhibited a love of killing, a fundamental breakdown in empathy and self-control veering toward the uncontained passions of the hardened heart. Alcott worried that, "whatever may be the apology, are not most of the animals around us, whether slain in one way or another—for food or for defence [*sic*]—are they not slain for sport? Where is the boy or young man to be found who hunts, entraps, fishes, &c., for any better reason . . . than because it is an amusement to him?"[53] Making "*game* of the suffering of God's creatures" hardened the heart and also graduated the practitioner to greater violence, in this case "rapine, murder, and war."[54]

It was also important not to take passionate revenge on "injurious" animals, or animals whose behavior worked at cross-purposes to human plans. Alcott worried about the "spirit of war and murder" with which children killed snakes that were simply acting "in accordance with their nature."[55] Harriet Beecher Stowe related the story of a man who set fire to a living rat to serve as a warning and example to the others in his house, and argued that it reflected "a very imperfect sense of right and wrong existing in the community": "Is there any wrong in rats wanting food? Does not the inconvenient pursuit of this, on their part, spring from the same source and the same craving that causes the work of man? If then, these interfere with man, have they not at least a right to the ordinary rules of civilized warfare? Has any human being a right to torture an animal, merely to avoid an inconvenience to himself?"[56]

The earliest advice on inculcating "kindness to all around" found in child-rearing advice books simply assumed the presence of small domestic animals around middle-class households and found it useful. Now, welcoming animals into the domestic circle was transformed from a personal indulgence to a positive good, and pet keeping was promoted. Affection for pets and the experience of caring for them were rational amusements, regarded as potent agents for cultivating "the love of natural history and intellectual improvement . . . thoughtful tenderness and moral sensibility." "Indeed, love in any form, and to any thing," noted Reverend A. B. Muzzy, "is an elevating motive. . . . [W]ho can doubt that many a heart, both of the happy and sad, has been made better by the multitudes of parrots, lap-dogs, canaries, &c., which have been objects of affection."[57]

In 1868, Sarah Josepha Hale published an essay titled "Pets and Their Uses" that actually argued in favor of pet keeping because it "humanized" boys and was "a great preventative against the thoughtless cruelty and tyranny they are so apt to exercise toward all dependent beings!" Like other advice authors before her, Hale presumed that boys, the future masters of their households, were inclined to casual cruelty. She did admit that girls should also be encouraged to have pets, if only because "as sisters and mothers, they must help and teach boys in whatever things are good, tender, and lovely."[58]

The relationship of the discourse of kindness to broader discussions of social violence took a particularly interesting turn in the case of songbirds and their broods. Although advocates of kindness viewed all boyhood hunting as suspect, songbird killing, smashing clutches of eggs, and bird-nesting (stealing young birds to raise, or nests and eggs for natural history collections) was by far the most troublesome because of the special status of birds in Victorian culture. Birds occupied a special place in the array of scriptural quotations used to promote kindness; children were warned repeatedly that God's eye watched birds just as it did children, and that his mercy extended even to them.[59] Birds had been praised for their apparent familial sentiments for centuries, but the conventions of middle-class domesticity in Victorian culture gave this formula extra resonance. Songbirds, both caged and wild, were appreciated not only because of the company their song provided in a world without recorded music. The "untiring devotion" of bird couples to one another, and their domestic habits, particularly the "care and tenderness they show to their young," were exemplary. The "ingenuity with which they construct their nests" was analogous to the care women were supposed to expend on the rooms of their own "nests."[60]

America's new humane organizations took up bird protection in the late 1860s; the very first issue of *Our Dumb Animals,* the children's magazine founded by George Angell of the Massachusetts Society for the Protection of Animals in 1868 (of which two hundred thousand copies were distributed gratis) included a poem called "Boys, Spare the Birds." These efforts have

1. "The Cruel Boys Robbing the Bird of her Little Ones." Colored lithograph; artist unknown. From the American Sunday-School Union, *Picture Lessons.* Philadelphia, ca. 1850. Courtesy The Winterthur Library: Printed Book and Periodical Collection.

been credited as effective promoters of bird study and appreciation.[61] However, the narratives under consideration here were not just about animals; they also were clearly intended as metaphors about the dangers of violence to human families, especially helpless women and children. Moralizing children's poetry, stories, and even picture Sunday school lessons often took destruction by boys of bird families as their text. *Picture Lessons, Illustrating Moral Truth,* a set of large colored pictures with didactic captions published by the American Sunday-School Union, included "The Young Robbers. The Cruel Boys Robbing the Bird of Her Little Ones" (figure 1). The caption chided, "Harken! My boys, Would a mother like to have a cruel robber come, and take her little ones out of the cradle or the crib while she has gone out, to get bread for them? Answer this question before you touch these helpless birds." And the lesson drove home the point by asking, "Has [the boy] a home, and would not he be grieved if robbers should break into it, and seize his little brothers and sisters and carry them off, and throw the house down or set it on fire? Why then does he not consider that the house of the little birds is dear to them, and that to break it up, is to distress, if not to destroy them?"[62] Birds deprived of their mother became "orphans," and their suffering as they slowly died of hunger and cold was catalogued in excruciating detail.[63] "The Robin," a poem included in *The Boy's Reading-Book,* and Harriet Beecher Stowe's "The Nest in the Orchard" both used the formula of a destruction of a bird family. Each began with the death of the male at the hands of a hunter and chronicled the eventual death of the female and her brood from starvation, grief, or further hunting.[64]

Poems where the birds spoke on their own behalf, expressing their sorrow—and sometimes their rage—at the humans who despoiled their nests or murdered their families also were directed to young readers from the beginning of the century, and the villains seem invariably to be male. *Pity's Gift: A Collection of Interesting Tales, to Excite the Compassion of Youth for the Animal Creation,* in circulation in Philadelphia by 1802, included verse in the voice of a nightingale mourning the death at the hands of a boy of her "husband" Elusino.[65] "Who Stole the Bird's Nest?" a widely reprinted mid-century poem for children on this same subject, used blame from a harmonious community of animals and little girls in concert against the boy who stole "five eggs I laid / And the nice nest I made." A dog, a cow and her calf, a ewe, and a hen, who had donated their own fur, fleece and feathers to help make the nest, and a chorus of the wronged bobolink, a dove, a crow, and two little girls who understand "how sad the bird would feel," speak to each other, agreeing "Let us find out his name / And all cry "For shame!" The result:

A little boy hung down his head,
And went and hid behind the bed
For *he* stole that pretty nest
From poor little yellow-breast;
And he felt so full of shame,
He didn't like to tell his name.[66]

Lucy Larcom's often-reprinted "The Brown Thrush" did urge both boys and girls to leave nests alone, but the anonymous "A Bird's-Eye View," reprinted by John Greenleaf Whittier in his popular edited collection *Child Life* (1871), ended with a "little Robin-bird" crying, "But a great mistake it is / Boys were ever made at all."[67] It should be noted that, as the literature promoting kindness developed, merciful boys began to appear with more regularity. Frederic, the protagonist of one of many stories titled "The Bird's Nest," is tempted to take a nest full of young robins to raise, but, recognizing how "unhappy" the adult birds are, he puts the nest back unharmed.[68]

Why was this element of the domestic ethic of kindness to animals clearly doubling a covert discourse about domestic violence? Myra C. Glenn's discussion of the antebellum crusade against corporal punishment points out that a campaign that successfully made physical punishment of convicts, seamen, and slaves controversial, and pushed executions out of public space, met its cultural limits in discussions of chastisement within families. While arguments against physical punishment for children were an important part of the new kinds of parenting advice discussed earlier in this essay, and while middle-class parents clearly tried to adhere to these limits, reformers were reluctant to restrict legally "a parent's traditional right to punish children. Various factors, particularly parents' allegedly God-given authority to discipline children and the increasingly private nature of middle-class domesticity, precluded efforts to regulate parental punishment within the home."[69] Ironically, the idealized views of women and marriage articulated as part of domesticity also effectively posed a "crucial ideological barrier to widespread discussion of wife beating." In 1854, free love advocates Thomas and Mary Gove Nichols complained, "Thousands of women suffer where one complains. The ministry, the physician, or sometimes an intimate friend, hears of these things; but there is a deep *hush*! Respectability, and the sacrednesss of the marriage institution demand silence."[70]

Pushing the violence graduation hypothesis to its logical limits, some argued that boyhood cruelty not only portended domestic tyranny, it even had worrisome implications for the future of the American republic. If "one who is known to have been a good father, son or brother, is relied upon as a good citizen also," the inverse could also be true.[71] Lydia Sigourney cited

American history as her evidence: "Those baleful dispositions, which desolate human happiness are often early developed. It was Benedict Arnold, the traitor, who in his boyhood loved to destroy insects, to mutilate toads, to steal the eggs of the mourning bird, and torture quiet, domestic animals, who eventually laid waste the shrinking domestic charities, and would have drained the life-blood of his endangered country."[72]

It seems likely that analysis of boys' "thoughtless" violence toward animals, and the subsequent argument that intervention during childhood was critical, was not inspired by parental observation of rough play alone. Violence enacted upon the bodies of animals in public spaces was routine in both rural and urban America throughout the eighteenth and nineteenth centuries—and almost all of it was committed by men. Such violence became newly "visible" to the eyes of sensitized people, and it may have inspired a new search for both the meanings behind the violent acts of boys and the connection between the violence of boys and the more dangerous violence of men. The deliberate quality of some of the most egregious behavior of adult men apparently could be explained only by postulating a "hardening effect" in presumed past behavior. It was a scary thought, especially since masculine brutality seemed to embrace so many kinds of victims, including women and children.

Much of the daily physical violence upon animals was what might be termed violence by community consensus, associated with the traditional demands of utility such as butchering for meat. Men were not the only perpetrators. Slaughter and butchering were processes that even city women still knew intimately. They often continued to kill small stock such as chickens with their own hands, and as good shoppers, they understood butchering and still saw their families' future dinners on the hoof in some city markets. However, men did the heavy work with animals in city and suburb almost exclusively; draft animals were a necessary source of power in an energy-poor world that relied heavily on animal muscle power until well into the twentieth century. And some workingmen seemed to express their passions with special freedom upon these dumb creatures who could not defend themselves.

The fledgling American anticruelty organizations of the late 1860s and 1870s were built upon a gradual breakdown in the unspoken community consensus about the permissibility of at least some of this treatment. As early as the end of the eighteenth century, a few sensitive folk had begun to object to treatment of livestock that had once been unworthy of comment. In 1794, Elizabeth Drinker was "really distress'd . . . to see the cruelty of the Dray-Men to their Horses, in forceing them to drag loads to heavy for them up the Hill" and her diary indicates that she had been troubled by this conduct for some years, although as a Quaker and a woman, she took no action against what she saw.[73] Caroline Earle White, founder of the Women's Auxiliary of the Pennsylvania Society for the Prevention of Cruelty to Animals, recalled late in life that she had been inspired to action against abuse of animals as a child. Described as "passionately fond of animals," she "avoided certain streets near her home because, in passing over them, she nearly always witnessed scenes of animal cruelty which depressed her for days afterward."[74] Some brave individuals even acted on their sympathy in an ad hoc manner. Mark Twain recalled his mother spontaneously upbraiding a carter for beating his horse—she even took his whip from his hand—on a visit to St. Louis in the 1840s.[75] Throughout the nineteenth century, statutes against cruelty to animals (laws introduced by men and passed by deliberative bodies of men) concentrated upon the kinds of abuse taking place in public spaces—transporting calves trussed and piled in wagons to market, beating an overtaxed animal carrying a too heavy load—but they did not question the use of animals either as valued food or as flexible power sources.[76] The issue was ameliorating the circumstances of their lives and deaths, rather than questioning the rightness of that exploitation.

However, other kinds of violence against previously "invisible" animals began to seem shocking; even Man's Best Friend was routinely victimized in public. Stray and feral dogs in cities were annually the object of a particular kind of officially sanctioned mayhem. In hunts during the "dog days" of summer to prevent the spread of rabies, mobs of working-class men and boys roamed city streets clubbing dogs for bounty payments. Writing from New York City in 1841, Lydia Maria Child deplored the practice:

> Twelve or fifteen hundred of these animals have been killed this summer in the hottest of weather at a rate of three hundred a day. The safety of the city doubtless requires their expulsion; but the *manner* of it strikes me as exceedingly cruel and demoralizing. The poor creatures are knocked down on the pavement, and beat to death. Sometimes they are horribly maimed, and run howling and limping away. The company of dog-killers themselves are a frightful sight, with their bloody clubs and spattered garments. I always run from the window when I hear them; for they remind me of the Reign of Terror. Whether such brutal scenes do not prepare the minds of the young to take part in bloody riots and revolutions is a serious matter.[77]

Cities' routine control of strays was also done with considerable cruelty. The contractors of city dog carts strangled their victims with wire lassos and sometimes clubbed or shot them on the spot, and frantic captured dogs often tore each other to pieces within the confines of the vehicle.[78]

Another type of violence against animals enacted by men was associated with entertainments in towns and cities. We have seen that blood sports were among the "desperate excitements of debauched men" singled out by Henry Ward Beecher as one of the avenues to exhaustion of the passions.[79] Setting packs of dogs on tethered bears and panthers ("baiting"), giving the starved lions in traveling menageries dogs, sheep, and goats to tear to pieces before an audience, fighting dogs and cocks for money, and shooting at tethered fowl for prizes were all part of the good fun around taverns, town commons, and other places men gathered, and they seem to have attracted audiences that cut across class, but not gender, lines.[80] Lewis Miller's sketchbooks of everyday life in York County, Pennsylvania, illustrate a "Bear-Beat" (*sic*) and "Panther beat" on the common in the town of York in the early nineteenth century. Miller was always careful in his depiction of crowds, and the audience for these events consists solely of men and boys, although women certainly heard about such events and may have seen them in passing. An itinerant trick rider, on the other hand, had an audience of both men and women. (In retrospect, Miller called the bear bait a "barbarous custom.")[81]

However, the violence perpetrated on animals that probably spoke most clearly to women about their own potential helplessness against men was individual and casual, reflecting a rough world of masculinity that regarded the suffering of the powerless as a joke. The remarkable autobiography of William Otter (1787–1856), a plasterer, tavern keeper, sometime slave hunter, and occasional horse dealer in southeastern Pennsylvania and Maryland, is built around a lifelong series of "scrapes" and cruel practical jokes perpetrated on both people and animals. Otter recalls with glee "sport" with a "very large baboon" owned by a tavern keeper in Lititz, Pennsylvania; he "let him have a slash of lime from my brush bip into his eyes, which burnt the poor devil equal to fire, it made him raving." The animal tore his chain from its moorings on the tavern signpost and attacked the tavern keeper's wife and a maid; Otter and his journeymen then subdued the animal by beating it over the head with a hod carrier and filling its eyes and mouth with lime wash before the tavern keeper returned and calmed it by washing and soothing its eyes and mouth with sweet oil (olive oil used medicinally). On another occasion, Otter deliberately frightened and ran a pastured herd of horses almost to death as a distraction so that he and his compatriots could steal pears from a nearby orchard. The "scrape" with an animal receiving the longest anecdote, however, was the plan of Otter and a man for whom he was plastering his house to "work" the death of a Scottish immigrant well digger's dog "for a peace of sport." First, they "turpentined" the dog's nether parts (a common practical joke on dogs and cats) in hope that Brown, the dog's owner, would be forced to shoot him as "mad." When this did not work, Otter cooked up a more elaborate scheme where he told the poor foolish man of a handbill advertising a fifty-dollar reward for a performing dog that could dive six feet under water and carry a twenty-five-pound weight to the surface. They rowed the dog out into the middle of a millpond and experimented with ever heavier weights until the animal finally drowned. Other "scrapes" with animals included almost scorching a monkey to death on a hot stove in a tavern and stuffing a live turkey down a chimney.[82]

Otter may or may not have been a singularly cruel man in his treatment of animals. He was, after all, fond of certain valued animals of his own. His autobiography includes references to a favorite horse named "Bonyparte," and he recalls how much he valued for slave hunting "a dog of the spaniel breed" named Ponto, "one among the finest of that species of dog."[83] He was not cruel enough to become an outcast among his peers, although performance of an increasingly marginalized masculine working-class identity was behind at least some of Otter's behavior. His scrapes exemplify long-standing uses of living and dead animals as the medium for a coarse humor that commented on the realities of physical and social inequalities between people or communicated displeasure at the conduct of others. Such commentary seemed always to rely on either cruelty or death dealing to make the point. Its perpetrators understood well the pain they administered; they did not in fact care about it and expected to suffer pain themselves if they were unable to assert their will.[84]

The middle-class advocates of kindness certainly recognized social and economic differences between themselves and the cart men and hack drivers they criticized and began to prosecute in the late 1860s. But while the fledgling American anticruelty movement clearly identified cruelty with immigrants (Henry Bergh, founder of the American SPCA in 1866, thought that the Irish were especially cruel) and as a typical behavior of the lower classes, the women and children's literature of the decades under consideration here resolutely blurred class distinctions.[85] If present at all, the class of some "cruelists" is registered in a kind of coded shorthand, where words like "ruffian" or "cartman" stand in for entire categories of workingmen.[86] Until the late 1860s, when middle- and upper-class men became active in chartering and running anticruelty groups, the notion that cruelty was fundamentally gendered seems to have trumped the overt connection of cruelty and class. Harriet Beecher Stowe's 1869 article "Rights of Dumb Animals," the most comprehensive statement she published on the subject, still discussed cruelty only in relation to the "heathen state of mind" of a generic man, who hunted animals for pleasure alone or did not care adequately for domestic animals in his possession.[87] By then, however, published cartoons against overloading cart horses or cropping the ears of fighting dogs made clear that the human subjects of the parodies were from the lower classes. In "Cruelty to Animals," published in the same journal as Stowe's essay, a calf whips two men who pull a cart of their trussed compatriots to the market; sheep dressed as fashionable city folk look on.

It is clear that some of the new concern for animals expressed by the advocates of the domestic ethic of kindness was actually a thinly disguised plea on behalf of some groups of people. At the same time, however, animals themselves were increasingly cherished around urban and suburban households for their own sake. One important way this new appreciation was manifested in everyday life was through the expanding practice of pet keeping. Changing sensibilities toward animals around the house can be charted through the changing use of the three metaphors that both described animals and shaped perceptions of their qualities: the animal as servant, the animal as child, and animal families as human families. In the conventions of discourse about pet keeping, the qualities of pet animals, and the feelings of owners toward their favorites, were often described using the same language attached to young children.

Before the nineteenth century, Americans did keep a select group of small animals in and around their dwellings primarily for pleasure, and they indulged some of the animal "workers" in their houses, the mousing cats and watchdogs essential to the good order and protection of the household. Museums have numbers of carefully preserved eighteenth-century birdcages and engraved brass dog collars that document the presence of pets; relatively few people seem to have written much about their charges, however. Diaries like that of Elizabeth Sandwith Drinker of Philadelphia, (kept from 1759 to 1807), which contains occasional observations on the behavior of the family's cats, dogs, and other small animals, are rare. During the busy years of her motherhood, brief entries noted in passing such events as the death of the old watchdog and his replacement by another. As Drinker's age permitted more leisure, however, notes about animals became more frequent and included more detail as she reminisced about pets kept by the family in the past: "a large mastiff dog" named Ranger, "a black cat—which we had for many years, and a white cock" called Chanticleer, "a favorite of our sons." When her house dog Tartar spent part of the night barking and woke her, she commented that "something there was to set him agoing, as he is too fat and lazy to exert himself for nothing." She noted with pleasure that Tartar and the cat were the "first to welcome us" when the family returned from some time spent in the country and commented on her grandchildren's pets, including unusual "English rabbits," chickens kept as "favorites," and an orphan kitten that had been suckled by a dog. Drinker's diary also suggests that pet keeping was not confined to the well-to-do. She noted when her son William "gave our little squirrel to a Negro lad over the way a barber who has several petts and appears fond of them"; she was glad to be rid of the creature but worried a bit about the quality of care it would receive. On September 20, 1800, Drinker noted the death from a "disorder among the cats" of "our poor old Puss . . . in the 13th year of her age." Her servant Peter "dug a grave 2 feet deep, on ye bank in our garden, under the stable window, where E.S., myself, and Peter saw her decently interred." Puss had been held in high enough regard to be transported in a basket to Germantown, where the Drinkers fled to escape the yellow fever epidemic in 1797; the next year, Drinker had noted with humor that "Our Cats progeny are much in demand whether it is her real merit, or the value her mistress sits on her, that gives her such consequence I cant say." Yet she closed her account of Puss's funeral (the only pet who apparently merited one) with the remark "I had as good a regard for her as was necessary."[88]

Drinker's comment can be used to suggest the difference in sensibility between pet keepers in this earlier period and at least some of their nineteenth-century counterparts, a kind of emotional restraint, even in the face of deep regard, bred of a clear sense of distance in the chain of being between human beings and nonhuman animals. Too much love for an animal was even a little embarrassing, as this journal entry from Virginia planter Landon Carter of April 15, 1758, suggests:

> I can't but take notice of the death of my little Canary bird, an old housekeeper having had it here 11 year this month and constantly fed it with bread and milk, and I wish the heat of this weather did not by Souring its food occasion its death, for it sung prodigeously all the forepart of the day. At night it was taken with a barking noise and dyed the night following, vizt, last night. I know that this is a thing to be laught at but a bruit or a bird so long under my care and protection deserves a Small remembrance.[89]

An advertisement for a lost dog, appearing in the *Pennsylvania Packet* on September 15, 1796, can also be used to suggest the emotional ambiguities of that time's animal-human relations:

> Lost. A Half bred pointer bitch, stands low and very square before—the colour is a kind of dusky white, her ears marked, a pale liver

> colour, and one of them a good deal cut or torn away—answer [*sic*] to the name of Romp. As she is valued only as an old servant, three dollars reward and no more will be given upon her being returned to the owner. Enquire at the office of the *American Advertiser.*[90]

Although Romp's owner was clearly worried about her, and had supported her long past the time of her utility as either a watch or hunting dog, he or she took pains to describe the dog only as a decrepit dependent of a household. This may have been at least in part to limit the amount of ransom a dog thief would have to be paid, but the defensiveness of the reward statement—"three dollars reward and no more"—suggests that the owner worried that his or her affection for the animal was potentially exploitable as a source of weakness.

The metaphor used in the advertisement for Romp, the animal as servant, encompassed much early thought about animal welfare and remained one of the organizing tropes for anticruelty work throughout the nineteenth century, too. It made good sense in a world where animals did so much of the physical work and provided so much material for human comfort in the products derived from their bodies. By the early nineteenth century, however, the old understanding of animals as servants to humans was augmented by two others. One, equating animal and human families, had two consequences. It suggested that middle-class family life was timeless and natural, rather than a construction with a relatively recent history. By comparing the family lives of animals and humans favorably, it also suggested that the integrity of animal families deserved respect, and the feelings of animal parents, consideration. A third metaphor paired selected animals, particularly household pets, with young children. This trope, the animal as child, grew out of the changing understanding of the natures of children that made childhood, and the relationships of mothers and children, a central organizing metaphor in a broader reconceptualization of the nature of dependent beings. Because cultural understanding of animals is always dynamic and often contradictory, the three metaphors are often found mixed together in stories and images about animals. Each had something different to say about the kind of regard appropriate for selected creatures, and each suggested different obligations on the part of humans, although few people seem to have considered the full implications of their social logic.

We have seen that household animals were important as objects for benevolence and affection. Yet the animals in and around households were not viewed simply as the passive recipients of this care. Although the era's moral philosophers insisted that whatever their capacity to feel pain, animals were "destitute of any moral faculty," and although they were no longer punished in the same ways as criminals, a strong strain of traditional thought survived that suggested domestic animals were indeed moral actors in their capacities as servants, parents, honorary siblings.[91] The apologists for kindness could draw on a long tradition of stories, folk belief, and popular natural histories when they argued that human beings, especially young ones, actually benefited from daily contact with certain animals because of *their* exemplary qualities.[92] Children could learn respect for the family dog, hearing their mother describe "the virtues of his race . . . their fidelity and enduring gratitude"; they also learned that these were valued qualities in people as well.[93] Cats were "very neat" and observant of changes in their environment (a quality "very useful for boys and girls and every body to imitate").[94] Stories of virtuous dogs were, of course, the most common account of the moral qualities of animals, appearing frequently in children's books and papers, and as brief anecdotes in magazines, gift books, and anthologies.[95] In 1848, the General Protestant Episcopal Sunday-School Union even published an entire volume of anecdotes titled *The Dog, as an Example of Fidelity,* urging young readers to consider the creature's "noble and disinterested nature" and, in effect, to be more like dogs themselves: "A faithful performance of duty implies a desire to be really useful in the world, and to act with the heart, as well as the head and hands. It gives an affectionate earnestness to the whole manner, and produces that willing and active service which is so delightful to witness."[96] Even the ubiquitous dooryard chicken provided a useful model for children, in part because of its direct connection to a scriptural metaphor that emphasized a "maternal" Jesus. The story "Thomas and his Chickens" reminded young readers that Jesus had compared his own efforts to gather his "children" "even as a hen gathereth her chickens under her wings . . . Matt. 23:37."[97]

Domestic animals were increasingly described as having feelings paralleling those of middle-class humans in both range and intensity—devoted love, happiness, grief, and mental as well as physical suffering. Pets' displays of affection could "awaken corresponding tenderness and care" in children, Catharine Beecher thought. When she credited animals with "intuitive perceptions of our emotions which we cannot conceal," Beecher used language which would have been perceived by her readers as appropriate to the delicacy of refined people, especially women.[98] And both domestic and wild animals provided a model for receiving benevolent treatment as well, since they had the capacity for gratitude.[99]

Popular media also promoted a vision of animal emotional life that had the idealized domestic life of human families as its point of comparison. As we have seen in the case of birds, middle-class people valued types of animals which seemed to exhibit the characteristics they found desirable in themselves. In popular narratives and through their actual presence in households, animals reinforced the cultural conventions of domesticity by, paradoxically, giving the virtues of middle-class family life origins in nature.[100] Denial of any gap in moral standards between people and certain animals also helped to open the door to a conventional acceptance of identity of feeling and shared sensibilities on the part of both parties.[101] Expressions of maternal grief were particularly compelling. An "Anecdote of a Cat," published in *Youth's Companion* in 1828 and *The Cabinet of Natural History and American Rural Sport* in 1832, recounted the efforts of a favorite cat to make her master follow her from a room. When this did not avail, the cat brought "the dead body of her kitten, covered over with cinder dust," to his feet. The stricken man "now entered into the entire train of this afflicted cat's feelings. She had suddenly lost the nursling she doated on, and was resolved to make me acquainted with it—assuredly that I might know her grief, and probably also that I might inquire into the cause . . . and divide her sorrows with her."[102] *Youth's Companion* repeatedly published anecdotes about the "maternal affection" of whales and cows and the "fervent" love of goats for their kids, among other animals, usually in settings where "savage" men were attacking their young; they expected their readers to understand the subtext.[103]

Prints of harmonious animal families, both wild and domesticated, were extremely popular throughout the nineteenth century precisely because of their metaphoric celebration of the naturalness of middle-class family life, especially maternity. Some invite comparisons between several sets and generations of mothers and infants, both biped and quadruped. The absence of any males suggests that the heart of family life is the relations between mothers and children (one assumes the menfolk are out "hunting" for their families), a point also stressed in the contemporaneous childrearing literature.[104] In fact, father figures seem to rarely turn up in these images at all. The 1868 Christmas catalog of chromolithographs offered by Boston printer Louis Prang invoked this reading in its description of a very successful edition after a painting by the celebrated wildlife and sporting painter A. F. Tait (figure 2): "*Maternal Love*. . . . The power of love, be it the love of the youth for the maiden, of friend for friend, of child for parent, or purest of all—of the mother for her offspring, has often been the painter's theme. Seldom, however, if indeed ever, has this subject been treated with so much tenderness and refinement as the picture under consideration, although it represents only a deer caressing a fawn in the cool shadow of the forest."[105]

As families partook of a set of literary and pictorial representations of animals as watchful parents, a convention that certainly received material support from their own experience of successive litters of kittens and pup-

2. A. F. Tait, "Maternal Love." Chromolithograph. Published by Louis Prang, Boston, 1868. Reproduced from the Collections of the Library of Congress.

pies, and broods of poultry or canary chicks, their selective recognition of the individuality of pet animals in their own households also drew heavily upon the conventional understandings of Victorian childhood. Domesticity made the relationship between mothers and children paradigmatic; hence it became an idealized description of the loving relationship between middle-class people and certain animals, particularly pets. An animal is defined as a pet by the act of human selectivity: according to the *Oxford English Dictionary,* a pet animal is "domesticated or tamed and kept as a favourite, or treated with indulgence and fondness." It is telling that "pet" was used interchangeably to describe both children and animals; by the late eighteenth century, it was applied to "an indulged (and, usually, spoiled) child"; by the early nineteenth century, the word denoted "any creature that is fondled and much indulged."[106]

Like the formulation that childhood cruelty inevitably led to even worse adult behavior, equating children and animals was nothing new. Seventeenth- and eighteenth-century parents had found their children uncomfortably close to animals in the kinds of sounds they made and their predilection for crawling on all fours. Infants represented all too vividly the beast within undisciplined, sinful humans. Because upright posture and walking were such fundamental signs of full humanity, parents laced babies to force them to sit erect or even stand, and discouraged infants from crawling by clothing them in long gowns that tangled around their legs, leaving them in walkers, or using leading strings.[107]

When more liberal philosophies of child rearing viewed children as innocent beings, more inclined to good than evil, the physical "animality" of babies was less troubling. If young children and animals shared many apparent physical characteristics, from crawling on all fours to the lack of speech, they also shared their innocence of the trials of the world and their need for gentle guidance in the routines and morals of family life. Thus the terms of the old comparison held, but its meaning was inverted: children and household animals now became innocent, good-hearted, dependent beings requiring "raising" into society but having special contributions to make of their own. The conventional metaphor seemed to work both ways; if the child was a pet, the pet was also like a child. Consequently, child-rearing and anticruelty literature sometimes paralleled each other in startling ways. While Lydia Sigourney described a boy who ran to his mother crying, "*My goodness grows weak . . . help me to be good,*" Charlotte E. B. Tonna informed her readers that the animal "wants the help of your reason to keep him from doing wrong; and he wants you to explain to him how he may please you."[108]

Somehow it is not surprising that Harriet Beecher Stowe was the most well-known proponent of the view that pets were like children. In the 1850s, 1860s, and 1870s, she published stories and didactic essays for children that introduced dogs as protagonists, and even used them as plot devices in several sentimental novels for adult readers.[109] Stowe published several fictionalized accounts of her own family's relations with their pets, including the serial story "Our Dogs," which was first published in the children's magazine *Our Young Folks* during 1866 and appeared in book form in 1867. The anecdotes of dog behavior in "Cunopolis" are notable for their depictions of pet dog behavior, from chewing shoes to frantic enjoyment of carriage rides. The children of the household and the dogs are described in the same tone and vocabulary: both boys and dogs are "rovers," both puppies and children commit "youthful errors." Each dog is described as a person with a distinctive nature: one mannerly dog is "a highly-appreciated member of society," another is "a friend and playfellow," a third "a little black individual . . . not very prepossessing in appearance and manners, but possessed of the very best heart in the world," a formulation that sounds suspiciously like descriptions of some of the slave characters in *Uncle Tom's Cabin.*[110] In the follow-up to the series, "Dogs and Cats," Stowe pointed out to her readers that each of her dogs "had as much his own character if he had been a human being," and that such individuality was a potential characteristic of every animal, given proper family care and love.[111]

Occasionally, authors even suggested that pet animals were an acceptable substitute for childless human beings to express their nurturing impulses because they could be so much like young children. Caroline Kirkland, author of several popular collections of stories of life in the Michigan frontier towns in the 1830s and 1840s, set one tale around the household of a childless couple who lavished care on many birds, squirrels, turkeys, guinea pigs, and a peacock. Pet keeping, she informed her readers, was "something that childless people are apt to do, if they are wise."[112]

This expanded conception of animals as "like children" also opened the way for new kinds of feelings toward beloved pets. The death of a favorite animal could even be a source of grief as deep as the death of a baby. Recall Elizabeth Drinker's diary entry about the death of Puss, and compare it to one by Anne Hampton Brewster (1818–92), a Philadelphia-born author and journalist. On May 9, 1859, she recorded the last moments and death of her small dog Beauty from a respiratory disease in as much detail as any account of a human death, and grieved, "No one can tell how I miss him and grieve too at night. I feel his little paws scratching me and his little tongue on my hand asking me to take him in my arms. . . . I come up to my now silent room & have no noisy loving bark to greet me—Oh I am very lonely."[113]

She mourned in the passionate voice of a bereaved mother. Compare Fanny Longfellow writing of the death of her daughter Fanny in 1848: "Sinking, sinking away from us. Felt a terrible desire to seize her in my arms and warm her to life again at my breast. Oh for one look of love, one mood or smile. . . ."[114]

Popular imagery also reinforced the linked metaphors "the pet is a child, the child is a pet" throughout the nineteenth century. This visual twinning of children and animals had distinctive limits, of course. Children were compared directly only to certain animals—lambs (which had a biblical basis), cats and dogs (particularly but not exclusively kittens and puppies), and some kinds of birds (songbirds and chicks).[115] Three common, generic representations, appearing repeatedly in prints, trade cards, and stereoscopic views, were "Play with Me," "Our Pets," and "Can't You Talk?" "Play with Me" (variations were titled "Playmates" and "Good Morning, Little Favorite") depicts the happy greetings of a small pet dog and a young child in a cradle or bed in anticipation of a day of play together. They are compatriots, about the same size, who look each other directly in the eyes (figure 3). "Little Pets" (variations were called "Our Pets," "The Pets," or even "Household Pets") depicts a child, usually a girl, holding one or more animals in her arms; several more may gather around. Sometimes all the figures look directly at the viewer. Holding the animals is also a maternal gesture, suggesting the place of quadruped pets in socializing biped ones. The most provocative of the three images, "Can't You Talk?" (alternatively titled "Can You Talk?" and "Why Can't You Talk?") not only appeared on paper but was printed on cloth in children's books, molded into majolica pitchers, and even sold as plaster-of-paris and ceramic figurines (figure 4). "Can't You Talk?" depicts the encounter of a crawling baby and a dog. Both are on all fours, and again, they look directly into each other's eyes, as equals. The image is made more poignant than the others, however, by its title, which intervenes and tries to shape the viewer's understanding of the picture. Although it is unclear which creature is, in fact, interrogating the other, the future resolution of the narrative is that the dog and the baby, who see each other as equals for a time and are equals in important ways, will grow away from each other as time passes; the child can grow "up" in a way the animal never can.

The shared assumptions underlying all this literary and visual rhetoric were part of a feminine recasting of the meaning of dependence that imagined not only animals but other beings as innocent children. A particularly striking correspondence can be found in the antebellum discourse called romantic racialism. This thread of antislavery thought, articulated in the late 1830s and fully developed in the 1840s and 1850s, argued that the

3. "God Morning! Little Favorite." Lithograph, artist unknown. Currier and Ives, New York, ca. 1850. Reproduced from the Collections of the Library of Congress.

4. "Can't You Talk?" Printed cotton fabric, ca. 1890. Courtesy The Strong Museum, Rochester, New York.

Negro race was blessed with a set of fundamental characteristics more in tune with the requirements of true Christianity than were those of the hard-hearted, world-beating Anglo-Saxon: it was "imitative and docile," "eminently gentle, submissive, affectionate, and grateful," despite the misery of circumstances. Blacks' "willingness to *serve,* the most beautiful trait in humanity, which we, from our innate love of dominion, and in defiance of the Christian religion, brand with the name of servility" was a trait shared naturally with domestic animals—middle-class infants had to be trained into it.[116] Bluntly put, young children, selected animals, and blacks, all liminal beings in their relation to the male world of money and law, were still better people than white men could ever be. In the realm of feeling, they were natural Christians, like middle-class women.

Like antislavery narratives, the domestic ethic of kindness also relied on the concept of speech—or, rather, the lack thereof—on the part of the oppressed. What made some categories of dependents especially worthy of defense was their inability to speak for themselves, as the title of *Our Dumb Animals,* the first anticruelty magazine directed to children, made clear. Eventually first-person slave narratives and speaking tours allowed an oppressed people to make its own case for freedom, whereas animals were a "helpless class who can neither speak, read, nor write, but who have no capacity for being taught any of these accomplishments."[117]

Anecdotes of the intelligence or devotion of individual animals, a genre discussed throughout this essay, made one case for new sensitivities, but the most powerful strategy for creating "voices for the speechless" was through allowing animals to "speak" either in poetry and fiction or through animal "autobiography," a genre that became quite similar to the slave narratives that have been credited with helping to "expand the definition of cruel behavior" by "delegitimizing the 'right' to private violence against persons." First-person slave narratives "established their authors as fully sentient and ardent in their desires for freedom, love, family; this contrast made their brutalization the more horrific."[118] The editor of *Youth's Companion* even made explicit the connection between narratives where slave families were broken up, and the destruction of bird families, urging reform:

> We hear much of slavery among the Colored People at the South—meetings are held—speeches are made—Anti-Slavery Societies are formed—the whole country is excited on the subject, and we hope some good will come out of it. But the above communication [another plea against bird-nesting] brings to light another species of Slavery, full as cruel to the sufferers as African Slavery, and that even in *free Massachusetts*! The peaceful home of the innocent bird is invaded—the helpless little ones are kidnapped—they are cruelly forced away from their parents, and shut up in wire cages, and often are left to suffer hunger and starvation, without the power of helping themselves—while their parents are left to mourn the loss of their beloved offspring, or are murdered for the mere amusement of the sportsman. We think it is high time our youth were called together, for the purposes of forming an ANTI-BIRD-SLAVERY-SOCIETY.[119]

Animals who either represented human beings, or were humans or otherworldly beings in disguise, had often been given voices in fables and fairy tales, but creatures who were not fabular already had clear "voices" of their own by the turn of the nineteenth century. We have seen that *Pity's Gift* included one poem in the voice of a grieving nightingale. It also contained "The Address of the Superannuated Horse to His Master" (subtitled "Who, on account of his (the Horse), being unable, from extreme old age, to live through the winter, had sentenced him to be shot.")

> And hast thou fix'd my doom, sweet Master, say?
> And wilt thou kill thy servant, old and poor?
> A little longer let me live, I pray,
> A little longer hobble round thy door.[120]

These kinds of poems and stories appear now and again through the first half of the nineteenth century; we have seen examples where birds and animals scold bad boys.[121] However, the animal "first person" was used most heavily and effectively after 1870, after the conventions of the domestic ethic of kindness were fully articulated. It became a standard tool in organized "humane education," especially following the publication in America of Anna Sewell's *Black Beauty* (first English edition, 1877). Once the book had been called to humane activist George Angell's attention by a woman sympathetic to the movement, he dubbed it "the *Uncle Tom's Cabin of the Horse.*" Angell's American Humane Education Society (founded 1889) saw three million copies into print before his death in 1909.[122] Other children's authors soon chimed in with tales like *Dicky Downy: The Autobiography of a Bird* (1899), a story that made a case not only against caging wild birds but also against bird ornaments for women's hats and hunting.[123] Eventually even Mark Twain got into the act with his pathetic *A Dog's Tale* (1903), where a mother tells the terrible tale of the blinding of her beloved only puppy by her master in a scientific experiment (this after the dog had saved the master's child from a fire and been terribly wounded).[124]

In the United States, organizational activity to pass and enforce effective anticruelty laws began in 1866 with the founding of the American Society for the Prevention of Cruelty to Animals in New York City. It lagged comparable British work by over forty years (the founding of the Society for the Prevention of Cruelty to Animals there dating to 1824). This delay reflected other preoccupations on the part of reform-minded Americans; the suffering of cart horses and fighting dogs seemed a minor evil compared to the blight of slavery in particular. Once the American Society for the Prevention of Cruelty to Animals was organized, SPCAs appeared in Philadelphia, Boston, Buffalo, and Chicago by 1869; dozens of other communities soon followed.[125]

When the ASPCA and its contemporaries were organized at last, they were able to draw extensive support from the constituency of middle-class men and women who had imbibed the formula of the domestic ethic of kindness as children. Further, by the 1870s, natural feminine tenderness had become a justification for women's involvement in the fledgling American humane movement, just as it had previously supported their participation in other reforms. "Our Dumb Dependents," an essay appearing in the August 1871 issue of *Godey's Lady's Magazine,* explicitly associated poor treatment of animals, particularly livestock, with the heedlessness of the "eager pursuit of wealth" by men: "The impulse of kindness towards the weak and helpless, which often exhibits itself in the love of pets, has doubtless been implanted in the hearts of women, like the love of children, for a special purpose; and this purpose we may believe to be that of impelling her to come between the hasty and thoughtless master and his dumb dependents, as the protectress of those who are unable to help themselves."[126] Cruelty to animals had become another disease of the male world of money that domesticity was supposed to counter. It was part of a wider vision of a domesticated social order.

Given the evident popularity of the conventional metaphor linking children and pet animals (including the iconography that paralleled the "speechlessness" of each) and the well-established argument that made kindness a fundamental characteristic of middle-class family life, one might assume that once organized anticruelty activity began, children and animals would again be linked. The first organized efforts to prevent cruelty to children were indeed related to animal protection efforts; in the last quarter of the nineteenth century, many communities used a single anticruelty organization to intervene on behalf of both animals and children; the Hartford, Connecticut, SPCA even included protection for the elderly in its original charter. However, anticruelty prosecutions on behalf of children grew out of the famous case of Mary Ellen, a child rescued by a social worker who appealed to the New York Society for the Prevention of Cruelty to Animals in 1874 after months of seeking relief unsuccessfully from the police and courts of New York.[127] Further, women did not assume leadership roles in the first groups, and their activism took forms that were related to the domestic ethic of kindness in interesting ways. They were unlikely to actively engage in enforcement of anticruelty laws against livestock or blood sport, prosecutions that usually pitted rural or working-class men against middle- or upper-class reformers. Women were involved in suitably gendered activities in the movement: organizing educational programs for schoolchildren, establishing "homes" for abandoned and abused animals, or arranging for creature comforts such as water fountains on city streets for horses and dogs.

Despite its significant limitations as a theoretical framework for addressing cruelty broadly in society, the domestic ethic of kindness did foster change in a world where violence against animals was widespread, casual, and often deliberate. It proscribed and stigmatized some cruel behavior that was once simply part of everyday life and encouraged greater emotional connection to selected animals, helping to reshape the old habit of pet keeping into the emotionally complex practice we recognize and engage in today. It laid the groundwork for organized anticruelty as well. But the domestic ethic of kindness also created unresolved new discomforts, as adults and children struggled with understanding the meaning of animal suffering in a world where animal life now meant different things. And the legacy of this ethic, which focused on the heart of each person, the private household, and the individual act of cruelty and kindness, may have limited popular discussion of cruelty and kindness by reducing human actions to a series of individual cases, and by refusing to consider the more far-reaching implications of the metaphors it used to describe human relationships to animals. In truth, the full implications of the metaphors of animal as servant or child, and animal families as like human ones, were never worked through. Despite their potentially radical implications for everyday practice (if animal families were indeed like human ones, how could a newborn bull calf be justifiably taken from his mother and killed?), the domestic ethic of kindness did not completely challenge the world of traditional animal-human relationships. Instead, it grafted its softened sensibility onto the range of utilitarian and cultural uses to which animals were already put, and it favored some animals at the expense of others.

Notes

1. The "Eden of Home," along with a score of other "synonyms" for "home," may be found in Rev. W. K. Tweedie, *Home; or, The Parents' Assistant and Children's Friend* (Norwich, Conn., 1873), pp. 34–41.

2. Charlotte B. Tonna, *Kindness to Animals; or, The Sin of Cruelty Exposed and Rebuked,* Revised by the Committee of Publication of the American Sunday-School Union (Philadelphia, 1845), pp. 8, 9, 12.

3. For a concise overview of middle-class culture in the nineteenth century, including a useful bibliography, see Mark C. Carnes, "The Rise and Consolidation of Bourgeois Culture," in *Encyclopedia of American Social History,* ed. Mary Kupiec Cayton, Elliot J. Gorn, and Peter W. Williams, vol. 1 (New York, 1993), p. 608. I use the descriptor "Victorian" for this culture in America, drawing upon Daniel Walker Howe's definition in "Victorian Culture in America," in Daniel Walker Howe, ed., *Victorian America* (Philadelphia, 1976), pp. 2–3.

4. For the principal arguments of the new liberal Protestant theology and an argument for its utility as an instrument of social control in one community, see Paul E. Johnson, *A Shopkeeper's Millenneum: Society and Revivals in Rochester, New York, 1815–1837* (New York, 1978), chap. 5, "Pentecost," pp. 95–115. See also Curtis D. Johnson, *Redeeming America: Evangelicals and the Road to the Civil War* (Chicago, 1993).

5. The most comprehensive discussion of the practice and meanings of gentility in America is found in Richard L. Bushman, *The Refinement of America, 1750–1850* (New York, 1992).

6. *A Manual of Politeness, Comprising the Principles of Etiquette, and Rules of Behaviour in Genteel Society* (Philadelphia, 1837), p. 276.

7. For the construction of domesticity and the early articulation of its conventions, see Nancy F. Cott, *The Bonds of Womanhood: Women's "Sphere" in New England, 1780–1835* (New Haven, Conn., 1977), chap. 2, pp. 63–100. For the antebellum conventions of domesticity, see Mary P. Ryan, *The Empire of the Mother: American Writing about Domesticity 1830–1860* (New York, 1982), esp. chap. 1, "From Patriarchal Household to Feminine Domesticity," pp. 19–44. For the interplay of domesticity and liberal Protestant theology, see Barbara Leslie Epstein, *the Politics of Domesticity: Women, Evangelism, and Temperance in Nineteenth-Century America* (Middletown, Conn., 1981), pp. 45–87.

8. Muzzy, *The Fireside,* p. 8. For changes in attitudes toward mothering, see Mary Lynn Stevens Heininger, "Children, Childhood, and Change in America, 1820–1920," in Mary Lynn Stevens Heininger et al., *A Century of Childhood 1820–1920* (Rochester, N.Y., 1984), pp. 1–32; Nancy Shrom Dye and Daniel Blake Smith, "Mother Love and Infant Death, 1750–1820," *Journal of American History* 73, no. 2 (September 1986): 329–53; Ryan, *The Empire of the Mother,* pp. 45–55.

9. The most comprehensive account of this genealogy remains Keith Thomas, *Man and the Natural World: A History of the Modern Sensibility* (New York, 1983).

10. Karin Calvert, *Children in the House: The Material Culture of Early Childhood, 1600–1900* (Boston, 1992), pp. 55–60.

11. Ibid., p. 105, quotation from "Childhood," *Godey's,* June 1832, p. 268.

12. Horace Bushnell, *Views of Christian Nurture, and of Subjects Adjacent Thereto* (Hartford, Conn., 1847), pp. 15, 6.

13. "Home Interests," *Youth's Penny Gazette,* November 20, 1850, p. 94. Published by the American Sunday-School Union.

14. Catherine C. Beecher and Harriet Beecher Stowe, *The American Woman's Home, or Principles of Domestic Science* (1869; repr. Hartford, Conn., 1987),

p. 18. Reverend A. B. Muzzy, *The Fireside: An Aid to Parents* (Boston, 1854), p. 109.

15. Thomas, *Man and the Natural World,* pp. 19, 24.

16. Francis Wayland, D.D., Ll.D., *The Elements of Moral Science,* rev. ed. (Boston, 1865, 1871), p. 395. This text was first published in 1837.

17. John Bigland, *A Natural History of Animals* (Philadelphia, 1828), pp. 187, 20.

18. A. L. Carroll, "An Apology for Dogs," *Harper's New Monthly Magazine,* June–November 1867, p. 192.

19. David Brion Davis, *The Problem of Slavery in Western Culture* (Ithaca, N.Y., 1966), chap. 7, "The Legitimacy of Enslavement and the Ideal of the Christian Servant: The Failure of Christianization," pp. 197–222.

20. Harriet Beecher Stowe, "Dogs and Cats," *Stories and Sketches for the Young,* vol. 14 of *The Writings of Harriet Beecher Stowe, Riverside Edition* (1896; repr. New York, 1967), p. 109.

21. Tonna, *Kindness to Animals,* p. 13. Harriet Beecher Stowe made a similar veiled threat to her young readers: "Our Lord says not even a little sparrow falls to the ground without our Heavenly Father, and we may believe that his eye takes heed of the disposition which we show toward those defenseless beings whom he thinks worthy of his protection." Harriet Beecher Stowe, "Dogs and Cats," p. 109. See also *The Dog, as an Example of Fidelity* (New York, 1848), p. 83: a dog is "not to be regarded as mere property to be disposed of at the will of the owner."

22. Lydia Sigourney, *Letters to Mothers,* 6th ed. (New York, 1846), p. 41, 35–36.

23. Horace Bushnell, *Essays on Christian Nurture, and of Subjects Adjacent Thereto* (Hartford, Conn., 1847), pp. 41, 14.

24. Jacob Abbott, *Agnes: A Franconia Story, by the Author of the Rollo Books* (New York, 1853), pp. v–vi.

25. Arnold Arluke, Jack Levin, Carter Luke, and Frank Ascione, "The Relationship of Animal Abuse to Violence and Other Forms of Antisocial Behavior" (paper presented to the Eighth International Conference on Animal-Human Interactions, Prague, Czech Republic, September 12, 1998).

26. Thomas, *Man and the Natural World,* 151.

27. Karen Halttunen, "Humanitarianism and the Pornography of Pain in Anglo-American Culture," *American Historical Review* 100 no. 2 (April 1995): 303.

28. James Turner, *Reckoning with the Beast: Animals, Pain, and Humanity in the Victorian Mind* (Baltimore, 1980), pp. 9, 12.

29. For a discussion of the circumstances surrounding the publication of these prints (Hogarth later commented that he was proud of the series since

he believed that it had helped to relieve cruelty to animals), and a detailed visual analysis, see Ronald Paulson, *Hogarth,* vol. 3: *Art and Politics, 1750–1764* (New Brunswick, N.J., 1991), pp. 17–36.

30. Mary Wollstonecraft, trans., *Elements of Morality for the Use of Children, with an Introductory Adress {sic} to Parents: Translated From the German of the Rev. C. G. Salzmann.,* 3d U.S. ed. (Wilmington, Del., 1796). The book was first published in English in 1790.

31. "Cruelty and the Gallows," *Youth's Companion,* April 22, 1836, p. 195.

32. Muzzy, *The Fireside,* p. 74. See also "Be Kind to Animals," *Youth's Companion,* June 2, 1830, p. 6; "Cruelty," ibid., July 25, 1832, p. 6; "Kindness to Brutes," ibid., September 21, 1838, pp. 74–75.

33. Lydia H. Sigourney, *The Boy's Book* (New York, 1845), pp. 54, 29. For Cowper as an inspiration to kindness, see "Stories about Cats," *Youth's Companion,* October 30, 1835, pp. 95–96: "A very sweet poet, named Cowper, who loved animals and used to keep hares, a dog, and a cat in his room, and loved to see them sport about so merrily, said, 'he would not call that man his friend who would needlessly tread upon a worm.'" The *Youth's Companion* often reprinted material from other periodicals with credit lines; this story was credited to the *Juvenile Miscellany.* This passage was paraphrased from Cowper's "The Task" (1784), a five-thousand-line work in blank verse that contained a lengthy discussion of cruelty to animals in book 6: "The Winter Walk at Noon," lines 560–67. John D. Baird and Charles Ryskamp, eds., *The Poems of William Cowper,* vol. 2: *1782–1785* (Oxford, 1995), p. 251.

34. Henry Ward Beecher, *Lectures to Young Men, on Various Important Subjects* (1844; 2d ed., New York, 1859), pp. 147–48.

35. Halttunen, "Humanitarianism and the Pornography of Pain in Anglo-American Culture," pp. 323–24. Antebellum arguments about juvenile delinquency similarly operated on the assumption that youthful disobedience led inexorably to future, more serious crime; see David J. Rothman, *the Discovery of the Asylum: Social Order and Disorder in the New Republic* (Boston, 1971), p. 76.

36. Dr. William A. Alcott, *Gift Book for Young Ladies; Or Familiar Letters on Their Acquaintances, Male and Female, Employments, Friendships, &c* (Auburn, N.Y., 1854), p. 271.

37. Mrs. Louisa Hoare, *Hints for the Improvement of Early Education and Nursery Discipline,* 3d ed. (Dover, Del., 1826), pp. 36–37.

38. Elizabeth B. Clark, "'The Sacred Rights of the Weak': Pain, Sympathy, and the Culture of Individual Rights in Antebellum America," *Journal of American History* 82, no. 3 (September 1995): 470, 476.

39. *The Youth's Companion,* one of the most popular children's periodicals of the nineteenth century, continued publication until 1929. Nathaniel Willis and Asa Rand intended their publication to "warn against the ways of transgression, error, and ruin, and allure to those of virtue and piety," while it entertained children with a wide variety of stories clipped "from the various publications which we receive and peruse." "The *Youth's Companion,*" in Frank Luther Mott, *The History of American Magazines,* vol. 2: *1850–1865* (Cambridge, Mass., 1938), pp. 262–74.

40. "The Puppy," *Youth's Companion,* August 17, 1838, p. 55. Reprinted from *Youth's Penny Paper.*

41. T. S. Arthur, "The Wounded Bird," ibid., April 27, 1848, p. 1.

42. "The Woodchuck," ibid., December 18, 1835, p. 123.

43. *Master Henry's Rabbit* (Troy, N.Y., n.d. [c. 1840]), p. 10.

44. Mary Wollstonecraft, *Thoughts on the Education of Daughters: with Reflections on Female Conduct, in the More Important Duties of Life* (London, 1787), p. 44.

45. This is a formula previously used by Salzmann in his *Elements of Morality for the Use of Children.* In the chapter on cruelty to animals, James's parents are warned by the entreaties of his sister of his intention to mutilate the mouse he has caught. *Elements of Morality for the Use of Children,* pp. 168–69.

46. "The Bird's Nest," *Youth's Companion,* June 16, 1833, p. 24 ("Reprinted from the *Cincinnati Journal.*") For a short story featuring the death of an "innocent" butterfly at the hands of a small girl, see "The Birth Day Present," ibid., March 12, 1829, p. 165.

47. William Lyon Phelps, *Autobiography with Letters* (New York, 1939), pp. 60–61.

48. For boys' taxidermy and amateur "cabinets" as an educational form of play, see *Charley's Museum: A Story for Young People* (Philadelphia, 1857); Phelps, *Autobiography with Letters,* pp. 54–55; Bellamy Partridge, *Big Family* (New York, 1941), p. 305.

49. H. H. Delong, *Boyhood Reminiscences: Dansville, N.Y., 1855–1865* (Dansville, N.Y., 1913), p. 56; Claudia L. Bushman, ed., "Life Along the Brandywine between 1880 and 1895 by Samuel Canby Rumford, Part II," *Delaware History* 23, no. 3 (spring/summer 1989): 168, 195; Henry Ward Beecher, "Country Stillness and Woodchucks," in *Eyes and Ears* (Boston, 1862), pp. 139–41.

50. Some commentators suggested that cruelty to animals was simply modeling on adult behavior that had gone awry. The healthy boy was constantly "tempted" to use behavior that had now been problematized as abuse against animals as "a trial of his skill" (at throwing rocks, for exam-

ple, or aiming an arrow) or as "proof of his strength" as he grew toward manhood. "Against the Abuse of Cattle," *Youth's Companion,* June 4, 1829, p. 6.

51. Alcott, *Gift Book for Young Ladies,* p. 274.

52. Mrs. L. G. Abell, *Woman in Her Various Relations* (New York, 1851), p. 53.

53. Alcott, *Gift Book for Young Ladies,* p. 274.

54. "Nauticus Agricola" [pseud.], "The Murdered Robin," *Youth's Companion,* June 23, 1837 (pages illegible).

55. Alcott, *Gift Book for Young Ladies,* p. 273.

56. Harriet Beecher Stowe, "Rights of Dumb Animals," *Hearth and Home,* January 1869, p. 24. See also "Be Kind to Animals," *Youth's Companion,* June 2, 1830, p. 6.

57. Muzzy, *The Fireside,* pp. 141–42.

58. Sarah Josepha Hale, *Manners; Or, Happy Homes and Good Society All the Year Round* (1868; repr. New York, 1972), p. 244.

59. "The Bird's Nest, " *Youth's Companion,* September 23, 1829, p. 70; "The Bird's Nest," ibid., April 12, 1834, p. 187; "The Robin's Complaint," ibid., August 11, 1837, p. 50.

60. "Attachment of a Swallow," ibid., September 2, 1836, p. 62 (reprinted from the *"Greenfield Adv."*); *Book of Household Pets, and How to Manage Them* (New York, 1866), p. 5. Keith Thomas notes that the family life of birds was occasionally the object of admiration in the seventeenth century. William Ramsay commented with pleasure on his pigeons' "mutual love, chastity, and constancy, obedience of the hen to the cock pigeon, care of the cock to the young," in *Man's Dignity* (1661). Thomas, *Man and the Natural World,* p. 121.

61. For a discussion of these efforts, in the context of the larger bird protection movement of the last quarter of the nineteenth and the early twentieth centuries, see Robin W. Doughty, *Feather Fashions and Bird Preservation: A Study in Nature Protection* (Berkeley, Calif., 1975), pp. 46–48.

62. *Picture Lessons, Illustrating Moral Truth: For the Use of Infant Schools, Nurseries, Sunday-Schools, and Family Circles* (Philadelphia, n.d.). The copy in the collection of the Winterthur Library is inscribed, "Zilpha Clarke/from LaFayette H.[(?)] S. School/Buffalo Apl 25, 1852."

63. "The Little Orphan Birds," *Youth's Companion,* August 7, 1835, p. 47.

64. "The Robin," Child, *The Boy's Reading-Book,* pp. 235–36; Harriet Beecher Stowe, "The Nest in the Orchard," *Stories and Sketches for the Young,* vol. 16 of *The Writings of Harriet Beecher Stowe, Riverside Edition* (Boston, 1898), pp. 334–38.

65. Anon., "The Nightingale," *Pity's Gift: A Collection of Interesting Tales, to Excite the Compassion of Youth for the Animal Creation.* Ornamented with vignettes from the Writings of Mr. Pratt. Selected by a Lady. 3d ed. (London, 1801), pp. 59–61. Elizabeth Drinker records reading *Pity's Gift* in her list of books for 1802; Elizabeth Drinker, *The Diary of Elizabeth Drinker,* 3 vols. (Boston, 1991), 2:1607.

66. "Who Stole the Bird's Nest?" was published as a picture book by Louis Prang in 1866; L. Maria Child, "Who Stole the Bird's Nest?" was also published as the cover story in *Godey's Lady's Book and Magazine,* March 1871, pp. 231–33. John Greenleaf Whittier collected it into his anthology *Child Life: A Collection of Poems* (Boston, 1871), pp. 70–73.

67. Anon., "A Bird's-eye View," in Whittier, *Child Life,* pp. 64–65.

68. "The Bird's Nest," *Youth's Companion,* May 11, 1843, p. 1.

69. Myra C. Glenn, *Campaigns Against Corporal Punishment: Prisoners, Sailors, Women, and Children in Antebellum America* (Albany, 1984), p. 30.

70. Ibid., p. 80. The quotation Glenn uses from Thomas and Mary Gove Nichols comes from their *Marriage: Its History, Character, and Results* (New York, 1854), p. 85.

71. "Home Interests," *Youth's Penny Gazette,* November 20, 1850, p. 94.

72. Sigourney, *Letters to Mothers,* pp. 37–38.

73. Drinker, *Diary,* 1:556 (May 1, 1794).

74. Sydney H. Coleman, *Humane Society Leaders in America, with a Sketch of the Early History of the Humane Movement in England* (Albany, N.Y., 1924), p. 179.

75. Mark Twain, *The Autobiography of Mark Twain,* ed. Charles Neiter (New York, 1966), pp. 29–30.

76. Turner, *Reckoning with the Beast,* pp. 45–46.

77. Lydia Maria Child, *Letters from New-York* (New York, 1843), p. 10.

78. *Twelfth Annual Report of the Women's Branch of the Pennsylvania Society for the Prevention of Cruelty to Animals, for the Year Ending December 31, 1880* (Philadelphia, 1881), p. 10.

79. Beecher, *Lectures to Young Men,* p. 223.

80. For a detailed discussion of cockfighting, see Peter W. Cook, "Cockfighting in North America and New England, 1680–1900," in *New England's Creatures: 1400–1900,* ed. Peter Benes and Jane Montague Benes, The Dublin Seminar for New England Folklife Proceedings, 1993 (Boston, 1995), pp. 64–82. Lewis Miller, *Lewis Miller, Sketches and Chronicles: The Reflections of a Nineteenth-Century Pennsylvania German Folk Artist,* introd. Donald A. Shelley (York., Pa., 1966).

81. Miller, *Sketches and Chronicles,* pp. 29, 48, 62.

82. William Otter, *History of My Own Times,* ed. Richard B. Stott (Ithaca, N.Y.: Cornell University Press, 1995), pp. 81–84, 146–48, 152–53, 167–68.

83. Ibid., p. 170.

84. Michael Kaplan discusses the strategic use of violence among working-class men in antebellum New York in "New York City Tavern Violence and the Creation of a Working-class Male Identity," *Journal of the Early Republic* 15 (winter 1995): 591–617.

85. Turner, *Reckoning with the Beast,* pp. 46–56.

86. "Cruelty," *Youth's Companion,* July 25, 1832, p. 39. In this article, the connection between torturing insects as a boy and whipping a cart horse as a man is made explicit.

87. Harriet Beecher Stowe, "Rights of Dumb Animals," *Hearth and Home,* January 1869, p. 24.

88. Drinker, *Diary,* 1:385, 386; 2:1245, 1586, 1833, 1936, 1175, 1007, 1341.

89. Jack P. Greene, ed. *The Diary of Colonel Landon Carter of Sabine Hall, 1752–1778* , 2 vols. (Charlottesville, Va., 1965), 1:216.

90. *Pennsylvania Packet (Claypoole's American Daily Advertiser),* September 15, 1796. Prime File, Winterthur Museum, Winterthur, Del.

91. Wayland, *The Elements of Moral Science,* p. 395. The most comprehensive account of animal "crime and punishment" remains E. P. Evans, *The Criminal Prosecution and Capital Punishment of Animals* (1906; reprint ed., London, 1987). An example of this older strain of thinking about animals survives in the *Chronicles* of Lewis Henry Miller (1786–1882). A lifelong resident of York County, Pennsylvania, Miller produced an illustrated remembrance of everyday life in his town. He told and illustrated the story of "Mrs. Cath. Weiser frying a sausage, and A hound came and Stole it out of the pan for his breakfast." Miller warned, "Woman Guard Your Kitchen. . . . teach a dog and put him in a way to fulfill his demands, and you make him a Moral Agent." Miller, *Sketches and Chronicles,* p. 82.

92. For a discussion of the tradition of drawing moral lessons from animals in popular natural histories, see Harriet Ritvo, *The Animal Estate: The English and Other Creatures in the Victorian Age* (Cambridge, Mass., 1987), pp. 15–30.

93. Sigourney, *Letters to Mothers,* p. 98.

94. "Stories about Cats," *Youth's Companion,* October 30, 1835, pp. 95–96.

95. In one typical story, a shepherd discovers that his dog has been feeding the man's lost son until the child could be rescued from a cave. Sacrificing his own small ration of bread, the "faithful dog guarded him like

a father, and fed him with a mother's tenderness." Sigourney, *The Boy's Reading-Book,* p. 37. See also "Canine Sagacity," *Youth's Companion,* October 17, 1828, p. 83; "Canine Affection," ibid., October 3, 1828, p. 75; "The Lives of Two Children Saved by a Dog," ibid., September 16, 1829, p. 68.

96. Anon., *The Dog, as an Example of Fidelity: From the London Edition.* (New York, 1848), pp. 10. 38. The copy I have examined was presented as a gift to a child: "George S. Payn / a token of affection from the Rector./ Grace Church, Christmas 1848."

97. "Thomas and his Chickens," *Youth's Companion,* July 6, 1827, p. 24. See also "The Hen," ibid., February 6, 1835, p. 150 ("Extract from the "INTELLIGENT READER" recently published by Messrs. G. & C. MERRIAM, Springfield"). The Reverend C. S. Lovell's *Young Pupil's Second Book* (1836) also included an essay on "The Hen." He took care to point out the utility of her eggs and meat but devoted most of his praise to her domesticity: "What motherly care does she take of her young. How closely and tenderly does she watch over them. . . . While the sight reminds you of the wisdom and goodness of her Creator, let it also remind you of the care which your own mother took of you, during your helpless years, and of the gratitude and duty which you owe to her for all her kindness." Rev. C. S. Lovell, quoted from Page Smith and Charles Daniel, *The Chicken Book* (Boston, 1975), pp. 212–13.

98. Beecher and Stowe, *The American Woman's Home,* p. 394.

99. "Animal Gratitude," *Youth's Companion,* March 14, 1827, p. 167 (from "London paper"); "The Grateful Lioness," ibid., May 19, 1830, p. 207 (from the *London Youth's Magazine*); "A Grateful Cow," *Youth's Companion,* October 5, 1831, p. 70 (from the *Cheltenham Chronicle*).

100. Richard Tapper notes that animal stories can serve three purposes: 1) "*through non-human metaphor* they allow teachers and learners to avoid articulating difficult or embarrassing truths about humanity"; 2) they *create a distinction* between human beings and other animals"; and 3) they can *reinforce morality* by giving it a 'natural' basis." Richard Tapper, "Animality, Humanity, Morality, Society," in *What Is an Animal?* ed. Tim Ingold (London, 1988), p. 51.

101. Ibid.

102. "Anecdote of a Cat," *Youth's Companion* 1, no. 47 (1828) 187; "Anecdote of a Cat," *The Cabinet of Natural History and American Rural Sport,* vol. 2 (Philadelphia, 1832), p. 246. In the latter source, the story is credited to *Good's Book of Nature.*

103. "The Cow and Her Calf," *Youth's Companion,* September 21, 1831, p. 72; "Barbarity of Whale Fishing," ibid., July 4, 1827, p. 23; "Remarkable Attachment of a Goat," ibid., April 6, 1831, p. 183.

104. Ryan, *The Empire of the Mother,* pp. 55–57.

105. Cited in Katherine Morrison McClinton, *The Chromolithographs of Louis Prang* (New York, 1973), pp. 222–23.

106. *The* word "pet" appeared as a description for favorite animals as early as 1539. *OED,* 2d ed., s. v. Pet.

107. Karin Calvert, *Children in the House: The Material Culture of Early Childhood, 1600–1900* (Boston, 1992), esp. chap. 1, "The Upright Child: Swaddling Clothes and Walking Stools," pp. 19–38.

108. Sigourney, *Letters to Mothers,* p. 39. Tonna, *Kindness to Animals,* p. 41. This language even gradually penetrated the rather hard-headed masculine world of dog training. The author of *Woman in Her Various Relations* (1851) noted that "The faculties of children often lie dormant for want of proper exercise, or pains taken to develop and draw them out. They are peculiarly creatures of imitation, but they can also be trained to a certain course of action, with judicious care and effort." The author of *Practical Training of the Shepherd Dog* (1891) informed his readers that "the shepherd pup is much like a child; he is a great imitator." And he sounded like Horace Bushnell when he counseled against physical coercion except in rare cases; "securing the respect and affection of the pupil" and exhibiting "kindness and imperturbable patience" were better educational techniques. Abell, *Woman in Her Various Relations,* p. 223; W. A. Wickham, *Practical Training of the Shepherd Dog* (Tipton, Iowa, 1891), pp. 72, 80.

109. Novels using dogs as plot devices included Harriet Beecher Stowe, *A Dog's Mission* (Boston, 1881), and *We and Our Neighbors* (New York, 1875).

110. Harriet Beecher Stowe, "Our Dogs," in *The Writings of Harriet Beecher Stowe, Riverside Edition,* vol. 16 (1896; repr. New York, 1967) pp. 65–67, 71, 78, 79. Compare Stowe's descriptions of canine personalities with the following passage from the chapter "The Feelings of Living Property," wherein Mrs. Shelby objects to her husband's plan to sell Uncle Tom and describes the plantation's slaves: "O, Mr. Shelby, I have tried . . . to do my duty to these poor, simple, dependent creatures . . . and how can I ever hold up my head again among them, if, for the sake of a little paltry gain, we sell such a faithful, excellent, confiding creature as poor Tom . . . ?" Harriet Beecher Stowe, *Three Novels: Uncle Tom's Cabin or, Life among the Lowly*; *The Minister's Wooing*; *Oldtown Folks* (New York: Library of America, 1982), p. 47.

111. Stowe, "Dogs and Cats," in *The Writings of Harriet Beecher Stowe, Riverside Edition,* 16:103.

112. Caroline S. Kirkland, "The Bee Tree," in *Western Clearings* (1845; repr. New York, 1969), p. 76.

113. Anne Hampton Brewster, Diary 1859–1860, entry for May 9, 1859. Library Company of Philadelphia, Philadelphia, Pa.

114. Longfellow Diary, September 11, 1848, quoted in Nancy Schrom Dye and Daniel Blake Smith, "Mother Love and Infant Death, 1750–1920," *Journal of American History* 73, no. 2 (September 1986): 342.

115. We commonly compare children to monkeys now, but in the nineteenth century the comparison of humans and primates almost always bore negative racial or class connotations. See Grier, "Material Culture as Rhetoric: Animal Artifacts as a Case Study," in *American Material Culture: The Shape of the Field,* ed. Ann Smart Martin and J. Ritchie Garrison (Winterthur, Del., 1997), pp. 65–104.

116. George M. Fredrickson, *The Black Image in the White Mind: The Debate on Afro-American Character and Destiny, 1817–1914* (New York, 1971), p. 103.

117. Stowe, "Rights of Dumb Animals," p. 24.

118. Clark, "The Sacred Rights of the Weak," pp. 463–64, 470.

119. "T. D.," "The Mourning Robin: A Word of Advice to Boys," *Youth's Companion,* July 31, 1835, p. 42.

120. *Pity's Gift,* pp. 69–71.

121. "The Captive Blue Bird," *Youth's Companion,* November 20, 1833, p. 103 (from *Parley's Magazine*).

122. Sydney H. Coleman, *Humane Society Leaders in America with a Sketch of the Early History of the Humane Movement in England* (Albany, N.Y., 1925), pp. 75–76.

123. Virginia Sharpe Patterson, *Dickie Downy: The Autobiography of a Bird* (Philadelphia, 1902).

124. Mark Twain, *A Dog's Tale* (1903; first U.S. edition, New York, 1905).

125. While eighteen states did have laws on the books against physical abuse of domestic animals before 1860, no mechanisms for enforcement existed, and there were few, if any, prosecutions. The Civil War also interrupted at least one early effort to organize an American SPCA, in 1860 by Philadelphian S. Morris Waln, who corresponded with the now Royal Society for the Prevention of Cruelty to Animals to obtain guidance on how to proceed with the humane cause here. The well-to-do men who founded, served as officers in, and financially supported the fledgling groups seem to have been interested in a variety of other reforms. Some, such as George Angell, founder of the Massachusetts SPCA, had ties to abolitionists; others served on the boards of asylums or other municipal institutions. And some of the earliest work by anticruelty groups represented public health reform as much as humane activity. One of the first campaigns of the American SPCA was against impure "swill milk" sold in cities, where tubercular cattle, often confined to underground stables, subsisted on the by-products

of the brewing industry. For discussion of the leaders of the early humane movement, see Turner, *Reckoning with the Beast,* pp. 35–36, 39–59. See also Coleman, *Humane Society Leaders in America,* passim. Coleman, who knew many of the men and women discussed in the book, not only provides particulars about organization building and early activities, but gives a sense of the social pedigrees of the early activists and the place of women in the first several decades of the movement.

126. "Our Dumb Dependents," *Godey's Ladies' Magazine,* August 1871, p. 182.

127. Coleman, *Humane Society Leaders in America,* chap. 3, "Elbridge T. Gerry and the Prevention of Cruelty to Children," pp. 65–88.

11

A Comparative Study of Women's Activities in the Domestication of Animals

Jacqueline Milliet

Anthropology, before it abandoned the quest for origins, took for granted the archeological and zoological discourses about domestication. According to archeologists, humans domesticated corn and animals at the same time, sometime during the Neolithic period, and domestication would have been a specific response to the need for food.[1] This postulate, in turn, is based on the conviction that domestication would not have been possible without an obligatory prior stage "taming," which involved hunting-gathering activities along with the taming of animals exclusively as pets.[2] Zoologists, for their part, considered domestication to be a state proper to animals that breed in captivity.[3] These scientists described animal morphology and endeavored to measure genetic consequences due to human actions. Archeological and zoological arguments were based on evolutionist and diffusionist preoccupations. These perspectives become blurred when the subtle relations between wildness, taming, domestication, domesticity, husbandry, training, exploitation, familiarization, etc. are considered. As we will see, these various relations reveal many different aspects in human intentions toward animals.

Using theoretical concepts such as alimentary determinism, survival of primitive societies and matriarchy, the change from hunting and gathering to agriculture, etc., however, Downs helped the comprehension of the domestication phenomenon to take a great leap forward.[4] He proposed to take into account simultaneously the animals' instinctive attributes, and the fundamental modifications generated by the new relationships between humans and animals. According to Downs, domestication should no longer be considered an isolated event, but rather a dynamic process that has continued through the entire course of human history, even to the present day. He noticed that social groups who lived next to one another did not necessarily domesticate the same animals, and this explained why the cultural and social potential of the human-animal relationship was a determining factor. At the same time, Haudricourt demonstrated, particularly in his article "Domestication des animaux, culture des plantes et traitement d'autrui,"[5] that the relationships humans entertained with plants and animals reflected relationships between humans. Thus, animal domestication may be clearly defined as a synchronic phenomenon, as a continuous process involving a great number of technical and cultural activities. These activities, constantly repeated and renewed, should be understood in relation to economic functions as well as to social, ritual, symbolic, and emotional values.

At the beginning, anthropology mostly focused on the domestication of plants.[6] This is probably due to at least three main reasons. First, the number of domesticated plants is far higher than that of domesticated animals. Second, behaviors toward the same animal species are manifold whereas each plant is usually treated the same way.[7] Third, general techniques of animal domestication require few tools or other materials; when needed, tools tend to be simple and multifunctional.[8] In what category should we put the stick used by the shepherd to both cudgel the dogs and lead the flock; or the stew-pan a woman uses to cook the food for her kin, as well as for animals? Simplicity of the tools used in domestication placed emphasis on techniques that take the form of a great many gestures and a great deal of know-how (using the voice, the hand, etc.) learned from the close observation of animal anatomy, physiology and behavior. The instruments utilized by women in the traditional sexual division/prohibition of labor have the same rudimentary quality. Women plant and harvest with bare hands; they fish without net or boat; they catch fish from the shore; they hunt ordinary prey without weapons and seldom with the help of dogs. In general, their activities require the use of their own bodies, and bare hands or, at most, of basic tools.[9] We will see how both domestication techniques and women's activities overlap because of the originality of their object: the animal, a living being.

We will also see that once the traditional definition of animal domestication is called into question and a new one constructed, it becomes possible to inventory its technological variants and to specify the role of women in this process. However, before describing the process, I should say a few words about the sources.

Some Elements of Methodology

Ethnographic literature deals only incidentally with domestication and even less with the particular role in it played by women. Traditionally, anthropologists liked—and still like—to express themselves through monographs, either thick volumes or articles. They tried to make a list of the patterns of the social group they studied, to combine the description of its various aspects such as social organization, material culture, religion, warfare, etc. In travel narratives, as well as in monographs and articles, information is usually scattered over thousands of pages. Sources present serious deficiencies—"second sexe oblige"—that precisely concern the key domain of task distribution according to sexes and the female engagement in the process of domestication. When data are found they take either allusive or affirmative. Women are "there"—an adverb from which Lofland created the operating concept of "thereness" that she uses in urban sociology:

> By thereness of women, I refer to a phenomenon rather similar to that of portrayal of the servant in the classic British mystery story. Such persons seem always present and yet, despite, or perhaps in part because of, their omnipresence, they remain, by and large, merely part of the scene. They are continually perceived, but rarely perceivers. Essential to the set but largely irrelevant to the action. They are simply there.[10]

The fact of being "simply there" fits exactly with what I found while investigating ethnographic materials. For what reason, indeed, should ethnographers be interested in a woman who, at sunset, calls her pigs back for feeding and shelter for the night? For what purpose, likewise, should an ethnographer describe a dog napping in front of a house? Ordinariness of this kind does not generally pique the curiosity of ethnographers. Rather, the history of anthropology shows that the discipline initially was, and still is, focused on facts that refer to differences, otherness ("altérité" in French), and to the exoticism of the observed people and phenomena.[11] Similarly, customs that varied from theirs were more fascinating to explorers, missionaries, and anthropologists than the ones that more closely resembled their own.

Since 1975, after the publication of Louise Lamphere and Michele Rosaldo's *Woman, Culture and Society,* a whole sector of anthropology has been dedicated to the study of women as social subjects/actors. This form of anthropology includes various subfields, such as the anthropology of sexes or of genders, the anthropology of women, feminine anthropology, or feminist anthropology. All these names demonstrate a more or less vigorous involvement of scholars in the feminist movement; however, specialized data on women (with a few exceptions) do not supply any interpretation of female roles in human-animal relationships.

My approach, pluralistic and interdisciplinary, involves an extensive and systematic examination of the Human Relations Area Files (HRAF),[12] as well as of other fields usually concerned with animal domestication, such as archeology, paleontology, primatology, zoology, ethology and veterinary medicine, ecology, cultural geography, history and art history, or mythology. I also delved into fiction, sculpture, even comic books, and found this digression a positive one, at least in the gathering of iconographical materials. In this way, I collected a great amount of data on almost three hundred diverse ethnic or social groups, including modern Western societies.[13]

This approach made it necessary to read over again the sources dealing with animals and their domestication, and with the sexual division of labor as well, in order to bring to light, and also to bring up to date, women's activities in animal appropriation.

Once the traditional definition of animal domestication was called into question and a new one constructed, it became possible to inventory its technological variants and to specify the role of women in this process.

The domestication processes applied in human societies correspond to a whole series of combined husbandry and utilization techniques.[14] Primarily, these are the products of countless combinations of the three basic conditions of domestication defined by Barrau:[15] (1) animal feeding; (2) animal protection and control of animal movements, and (3) reproduction (three conditions animals have to meet by themselves when they live in a natural environment). The phenomenon is all the more complex since humans do not intervene uniformly at these three levels of domestication, and since their interventions depend on the function of the animal as well as on its innate behavior. This is why it is essential to correlate husbandry techniques with utilization techniques, that is to say, with the practical values that correspond to utilitarian, nonmaterial, specialized, or multifunctional animal roles and statuses in diverse social systems. A utilization technique consists in what is used from the bodies of live, as well as dead animals. Besides the exploitation of the animal strength (traction), and physical aptitudes (watching, scavenging, etc.), natural bodily productions such as bristles, skins, teeth, excrement, sacrificial flesh, or meat, are used and more or less altered by human interventions.

To elaborate these ideas, I have limited the zoological scope of this research to pigs (including here true pigs and peccaries) and dogs. These animals belong to the most ancient groups of domesticated species;[16] and they can be found in the four corners of the world, observed at the same time in wild and domesticated varieties, and seen in the most varied societies.

However, sources dealing with pigs and dogs are extremely uneven, in quality as well as in quantity. When explorers, missionaries, and anthropologists focus on dog meat and on its possible corollary, the rejection of pork, clearly they are influenced by the reverse of their own practices. Nothing is more human than comparing or differentiating oneself from others to protect one's own habits. A butcher from Ghat (South Algeria), himself very fond of dog meat, justified his choice quite simply: "J'aime la viande de chien comme les chrétiens aiment la viande de cochon."[17]

Eberhardt, a specialist in Asia and the Near East, supplied one more justification for confining my research to pigs and dogs. He made the assumption that these two species were cultural equivalents: the first would occupy in certain societies the same place as the second would in others. Downs took up and expanded on this hypothesis: "This position appears sound, particularly in Asia and the Near East where the conditions under which both animals exist are similar and may, as suggested, stem from a parallel domestic history. It also permits a functional analysis of the dog's role in areas such as pre-Conquest Mexico where certain races of dogs where bred for eating and others for hunting, as pets, etc."[18] If Downs acknowledged that the meat potential of animals often led to their domestication, he seemed, however, to suppose that pigs and dogs were subjected to comparable domestication processes, which mostly depended upon the domestication system elected by each society. Wild, tamed, or domesticated, both animals mostly live close to human settlements and, depending on social mechanisms, both are allowed to enter dwellings. We shall see how animals follow spatial rules that most of the time regulate domestication techniques.

We know now that so-called primitive societies resolutely chose to remain hunters and gatherers even though they had the opportunity to make another choice. In tropical America, numerous animals tend to gather around Indian settlements, whose differences in status vary not only because of the variety of species. The dog, the domestic animal par excellence, is never eaten, while poultry, also domestic, is raised for its flesh. Wild game animals are killed where caught and brought back home to be cooked and eaten. Orphaned offspring are taken in alive, breastfed and given premasticated food, and become pets for the remainder of their lives. The Barasana (a people of eastern Colombia) keep a great number of small rodents, dogs, parrots, and other kinds of birds and to a lesser extent, tapirs, peccaries, ocelots, and jaguars.[19] Among these animals, the peccary is a good example of a species that would be advantageous to domesticate. It is certainly not because they do not know how to domesticate peccaries that the Amerindians seem not to be interested in raising them; it is rather because domestication of peccaries would throw their various relations with nature into confusion. People knowingly manipulate various states of animals. Thus, it is possible to act differently according to the animals they select. Modern urban people do not behave much differently when they pamper cats, dogs, guinea pigs, goldfish, snakes, dwarf hens, miniature pigs, and horses at home, while at the same time concealing behind the walls of animal factories the pigs and poultry that they intend to eat.

In this paper, I will survey the techno-cultural activities that control the existence of animals. To do so, I will reconsider the traditional definition of animal domestication, and resituate the characteristics of the social dimension women personify into distinctive ethnographical contexts. I intend to show to what extent the distribution of animals into wild, tamed, and domesticated categories reflects the hierarchical organization and the differentiated distribution of individuals in human societies.

PIG HUSBANDRY TECHNIQUES

Pigs search for food in three different ways. One is to find edibles in the forest. Their pasture involves short-distance movements and transhumances that are usually under women's responsibility. For example, in Greece, women lead the pigs under oaks that they have previously beaten to shake loose the acorns.[20] Secondly, pigs are integrated into the agro-pastoral system. They graze with herbivores and glean the roots overlooked in harvested fields. In Manchuria, among the nomadic Lolo, the practice was to use women captives to herd the pigs.[21] Both examples indicate that women's role and pig husbandry are not strictly tied to a settled way of life.

Pigs also scavenge in villages or are fed kitchen, farm, and industrial wastes. Garbage and excrement are consumed. Wastes are also gathered together and cooked to complement the pigs' own search for food. In Thailand, for example, pigs are fed peels, weeds, bananas, and rice brew, which women cook in a big earthenware pan over a fire set in front of the house.[22] On the whole, from household economy to animal factories, wastes mix together differently in the trough as the pigs progressively lose their freedom of movement, and as the importance of women's activities lessens. In short, while the contents of the pigs' meal can be made up of either kitchen, farm, or industrial wastes, more often it is a combination of all three.

A full stomach is not enough to develop the animal's loyalty: feeding an animal also includes establishing a relationship with it in order to control its movements. On Hainan Island (China), young women call the pigs back twice a day with a piercing "you-ou," leaving the food inside their overnight shelters.[23] Women exert a centripetal force over animals' movements. In addition to calling, a few specific techniques keep animals in proximity to habitations. Leashes or triangular collars are placed on pigs to prevent them from escaping. Very seldom do women intervene directly on the animals' bodies. Rather, as among the Irakia Awa of New Guinea, it is the men who clip the ears, tail, or testicles.[24]

In some Tzelzal villages (Yucatán, Mexico), Indian women purchase piglets from *ladinos* (mestizos) of other communities. After eight to ten months of fattening, they resell them to the same people who will then slaughter the animals and get money from the Indians who buy the meat.[25] Twice in their pig-raising activities, Tzelzal women are required to make use of persons who are strangers to them: males, belonging to another community than their own, do the breeding, spaying, castrating, and slaughtering.

Whatever social group women belong to, they do not castrate or spay, nor are they responsible for trading the boar or selecting offspring from the litter. On the other hand, they are expected to know when sows are on heat, and occasionally they bring them to the boar. They may also assist birth, but above all, they take care of the sow and her litter.

PIG UTILIZATION TECHNIQUES

The function of pigs as food is treated almost exclusively in the sources, and this has caused some interpretative difficulties. We saw that animal domestication in general, and pig domestication in particular, is generally explained by the human need for food. Works on pigs definitely focus on the production of meat, and therefore consider any other utilization of the animals as marginal, if not nonexistent.

It is not easy to separate husbandry and utilization techniques because they intersect in many areas. Where should we set the limit between fattening (meat production and waste recycling) and confinement (protection, control of movement, and collection of manure for the fields)? Even though it may seem artificial sometimes to distinguish between utilization and husbandry techniques, this nevertheless remains an effective way of understanding the various roles and functions animals play inside social and sexual groups. We will see that women do not have the same access to animal products as men.

When alive, pigs very often have diverse functions that include food production in general, and the individual's search for food in particular: gathering roots and acorns in harvested fields not only feeds the pigs, but also makes possible plowing, manuring, and land clearance. When dead, pigs yield numerous useful products: meat, fat, bristles, leather, and teeth, and can also be used as ritual offerings, or even as currency. Therefore pig products can disperse in many directions. They fit into an extended market and can cover a wide area of social interactions. Women find themselves at the center of such an area, through feeding, sheltering, buying, and selling the animals; and men act at both ends through control of reproduction, of slaughtering, and of the whole market process. Poor country people of Puerto Rico raise pigs for money and festivals. For families who cannot keep this capital alive, a visiting butcher makes the rounds. For destitute families, pig husbandry represents a way of saving money rather than making money. Haitians regard their small black pigs in the same way, calling them "la banque du pauvre."[26] As one climbs the social scale, animal husbandry becomes more specialized: that is to say, as soon as there is capitalization of the dead body, the animal fulfils its sole function of being a meat reserve. As pig husbandry gains in status, the meaning of the pig is reduced to the profit it brings when sold as meat.

The connections between life and death play an important but not determining role in female techniques of domestication. We should point out here that the failure of women to participate in slaughtering has been too often associated with the female reluctance—imposed or intentional, but always supposed "natural"—to put an end to a life and then to eat the product of this action. Women raise animals they know will be slaughtered sooner or later. Furthermore, women do not stay away from animal blood; they unhesitatingly dip energetic hands in the victims' warm blood. On traditional French farms, just after they have rinsed the guts, women first drain the blood to make blood sausages and exchange recipes involving blood.[27]

In farm husbandry, female responsibilities are very important, but effort is also made to distance oneself from the animal about to be slaughtered. In Hungary, women periodically pull the animal's tail to check whether it is time to proceed with slaughtering: as long as they can hold it, the animal has not grown enough.[28] In France, the words used to describe the animal show the need of turning away progressively from it. First, as a neonate, it is a pampered and cherished "bébé"; then, when the pig begins to flesh out, it is respectfully called "monsieur"; finally, when the day of slaughter arrives, it has become "méchant."[29]

Young Zulu women will never eat pork under any circumstances: they find the animal disgusting, its mouth hideous, and its snout too long, and they fear that they might give birth to children who would resemble it.[30] This reaction might be a residue of the influence of ancient Egypt in South Africa.[31] Religion—and a fortiori Islam—is the principal reason adduced by travelers and ethnographers to explain pork avoidance. This is too large a question to be dealt within this paper. Instead, we shall review some ethnographic data from cultural areas where monotheistic principles are not involved. Although Buin women of Bougainville (Solomon Islands) are responsible for pig husbandry, they have access to pork only with the permission of their husbands.[32] In the Hawaiian and the Marquesas islands, pork was available to women only on certain occasions or during specific periods of the year.[33] De Coppet observes that in Oceania pigs and women share the same dependent status: "Si, dans la réalité quotidienne, le porc est tout à la fois le seul animal domestique et la victime par excellence, les femmes sont dans une relation semblable, unilatérale, avec les hommes: elles ne sacrifient jamais le porc et ne peuvent le consommer."[34] Confronted with a very segregated society, de Coppet thinks he has discovered why women are kept out of things, but in reality, he just rephrases his informants' arguments. Indeed, a few lines further on, he opines regarding the sacrificed pig that "cette chose victimisée est de nature féminine," thus allowing biological determinism to reassert itself in his own argument.

DOG HUSBANDRY TECHNIQUES

The decision to feed or starve a dog depends upon the status of the animal, which derives directly from its function. The privilege of being pets or hunting auxiliaries brings a certain degree of food security. Whatever the form of the daily meal, it is never an established right: the age, the sex, and the function of the animal, as well as the rhythm of the seasons, are parameters that we must also take into account. Dogs, like pigs, will eat decaying carcasses, feces, and vegetal waste; they will ingest almost anything that lies around dwelling houses, stables, and harvested fields. They also get household refuse, kitchen scraps, and meal leftovers. They profit from butchery and dairy farming remnants.

Human interventions in the lives of animals grow proportionately to the extent that people ensure them a daily food intake. At first glance it seems that no human effort is expended when animals scavenge. In fact, some care must be taken to channel scavenging dogs away from stored food provisions and even leather. Women, the home keepers, have these tasks as their responsibilities. Montagnais Indians (Québec) put harnesses, snowshoes, and bags out of reach of dogs.[35] Among the Fali (Cameroon), "Tout est suspendu au-dessus du sol afin de ne pas être sali, ni mangé par les insectes ou dévoré par les chiens."[36] Women from Bali cannot perform their devotions quietly when they are harassed by dogs that attack their rice offerings.[37] In Burmese villages, dogs dash into kitchens and pounce on offerings in pagodas and on the meals set for mendicant monks.[38]

When dogs are free scavengers, their security is minimal without specific shelters. Among the Nyangatom nomads of the Omo Valley, Ethiopia, settlements are surrounded with thornbush enclosures, made of heavy plaited bundles of branches. Kept outside that perimeter are dogs that bark at the approach of jackals, hyenas, or human prowlers.[39] The lives of these scavenger dogs are constantly threatened, and they tirelessly try to crawl underneath the fences to enter the settlement enclosure and protect themselves from night predators.

Animal confinement occurs whenever dogs share in family life, or use their own specially created living arrangements (hammock, litter, kennel, etc.), and has two purposes: protection and humanization. The voice is one of the principal instruments used to supervise dogs: they are cheered or rejected through the granting of names or nicknames, or by jeers and laughter at their behavior. Words directed at dogs also often serve the end of indirect communication between people. They are ways to express unspoken thoughts and to criticize deviant behavior. For instance, among the Jankuntjara (Australia), dogs embody human character traits, such as hypocrisy, greed, and lust. Some of the names given to dogs are insults that women say to each other; among these epithets, one of the most frequent refers to copulating with dogs.[40]

Mobility of dogs may be restrained by objects tied around the neck, such as a collar, leash, shackle, or muzzle. Clipping the ears or cutting the tail are both thought to improve natural canine aptitudes, such as fierceness, auditory acuteness, and the guard instinct. These measures are also supposed to protect dogs from sickness.

As with pigs, it is the men who intervene when a surgical act is required: they castrate males and spay females. Men also exchange and select pups from the litter. Women watch a bitch in heat, lock it away from inopportune couplings, and take care of the puppies.

Dog Utilization Techniques

Uses of dogs fluctuate between two tendencies: specialization and versatility. Dogs are either idle pets that keep women company, or animals that help men in their work, such as protecting and herding flocks, and hunting. Canine functions are maximized in societies such as those in Arctic and sub-Arctic regions, where they are the only animals kept. There they perform "female" and "male" tasks simultaneously or successively.

Everywhere that dogs are kept more or less vagrant, women select individuals from the scavenging pack to feed in front of their houses, around their tents and fields, or harnessed to the sleighs, sleds, and toboggans. The women expect help from these dogs in amazingly contradictory ways. Among the Beng (Ivory Coast) the same animal that keeps dangerous warthogs away from the fields will also babysit in the absence of the mother: it puts a sleeping baby back in the right position with a very gentle nudge of its muzzle.[41] Dogs of the Ona in Patagonia both hunt pumas and chase flies away from a baby's face with a slight move of the paw.[42] Poorly nourished stray dogs of the Jakuntjara like feces so much that a stick is needed to keep them at arm's length when relieving oneself. Yet, women will allow one of them indoors to wipe an infant's bottom with its tongue.[43] Tibetan nomad children regularly play with the very aggressive family guard dog, which is trained to react to the presence of strange people or other dogs. These mastiffs mellow in the company of women, and not only the one who feeds or takes care of them: a female friend or stranger is not nearly as liable to be bitten as is a man.[44]

The transition from fierceness to gentleness does not happen by chance. A drastic diet, ear cutting, and rough relationships with humans exacerbate savagery. We know less about the development of gentleness, but we must suppose that women use specific techniques to pacify dogs, to compensate for rough treatment they get elsewhere. Practically speaking, how do they manage to mollify these fierce mastiffs? Is it enough, for instance, that during the cold nights of Patagonia, Ona Indians curl up against them,[45] to make animals understand that, under certain circumstances, coming close to humans is not dangerous? Training to be near or to keep at a distance is a form of managing social and animal territories that is incumbent on women.

Among the Hare Indians of sub-Arctic Alaska, adult women support the whole family and own both the dog team and the harness. But if female duties are equivalent to those of men, nonetheless the range of their interventions differs from men's, and is limited to the vicinity of the settlements.[46] The Hare expect girls and boys eventually to do the same tasks, but they do not start equal in life.[47] Infant mortality among boys is lower, for they are better dressed and better fed than girls. Boys are more spoiled and get more affection for a longer time than girls, who enter real life sooner. Rough handling sets the limits of sexual identity through a succession of acts of negligence that may even lead to selective infanticide of baby girls. A couple of elements should here be pointed out in light of the comparison between upbringing and domestication techniques. These concern the precise correlation between the joint socialization of females and males, children and dogs, that is to say, in reality, gender role learning and sexual selection. First, given this assumption, it is not too fanciful an idea to draw a parallel between Hare child-raising methods and the elimination of most female puppies at birth. Killing female rather than male puppies fulfils a regulatory rather than a breeding function. Secondly, while physical contact with both children and puppies oscillates between gentleness and cruelty, any closeness disappears as all of them grow older. Here, we should recognize one of the conditions of domestication: the establishment of a distance towards animals—alternating proximity and taking away—according to age, sex, status, and role.

In dog slaughter, both the sacrificial knife and the products of the animal remain in masculine hands. As a general rule, one does not kill or eat one's own dog. One does not sacrifice a pet, but rather an animal for which one does not show much affection. That is why the former young pet is taught to leave the vicinity.[48] For Vietnamese peasants the perfect victim is a village stray, which belongs to no one. They may use any kind of misbehavior as a pretext for endless arguments about the stray dog, then punish it by killing and eating it.[49]

Men immolate animals for themselves and for the whole community, most often in sacrifices dedicated to the dead. Among the Bambara (Mali), for instance, some women raise dogs that they expect to be killed for their funeral.[50] The Massa (Chad) believe that sickness during pregnancy, or infant suffering and death, express the revenge of a dog whose puppy has been killed, and that another dog must be sacrificed to undo the harm.[51]

Ethnographic sources say little about the products of dog corpses. However, we may deduce that in societies where the visible epidermic parts (hair, claws, nails, and teeth) are used, women very likely weave, sew, and assemble from them everyday objects such as blankets, or clothes for the family and community. As the market becomes wider, men sell or exchange raw materials from dogs such as wool, skin, fur, and fat, as well as artifacts made by women.

Meat, another product of the animal corpse, is subjected to many ambivalent social practices. In general, eating meat is an act that is never commonplace: it is a symbolic act, all the more so since others do not have access to this meat. Eating dog meat is an initiation ritual that consolidates the feeling of membership and reinforces sexual identity.[52] Female Tallensi (Ghana) will never cook or eat the dogs and the chickens that men slaughter during the New Year festival.[53] A similar case of exclusion has been observed among the Arapayo (Philippines), where men occasionally eat dogs, but women never. For the Arapayo, as well as for the Bontoc Igorot (Philippines), this segregation from ritual meat eating is a serious matter as it takes place at weddings and funerals.[54] Kabyle women (Algeria) were permanently subjected to cynophagic prohibition, while it was only temporary for boys and male adolescents, who were allowed to eat dog meat during their initiation.[55]

THE THRESHOLD WOMEN WOULD NOT STEP ACROSS

The ethnographic record reveals quite clearly that the control of animal reproduction is beyond the responsibility of women. Men usually select young animals and kill them right after birth; they also castrate males, neuter females, and exchange stud animals. Women become engaged in the process when female animals come into estrus and drop their litters, and when the young require care. Men build shelters, stables, and fences; they cut ears and tails to recognize or punish the animals. Women, on the other hand, just kick them or knot a piece of cloth around their necks. They allow animals entry into the house, and encourage them to come into the family circle. They call them back for eating and sleeping. Women also carry and convey them. Feeding appears to depend almost completely upon women.

At the level of utilization, women's activities differ from those of men in two domains. When the animal is alive, it plays different roles: the dog, for instance, may be a pet, scavenger, guardian, or draft animal. These roles may help women to do their work, but may also be limited by the traditional specialization of labor. When the animal plays a single role, like the hunting dog, women may feed and protect it, but cannot use it. Women may hunt, but when they do so, they are not allowed to use a dog, or at least not the same dog used by men. When the animal is dead, its corpse and the products of its corpse will transit through male hands, while the meat serves to strengthen the social differences.

Women do not step across this threshold that seems to exist in the process of production. As a border between the domestic mode of production and extended exchanges, the threshold clearly defines general work from specialized work. Across the threshold, the more labor is specialized, the more tools are also specialized, and the more the work becomes solely a male concern. The boundary between living and dead animals is usually the moment of slaughter. The accompanying collective rituals of the slaughtering process mark the absolute limit between the living and dead animals. These rituals empower the restriction of women to the private or domestic sphere. However, aside from these sociocultural strictures, all the evidence indicates that women are not by nature more afraid of blood and death than are men.

These contrasts—between women and men, life and death, numerous functions and specializations—should be related to both general techniques of domestication and the work women usually accomplish. Women are under-equipped and so must use their bodies as instruments. It stands to reason that breastfeeding, feeding, nursing, petting, coddling, and calling are so many techniques, designed to help young animals adapt to the human world. Though these techniques concern very young animals, we know that women do not stop domesticating as pigs and dogs grow older. On the contrary, they play an important role in modifying the animals' statuses. If they do not remain pets, animals become potential source of profit and therefore must be moved away from and by humans, practically as well as symbolically. These domestication techniques bring into play many diverse and subtle levels in both taming and keeping wild. This means that there are numerous ways of managing distance with animals, from closeness to the greatest distance possible; in other words, at the symbolic level, from integration to rejection. These various opposing elements are centered in the area where women practice their know-how, and are based on ethological observations, as well as on household work and education of children.

To breed poultry and pigs in animal factories, none of the above skills are required. Frequency and intensity of contacts with animals do not have to be cautiously measured any more. The main technique of industrial livestock husbandry consists in locking up and keeping the animals away from any human proximity. In other words, animals raised in factories are no longer linked to human daily life, and this is where they really differ from pets. However, women still remain active in animal factories: on one hand, they work as female unskilled laborers; on the other hand, they occupy an important place in the production and the use of pets as consumer goods.

DOMESTICATION TASKS AND DOMESTIC TASKS

According to utilization techniques, we may conclude that in "traditional" societies women train dogs to accomplish selective actions that, while not substitutes for their own work, do make it easier. North American Indians hitched up dogs to the travois, a sled made of long tentpoles between which they fixed crosspieces to hold a load.[56] As the load could not be heavier than twenty-five to thirty kilograms, the Indians needed large numbers of travois. They kept packs of several hundred dogs to pull the sleds, and "seules les femmes s'(en) occupent, et ce sont elles—lorsqu'elles en ont le droit—qui prennent soin des chiots et de leur mère."[57] Women, who were in charge of moving the tents and all the household goods, also managed the travois dogs and took care of the litters. Not surprisingly, these tasks fit the conventional limits of labor distribution according to sex. However, when Poux argues that women "have the right" to take care of dogs, he probably means that it is more an honor than a duty. In a roundabout way, he seems to be saying that it would be wrong to assume that connections between household work and animal domestication are mechanical because there is a power relation in the sexual division of labor.

Regarding current views of women's roles in domestication, two kinds of sources in the corpus of ethnographic data depend upon one another: brief accounts and prolix interpretations. The dearth of concrete and detailed information, indicating a decided lack of traveler interest in activities peculiar to females in general, contrasts significantly with the superabundance of images and discourses concerning women's role in animal domestication. Both the impressionistic evidence and the lengthy interpretations are forms of rhetoric that express metaphorically the foundations of the relationships that humans maintain with animals and nature, as well as with each other. The simple fact of alluding to women's activities in animal domestication becomes self-sufficient, as if the authors shared in advance with the readers the myth of the "natural" connection of the female body with taming. This myth is still firmly rooted in contemporary mentalities. A young Swiss artist who worked with me on archeological excavations during the late 1980s, and who was aware that my research field was human/animal relationships, once gave me one of his watercolors. It represented a squatting woman nursing a piglet—a typical genre painting illustrating the pregnant myth of a specific link between women and animals, a civilizing relationship whose quintessence might be the breastfeeding of animals.

My interest in the role of women in animal domestication was born while I was the curator of the exhibition "Animals and Humans" at the Museum of Anthropology in Neuchâtel.[58] In the course of my research, I came to understand that dogs were the first animals to be characterized as domestic. The domestication of dogs also seemed to be a female concern, as long as the woman was considered to be the "nurse-mother-woman," the only regular source of milk before the domestication of herbivores.[59] The woman, regarded as the Egeria of civilization, adopted dogs into the human family, where dogs and children became relatives. Forty-five years ago, Sauer was convinced that the prehistoric hunters of Southeast Asia had been the first to domesticate pigs and dogs by bringing them back home to be breastfed by women. He compared these women with the current natives of the same region, as well as with the Amerindians. Many scholars, impressed by what they see as a prehistoric survival, were convinced that animal nursing might be a possible—if not obligatory—prior stage of domestication.[60] It was thus that the idea of "taming" arose. In this literature, females' bodily functions, raised to the status of determining factor, seem to endow women with a specific role in the linear definition of animal domestication. I propose to call this specific role the taming breast myth. Taming, associated with breastfeeding, refers to the diverse orders of reality, to the wild and the domestic, to the feminine and the masculine. Considered together, these show the difficult management of the imprecise boundary between nature and society. Do women's bodies remain the single material element leading to actions towards animals?

I raise two points for consideration in this respect. One emanates from the concrete results of my research. Considered as a synchronic phenomenon, animal domestication reveals specific techniques that have remained hidden behind the mask of the sexual division/prohibition of labor and have escaped the observation of various scholars. It is now possible to inventory an almost exhaustive list of the technological variants that pertain to women in domestication, and to bring to light the stubborn permanency of a social mechanism. Secondly, the interpretation of these facts leads one to question the logic of the assumption that the role of women in domestication appears to be universal, and by consequence, "natural."

To a great degree these questions illustrate the imprecise boundary between nature and society. In the constant movements humans make to get closer or further away from nature, animals and women are significant protagonists. I do not deny differences between sexes. Females obviously give birth because of their biological makeup and for the same reason they are the ones to be concerned with breastfeeding. But biological makeup stops here, and there is no irrefutable proof that women are more prepared to be parents than men. Murdock attributes different predominant tasks to women and men: female tasks include gathering food and various kinds of fuel, grinding grain, carrying water, preserving food, cooking, making pottery, weaving, making clothes and baskets, making dairy products, and doing washing.[61] Many of these tasks, however, have also been carried out by men, particularly at decisive turning points in history, such as the introduction of the potter's wheel or the plow. Fieldwork conducted in Egypt after Nasser's agrarian reform and its reorientation towards a market economy by Sadat showed that men become involved in mechanized agriculture while women have continued to plant and harvest with their bare hands.[62] In agriculture and in animal husbandry, the introduction of more sophisticated mechanical devices causes these activities to come under the control of men. Sexual differences are applied in various ways, for practices do not strictly conform to the rules dictated by normative systems. That is why the sexual division of labor, the technical choice, and the type or the value of the products should be analyzed in conjunction. For these structural reasons, after birth it is the mother, or another woman, who bears the main responsibility for taking care of the newborn infant. That nurturing experience may compel her to continue fostering the child as it becomes older. If women are, in fact, the overwhelmingly predominant defenders and handlers of animals—in traditional societies, as well in modern urban settings—it may be because of the experience they acquired with infants, experience that generates historically constructed attitudes and values.

An alternative explanation is that the division of labor should be understood more as a prohibition than a mere distribution of tasks. We should always keep in mind Lévi-Strauss's observation that "when it is stated that one sex must perform certain tasks, this also means that the other sex is forbidden to do them."[63] Lévi-Strauss attempted to answer the intricate question of whether there exist natural and/or universal characteristics for some social patterns, and how these are integrated by each peculiar ethnic or social group. Despite his brilliant argumentation proving the essential artificiality of the human condition, he uses—like many other anthropologists—biological and moralistic presuppositions when he analyses women's activities: periods, pregnancy, delivery, breastfeeding, etc. underlie the sexual division/prohibition of labor and its considerable weight in the social process. In other words, sex—nature—and gender—social matters—should correspond in a symmetrical relation.[64] When Lévi-Strauss grants to maternal filiation a larger influence than to the paternal one, he defines the former as the actual pattern of filiation.[65] In some ways, the Lévi-Straussian approach meets the essentialist trend of some current feminists. This trend claims the precedence of protocultural (presocial) factors in order to ground human society in nature by means of the mother-child link.[66] Consequently, maternity should generate a feminine essence (a cooperative, peaceful, nonincestuous, altruistic one), that contrasts in a Manichean way with a masculine essence (a competitive, violent, incestuous, egoistic one).

SOME CONSTRUCTIVE HYPOTHESES OF THE ANTHROPOLOGY OF ANIMAL DOMESTICATION

It may seem paradoxical to have taken as a starting point the phenomenon of breastfeeding of animals,[67] since nothing could be found that better illustrates women's closeness to nature and to animals, or less favorable to any attempt to prove the contrary. According to the observers, what was more "natural" than an Amazonian woman breastfeeding a young wild mammal? It was an obvious biological reality that the orphaned animal, so young, could not ingest anything but milk. However, to take place, breastfeeding of animals requires a biological conjunction of a woman having milk and a newborn animal. But this obvious biological reality does not explain why people opt for integrating animals in the human world in this way. Breastfeeding, or even domesticating in general, comes down to making a choice, to integrating or rejecting, to transforming and selecting among the available elements existing in the physical world those that may be compatible or incompatible with other choices: these choices form social relations, i.e., a culture.

We have seen that activities forbidden to women depend upon the level of technical sophistication of the tools they may use. Domestication, however, involves as much know-how and "body techniques" as basic material implements. The traditional sexual division of labor limits women to the use of their own bodies, as well as to rather rudimentary instruments of production. This is precisely how female techniques in general and those that relate to domestication in particular intersect; underequipment in animal appropriation affects the underequipment of women's part in domestication, and vice versa. Thus, instead of trying to understand domestication by women as a whole, we should rather define women's roles as the feminine parts of domestication.

Another constructive hypothesis of the anthropology of animal domestication is based on the observation of animals, above all on the observation of the instability of their statuses. This hypothesis helped us to understand that whatever the species, and whatever the origin of domesticated animals—whether in nature or in husbandry—they are not raised either naturally or in an identical way. Actually, techniques of domestication represent various ways of playing with the diverse statuses of animals. When the cultural anthropologist focuses on animals, insofar as these reveal something about human relationships, her or his thought processes consist in dealing with animal distribution according to the distribution of human tasks, which in turn reflects the hierarchical organization and differentiated distribution of individuals in the people that the anthropologist studies. As soon as an explanation considers the effects of biology and society apart, it becomes possible to relieve women of the weight of the biological determinism.

From the methodological point of view, HRAF data need to be updated. It is also urgent to put the described phenomena back in their context, in order to place the problematic of the choices involved in domestication activities into the context of more concrete social systems. Then it becomes possible to define new perspectives using other criteria than gender, such as the kind of production (pastoralism, horticulture, hunting, and also capitalism); the relationship between the biological and the social; the articulation between normative systems and social praxis; consubtantiality and transfer of substances; and finally the relation between animal and human reproduction. Henceforth, these will be the main anthropological themes that open comparative leads. They will supply a mass of qualitative data into which the topic of this research may be inserted, and may lead to the discovery of domestication formulas.

Notes

1. P. J. Ucko and G. W. Dimbleby, eds., *The Domestication of Plants and Animals* (Chicago, 1969); A. F. Mourant and F. E. Zeuner, eds., *Man and Cattle* (London, 1963).

2. G. Richard, "Domestication," in *Encyclopedia Universalis,* 5:748–52; Michel Brézillon, *Dictionnaire de la préhistoire* (Paris, 1969).

3. I. L. Mason, *Evolution of Domesticated Animals* (London, 1984); René Thévenin, *Origine des animaux domestiques* (Paris, 1947).

4. J. F. Downs, "Domestication: An Examination of the Changing Social Relationships between Man and Animals," in *Kroeber Anthropological Papers* 22 (1960): 18–67.

5. André-Georges Haudricourt, "Domestication des animaux, culture des plantes et traitement d'autrui," *L'Homme* 2, no. 2 (1962): 40–50.

6. André-Georges Haudricourt and Louis Hédin, *L'homme et les plantes cultivées* (Paris, 1943); Jacques Barrau, "De l'homme cueilleur à l'homme cultivateur: L'example océanien," *Cahiers d'histoire mondiale* 10, no. 4 (1967): 275–92.

7. François Sigaut, "Un tableau des produits animaux et deux hypothèses qui en découlent," *Production pastorale et société* 7 (1980): 21.

8. Jean-Pierre Digard, *L'homme et les animaux domestiques: Anthropologie d'une passion* (Paris, 1990).

9. Paola Tabet, "Les mains, les outils, les armes," *L'Homme* 19, nos. 3–4 (1979): 5–61.

10. L. H. Lofland, "The 'Thereness' of Women: A Selective Review of Urban Sociology," in *Another Voice: Feminist Perspectives on Social Science and Social Life,* ed. M. Millan and R. Moss Kanter (New York, 1975), pp. 144–45.

11. Francis Affergan, *Exotisme et altérité* (Paris, 1987).

12. Following the comparative researches of H. Spencer and E. B. Tylor, G. P. Murdock created the HRAF in 1937 at Yale University. The HRAF is a system of files based on primary ethnographic sources and divided into eight main geographical areas: Africa, North America, South America, Asia, Europe, the Middle East, Oceania, and Russia. The data are organized and

classified under seventy-nine general headings, such as "23. Animal Husbandry," for instance. Each of these headings is subdivided into 637 comparative items, such as "23. Animal Husbandry; 231. Domesticated Animals—captive and tamed animals; domesticated insects (e.g., silkworms, bees); pets (e.g., dogs, cats); domestic animals; special treatment and use of each; taming and training; care and feeding; grooming; slaughtering; associated beliefs and rituals; raising of pig herds; raising of fur-bearing animals; economic and social importance of animal husbandry; diseases; etc."

13. Jacqueline Milliet, "Nature et enjeu du rôle des femmes dans la domestication des animaux: L'example du porc et du chien" (Diss., Ecole des Hautes Etudes en Sciences Sociales, Paris, 1993).

14. Haudricourt, "Domestication des animaux"; François Sigaut, "Critique de la notion de domestication," *L'Homme* 108 (1988): 59–71.

15. Jacques Barrau, "Domesticamento," in *Encyclopedia Einaudi,* 5:49–71.

16. Domestication of dogs was conventionally dated to the Palestinian Mesolithic around 9000 B.C., or 8400 B.C. in the Americas; pigs in Thessaly around 9000 B.C. Robert Delort, *Les animaux ônt une histoire* (Paris, 1984), p. 356. Of course, these dates have since been revised.

17. R. Thiriet, "La cynophagie à Ouargla," *Bulletin de liaison saharienne* 16 (1954): 14.

18. Downs, "Domestication," p. 59, n. 8.

19. James Serpell, *In the Company of Animals: A Study of Human-Animal Relationships* (Oxford, 1986), p. 51.

20. Irwin Taylor Sanders, *Rainbow in the Rock: The People of Rural Greece* (Cambridge, Mass., 1962), p. 148.

21. Sergei Mikhailovich Shirokogorov, *Social Organization of the Manchus: A Study of the Manchu Clan Organization* (Shanghai, 1925), p. 133.

22. John E. de Young, *Village Life in Modern Thailand* (Berkeley, Calif., 1955), pp. 31, 94, 98.

23. K. Odaka, *Economic Organization of the Li Tribes of Hainan Island,* Yale University, Southeast Asia Studies, Translation Series (New Haven, Conn., 1950), p. 40.

24. D. J. Boyd, "The Production and Management of Pigs: Husbandry Options and Demographic Patterns in an Eastern Highlands Herd," *Oceania* 55, no. 19 (1984): 35.

25. A. Villa Rojas, "The Tzelzal," in *Handbook of Middle American Indians,* ed. Robert Wanchope (Austin, Tex., 1969), p. 203.

26. Alain Campiotti, "Haïti: La Suisse entre deux cochons," *L'Hebdo,* November 12, 1987, p. 26.

27. Alice Peeters, "Tuer le cochon à Brantes," *Ethnologie française* 10, no. 3 (1980): 247–56; Yvonne Verdier, "Un temps du cochon . . . ," *Ethnozo-*

otechnie 16 (1976): 130–39, and "Le langage du cochon," *Ethnologie française* 7, no. 2 (1977): 143–54.

28. Karoly Viski, *Hungarian Peasant Customs* (Budapest, 1932), p. 141.

29. Verdier, "Le langage du cochon," p. 144.

30. Henry Callaway, *Nursery Tale Traditions, and Histories of the Zulus, in Their Own Words, with a Translation into English and Notes* (Pietermaritzburg, 1868), p. 281.

31. Alfred T. Bryant, *The Zulu People as They Were before the White Man Came* (Pietermaritzburg, 1949), pp. 342–43.

32. Robert Harry Lowie, *An Introduction to Cultural Anthropology* (New York, 1940), p. 441.

33. Franz Steiner, *Taboo* (London, 1956), p. 88; Hutton Webster, *Taboo: A Sociological Study* (Stanford, Calif., 1942), p. 118.

34. Daniel de Coppet, "Des porcs et des hommes," *Traverses* 8 (1977): 65.

35. Georg Henriksen, *Hunters in the Barrens: The Naskapi on the Edge of the White Man's World* (Saint John's, Newfoundland, 1973). p. 22.

36. J. P. Lebeuf, *L'habitation des Fali, montagnards du Cameroun septentrional: Technologie, sociologie, symbolisme* (Paris, 1961), p. 540.

37. Miguel Covarrubias, *Island of Bali* (New York, 1937), p. 98.

38. James George Scott, *The Burman: His Life and Notions,* 3d ed. (London, 1910), p. 83; Cecil John Richards, *The Burman: An Appreciation* (London, 1945), p. 42.

39. Serge Tornay, "Vies de chiens ou des excréments comme bouillon de culture," *Production pastorale et société* 8 (1981): 35–42.

40. Annette Hamilton, "Aboriginal Man's Best Friend," *Mankind* 8, no. 4 (1972): 287–95.

41. Alma Gottlieb, "Dog: Ally or Traitor? Mythology, Cosmology, and Society among the Beng of Ivory Coast," *American Ethnologist* 13, no. 3 (1986): 481.

42. Martin Gusinde, *Die Feuerland-Indianer,* vol. 1: *Die Selk'nam: Vom Leben und Denken eines Jägervolkes auf der grossen Feuerlandinsel* (Mödling bei Wien, 1931): p. 376.

43. Torney, "Vies de chiens," p. 38.

44. R. V. Ekvall, "Role of the Dog in Tibetan Nomadic Society," *Central Asiatic Journal* 8 (1963): 172.

45. Gusinde, *Die Feuerland-Indianer,* p. 200.

46. Joel Stephen Savishinski, "Stress and Mobility in an Arctic Community: The Hare Indians of Colville Lake, Northwest Territories" (Diss., Cornell University, 1970), p. 417.

47. Joel Stephen Savishinski, *The Trail of the Hare: Life and Stress in an Arctic Community* (New York, 1974), p. 175.

48. Jacqueline Milliet, "Manger du chien? C'est bon pour les sauvages!" *L'Homme* 136 (1995): 75–94.

49. Frederick J. Simoons, *Eat Not This Flesh: Food Avoidance in the Old World* (1961; repr. Westport, Conn., 1994), p. 100; James George Scott, *France and Tong-King: A Narrative of the Campaign of 1884 and the Occupation of Further India* (London, 1885); Pierre Gourou, *Les paysans du Delta tonkinois: Etude de géographie humaine,* Publications de l'Ecole Française d'Extrême-Orient, vol. 27 (Paris, 1936), pp. 458–59.

50. Joseph Henry, *L'âme d'un peuple africain: Les Bamba, leur vie physique, éthique, sociale, religieuse* (Münster, 1910), p. 245.

51. Françoise Dumas-Champion, "Sacrifice et homicide en pays massa (Tchad)," *Systèmes de pensée en Afrique noire, le sacrifice* 4, no. 5 (1981): 175–93.

52. Milliet, "Manger du chien?"

53. Meyer Fortes and Sonia L. Fortes, "Food in the Domestic Economy of the Tallensi," *Africa* 9 (1936): 272.

54. Morice Vanoverbergh, "The Isneg Life Cycle," *Publications of the Catholic Anthropological Conferences* 3, nos. 2–3 (1936–38): 206, 111.

55. Raymond Verdier, "Malheurs de l'homme et mise à mort rituelle de l'animal domestique dans la société Kabyle," *Systèmes de pensée en Afrique noire, le sacrifice* 4, no. 5 (1981):164–65.

56. André Leroi-Gourhan, *Evolution et techniques,* vol. 1: *L'homme et la matière* (1971; 1st ed., Paris, 1943), p. 139.

57. Jacques Poux, *La traction canine* (Paris, 1945), p. 150.

58. *Des animaux et des hommes,* May 10, 1987–January 3, 1988, Musée d'ethnographie de Neuchâtel, in the French-speaking part of Switzerland. I contributed an article to the book that accompanied the exhibition: "Un allaitement insolite," in *Des animaux et des hommes,* ed. Jacques Hainard and Roland Kaehr (Neuchâtel, 1987), pp. 87–118. In addition, an anthropological survey was carried out among the visitors and published two years later as Jacqueline Milliet, "Une enquête sur les nouvelles sensibilités à l'égard des animaux domestique," in collaboration with Jean-Pierre Digard, *Anthropozoologica* 10 (1989): 23–30.

59. C. O. Sauer, *Agricultural Origins and Dispersals* (New York, 1952).

60. See Frederick Simoons and James A. Baldwin, "Breast-Feeding of Animals by Women: Its Socio-Cultural Context and Geographic Occurrence," *Anthropos* 77 (1982): 421–48; Brézillon, *Dictionnaire de la préhistoire*; Downs, "Domestication"; Richard, "Domestication"; Serpell, *In the Company of Animals*; F. E. Zeuner, "The History of the Domestication of Cattle," in Mourant and Zeuner, *Man and Cattle,* pp. 9–19.

61. George Peter Murdock, *Social Structure* (New York, 1949).

62. Lucie Wood Saunders, "Unseen Hands: Women's Farm Work in an Egyptian Village," *Anthropological Quarterly* 59, no. 3 (1986): 105–14.

63. Claude Lévi-Strauss, "Family," in *Man, Culture and Society,* ed. Harry L. Shapiro (1956; repr. London, 1971), pp. 347–48.

64. Nicole-Claude Mathieu, *L'anatomie politique: Catégorisations et idéologies de sexe* (Paris, 1991), p. 231.

65. *Anthropologie structurale deux* (Paris, 1973).

66. G. Gaard, ed., *Ecofeminism: Women, Animals, Nature* (Philadelphia, 1993).

67. Milliet, "Un allaitement insolite," "Enquête sur les nouvelles sensibilités," and "Le mythe scientifique de l'apprivoisement des animaux par les femmes," *Ricerca folklorica* (forthcoming).

Part Five

OF MICE AND MEN: MODELS FOR THE LABORATORY

12

The Multiple Meanings of Laboratory Animals

Standardizing Mice for American Cancer Research, 1910–1950

Karen A. Rader

Introduction: Why Mice?

When is a mouse equivalent to a human being? When is a so-called "common house mouse" an animal like any other member of its taxonomical class, genus, and species (Mammalia, *mus musculus*)? And when is it a creature that, by its very existence, redefines or challenges conventional zoological and social understanding? Such questions lie at the heart of debates about the validity and meaning of mouse models in both American biological research and American culture at large. Nowhere are they more hotly contested than in the continuing controversy over Walt Disney's premier rodent, Mickey Mouse. Stephen Jay Gould, for example, has argued that Disney artists have gradually neotenized Mickey to reflect the trend towards youthful features characteristic of human evolution.[1] Elizabeth Lawrence has claimed that Mickey embodies a much larger paradox, related to the mouse's use in scientific experiments: "[B]ecause his animality is camouflaged, Mickey also represents that uncertainty which we as human beings feel toward exploited species of animals. For we celebrate the innocence and charm of friendly and appealing mice . . . even as we almost unquestioningly exploit (and exterminate) millions of the same species for the benefit of our own kind."[2] Ultimately, Marc Holthof characterizes Mickey as a product of co-constructed social worlds: "'Disneyland exists to conceal the fact that America is Disneyland,' wrote Jean Baudrillard. "We might also add: Mickey Mouse exists to conceal the fact that Man is a mouse."[3]

Analogous disputes over mouse models also abound in scientific work, specifically with regard to the use of "transgenic" mice (mice with genomes containing human DNA), "knockout" mice (mice with a single gene, responsible for production of a known biochemical product, "knocked out"), and otherwise genetically controlled rodents. The series of transgenic mice designed to mimic features of human Alzheimer's disease and knockout mice models for studying parental nurturing and sexual aggression have raised ongoing questions about whether the biochemical and behavioral patterns displayed by such rodents are (in the words of one scientist) "comparable to what's going on in humans."[4] For the activists of the German organization *Kein Patent auf Leben* (No Patents on Life), legal contests over patenting these and other transgenics—especially Harvard's cancerous OncoMouse™—carry even more deeply profound moral consequences beyond the laboratory.[5]

What these seemingly varied debates have in common is that they compel scientific researchers and laypeople alike to come to terms with the boundaries between "humans" and "animals," and "natural" and "technological"—as well as with the proper relationships among the resulting fields. Just as importantly, they also highlight the ways in which such boundary making is neither a trivial nor an intangible activity. Donna Haraway has repeatedly insisted that the power of scientific animals lies primarily in their "cyborg" status: "mice and men of technoscience are simultaneously research models, cultural metaphors, and potent jokes. . . . We inhabit their narratives and they inhabit us."[6] This is the very power that Harvard-trained geneticist C. C. Little (1888–1971) self-consciously sought to wield when he wrote to his lawyer in 1954 about the possibility of his Jackson Laboratory—then the world's largest mass producer of mice for research—obtaining some publicity by "arousing Disney's interest in [a] film, to tell the story of our mouse (which might easily be a brother or some other relative of Mickey). . . . "[7] In the case of the mouse, then, the drama of its laboratory success has played out in terms of traffic across culturally constructed borders, but the historical actors themselves have staged it this way, and their stake in the outcome reflects local goals and material enterprises.

For at least the last three decades, *mus musculus* has ranked as one of the most—if not *the* most—widely used animal in contemporary American biological and medical science. By best estimates, the numbers are quite staggering. In 1965, a total of nearly thirty-seven million mice were consumed in U.S. laboratories, and by 1984, that figure had risen to an estimated forty-five million—or 63 percent of the total number of (counted) animals used by U.S. scientists.[8] In a draft of his 1965 testimony before Congress, Little himself raised a series of pointed questions about this transformation: "I have spoken of mice as the servant of man. Why is this true? What made its truth evident? In other words, why mice?"[9] Little was then nearing the end of a long career dedicated to "building a better mouse" for science, so there was perhaps no one for whom these questions held more personal urgency.[10] Still, taken as broader reflections, his queries frame a compelling problem: how and why did mice come to be standard animal players in the making of biological and medical knowledge?

Most scientists who work with the mouse respond to this question by citing the creature's inherent biological properties. As Patricia Lauber reported this explanation: "*Mus musculus* is a small animal. Its size makes it easy to handle, easy to house, and cheap to feed. The mouse breeds readily and often—several times a year. The young are born only three weeks after the female has mated, and they are born in good size litters. Like us, *mus musculus* is a mammal, and it happens to be highly susceptible to many of the diseases that afflict human beings." Similarly, in her January 1997 *New York Times* feature on model organisms, Gina Kolata wrote: "Scientists wanted something small, tame and highly breedable, something that would quickly reach reproductive age. Otherwise, they might have to wait decades to see generational results."[11] But such answers presume that the qualities which demonstrate the suitability of mice for science are universal and timeless. More specifically, they decontextualize these animals from the places and the circumstances under which they were developed into useful experimental tools, and thus render invisible the very nexus of politics and practices that defines what counts as their success.[12]

Alternatively, paying attention to the process through which mice were introduced into early twentieth-century biological and medical research laboratories and developed into model organisms[13] suggests a very different understanding—one in which the suitability of the mouse for research is the product of, rather than the cause of, its mobilization as a tool for solving experimental, as well as organizational and larger cultural problems. In particular, when one focuses on cancer research—one of the two prominent scientific fields of large-scale mouse use in America[14]—the standardized mouse strains and models that became laboratory fixtures in cancer studies from 1910 to 1950 emerge as creatures embedded with multiple, coexistent meanings. These were animals whose materialities and representations were deliberately engineered by Little and his Jackson Laboratory to simultaneously further the local goals of particular research communities and the social aims of those people and institutions that surrounded and supported laboratory work.[15] Only by articulating these developments, then, can we come to understand the historical significance and the contemporary power of standardized mice as both scientific and cultural artifacts.

CREATING MOUSE STANDARDS: SCIENTIFIC CONTROL, PROFESSIONAL COOPERATION, AND SOCIAL RELEVANCE

In 1963, the National Cancer Institute's Walter Heston lauded the development of inbred animal strains as "probably the greatest advance in all cancer research." Near the end of his life only a few years later, C. C. Little told a reporter that he had envisioned this outcome when he initially made the decision in 1909 to systematically inbreed mice for experimental use.[16] Doubtless by 1910—his senior year at Harvard—Little had developed a consuming (though not altogether explicable) interest in the scientific possibilities of the murine species. Friends confirm that Little's conversion—from amateur dog breeder and kennel club hobbyist to experienced mouse caretaker and mammalian genetics researcher in training—dated to his first genetics class, when instructor W. E. Castle skidded a live mouse at him across a lab benchtop and instructed him to learn everything about it. Also, while Little's introductory zoology notebook was filled with lecture notes on the genetics and breeding of mammals as varied as dogs, horses, cattle, and sheep, frequent mouse drawings and reflections on mice examples in the margins testify to his abiding intellectual passion.[17] Nevertheless, Little was not the first person to think of inbreeding mice as a strategy for eliminating their natural variability—this method had been loosely employed by mouse fanciers since the eighteenth century for perpetuating exotic rare specimens[18]—and the successful use of the inbred mouse as a standard tool for cancer research was hardly a foregone conclusion. In fact, in the period from 1910 to 1925, the various meanings this creature acquired in cancer studies emerged from Little's own still developing scientific interests, and his limited ability to align these with the goals of the larger cancer research community.

Little began his scientific career as a traditional geneticist, but within five years he had made a crossover to medical research—a shift sanctioned by a marked change in his programmatic interests. As a doctoral student (1910–14) of W. E. Castle at Harvard's Bussey Institution, Little chose to focus on developing a Mendelian understanding of mouse coat color inheritance. This entailed first obtaining well-established varieties of his chosen organism—primarily from mouse fanciers, who kept them for show purposes.[19] Methodologically, Little then brother-sister inbred them to insure their genetic homogeneity,[20] and performed cross-breeding experiments to demonstrate predictable similarities and differences in their heritable "factor" composition and transmission. In this experimental scenario, mice were defined as suitable in relative terms. They were not as cheap and easy to house, or as fast to generate genetic data from, as *Drosophila* fruit flies, the paradigmatic mutant "breeder reactors," but mice were the easiest and fastest means through which to obtain mammalian genetic data—information then considered key for understanding physiological effects on the expression of Mendelian factors. Other possibilities considered included cats, dogs, rabbits, guinea pigs, and even cattle, but mice were more immediately convenient than these animals precisely because their fanciers had preserved so many different coat color varieties.[21]

By the time Little was finishing up his doctoral work in 1914, he had become preoccupied with more complex genetic problems—specifically, the question of multifactorial inheritance or the idea (as he put it) that "certain characters of an organism depend for their visible manifestation . . . upon the simultaneous presence of more that one mendelizing factor." Little equated the solution of this problem with opening up an entire new range of phenomena for genetic study, and he subsequently published a theoretical paper that tried to convince his genetics colleagues of this point: even the more complicated patterns of inheritance shown in higher animals like mice could be explained and experimented on in the Mendelian framework.[22] Simultaneously, he encountered one such pattern in his own mouse colonies and in the medical research literature—the frequent occurrence of cancerous mammary tumors in the "dilute brown non-agouti" (*dba*) coat color stock.[23] Abbie Lathrop, a mouse fancier supplier, and medical researcher Leo Loeb were just publishing a co-authored series of articles claiming that certain "families" of mice were more prone to develop spontaneous cancers.[24] Similarly, Harvard Cancer Commission investigator E. E. Tyzzer had begun to explore the "inoculability [i.e. transplantability] of certain tumors in animals" as "an experimental basis . . . for the investigation of problems related to tumor growth." Tyzzer openly wondered "if susceptibility to an inoculatable tumor" was "transmitted in accordance with . . . Mendel's laws."[25] Such developments combined to distract Little from his initial scientific path. In 1913, for example, his dissertation on coat color research was published as a Carnegie monograph about the same time that he failed his Bussey doctoral exams. While waiting to finish his degree, Little thus took his mouse colonies from the Bussey to the Harvard Medical School, and obtained transitional employment as Tyzzer's research associate.[26]

Over the next four years, Little and Tyzzer moved to simultaneously redefine the phenomenon of cancer inheritance and the mouse materials used to study it. In contrast with the broad hereditarian terms like "family constitution" still popular among medically trained workers, Little explicitly conceived the etiology of cancer as a series of "problems of the factors causing, limiting, and directing cell division." Accordingly, susceptibility to a spontaneously arising cancer or to the growth of a transplanted tumor became a "trait" or character, and Little argued that the way to obtain knowledge of its heredity was to examine the Mendelian factors responsible for its expression in populations. The problem thus defined, inbred mice were the required standard, for inbred mice were the only cancer-bearing mammals that were "homogeneous" and fast-breeding enough to produce data amenable to Mendelian generalizations, and genetically controlled enough to overcome objections that the experimental system was burdened with irrelevant variables.[27] Little and Tyzzer thus initiated a series of crosses between Tyzzer's cancerous Japanese waltzing mice and Little's *dba*'s—two highly inbred strains—in order to determine whether tumors arising in the Japanese mice could be successfully transplanted to first (F1) and second (F2) generation hybrid mice. This work included some traditional clinically based research methods, such as taking pathological charts of individual mice and making sketches of the physical progress of their tumors over the course of an experiment. But Little and Tyzzer ultimately opted to express their findings in a Mendelian framework: though all of the F1 hybrids accepted the waltzer tumor, only 3 of the 182 F2 animals in their Japanese waltzer/*dba* crosses were susceptible. Breaking new ground, they argued that this was not a negative result because as the number of factors responsible for a Mendelian trait increased, so would the ratio of observed F2 animals which *do not* show the trait to those who *do.* Calculating backwards, then, they estimated that mouse tumor susceptibility was a Mendelian trait involving twelve to fifteen separate factors. Little and Tyzzer had effectively used the concept of the Mendelian gene to define both the problem of cancer as a heritable disease and the suitability of mice as research materials.[28]

Simultaneous with this work, Little independently began to articulate and promote the larger professional goal that he hoped such genetic studies of mouse cancers would achieve—namely, a state of enlightened cooperation between medical and genetic practitioners. In a 1918 letter to Harvard Medical School professor W. T. Bovie, Little declared that successful mouse genetics work on cancer and other "subjects of interest to medical men would lead to their conversion more rapidly" to the genetic point of view for other human ailments. As it stood now, he complained, "unless my work led to good advertising material for [the Cancer Commission], I should have to walk the plank."[29] Two years earlier, in a review essay on "The Relation of Heredity to Cancer in Man and Animals," Little conceded that the complexity of cancer genetics in mice likely meant that medico-genetic research work would be "of negative importance as a practical preventative in man." Nevertheless, he argued that it would be an "unfortunate" social consequence if neither medical researchers nor geneticists could recognize cancer as a problem "essentially biological in nature," and make it their priority to cooperate through research to solve it.[30] Little's ultimate goal, then, was to realign the disciplinary boundaries of the cancer research enterprise. His inbred mouse now embodied both a practical and a rhetorical symbol of the vision that genetics was the strong scientific basis on which medical investigation (if not cure or treatment) of this disease should be founded.

But however quintessentially Progressive this plan may appear in retrospect, Little's idea of a "cooperative rationality" (as he later called it) was not embraced as such by all of his contemporary colleagues.[31] Clinically oriented cancer researchers, eager to demonstrate the usefulness of increasing American investments in experimental work,[32] focused on the problem of extrapolation from the controlled mouse data to the uncontrollable human condition and concluded the genetic results were both unwieldy and inconclusive. A. H. Thaites, for example, wrote that early studies on spontaneous tumor mice (such as Loeb and Lathrop's) showed the "hereditary factor . . . plays an important part" in cancer, but he concluded that overall mouse genetic work provided "contradictory and unreliable" evidence. Similarly, Francis Carter Wood told the American Society for the Control of Cancer that experimental mouse studies showed "in a sense, cancer is hereditary," but "under ordinary conditions" of human mating (i.e. non-inbreeding) "we may safely assume that heredity plays a very small part, if any, in the production of cancer."[33]

Also, beginning in 1915, Little and his inbred mice faced an unrelenting methodological battle with University of Chicago researcher Maud Slye over how animal research should be used to determine the proper social and scientific meaning of human cancer. This former psychology teacher worked in relative isolation, like other woman medical researchers before her,[34] from 1911 to 1913 on a study of spontaneous tumor occurrence within and between mouse "families." Unlike Little, she framed her work entirely in traditional methods of clinical research—namely, she kept extensive pedigrees of individual mice, kept pathological charts for each descendant, and autopsied each individual dead mouse to assure that an accurate cause of death had been recorded. For Slye, these decisions were self-conscious: she rejected outright the complexities of "authoritative" genetic theory, and openly "deplored" the "artificial" nature of Little's inbred strains in which, she said, "variation becomes wholly a matter of the environment."[35] Despite these differences, however, Slye announced in 1914 that her results proved a simple Mendelian conclusion: cancer susceptibility was transmitted in mice as a single recessive factor. Because her use of mice more closely matched their intuitive and practical understanding of what it meant to study heredity in the clinic, physicians and many medical researchers lauded Slye's work.[36] Little reacted by attacking her credibility and reiterated his position that the complex behavior of cancer inheritance necessitated inbred mouse tools and a cooperative approach. Slye's non-inbred mouse material and lack of proper statistical, population-oriented experimental design, he said, indicated that she could never possibly answer *genetic* questions[37] and her work needed an "extremely important postscript" to correct the impression it made of an easy eugenic fix.[38] In terms of practical changes, Slye sought to use her mice to convince physicians that better record keeping for human cancers might eventually help them control cancer in their clinical practices.[39] Little sought to use his mice to achieve a comprehensive organizational reform that would empower geneticists alongside medical researchers in a new scientifically based cancer control movement.

The general clinical utility critique of mouse work, in combination with their increasingly public conflicts,[40] temporarily neutralized both Slye's and Little's effectiveness in defining the role of mice in cancer research. For Slye, this meant she was not able to win the AMA's sanction for her human cancer statistics bank.[41] For Little, it meant that by 1920, though roughly a quarter of the research articles in the *Journal of Cancer Research* were on mouse topics, and several dealt with tumor transplantation, none of these employed inbred mice or Little's transplantation genetics methods. By 1925, the situation remained roughly the same: the percentage of mouse articles was about 20 percent and inbred animals were used in only one study among this group.[42] In 1918, already expressing frustration that it no longer seemed possible for him to "prove the value of genetics to a medical school as well under existing circumstances,"[43] Little told Tyzzer he would not return to the Cancer Commission after completing his World War I service.

That same year, Little accepted a position at the Station for Experimental Evolution (SEE) in Cold Spring Harbor, New York—an institution that had developed a reputation for research that was both genetically rigorous and socially applicable. This place was, in many ways, the ideal place for him to retrench and consolidate his mouse research at the time.[44] While only part of Little's mouse work at Cold Spring Harbor directly addressed the cancer genetics problem, this part nevertheless proved especially significant for rearticulating the experimental and organizational meanings of inbred mice. Little first devised a new strain of spontaneous tumor mice—the *C57B*'s—based on a line of black and brown mice observed to show a less than 1 percent chance of developing mammary cancer,[45] and then used them to refine his earlier transplantable tumor genetics studies. In 1920, he suggested that even fewer genes (four to five) controlled susceptibility to Japanese waltzer tumor grafts. A year later, he also presented a paper at the Second International Eugenics Congress at the New York Museum of Natural History, in which he analyzed data on cancer incidence drawn from the human pedigrees of the Eugenics Record Office. Perhaps learning from his clashes with Slye, Little now cast inbred mice in a more conciliatory combination of medical and genetic discourses. He argued that human pedigrees showed that "inheritance in the direct sense of transmission is *not* involved," but that inbred mouse research revealed genetic "predisposition" to cancer to be "beyond question" and that "the case of man is clearly parallel" to that of the mouse.[46] This conciliation bore fruit in terms of two collaborative cancer projects while at Cold Spring Harbor, one with medical researcher Halsey J. Bagg, on X-ray induced heritable cancerous mutations in Little's *dba*'s,[47] and another with graduate student L. C. Strong, on the genetics of tumor transplants.[48] In turn, Little was subsequently promoted to assistant director of the SEE, and began to be recruited to speak on cancer inheritance in more popular and medical forums.[49] These SEE projects thus went a long way towards practically convincing at least some geneticists and medical researchers that if they worked together and used inbred mice, cancer inheritance could be experimentally understood.

About the same time, Little himself began to address a nonspecialist audience, advocating social reform based on an increased recognition of the "interrelatedness" of various branches of scientific work—especially genetics and medicine. In a 1922 *Scientific Monthly* essay, Little called for an alliance between experimental geneticists—who provided "immediate knowledge" of "sufficient accuracy"—with those better trained in the long-term observation of human material: teachers, YMCA directors, ministers, and especially doctors. He argued that if more precise observational data on human material could be obtained, experimental geneticists would naturally be more willing to study human problems and thereby render their "unfulfilled duty" to society. Furthermore, this practice—in conjunction with work on inbred mice—would benefit both science and society because genetic research would act as a rational check on hastily conceived eugenic advice.[50]

Such arguments, which originated in Little's inbred mouse research program, now attracted the attention of the reform-minded trustees and administrators at the University of Maine, who hired him as their president in March 1922. That summer he transferred his mouse strains to a new laboratory in the Science Building at Orono, where he continued to expand both on the experimental possibilities of inbred mouse cancer work and on its ultimate medical and social value. Here a slightly increased research budget enabled him to support a few official students of his own.[51] By 1923, one of these, Beatrice Johnson, had begun to extend the transplantation studies to normal tissues (specifically, spleen), and in 1924, Little and Strong finally completed a study that explored chromosomal linkage (a classical genetics problem) of factors responsible for certain mouse cancers and mouse coat colors.[52] While this research itself practically stabilized the experimental meaning of inbred mice, Little also worked to stabilize their social meaning by reiterating the general value of a research-centered approach to the cancer problem. For example, in a 1923 review on "Congenital and Acquired Predispositions and Heredity" done for *Abt's Pediatrics,* Little suggested that since the "genetic constitution" problem was clearly of interest to medical men, geneticists would need the M.D.s' support to convince the American public that there was a truly scientific means of studying it—namely, through inbred animal research.[53]

By the time Little left the University of Maine in 1925—this time, to take on the presidency of the University of Michigan—the inbred mouse had come a long way from its humble beginnings as a hobbyist's pet. Little had co-constructed the problem of disciplining experimental studies of cancer heredity and inbred mice as the standard tools for such research. Meanwhile, he also promoted such animals as solutions to the American biological research enterprises' worst social and organizational difficulties: lack of coordination between medical and genetic workers and lack of relevance to human problems. But in the mid-1920s Little's professional interests in research were still only tenuously aligned with those of more clinically inclined cancer workers and those of the American public. Finally, and perhaps most importantly, there was no well-established infrastructure through which mouse material of the standard Little was advocating could be easily obtained by the cancer research community. Short of raising their own inbred mice—which meant huge investments of time and money—or obtaining creatures directly from Little—who didn't have a large enough supply—such work literally could not be realized. These problems would form the basis of his subsequent efforts to institutionalize his scientific organism of choice.

INSTITUTIONALIZING THE ORGANISM: STABILIZATION, STANDARDIZATION, AND A "NEW DEAL" FOR MICE

When Little spoke of the founding of the Jackson Lab in 1929 as "the fulfillment of a lifelong ambition,"[54] he was not merely posturing for the press. Indeed Little's professional activities in the five years from 1925 to 1929—from Ann Arbor to Bar Harbor—reflect a significant evolution in his commitments to the social, material, and cultural meanings of the inbred mouse. When given the opportunity at Michigan to further expand his research, he initially directed his efforts at fixing narrowly defined infrastructure problems of mouse breeding and student support. But Little and his inbred mice were more successful when he promoted both the animals themselves as valuable research tools and the cancer genetics research they made possible, as the embodiments of a reform-oriented alliance between physicians, medical researchers, geneticists, and public health advocates. Unlike in previous scenarios, here the success of inbred mice dovetailed with the promotion of the emerging social concerns and industrial sensibilities of several upper-class Michigan industrialists. Depression-induced financial pressures soon prompted Little to literally transform inbred mice into multiple-user commodities—market-driven entities which were suitable for addressing a range of organizational and experimental concerns among biological researchers and policymakers. Still, by 1935, Little's articulation of inbred mice had come full circle. His definition of the mouse as suitable for conquering disease and doing good laboratory genetics was what drove the founding of the Jackson Lab. In turn, the Jackson Lab now provided an institutional infrastructure for perpetuating and distributing the mice thus defined. Such an arrangement represented, as Little would put it, a "New Deal" for mice, in that it created stabilizing multiple constituencies for the scientific knowledge obtained from inbred mouse research and for the experimental animals this work could provide.

In 1925, Little began his term at Ann Arbor by demanding a significant increase in infrastructure support for his mouse work. By an unprecedented concession, he was granted the five thousand dollars he requested for the maintenance of what was simply to be called "the President's Laboratory," and the Michigan trustees lauded their new leader as a "man who knows the value of scientific research through personal experience."[55] But in fact, Little directed this initial influx of funds towards stabilizing living conditions for his experimental animals—now an especially important goal in light of his public efforts to convince medical research workers of the inbred mouse's suitability for cancer studies. The *dba* strain, for example, barely survived a mouse typhoid epidemic at Cold Spring Harbor in the early 1920s, and meager space and caretaking arrangements at Maine had further weakened their fecundity.[56] Thus in Little's first full year at Ann Arbor, nearly three thousand dollars—an amount equal to what he approved for the total Michigan faculty research fund—went for the salaries of two full-time caretakers, mouse food (canary and hemp seeds), cotton batting, wood shavings, and other miscellaneous laboratory equipment such as cages.[57] Remaining funds went to pay the salaries of graduate students Joseph and William Murray. The Murray brothers had been in Little's Maine laboratory for several years, and they already had valuable hands-on knowledge of his desired work arrangements. The presence of more students increased the lab's research output somewhat. A few more mouse tumor transplant studies began to emerge from Michigan in 1926–27, as did a few anatomical and histological studies.[58] But just as importantly from Little's perspective, the larger number of students provided the added security of strict colony supervision while his administrative duties made daily visits to the mouse room impossible.[59] The survival of individual strains was thus linked primarily to individual researchers and their projects. Bill Murray, for example, looked after the *dba* stock (in which he studied spontaneous tumor rates and X-ray–induced developmental anomalies) and Joe, the *C57B*s (in which he studied the histological damage wrought by X-rays to females' ovaries).[60] Together, all of these developments combined to stabilize the actual production of animals as well as the meaning of inbred mice as controlled, reproducible commodities within the Little lab's own economy of research.

Yet Little would soon learn that his concern with animal material stabilization was not unique. In fact, so many biological researchers now complained of the problem of inadequate plant and animal supplies that in the mid-1920s, the National Research Council convened a Committee on Experimental Animals and Plants (CEAP) and charged this group with making suggestions about how to improve research material production and distribution. Perhaps because of his reputation as a "mouse advocate," Little was called to join CEAP; the remaining members were naturalists, geneticists, and some medically affiliated researchers. L. C. Dunn, mouse geneticist and CEAP chair, initially conceived of the supply problem as one of poor planning and inefficient organization, tied to the imperatives of a scientific career system that rewarded results but not infrastructure building: "[The] plants and animals which have been used in genetic or other experiments . . . in the absence of special provision usually are discarded by the investigator or the laboratory which used them. Some laboratories have taken upon themselves the preservation of stocks. . . . But this takes practically the whole time of one assistant. . . . [F]or animal material in general there has been no attempt to create a service or preserve a supply." Dunn suggested solving these predicaments by imitating bacteriologists, who had created the American Type Culture collection as a centralized repository for their research material.[61] Others who subsequently wrote to the committee acknowledged the convenience of such centralized sources, but many objected that the centralization solution was predicated on the assumption that biologists could agree beforehand on how to define a necessarily finite group of "strains" or "stocks" as standard—apart from an individual scientist's projects. As Wisconsin's L. J. Cole put it: "It is somewhat difficult to say just what one means by 'stocks'. . . . We cannot, of course, expect that everything we should like to use should be available any more than can the chemist. When he wants some special reagent, he often has to make it up. . . ." Alternative solutions posed thus ranged from forming cooperative arrangements between local labs (a suggestion of which Little approved) to the NRC providing financial compensation to those researchers willing to act as animal or plant "mutant custodians."[62] Ultimately, what was at stake in these discussions was not simply the establishment of a materials supply network within the existing practice of research but also the proper organization of the entire American biomedical research enterprise. And this was the problem to which Little had already begun to construct inbred mice as a solution.

Still, though such issues were potent for biologists like Little, they remained esoteric for clinically oriented medical workers and members of the nonscientific public interested in scientific endeavors. Among these groups of historical actors, sympathy toward the idea of research was growing, but only in so far as such research could be demonstrated to address pressing social problems like the elimination of diseases.[63] Cancer, in particular, had begun to gain media momentum as one such public health problem to which research might provide a long-term solution. In 1900, cancer was ranked as the ninth leading cause of death in the United States, but by the 1930s government statistics ranked it second only to heart disease. Critics disputed the interpretation of these statistics, but publicly agreed: cancer was a threat to the lives of many Americans, as well as a threat to the American way of life. West Virginia Senator Matthew Neely, for example, proclaimed in 1928 that cancer was "a monster more insatiable than the guillotine," which had "inflicted more suffering and agony upon the American people than all the other diseases known to humanity." Magazines and newspapers from the period further underscored that cancer had captured many Americans' imaginations as a "dread disease." Beginning in the late 1920s, the "conquest of cancer" challenge, constructed primarily by the American medical profession in response to public fears, had attracted significant interest among wealthy upper-class citizens and philanthropists, like the Rockefellers, looking for noble and popular causes to support. This movement, unlike previous efforts to address the cancer problem, had a research component.[64]

Little's Michigan lab, though continuing to invest its financial resources primarily in the problems of mouse material stabilization, was not immune to such broader cultural forces. In March 1926, Roscoe B. Jackson, the president of Detroit's Hudson Motor Car Company and Little's friend from summers in Maine, asked Little to further apprise him of possible cancer research plans. This was the inbred mouse work, Jackson reminded him, "which you mentioned when you were last in Detroit of offering talks to the community of popular interest." Jackson also noted that if this work could be linked more directly to the problem of curing cancer, then "it would make it easier to present the case [for funding] to some of my friends."[65] This connection clearly already existed in Little's own writings and experimental projects. In a 1926 *Commonwealth Review* article, Little once again called for an urgent program to study "the general principles of research on the cancer problem." He noted that since cancer was, in humans, a disease of old age, such work could only be done with "some mammalian material which ages more rapidly than does man." Only knowledge such as that obtained from controlled laboratory rodents, he argued, would be precise enough and arrive quickly enough to address the problems of cancer's true causes. Meanwhile, as his many invited speeches from this period testify, Little's recognition and acceptance in the medical community for his cancer work and genetic ideas now matched his media profile as the country's youngest college president.[66] Cancer was a powerful cultural resource of which Little was already an adept mobilizer.

Jackson's direct prompting and increased financial incentive, however, led Little to propose a special inbred mouse research laboratory as a valuable organizational and institutional contribution to the applied cancer effort. By April 1926, Jackson convinced his brother-in-law Richard Webber, the head of Michigan's J. L. Hudson Department Stores, and Edsel Ford, son of automaker Henry Ford and head of Ford Motorworks, that funding Little's work on cancer would be a worthwhile corporate investment towards this same end. The three industrialists subsequently crafted a proposal for "the creation and financing . . . of a laboratory for scientific research and investigation of the biological phenomena of cancer . . . ," with an ultimate "end in view" of "discovering and promoting scientific knowledge of the causes, prevention and cure of cancer."[67] Michigan's Regents accepted the accompanying forty-five-thousand-dollar endowment for the next year, when they inaugurated the University's "Laboratory of Mammalian Genetics, in Co-Operation with the Cancer Research Fund."[68]

In a letter shortly thereafter to John Kellogg, director of the Battle Creek Sanitarium, Little described his new laboratory's inbred mouse work as "concerning . . . the principles of human biology and eugenics." Yet echoing his previous arguments, he cautioned Kellogg that such applications would only be possible to the extent that geneticists and physicians could learn to cooperate through research: "we [at Michigan] are making a real effort, although an admittedly inadequate one, to place a sound foundation of biology under the training of many people who now exercise a great influence in shaping the training of humans. . . ." Furthermore, Little also mobilized the inbred mouse's ability to function as a mediator—between the detailed, but useless genetic knowledge of simple *Drosophila* flies and the inaccurate, but useful genetic knowledge of complex humans—as a means of demonstrating the animal's dual social and scientific suitability. For example, in a review essay called "Opportunities for Research in Mammalian Genetics," Little argued:

> It is true that there are chromosomes in insects and in mammals and that there is evidence for Mendelian inheritance in both forms. To draw a close homology, however, between the detailed genetics of man and of insects would be exceeding our present knowledge. There is a real gap between the two forms. There is, however, already in position the foundation of the main support of the bridge over [this] gap: [this is] the steadily-growing mass of information . . . in the genetics of laboratory mammals. . . . [To] help bridge that gap . . . the problems to be investigated must be of importance to man. . . . Fortunately, the great majority of problems . . . capable of study in the laboratory mammals are of just that sort. . . .

Thus for Little, inbred mice now also embodied a kind of new eugenic vision—one in which the application of laboratory genetic knowledge was sanctioned by medically desirable goals (e.g. curing cancer) and supervised by practicing research geneticists. Though there were severe fractures in the U.S organized eugenics movement in the late 1920s, enough active researchers agreed with Little's vision to subsequently elect him to a one-year term as president of the American Eugenics Society.[69]

Little invested the Detroit windfall in his existing operation, though as his Michigan presidency began to fall apart, he began searching for new ways to expand this agenda for biomedicine. The bulk of the initial money went towards assistants, and Little anticipated increases in this number—corresponding to increased mouse production—during the next academic year.[70] Research continued on a larger scale, too, although the lab's university location eventually imposed space limits on this growth. Strong, whom Little was finally able to hire as a postdoc, expanded the series of *dba* factor-ratio transplant experiments that he had begun at the SEE, and in 1926, he published a case study claiming one-factor mammalian tumor susceptibility inheritance.[71] Then in 1928, after two years of bitter clashes with the Michigan trustees over campus and curricular reforms, Little resigned his administrative office and immediately shifted to making alternative institutional plans for the mouse research.

He met with Jackson to explain what he now saw to be the most promising possibility: "to establish an independent center of biological research on cancer and other constitutional diseases as a continuation of the work which I had begun twenty or more years before."[72] Earlier he had summarized his social and scientific vision of cancer research on inbred mice in familiar terms: "The complicated nature of cancer, its long and successful history of menace to the human race, and its social importance, together with the need of increasing the confidence of the medical profession in genetic methods of analysis, justify us in deciding definitely at this time that . . . continued study with more careful genetic analysis than [is needed]"[73] But as Little later recalled it, he was now asking his Detroit patrons to take a risk, grounded as much in their social conscience as in their belief in the scientific value of the lab's ideally suited research materials: "Jackson . . . , being a good businessman, saw . . . that they added efficiency, accuracy and repeatability to the biological work." [74] This well-timed and powerful articulation of mice, both as humanity's chosen disease fighters and as efficient, controlled systems for undertaking this task, proved uniquely effective for propelling the animals to a new level of institutional support. Jackson, Webber, and Ford together pledged more than seventy thousand dollars a year over three years, in order to transform Little's mice colonies and his Michigan cancer researchers into a freestanding cancer research unit.[75]

What this development subsequently entailed for the successful scientific use and social mobilization of inbred mice, however, stemmed largely from a series of unexpected contingencies. First, while Little was negotiating with George Dorr, another Maine summer colony resident, for land on which to build his lab, he learned that Roscoe Jackson had died suddenly of influenza while vacationing in southern France.[76] Jackson's widow, as well as Edsel Ford and the Webbers, confirmed their standing financial commitments to Little and his project, but now requested that the resulting facility be named in Jackson's honor.[77] On October 7, 1929, the *New York Times* thus glowingly proclaimed the beginning of scientific work at the "Roscoe B. Jackson Memorial Laboratory of Cancer Work" in Bar Harbor, Maine. This article further noted that soon "the structure will house 25,000 mice on which cancer experiments will be made" in "search for a possible cure."[78] Here members of Little's Michigan lab set up a new series of transplantation studies designed to use tumor susceptibility for understanding more simplistic genetic problems. Strong, for example, set out to determine if one of the genes involved in mammary tumor susceptibility was sex-linked, and John Bittner, a recent Michigan Ph.D., published work on the genetic connections between mammary tumors and mouse coat color.[79] Around the same time, Little was given an unexpected boost in the medical research and policy community when he was simultaneously elected president of the American Association of Cancer Research and managing director of the American Society for the Control of Cancer (ASCC). For the record, ASCC leadership noted that Little had been chosen for his national reputation, as well as for his expertise on the experimental study of cancer—a quality that better resonated with what they saw to be a broadening public consensus in support of medical research more generally.[80] Then, on October 27, 1929—two weeks before the Jackson Laboratory officially opened its doors to the public—the U.S. stock market crashed and the nation entered into the most severe economic crisis in its history. Just as Little and his Jackson Lab group appeared to be at the height of their powers for convincing both scientists and society at large of the multiple values of inbred mice, financial disaster—in the form of the Detroit sponsors near-total withdrawal—threatened to destroy their fledgling enterprise.

In response to these circumstances, Little unsuccessfully tried to salvage his institution based on his arguments about the suitability of inbred mouse genetics research for encouraging a cooperative approach to the cancer problem. In his 1931 AACR presidential address, he resurrected the rhetoric of his early career and praised the "growing interest of medicine in genetic research" as well as "the continued growth of that cooperative spirit." Still, he emphasized the field of tumor transplantation in "genetically-controlled" mice as the key to the cancer problem. Previous transplantation data, Little argued, were full of "statistical artefacts due to the mixed genetic nature of the mice used." Use of inbred mice, he concluded, would serve to accomplish more quickly and carefully both geneticists' and physicians' goals.[81] Along these same lines, Little also told an executive of the Fuller Foundation in early 1933 that the primary scientific value of the Jackson Lab was its interdisciplinary cancer and genetics research: "The group at the Laboratory is one which has worked together for years. It has been built up with a specific program and purpose in mind."[82] "Of course," Little confided to Richard Webber that same year, "the same general conditions that are shaking the foundations of all industry will make it difficult for us to do as well as possible in normal times."[83]

Ironically, however, it was precisely Little's adoption of an industrial strategy that saved the Jackson Lab, and in the process, significantly affected the scientific meaning of the mouse materials his institution produced. At a plenary meeting of the Sixth International Congress of Genetics in August 1932, Little proposed a formal resolution that "In all experimental work involving investigations of a pathological . . . or of a physiological or psychological nature with animals or plants, the importance of genetically pure strains be emphasized."[84] Though this motion was ultimately tabled, in 1933 Little appealed directly to members of the genetics and medical research communities. Now his argument no longer relied on the Jackson Lab's research but on its inbred mouse materials as an efficient solution to the research material supply problem. Little and his group first began to focus on creating and stabilizing a wider variety of spontaneous-tumor inbred mice. Though the practice behind this enterprise was the same (i.e. inbreeding) its immediate goal was significantly different: not the production of mice strains with stable, homozygous genes (as it was for their own transplantation studies) but the production of mice with *predictable* tumor incidence.[85] Thus, alongside strains *dba* and *C57Black,* Jackson Lab workers developed more specialized low and high mammary cancer strains like *CBA* (13 percent) and *C3H* (90 percent). They also began inbreeding to stabilize strains with predictable spontaneous lung tumors: strains *A* (80–85 percent) and *X* (8 percent).[86] Then, in mid-March, he drafted a primitive catalog of inbred mouse strains, which created universal specifications for each of the ten "stocks" available at JAX (a new abbreviation taken from the lab's cable address). These specifications had multiple origins. Primarily, they were a combination of the Jackson Lab researchers' own configurations of the useful features of inbred mice, and initial feedback obtained from the U.S. Public Health Service, which had bought a small but significant number of mice from Little's lab in 1932. But Little also drew on the more general lessons of the NRC's CEAP Committee. For example, the list noted each stock's known gene makeup, including the number of inbred generations as a sign of homogeneity; percentage of tumor incidence; and a standardized name. All mice were priced equally: ten cents each plus the cost of delivery, with additional charges for specific sexes. And in the true spirit of American consumer culture, researchers were offered a money-back guarantee in the rare event that mice might arrive dead. The target market: "heads of all medical schools and various biological departments in schools and universities."[87]

This inbred mouse mass production and sales venture represented an institutional success for JAX. Cash orders for these animals quickly came in from medical researchers—generating enough money to obtain food and bedding materials for more mouse production, and nearly exceeding available supply.[88] But more importantly, when Little made the decision to sell the Jackson Lab's inbred mice to outside experimenters, the standards for successful scientific mouse production and use in research changed. Inbred mouse suitability became less a matter of these animals' place in the economy of a particular research program and more a matter of their place in the economy of an institution. In turn, the Jackson Lab itself was now defined by its ability to meet a wide variety of user needs in different disciplinary and institutional contexts.

This transformation was not lost on Little, who struggled to retain support for inbred mouse cancer research at JAX , only to find that scaled-up mouse production had created a new research momentum of its own. In 1933, for example, "The Staff of the Jackson Laboratory" published a landmark paper detailing the results of a series of reciprocal crosses between two high and two low inbred mammary tumor strains. These experiments suggested maternally influenced inheritance: that is, the females of the four F1 mouse hybrid types derived from high cancer females were high cancer, while those derived from high cancer males were not—despite their presumably identical genetic constitution. The JAX workers concluded that they had established "the transmission of extra-chromosomal influence [although] this should not be taken as a denial of the existence of chromosomal influence," [89] and began to plan a major research program exploring the *relative* importance of chromosomal and nonchromosomal factors. Bittner subsequently studied the impact of foster nursing on tumor development of the F1 strains: he found that female progeny of the high cancer strains could be transformed into low-tumor strains by nursing them with low-tumor mothers, an effect he labeled "milk influence."[90] But this work carried almost no weight in Little's last-ditch negotiations for a grant from the Rockefeller Foundation's Medical Sciences division, which was more interested in the other, more traditional mammalian genetics questions that JAX's new production arrangement promised to solve. RF program officer Frank Blair Hanson visited JAX Lab, and took note of the factory-like quality of its mouse breeding, stressing its value for producing more mutants for linkage and mapping studies like those with which Little began his career. "The whole arrangement," Medical Sciences head Alan Gregg later summarized, "is . . . probably the most complete plant in America for mammalian genetics." In turn, Little's pitch to Gregg underscored this aspect of the lab's work, by comparing the carefully recorded "pedigreed" mouse material at JAX to Morgan's *Drosophila* stocks and noting (correctly but unrepresentatively) that some JAX workers' published papers had "been on genetics purely, without any reference to cancer research." Little also described the Lab building as "three and one half stories of factory construction especially designed for the maximum utilization of living materials." That December, Rockefeller agreed to appropriate eleven thousand dollars in natural sciences emergency funds to Little's group "for research in mammalian genetics." And again, this money went first for buying enough supplies (food, bedding, and cages) to sustain the ever-growing inbred mouse colonies, and second for supplementing researchers' salaries.[91]

One result of these developments in the cancer research community was that while inbred mouse use did increase (in terms of sheer numbers in circulation), such increases were not in practice tightly linked either to JAX or to Little's cancer genetics research program. In 1930, for example, half of the total mouse articles in the *Journal of Cancer Research* utilized inbred strains for research directed at genetics problems—but virtually all of these articles were published by JAX researchers using their own mouse materials. By 1940, significantly more than half of the total mouse articles in this same journal[92] utilized inbred mouse strains—of which 65 percent originated with Little or at JAX—but only one-quarter of these articles dealt with the genetics of cancer. Thus, according to this crude index, JAX research was now less well represented in laboratory cancer studies than were JAX research materials. In contrast—and according to the Rockefeller Foundation's prediction—JAX's mouse production enterprise did seem to be increasing the amount of raw mutant material, which led to more mammalian genetics linkage and gene mapping research. Between 1932 and 1939, for example, JAX distributed a total of approximately 170,000 inbred mice to various research institutions, during which time twenty-seven new mouse mutations were reported in various scientific journals. This represented an increase in mutant reports of more than 40 percent over the 1920s, and most of these mutants were observed in mouse stocks that originated at Jackson Lab.[93] Such effects were, in turn, reinforced by the attrition of several of JAX's initial cancer research workers, including Strong (who left in 1933) and Joe Murray (who left in 1934),[94] and the subsequent hiring of mouse geneticist George Snell, a Bussey alumnus who had just completed postdoctoral work with H. J. Muller at the University of Texas.[95] In short, by the late 1930s, JAX's identity as a cancer research laboratory had been usurped by its identity as a mouse materials producer. At the same time, the inbred mouse's meaning as a tool for curing cancer was usurped by its meaning as a controlled, predictable commodity for use in a range of biological and medical research projects—including (but not limited to) mammalian genetics.

In 1935, in a move that simultaneously recovered some of his earlier programmatic vision for inbred mice and heralded the changes JAX's new industrial production of these creatures had wrought, Little penned a *Scientific American* article that he called "A New Deal for Mice." Here Little juxtaposed the changes the mouse had undergone in science with prevailing negative cultural images of the animal: "Do you like mice? Of course you don't. 'Useless vermin,' 'disgusting little beasts,' or something worse is what you are likely to think as you physically or mentally climb a chair." Against this background, he cast himself as "attorney for the defense," and argued that through science, mice have been positively transformed. Inbred mice—as opposed to their "not very convenient" wild mice relatives—

"provided a particular service" both to science and to humanity. Little implored his readers to visit one of JAX's "mouse laboratory 'cities' with its cleanliness, orderly arrangement, and activity"—all of which would testify that "thoroughbred" mice (though "amusing" to some) had become "an integral part of man's helpers." Inbred mice, he wrote, were "the troops which literally by the tens of thousands occupy posts on the firing line of investigation" into the "nature and cure of cancer." Also, in contrast with those medical researchers who wrote in defense of dogs, Little included photographs of inbred mice undergoing laboratory procedures. This rhetoric and these images combined to express what had previously been a tacit moral stance regarding the suitability of mice for research: mouse manipulation (in the form of the inbred strain creation and production at JAX) and mouse suffering (in the form of such experiments as tumor transplants) was a small price to pay for future human well-being. "Scientizing" mice was the only way to redeem these otherwise socially useless organisms. "Under these circumstances," Little concluded solemnly, "perhaps mankind will accept and develop his relationships with mice in a different light." [96]

Coming at the end of a period of dramatic change, such a message eloquently captured the many meanings that had defined the inbred mouse's successes during the previous decade, and it foreshadowed developments that lay ahead. Inbred mice were no longer merely narrow solutions to the problems of Little's cancer genetics research group, or potential solutions to the organizational problems of research coordination and experimental material supply. By virtue of the founding of the Jackson Lab and the development of its inbred mouse sales and production enterprise, inbred mice had become standardized social and scientific commodities. Yet this standardization, while institutionally embodied, was achieved not through a consolidation of these animals' previous meanings but through a multiplication. Little's literal circulation of inbred mice among laboratories and his rhetorical circulation of their broader social utility combined to define them as suitable experimental animals. This combination was so potent precisely because it allowed for a stable, but not rigid, alignment of those interested in knowledge obtained from mice and those interested in the mouse materials themselves. This "New Deal" for mice, like the Rooseveltian plan that had come before it, presupposed some points of tacit agreement among the various parties involved in its execution—in this case, the need for some level of control, coordination, and timeliness in experimental research and the moral correctness of animal use in science. But these, too, would soon be mobilized as resources over the next ten years by those who sought to fulfill one of the U.S. government's New Deal promises: increased federal involvement in national health issues. By virtue of changes in the political climate around 1937, inbred mice would embody exactly the right combination of experimental, cultural, and organizational meanings for achieving this goal.

Constructing Policies: The NCI Act, Animal Supply, and Meanings of Mouse Research through 1950

In retrospect, 1937 clearly marked a turning point for the American political fortunes of cancer. Anticancer sentiments reached fever pitch both in the media and on the floor of the U.S. Congress, and earlier deadlocks between the executive and legislative branches gave way to a near unanimous agreement among federal officials that cancer was a problem requiring immediate government attention. Viewed in context, however, this tide of events—which culminated in the passage of the first National Cancer Institute Act—also uncovered a persistent policy anxiety: How exactly should the U.S. government go about constructing a program to fight this chronic disease?[97] Little went before Congress in 1937 to offer inbred mice as one powerful solution. Against the predominant Congressional rhetoric of moral outrage over cancer's human death toll, he juxtaposed an image of controlled animals in the laboratory: "Let me point out that research in the cause of cancer is not entirely a medical problem in any sense of the word. . . . The emphasis is entirely shifted from working with the slow, unsatisfactory human material to the material that is easy to handle, rapid breeding and conveniently controllable. That has been the biggest change in cancer research."[98] But this was not an isolated event. In fact, by underscoring to both scientific and public audiences their usefulness in research and their moral status as stand-ins for humans, Little systematically mobilized inbred mice in a well-timed effort to help pass the NCI Act. In turn, the inbred mouse's suitability for experimental cancer studies became instantiated through the construction of an initial NCI policy that promoted its use for ensuring coordination and efficiency in research. This policy stabilized inbred mouse use in NCI intramural cancer research over the next ten years, and it justified the expansion of production infrastructure at JAX in order to meet these material needs. But by 1950, the suitability of mice as cancer research organisms was not entirely defined by this industrial arrangement. Instead, it was located in a multiplicity of experimental and organizational practices.

Little's promotions of inbred mice in his official capacity as JAX director in the years leading up to the 1937 NCI Act drew on his existing repertoire of familiar motifs, but with greater intensity and increasing political sensitivity. To medically trained audiences, Little once again emphasized that only a cooperative research program between experimental genetics and clinical medicine could ultimately find a cure for cancer, and inbred mice would provide the material link for such a partnership. In a 1936 presentation before the Mayo Clinic staff, for example, Little remarked that although "most medical men . . . are skeptical concerning its actual practical value . . . experimental genetics is one of the most promising of all the lines of cancer research in the laboratory. The reason for this is to be found in the remarkable extent to which such rodents as mice are now adapted for work of this sort."[99] Six months later, in the *Journal of the American Medical Association,* Little would expand on this point by noting that "the technic of keeping genetic constitution constant" in a research material—a control embodied by the JAX strains—was far superior experimentally to "the accepted method of making human matings, viz. by uncontrolled outcrossing combined with inadequate records and small numbers of progeny . . . in human families."[100] In short, he told a 1937 medical congress, the usefulness of inbred mice can be analogized to the usefulness of pure-grade chemicals:

> The chemist would be helpless in his attempt to analyze unknown material if he did not have known chemical agents on the shelves of his laboratory to which he could turn for assistance in analyzing the unknown. The biologist, in the past fifteen years, has been given the tools by which he could approximate [his unknowns]. . . . In experimental medicine today, . . . the use of in-bred genetic material . . . is just as necessary as the use of aseptic and anti-septic precautions in surgery.[101]

Thus defined, inbred mice were the methodologically responsible tools for investigating a socially responsible research problem—the causes of human cancers.

Simultaneously, however, Little's arguments took an increasingly public turn, as he led the ASCC into the media fray surrounding efforts to push a compromise bill through Congress. In these cultural forums, the usefulness of inbred mice was recast into the more dramatic and popular rhetoric of warfare. Cancer research was, as Little put it in a *Time* cover story, "a hard task requiring patience: trench warfare against a ruthless killer."[102] A few weeks later, 274 of JAX's spontaneous cancer mice appeared on the cover of *Life* magazine, introducing readers to the story "U.S. Science Wars Against Unknown Enemy: Cancer." Tapping into existing cultural fears of scientific cancer treatments, the "biggest gun" in the clinical treatment of cancer was presented first: Crocker Laboratory's one-and-a-quarter-million-volt X-ray machine. Readers were cautioned that X-ray machines such as the one pictured were only in "experimental" use so far, and warned that, in clinical terms: "Only the Surgeon's Knife or Radiation Can Cure Cancer." By comparison with radium and its associated heavy machinery, mice emerged as far more humane and more effective weapons for the cancer war: they first appeared on a page headed "Mice Replace Men on the Cancer Battlefield." This animal, wrote the author, is "an ideal laboratory for the propagation and study of cancer," and that is why tens of thousands of them a year must "lose their lives on the battlefields of cancer research."[103] The only explicit connection of Little to this piece was a brief mention of the Jackson Lab's "valuable discoveries" about cancer through mouse research. But similar stories involving some combination of Little, inbred mice, and cancer research appeared (often with some of the same pictures) in several Henry Luce–owned magazines, such as *Newsweek* and *Fortune,* at around the same time.[104] Ultimately, the ASCC media campaign articulated the meaning of inbred mice as acceptable animal stand-ins for humans in the cancer laboratory, and in the process, generated broader social support for the NCI act.

Immediately after the NCI bill passed, Little himself resurfaced at a joint hearing on cancer research. Shedding all such moralistic rhetoric, Little drew again on the image of industrially produced pure-grade chemicals[105] to illustrate the organizational advantages these animals promised the cancer research enterprise. He told his audience of congressmen and Public Health Service policymakers:

> [T]he first thing you should do . . . in spending your money would be to insure a constant and adequate supply of controlled, known animal material on which investigations could be carried out. . . . we can produce as nearly a chemically pure animal . . . as it is possible to produce. . . . during this past year the little laboratory where I work has sent out over 65,000 such animals . . . for research in cancer and in other experimental medicine.

Inbred mice, Little was arguing, should be the practical cornerstones of a coordinated basic research effort: the biological equivalents of standardized, interchangeable parts in a well-oiled machine of biomedical research production.[106] More self-interestedly, he was also suggesting that JAX could provide the infrastructure to make this possible, though he was notably not lobbying that genetic research on inbred mice also be incorporated into this scheme.

This argument resonated with other NCI policymakers, whose subsequent organizational decisions did bear out Little's vision. The NCI act initially authorized seven hundred thousand dollars annually in federal funds to support cancer projects, and it stipulated that Surgeon General Thomas Parran assemble a group of "leading scientific and medical men" to determine how this money should be spent.[107] This National Advisory Cancer Council (NACC), of which Little was a two-year charter member, decided in its initial round of 1937–38 meetings that coordination of federally sponsored research should be a top NCI priority. As NACC member James Conant saw it, this was precisely what should distinguish fundable projects from unfundable ones: "is this work that could not be done except by government money, by virtue of its being mobile and centralized?" And committee members agreed that past experience showed well-coordinated research to hold the greatest potential for applied human cancer treatment or cure.[108]

What the NACC ultimately decided coordination should mean, however, rested not on a theoretical discussion of what kinds of research should be done but on a practical determination about how such research should be managed. At the center of their discussion was the problem of animal supply—specifically, ensuring that all NCI-sponsored researchers would be working with "controlled" and "known" materials. Little submitted a proposal that requested federal support for the "stabilizing" work that JAX had already done with inbred mice—"the animals that are generally used . . . in the American cancer field" and "are now recognized as standard material." In exchange for (as Conant put it) "the NCI . . . subsidizing this industry," Little promised to double inbred mouse production at JAX (to 120,000 mice annually) and to give government intramural and extramural grantees first priority on these materials. Drawing on familiar analogies, the NACC committee framed Little's plan for making inbred mice available "with greater efficiency and . . . much *more economy*" according to their general goals of achieving organizational efficiency in the NCI. Carl Voegtlin and Arthur Compton labeled JAX's effort "essential" to the federal cancer research as a "standardizing job." Francis Carter Wood argued that it would "meet the situation we are in, like chemists doing work where everyone had to purify their own chemicals," and William Middleton seconded this opinion, noting "we do not want to waste time" in either research material preparation or the unintentional duplication of results. Ultimately, the NACC approved a three-year, ten-thousand-dollar annual grant for JAX's mouse production enterprise. Through this arrangement, then, the suitability of inbred mice for experimental cancer research became a matter of official NCI policy. Their known genetic composition and their fecundity were useful, according to the NACC, because these characteristics ensured accuracy and efficiency at the benchtop, while also enabling better coordination among all the various NCI-sponsored projects.[109]

This decision sanctioned an immediate expansion of the JAX physical plant—a development that even further increased the inbred mouse's availability for use in cancer studies and other kinds of medical research. By 1937, JAX Lab's production facilities had grown increasingly inadequate to meet what RF's Alan Gregg noted was a very high demand.[110] Meanwhile, Little's efforts to seek out small but stable markets for inbred mice in medical research had been more successful than he anticipated, so JAX Lab was already committed to even more production. The original JAX building had been designed to accommodate only sixty to eighty thousand mice, and was already operating at 100 percent of capacity.[111] Little thus turned to the Rockefeller Foundation, and requested funds to build a "production unit," which would be physically separate from the research labs in the main building. He worded JAX's needs in blunt economic terms: "The demand [for mice] which is steadily increasing greatly exceeds the supply. . . . [JAX] needs to insure the maintenance of the stocks under circumstances and in sufficient quantity to stabilize and continue the present supply to various institutions at approximately the present cost." The RF's Warren Weaver visited the JAX facilities and immediately grasped the material problem: the suitability of inbred mice for research depended on their sheer availability as well as quality control in production. Ultimately, he agreed to provide forty thousand dollars in funds from the Division of Natural Sciences.[112] Within a year, construction of the first JAX "mouse house" was complete, and Little sent RF's publicity office photos for their *Annual Review.* These images foregrounded the scaled-up, industrialized nature of inbred mouse breeding that the new building embodied. Scientist-workers were shown routinely marking and recording litters of mice, in juxtaposition with an external view of the modernistic glass and steel production wing. During 1939, JAX used this new facility to produce and distribute 110,000 inbred mice—a 175 percent increase in two years. In the first six months alone that the building was operational, animal production soared 53 percent.[113]

Around the same time, the NACC also crafted a program of "Fundamental Cancer Research"[114] that maximized intramural laboratory consumption of these mouse materials by highlighting their special suitability for particular experimental projects. For example, Howard Andervont of the Public Health Service's Harvard Cancer Laboratory moved to the new NCI Bethesda, Maryland, compound in 1939, where he began a series of experiments designed to explore the "milk influence" in *C3H* and *C57Black* mice.[115]

Recall that JAX workers had themselves discovered this phenomenon and that throughout the 1930s, Bittner continued to receive small federal grants to support his continued researches on the problem. But ironically, Little never allowed Bittner to completely consolidate these efforts.[116] Articulating the meaning of nursing-induced mouse mammary tumors at JAX stood to threaten JAX claims to be producing genetically stable commodities, and would link JAX mice too closely to one narrow and applied cancer research problem. At the NCI, however, milk factor and mouse mammary tumor studies flourished precisely because they were designed around the now shared assumption that JAX inbred mice embodied the standards for tumor and milk influence stability. Thus when Andervont failed to reliably (i.e. in a way that could be replicated) convert a low-cancer *C57* mammary tumor strain into a high tumor strain by nursing its young on *C3H* high-tumor mothers, he postulated that it was because his animal materials were not sufficiently "pure." Alternatively, he could have suggested that this effect was caused by some variability in the milk effect itself. As Jean-Paul Gaudillière has observed, "[JAX] ready-made mice . . . facilitated the convergence of laboratories" around the milk influence problem.[117] But such a convergence, while important for defining the suitability of inbred mice for local research communities, could no longer occur at JAX itself. JAX, like the animal commodities it produced, was now defined primarily by its ability to sustain diverse user communities. Bittner, in turn, left Bar Harbor in 1943 for the University of Minnesota Medical School.[118]

In the early 1940s, Little renewed his publicity efforts to highlight the benefits of NCI support. Now, however, he and others covering JAX's work cast the usefulness of inbred mice in multiple terms—alternating freely between the animals' research roles, their status as industrially produced commodities, and their role as laboratory stand-ins for humans. In 1940, for example, *Life* once again showcased JAX, this time under the heading: "Dr. Little's 1 Million Mice Have Proved Heredity Is a Factor in Cause of Cancer."[119] And in early 1941, *Collier's Weekly* ran an article touting the Jackson Lab as the "front line against cancer." In it, the reporter praised Little, "the major-domo of this Mousetown," and noted that "when the problem of cancer is finally solved we'll owe a gold statue to the mouse."[120] Only two years earlier Little himself published the book *Civilization against Cancer,* which was aimed at educating the lay public about recent developments in what he again called "America's cancer war." This book contained two full chapters that presented Little's familiar line on the productive and efficient use of inbred mice as materials for cancer research, and mouse data and examples are inextricably woven throughout the other eight chapters.[121]

Despite this positive press, the industrial character of both the JAX commodification of inbred mouse materials and the NCI's vertically integrated intramural research utilizing these animals did not entirely escape criticism. In the decade immediately following passage of the NCI Act, for example, several members of the nonscientific American public complained that the government's basic research–oriented plan of attack was so focused on the problem of cancer causes that it held no promise for helping those already ill with the disease.[122] Furthermore, Peyton Rous, of the Rockefeller Institute for Medical Research, led a faint chorus of voices within the cancer community that disparaged the NCI as a "GM approach to science." Rous thought the NCI's research suffered for being too narrowly conceived in terms of both its materials and methods—neither were relevant beyond its own institutional boundaries.[123]

But because the NCI did not yet represent the sole significant source of cancer funding, researchers who still did not consider inbred mice the most suitable experimental organisms could continue pursuing their alternative lines of work virtually unabated. Thus Rous himself induced cancer in healthy chickens with injections of a subcellular sarcoma extract and argued that the malignancies were likely caused by some sort of "minute parasitic organism" or virus.[124] And Slye, though no longer actively pursuing the problem of cancer inheritance, continued her murine cancer work through studies of the metastases of mouse lung and breast tumors. In this regard, her almost pastoral depiction of her breeding and observational procedure contrasted with the industrial commodification and production ethos embodied by JAX and the NCI: her lab, she repeatedly emphasized, was a place "where every mouse is simply allowed to live out its span of life."[125]

Even brief surveys of JAX's own activities and of the articles published in cancer research journals also suggest that, at least through 1950, a wide range of definitions for the laboratory suitability of inbred mice peacefully coexisted with the NCI's vertically integrated mammary tumor work. For example, throughout the 1940s, JAX Lab contracted with area Maine physicians to conduct Ascheim-Zondek (A-Z) tests for pregnancy on JAX inbred mice. Not only could these tests provide faster results for the doctor's patients than similar tests done with higher mammals, but also JAX's economies of scale enabled Little to charge 60 percent below commercial dealer fees for this service. In 1944, A-Z tests netted JAX Lab about thirty-two hundred dollars, which did not compare to the sixty thousand dollars netted in mouse sales. Still, this extra source of income was a welcome addition to the budget during wartime, and right after the war, such non-NCI-centered activities formed the basis of Little's second successful RF "mouse house" grant.[126] Likewise, by 1950, although 70 percent of the mouse researchers who published in *Cancer Research* employed inbred strains, much of this work ignored genetics and the milk factor problem and focused instead on the biochemical basis of tumor induction and/or spontaneous formation.[127] Ultimately, then, the experimental success of inbred mice did not completely eclipse the use of other biomedical research organisms or methodologies. Yet the utility of inbred mice was uniquely stabilized because it drew simultaneously on the consolidation of one narrowly defined cancer research program at the NCI and on a persistent flexibility of practical meanings outside of that program.

Perhaps the most striking affirmation of JAX Lab's system of production and distribution came in 1947, when the community of researchers was faced—literally—with its destruction. In October of that year, JAX Lab burned to the ground in the great Bar Harbor Fire. Flames completely leveled the original laboratory building, and the two Rockefeller "mouse houses" for production—the second of which was then still under construction—were also seriously damaged. Except for a few hundred mice readied for shipment in a corner isolation room, all ninety thousand then resident (housed primarily in wooden mouse boxes) perished in the blaze.[128] Even Little was surprised by the magnitude of the biological community's response to his own professional tragedy. Most of the earliest offers for help were extended especially to the JAX mice. The very day of the fire, for example, Little received telegrams from J. B. Murphy of the Rockefeller Institute and Miloslav Demerec of the Carnegie Institute—both offered facilities for maintaining the mouse strains. Bill Murray, who was away on business in Michigan, also wired Little later to say that he had made the American Cancer Society and the National Institute aware of the JAX crisis. The boards of both institutions held special meetings, and decided—before they even knew the full extent of the damage—to offer Little a new building for mouse production. Little told Weaver that there was "hardly a genetics or cancer research institute east of the Mississippi" that didn't offer JAX workers temporary laboratory space. And geneticists and medical researchers who had formerly received stocks of mice from JAX Lab now began to send back breeding pairs of the same stock, even though JAX had not yet solicited them.[129]

Little described return of mouse material to JAX Lab using a melodramatic religious analogy: "the bread which we cast upon the waters several years ago, is now returning to us." But the spiritual overtones of inbred mouse users' response to the JAX Lab destruction should not be overlooked. Their desire to rebuild JAX Lab as a center of inbred mouse production and supply ultimately reflected a strong communal belief in the value of both JAX's unified system of standardization, production, and distribution, and the research materials it created. In the end, regardless of where each individual researcher or institution stood on precisely why inbred mice were useful, as a group they agreed with Little. The JAX Lab and its standardized animals should be maintained "not only as a physical entity but as a spiritual force" to unify diverse inbred mouse users.[130] Several years later, in a foundation endorsement letter, JAX's own board of trustees enthusiastically noted that the institution had rebuilt and reclaimed its original status as the "Bureau of Biological and Medical Standards."[131]

CONCLUSION: MICE AS SCIENTIFIC AND CULTURAL ARTIFACTS

In 1955, the mutually reinforcing industrial relationship between NCI research and JAX inbred mice got a dramatic boost when a powerful postwar ASCC business leadership pressured the federal government to achieve better coordination of NCI cancer research labs and cancer treatment clinics. This led to the establishment of the Cancer Chemotherapy National Service Center (CCNSC), for which Congress allotted nearly seventy million dollars over the period from 1956 to 1958. The CCNSC's sole purpose was the screening of anticancer drugs, an enterprise that relied heavily on standardized tumors and standardized mice. Inbred mouse breeding for the CCNSC began under the supervision of the National Research Council at the Jackson Lab, but because of the enormous numbers of animals needed, breeding was soon also contracted out to commercial animal suppliers who agreed to follow the established protocols. As a result of the CCSNC, then, inbred mouse production literally became part of what one NCI leader called "an industry-government cooperation . . . as effective in the pharmaceutical industry as it is in some of the defense areas."[132]

These developments, while driven by a different set of actors with different priorities, relied heavily on the preexisting strategies for organizational and cultural justification that had led to the inbred mouse's success in the prewar period.[133] In the mid-1930s, NCI policy sanctioned the inbred mouse's suitability for cancer research by underscoring its potential for coordinating work across laboratories, and by sponsoring the industrial production of animals at JAX that made the desired research material circulation possible. NCI researchers demonstrated this animal's local experimental potential in a focused research program on the behavior of mammary tumors. These studies, though new, also drew methodologically on the earlier practices and inbred mouse meanings developed in work on the genetics of tumor transplantation. And perhaps most importantly, throughout the period from

1910 to 1950, C. C. Little consistently articulated the otherwise unspoken scientific and cultural assumptions of this enterprise: these animals were at once like technical objects (i.e. controlled, standardized research materials) and like humans (i.e. morally acceptable animal proxies in the laboratory).[134]

Inbred mouse suitability for American cancer research was thus defined in the early twentieth century by a multiplicity of criteria—experimental, organizational, and cultural—all of which were inescapably historically situated. Ultimately, the persistence of this multiple identity remains an open historical question: were, for example, the criteria of mice suitability narrowed in response to new and different social or biological understandings of the problems they were used to investigate? At the very least, however, this analysis suggests that to understand the roles played by standardized animals in science we must also understand the context under which those roles were defined and played out. "The bioengineered mouse," Donna Haraway recently wrote, "is simultaneously a metaphor, a technology, and a beast living out its many-layered life as best it can."[135] Indeed, in order for this creature to be able to live its life at all, it must inhabit a technoscientific reality that is perhaps inevitable, perhaps not—but is most certainly a world of our own making.

NOTES

1. Stephen Jay Gould, "Mickey Mouse Meets Konrad Lorenz," *Natural History* 88 (1979): 30–36. More specifically, Gould notes that Disney artists made Mickey's nose thicker and shorter, his limbs softer and puffier, his forehead more rounded than sloping, and his eyes far larger and rounder—in response to an intuitive understanding of what changes would make the mouse cuter and more appealing to his human audience. On the general scientific problem of homology and kinship between species, see Gould, "Of Mice and Mosquitoes," *Natural History,* July 1991, pp. 12–20.

2. Elizabeth A. Lawrence, "Neoteny in American Perceptions of Animals," in *Perceptions of Animals in American Culture,* ed. R. J. Hoage (Washington, D.C., 1989), pp. 57–75. Cf. also her "In the Mick of Time: Reflections on Disney's Ageless Mouse," *Journal of Popular Culture* 20, no. 2 (1986): 65–72.

3. Marc Holthof, "To the Realm of Fables: The Animal Fables from Mesopotamia to Disneyland" (translated from the Dutch by Milt Papatheophanes), in *Zoology on (Post) Modern Animals,* ed. Bart Verschaffel and Mark Verminck (Dublin, 1993), pp. 37–55.

4. "Transgenic Mouse Model for Alzheimer's Disease?" *Alzheimer News,* August 1995; cf. Gina Kolata, "Landmark in Alzheimer Research," *New York Times,* February 9, 1995, p. A20; "New Breed of Mouse Aids Alzheimer's Work," ibid., October 4, 1996, p. A22 (from which the quote comes). Natalie Angier, "Gene Defect Tied to Violence in Male Mice," ibid., November 23, 1995 (thanks to Elizabeth Hanson for this reference); and Douglas Birch, "Scientists Link Missing Gene in Mice to Violent Behavior," *Baltimore Sun,* November 23, 1995, p. A16. "Study of Mice Links a Gene to Nurturing," *Boston Globe,* July 26, 1996, p. A20.

5. Cf. Soren Germer, "Life and Evolution in the Case of the Harvard Oncomouse" (paper presented at the Society for the Social Study of Science Meeting, New Orleans, October 13, 1994; my thanks to the author for forwarding his work in progress).

6. Donna Haraway, "Where Man™ is on the Menu," in *Incorporations,* ed. J. Crary and S. Kwinter (New York, 1992), pp. 38–43. Cf. also her *Primate Visions: Gender, Race, and Nature in the World of Modern Science* (New York, 1989). On boundary definition associated with the general question "what is an animal?" see Tim Ingold, "From Trust To Domination: An Alternative History of Human-Animal Relations," in *Animals and Human Society: Changing Perspectives,* ed. A. Manning and J. Serpell (New York, 1994). For a sociological perspective, see Arnold Arluke and Clinton Sanders, *Regarding Animals* (Philadelphia, 1995).

7. C. C. Little to Roy Larsen, November 5, 1953, Box 12, Folder L, Jackson Laboratory Archives, Joan Staats Library, The Jackson Laboratory, Bar Harbor, Me. (hereafter JLA-BH).

8. By "animals" here, I refer to the technical scientific classification—members of the kingdom *Animalia*—rather than just "organisms." See Andrew Rowan, *Of Mice, Models and Men* (New York, 1984), chap. 5. Any estimate of mouse use will be problematic: Rowan's sources, for example, are unclear and he does not break down the kinds of mice included in that percentage (namely, percentage of inbreds versus outbreds). But the best statistics on animal use at the U.S. National Institutes of Health confirm his conclusions: mice and rats together account for 60–70 percent of all animals used, and approximately 90 percent of all mammals used. See Office of Technology Assessment Report on Lab Animal Use, 1990; I am grateful to Dr. Louis Siebel for this material.

9. C. C. Little typescript, draft of his 1965 testimony on the Cigarette Labeling Hearings (n.d.), Box 732, C. C. Little Papers, Raymond Fogler Library, University of Maine at Orono (hereafter CCL-UMO).

10. On Little's own personal and professional life history, see my sketch in the forthcoming *American National Biography* (Oxford University Press);

cf. also George Snell, "Clarence Cook Little," in *Dictionary of Scientific Biography* (suppl., 1990), 17:562–64.

11. Patricia Lauber, "Why Mice?" in *Of Man and Mouse: How House Mice Became Laboratory Mice* (New York, 1971), p. 49. Gina Kolata, "A Star Is Born: Even a Lab Mouse Needs an Agent," *New York Times,* January 26, 1997, p. E5.

12. Such explanations frequently also neglect the basic sociological dictum that most working scientists freely admit has always governed their choice of model organisms: namely, that the predominant use of particular experimental systems has (in the words of scientist Harold Hewitt) "a cumulative effect in recommending their use." Kolata's piece (cited in note 11) is exceptional in this regard, as she notes: "To get other scientists to accept a new lab creature is, in a sense, to create a new research library. Success depends on how famous, and how persuasive, the creature's champion is."

13. For an overview of recent work in this area, see the essays in Adele E. Clarke and Joan H. Fujimura, eds., *The Right Tools For the Job: Materials, Instruments, Techniques, and Work Organization in Twentieth-Century Life Sciences* (Princeton, N.J., 1992), and especially the bibliography to the authors' introduction, "What Tools? Which Jobs? Why Right?" In the collection itself, see Barbara Kimmelman, "Organisms and Interests in Scientific Research: R. A. Emerson's Claims for the Unique Contributions of Agricultural Genetics," and Gregg Mitman and Anne Fausto-Sterling, "Whatever Happened to *Planaria*? C. M. Child and the Physiology of Inheritance." Robert E. Kohler, *Lords of the Fly: Drosophila and the Experimental Life* (Chicago, 1994). Bonnie Tocher Clause, "The Wistar Rat as a Right Choice: Establishing Mammalian Standards and the Ideal of a Standardized Mammal," *Journal of the History of Biology* 26 (summer 1993): 329–49; Doris Zallen, "The 'Light' Organism for the Job: Green Algae and Photosynthesis Research," ibid., pp. 350–65; Frederic L. Holmes, "The Old Martyr of Science: The Frog in Experimental Physiology," ibid., pp. 233–67. For philosophical and sociological perspectives on model systems, see respectively Hans Jorg Rheinberger, "Experiment, Difference, and Writing (I. Tracing Protein Synthesis and II. The Laboratory Production of Transfer RNA)," *Studies in the History and Philosophy of Science* 23 (1992): 305–31, 389–422; Joan Fujimura, "Tools of the Trade" (draft provided by author); and Angela N. H. Craeger, *The Life of a Virus* (Chicago, 2002). Several more historical studies of model organisms are also in progress, e.g. on paramecia (Judy Johns Schloegel, "Herbert Spencer Jennings, Tracy M. Sonneborn, and the Career of American Protozoan Genetics" [Ph.D. diss. in progress, Indiana University]), and postwar mouse research (Jean-Paul Gaudillière

and Ilana Lowy, "The Makers of the Laboratory Mouse," book in progress). Cf. also Adele E. Clarke, "Research Materials and Reproductive Science in the United States, 1910–1940," in *Physiology in the American Context, 1850–1940,* ed. Gerald L. Geison (Bethesda, Md., 1987), pp. 323–50, for an elegant and insightful early treatment of biological materials.

14. The other field—radiation genetics from 1947 to 1965—will be the subject of another chapter of my book, *Making Mice: Standardizing Animals for American Biomedicine, 1900–1965* (Princeton, N.J., forthcoming).

15. Historiographically, this set of claims represents a synthesis of Susan Leigh Star and James Griesemer's sociological work on "boundary objects" and Norton Wise's historical work on "mediators." Star and Griesemer are concerned with explaining the management of collective scientific work, i.e. how members of different social worlds come to cooperate in constructing scientific representations despite their different viewpoints and agendas; Wise is concerned with explaining the productive effect of social context on scientific work, i.e. how categories of a local scientific community are interdefined and mutually reinforced by larger cultural categories from such areas as politics and economics. In the case of establishing the mouse as a standard organism, both things need explanation: the larger socio-moral consensus of the actors, as well the varying motivations by which they come to this particular scientific project. See S. L. Star and J. Griesemer, "Institutional Ecology, 'Translations,' and Boundary Objects: Amateurs and Professionals in Berkeley's Museum of Vertebrate Zoology," *Social Studies of Science* 19 (1989): 387–420; and N. Wise, "Mediating Machines," *Science in Context* 2, no. 1 (1988: 77–113.

16. W. E. Heston, "Genetics of Neoplasia," in *Methodology in Mammalian Genetics,* ed. W. J. Burdette (San Francisco, 1963), p. 247–68; CCL, "Your Question #2" (one of a set of typescript answers to a CBS reporter's questions, numbered 1–8, not dated but from content appears to be early 1960s), Box 737, CCL-UMO.

17. Cf. Roberta Clark, "The Social Uses of Scientific Knowledge: Eugenics in the Career of Clarence Cook Little" (M.A. thesis, University of Maine, 1985), p. 59; CCL's Zoology Notebook (c. 1910), pp. 80–81, Box 726, CCL-UMO.

18. Mitosi Tokuda, "An Eighteenth Century Japanese Guide-Book on Mouse-Breeding," *Journal of Heredity* 26 (1935): 481–84; cf. C. J. Davies, *Fancy Mice,* 5th ed. (London, 1912). Around the same time as Little began inbreeding mice, Helen Dean King at the Wistar Institute in Philadelphia began inbreeding rats—primarily for anatomical and nutritional research but also for genetics: see Clause, "The Wistar Rat."

19. 1958 letter from Castle to Dr. Michael Potter, reproduced in Herbert Morse III, ed., "Introduction," in *The Origins of Inbred Mice* (New York,

1978). Also on Lathrop, see Michael B. Shimkin, "A. E. C. Lathrop (1868–1918): Mouse Woman of Granby," *Cancer Research* 35 (June 1975): 1597–98. Cf. "Mice Beautiful," *Time,* July 19, 1937, p. 50, for a brief discussion of the American Mouse Fanciers Club, based in Stockport, New York, and its British counterparts.

20. On inbreeding as a scientific methodology, see M. F. W. Festing, *Inbred Strains in Biomedical Research* (New York, 1979); and Lee Silver, *Mouse Genetics: Concepts and Applications* (New York, 1995).

21. Cf. Karen Rader, "'The Mouse People': Murine Genetics Work at the Bussey Institution, 1909–1936," *Journal of the History of Biology,* no. 3 (fall 1998): 327–54. On *Drosophila* as "breeder reactors," see Kohler, *Lords of the Fly.* On the field of mammalian genetics c. 1910, see William Provine, *Sewall Wright and Evolutionary Biology* (Chicago, 1986).

22. C. C. Little, "A Possible Mendelian Explanation for a Type of Inheritance Apparently Non-Mendelian in Nature," *Science* 40 (1914): 904–6.

23. Little, "Your question #2."

24. A. E. C. Lathrop and Leo Loeb, "The Incidence of Cancer in Various Strains of Mice," *Proceedings of the Society of Experimental Biology and Medicine* 11 (1913–14): 34–38; "The Influence of Pregnancies on the Incidence of Cancer in Mice," ibid., pp. 38–40; "Further Investigations on the Origin of Tumors in Mice, I. Tumor Incidence and Tumor Age in Various Strains of Mice," *Journal of Experimental Medicine* 22 (1915): 646–73; "Further Investigations on the Origins of Tumors in Mice, V. The Tumor Rate in Hybrid Strains," ibid., 28 (1918): 475–500. At about the same time or slightly earlier, other scientists such as Haaland published observations that tumor transplants took more easily among mice of the same "race," but these scientists did not describe this "race effect" in genetic terms: I am grateful to Jean-Paul Gaudillière and Ilana Löwy for pointing this work out to me.

25. E. E. Tyzzer, "The Inoculable Tumors of Mice," *Journal of Medical Research* 18 (November 1907): 137–53; "A Study of Heredity in Relation to the Development of Tumors in Mice," ibid., 17 (1907): 199–211; "A Study of Inheritance in Mice with Reference to Their Susceptibility to Transplantable Tumors," ibid., 21 (1909): 519–74; "A Series of Spontaneous Tumors in Mice with Observations on the Influence of Heredity on the Frequency of Their Occurrence," ibid., pp. 153–92, where Tyzzer suggested that successful transplantation depended as much on the method of inoculation and the "character" of the tumor as on "the nature of the soil upon which the tumor is transplanted." For an explanation of Tyzzer's experiment, see also Kenneth Paigen, "Seventy-Five Years of Mouse Genetics: Some Perspectives and Lessons" (unpublished manuscript, 1983).

26. Although Little appears to have bounced back quickly in the professional sphere, failing his exams was an extremely traumatic event personally: see his description of this event in his entries for the 1917 (third) and 1935 (twenty-fifth anniversary) Harvard College Class Notes, Harvard University Archives, Pusey Library, Harvard University, Cambridge, Mass.; Little's dissertation, supervised by Castle, East, and G. H. Parker, was entitled "Experimental Studies of the Inheritance of Coat Color in Mice and Their Bearing on Certain Allied Problems in Genetics" (Sc.D, Harvard, 1914) but it was published as *Experimental Studies of the Inheritance of Coat Color in Mice,* Carnegie Institution of Washington, publication no. 179, paper no. 19 (Washington, D.C., 1913).

27. C. C. Little, "The Relation of Heredity to Cancer in Man and Animals," *Science Monthly* 3 (1916): 196–202.

28. C. C. Little and E. E. Tyzzer, "Studies on the Inheritance of Susceptibility to a Transplantable Sarcoma (JWB) of the Japanese Waltzing Mouse," *Journal of Cancer Research* 1 (1916): 387–89. See also their "Further Experimental Studies on the Inheritance of Susceptibility to a Transplantable Carcinoma (JWA) of the Japanese Waltzing Mouse," *Journal of Medical Research* 33 (1916): 393–427, where Little and Tyzzer also take on the seemingly contradictory experimental result of Cuenot and Mercier.

29. CCL to W. T. Bovie, December 18, 1918, Box 727; CCL to E. E. Tyzzer, December 11, 1918; also, E. E. Tyzzer to CCL, December 9, 1918, both in Box 739, all CCL-UMO.

30. Little, "The Relation of Heredity," p. 198.

31. "In this period [1900 to 1930] of transition from hopelessness to courage, from fear to cooperative rationality, and from isolated research to a planned and continuing attack on the disease [cancer], idealistic and yet practical leadership was needed." Little, "James Baumgardner Murphy," *National Academy of Sciences Biographical Memoirs* 34 (1960): 183–203, quote p. 191. On the cooperative ethos in the early history (1930–50) of molecular biology, see Lily E. Kay, *The Molecular Vision of Life: Caltech, The Rockefeller Foundation, and the Rise of the New Biology* (New York, 1993). On the problems with the historical label "Progressive," see Daniel T. Rogers, "In Search of Progressivism," *Reviews in American History* 10 (1982): 113–32.

32. On the rise of a research tradition in American medicine, see R. L. Numbers and J. H. Warner, "The Maturation of American Medical Science," in *Sickness and Health in America,* ed. J. W. Leavitt and R. L. Numbers, 2d ed. (Madison, Wis., 1985), p. 113–25.

33. A. H. Thaites, *Report on Cancer Research,* Commonwealth of Australia Quarantine Service, Service Publication No. 11 (Melbourne, 1916), p. 27;

"Cancer as a Social Problem," American Society for the Control of Cancer, Bulletin 3, December 1914, p. 5. All obtained at the Countway Library, Harvard University Medical School, Cambridge, Mass.

34. Margaret Rossiter, *Women Scientists in America: Struggles and Strategies to 1940* (Baltimore, 1982).

35. Quote from Maud Slye, "A Reply to Dr. Little," *Science* 42 (1915): 246–48. J. J. McCoy, *The Cancer Lady: Maud Slye and Her Hereditary Studies* (Nashville, Tenn., 1977); and John Parascondola, "Maud Caroline Slye," in *Notable American Women: The Modern Period* (Cambridge, Mass., 1987): 651–52.

36. Maud Slye, "The Incidence and Inheritability of Spontaneous Cancer in Mice," *Journal of Medical Research* 32 (1915): 159–72: "Whether or not my strains of house mice have behaved in hybrid crosses in accordance with the established canon has no bearing whatever upon the behavior of cancer. . . . it is an academic debate which lies in quite another field." In 1928, as a result of new observations on over one hundred thousand animals, she slightly modified this conclusion to say that two genetic factors were involved, one controlling the type of cancer and another controlling the location of the tumor: see Slye, "The Relation of Heredity to Cancer," *Journal of Cancer Research* 12 (1928): 83–133.

37. Little, "Cancer and Heredity," *Science* 42 (1915): 1076–77.

38. The pinnacle of the journal exchanges came in 1928, in Little, "Evidence that Cancer is Not a Simple Mendelian Recessive Disease," *Journal of Cancer Research* 12 (1928): 30–46, where Little systematically lists and attacks all twenty-six of Slye's publications up to that time. Cf. C. C. Little, "The Inheritance of Cancer," *Science* 42 (1915): 494–95.

39. McCoy, *The Cancer Lady,* pp. 88–90; see also Ludwig Hekoten (and other, anonymous scientists' reflections) on Slye in Greer Williams, *The Virus Hunters* (New York, 1961), pp. 419–21. It is interesting to note that while the points of disagreement between Little and Slye reflected earlier debates between the Biometricians and Mendelians, neither Slye nor Little drew explicitly on these rhetorical resources. On the conflict between Biometricians and Mendelians, see Lyndsay Farrell, "Controversy and Conflict in Science: A Case Study—The English Biometric School and Mendel's Laws," *Social Studies of Science* 5 (1975): 269–301, and Giorgio Israel, "The Emergence of Biomathematics and the Case of Population Dynamics," *Science in Context* 6 (1993): 469–507.

40. Little made the cover of *Time* in March 1937 (article: March 19, pp. 40–41); three weeks later, Slye made the cover of *Newsweek* (article: April 10, 1937, pp. 26–28).

41. McCoy, *The Cancer Lady.*

42. I am very grateful to Gail Schmitt and Jamie Thompson for research assistance with regard to the cancer journal article counts that appear here, and elsewhere throughout this paper.

43. CCL to E. E. Tyzzer, December 11, 1918; E. E. Tyzzer to CCL, December 9, 1918, both in Box 739, CCL-UMO.

44. CCL to E. E. Tyzzer, December 1918; E. L. Thorndike to CCL, November 25, 1918; both in Box 739, CCL-UMO; The Eugenics Record Office (ERO), a national headquarters for the maintenance of data on human heredity, was located at Cold Spring Harbor, and during summers, the SEE hosted a large number of distinguished visiting experimental geneticists. See Garland Allen, "The Eugenics Record Office at Cold Spring Harbor: An Essay in Institutional History," *Osiris* 2 (1986): 225–64.

45. See "Your Question #2." Leonell Strong also remembers that it was Little who started the *C57B*'s: L. C. Strong, "Inbred Mice in Science," in Morse, *Origins,* pp. 45–67.

46. On Little's work at the SEE, see excerpts from the Carnegie Institution Annual Reports: 1919 ("Department of Experimental Evolution," pp. 123–41) and 1920 ("Department of Genetics," pp. 101–34), in Box 738, CCL-UMO; Little, "Factors Influencing the Growth of a Transplantable Tumor in Mice," *Journal of Experimental Zoology* 31 (1920): 307–26; "The Heredity of Susceptibility of a Transplantable Sarcoma (J.W.B.) of the Japanese Waltzing Mouse," *Science* 51 (1920): 467–68; "The Inheritance of a Predisposition to Cancer in Man," in *Eugenics, Genetics and the Family: Second International Congress of Eugenics,* vol. 1 (Baltimore, 1923), pp. 186–90, where Little also highlighted the Mendelian meaning of cancer as a genetic disease: "It is quite possible to have true inheritance of a tendency to cancer, and yet little or no available evidence from a statistical point of view if only the relationship of parent to offspring is considered" (p. 187).

47. Obituary: Halsey Bagg, *New York Times,* April 15, 1947; C. C. Little, "Halsey Joseph Bagg," *Anatomical Record* 100 (1948): 397. On maintenance of the "Bagg albinos" at the SEE, see last page of an untitled "Mouse Club Newsletter," c. 1922, Box 732, CCL-UMO; *Carnegie Institution Annual Report* 1921 reports that Little and Bagg began their joint experiments in fall 1920: see pp. 112–13 of this report, Box 738, CCL-UMO. See H. J. Bagg and C. C. Little, "The Occurrence of Four Inheritable Morphological Variations in Mice and Their Possible Relation to Treatment with X-rays," *Journal of Experimental Zoology* 41 (1924): 45–91; "Hereditary and Structural Defects in the Descendants of Mice Treated With Roentgen-ray Irradiation," *American Journal of Anatomy* 33 (1924): 119–45; on work with mye-

lencephalic bleb mice, see Hans Gruneberg, *The Genetics of the Mouse* (Cambridge, 1943), p. 229–31.

48. Cf. C. C. Little and L. C. Strong, "Tests for Physiological Differences in Transplantable Tumors," *Proceedings of the Society for Experimental Biology and Medicine* 18 (1920): 45–48; L. C. Strong, "Indication of Tissue Specificity in a Transplantable Sarcoma," *Journal of Experimental Medicine* 39 (1924): 447–56.

49. Little became assistant director when the SEE was reorganized into two sister institutions, the Department of Genetics and the Eugenics Record Office: see Clark, *Social Uses*, p. 77. The Davenport-Little clash is also described in detail ibid., pp. 83–87; CCL to Davenport, May 17, 1922, and Davenport to CCL, May 18, 1922: both Charles B. Davenport Papers, American Philosophical Society (hereafter CBD-APS). C. C. Little, "Some Factors Influencing the Human Sex Ratio," *Proceedings of the Society for Experimental Biology and Medicine* 16 (1919): 127–30; "Relations between Research in Human Heredity and Experimental Genetics," *Scientific Monthly* 14 (1922): 401–14; *The Relations of Genetics to Problems of Cancer Research*, Harvey Lectures, 1921–22, ser. 17 (Philadelphia, 1923), pp. 65–88.

50. C. C. Little, "The Relation Between Experimental Research in Human Heredity and Experimental Genetics," *Scientific Monthly* 14 (May 1922): 401–14, quotes pp. 403–5, 407–9.

51. Little brought with him guaranteed annual support of fifteen hundred dollars from the Carnegie Institution, and he negotiated with Maine's trustees for another three thousand dollars in research money as a condition of his appointment. David C. Smith, *The First Century: A History of the University of Maine, 1865–1965* (Orono, Maine, 1979), pp. 120–22; H. J. Thorkelson, "Memorandum: University of Maine," General Education Board Collections, Record Group 2441, Box 680, Folder 7032, Rockefeller Archive Center (hereafter RAC-NY). Little also took on two more students once in Orono: the brothers Joseph and William Murray, former Maine undergraduates who began new research on X-rays and mouse genetics. On this, and Little's funding situation, see Karen Rader, "Making Mice" (Ph.D. Diss., Indiana University, 1995), chaps. 1 and 2.

52. Cf. C. C. Little and Beatrice W. Johnson, "The Inheritance of Susceptibility to Implants of Splenic Tissue in Mice," *Proceedings of the Society for Experimental Biology and Medicine* 19 (1922): 163–67; Little and Strong, "Genetic Studies on the Transplantation of Two Adenocarcinomata," *Journal of Experimental Zoology* 41 (1924): 93–114.

53. C. C. Little, "Congenital and Acquired Predispositions and Heredity," *Abt's Pediatrics* 1 (1923): 171–256, esp. pp. 173, 185, 246, 249, and

251 (though on p. 248, Little indirectly implies that genetics is the dominant discipline).

54. "To Begin Cancer Research," *New York Times,* October 7, 1929, p. 10.

55. Editorial: "New U. of M. President Seems an All-Around Man," *Detroit Saturday Night,* July 11, 1925 (which reported that the only condition Little attached "to the acceptance of his appointment was that he be provided with a private laboratory for research work, which he wishes to continue during his leisure from administrative work"), Box 742, CCL-UMO; Howard Peckham, "President Little Embattled," in *The Making of the University of Michigan, 1817–1967* (Ann Arbor, Mich., 1967), p. 157; cf. Allen Shoenfield, "'U' Pins Faith in Dr. Little," *Detroit News,* July 7, 1925, Box 742, CCL-UMO.

56. Strong, "Inbred Mice," in Morse, *Origins,* pp. 45–67; CCL to Strickland (a Maine trustee), March 7, 1922, CCL to Davenport, March 21, 1922 (cf. CCL to Davenport, October 26, 1921, and Davenport to CCL, October 31, 1921) (all CBD-APS); and a fragment of a "Mouse Club Newsletter" by Little about transporting the mice to Maine, [n.d.], Box 730, CCL-UMO.

57. President's Laboratory, Department of Cancer Research, Budget 1926–29, Box 735, CCL-UMO.

58. "Bibliography of Papers published 1926–29 inclusive," Box 740, CCL-UMO. Cf. Little bibliography appended to G. D. Snell, "Clarence Cook Little," *Biographical Memoirs of the National Academy of Sciences* 46 (1975): 240–63.

59. The only printed biographical sources on William S. Murray and Joseph M. Murray of which I have knowledge are their brief sketches in the 1938 edition of *American Men of Science.* On the shift of mouse workers from Maine to Michigan, see Elizabeth Russell, "Origins and History of Mouse Inbred Strains: Contributions of Clarence Cook Little," in Morse, *Origins,* p. 38.

60. On the Murray brothers' husbandry tasks, see Russell, "Origins": cf. "Bibliography of Papers." William Murray, "Studies of Developmental Anomalies in Descendants of X-Rayed Mice" (Ph.D. diss., University of Michigan, 1928); J. M. Murray, "A Study of the Histological Structure of Mouse Ovaries Following Exposure to Roentgen Irradiation" (Ph.D. diss., University of Michigan, 1929).

61. Quote from L. C. Dunn to W. Crocker, March 31, 1928 (in this letter, Dunn also complains about problems he is having getting mutant strains of sheep from Norway for genetics research); Crocker to Dunn, April 4, 1928, and May 9, 1928; all from L. C. Dunn Papers, American Philosophical Society (hereafter LCD-APS). On the ATCC and standardization, see R. E. Buchanan, "History and Development of the American Type Culture Collection," *Quarterly Review of Biology* 41 (June 1966): 101–4; and Eric Kupferberg, "Fixing a Name:

Bacterial Systematics and the Role of *Bergey's Manual of Determinative Bacteriology*" (paper delivered at the International Conference for the History, Social Study, and Philosophy of Biology, Northwestern University, July 1991). For a summary, see Karen A. Rader, "The Origins of Mouse Genetics beyond the Bussy Institution. II. Defining the Problem of Mouse Supply: The 1928 National Research Council Committee on Experimental Plants and Animals," *Mammalian Genome* 13, nos. 2–4 (2002).

62. The mouse researchers whose replies to Dunn are extant are: William Gates of Louisiana State University (May 26, 1930); L. J. Cole of Wisconsin (May 26, 1930); H. R. Hunt of Michigan State (May 29, 1930); Lee Dice of the University of Michigan's Museum of Zoology (June 5, 1930); C. H. Danforth of Harvard (June 28, 1930); and Horace Feldman of Michigan (February 10, 1932); all in LCD-APS.

63. Cf. "Reductions in the Appropriations for Scientific Work under the Federal Government," *Science,* July 29, 1932, p. 76: and Frederick Schlink, "Government Bureau for Private Profit," *The Nation,* November 11, 1931, p. 508. On the 1933 "moratorium on research" in physics, see Daniel J. Kevles, "Revolt Against Science," in his *The Physicists* (New York: Knopf, 1971), pp. 236–51.

64. James T. Patterson, *The Dread Disease: Cancer and Modern American Culture* (Cambridge, Mass.: Harvard University Press, 1987), chaps. 3 and 4, esp. pp. 76–77 and 97–98 on ASCC "Cancer Weeks" (an early 1920s invention) and increasing media attention for cancer. Rockefeller philanthropy, for example, supported cancer research on a large scale, but primarily through its own Medical Institute and the individual initiatives of John D. Rockefeller, Jr. For Neely's full remarks, see *Congressional Record,* 70th Congress, 1st session, 1928, 69 (pt. 8), pp. 9048–50. "Statement from Thomas L. Parran," U.S. Congress, Committees on Commerce of the Senate and House, *Joint Hearings on Cancer Research,* 75th Congress, 1st session, July 8, 1937, pp. 44–45. Historian Robert Bud also mentions F. L. Hoffman, *The San Francisco Cancer Survey,* 9 vols. (Newark, N.J., 1924–34) as the classic interwar source for cancer statistics: see Bud, "Strategies in American Cancer Research after World War II," *Social Studies of Science* 8 (1978): 425–59, esp. p. 449 n. 9. Stephen Strickland, *Politics, Science, and Dread Disease: A Short History of U.S. Medical Research Policy* (Cambridge, Mass., 1976), chap. 1, also cites several examples of congressmen challenging the medical statistics during the cancer hearings.

65. R. B. Jackson to CCL, March 3, 1926, and CCL to R. B. Jackson, March 10, 1926; also, CCL to R. B. Jackson, June 1, 1926; R. B. Jackson to CCL, June 8, 1926: all Box 739, CCL-UMO. Although it is mentioned in these letters, I have found no extant copy of Little's description of his lab's cancer-related activities. When Little was president of the University

of Maine, Jackson funded his summer biological field class on Mount Desert Island: see chap. 2 of Rader, "Making Mice."

66. C. C. Little, "A Discussion of Certain Phases of Sterility," *Annals of Clinical Medicine* 5 (1926): 1–4; "Agents Modifying the Germ Plasm," *Surgery, Gynecology and Obstetrics* 46 (1928): 155–58.

67. CCL to Jackson, January 18, 1927, and February 17, 1927; Jackson to CCL, February 19, 1927; see also "Preliminary proposal of R. H. Webber, E. B. Ford and R. B. Jackson . . . for the creation and investigation of the biological phenomena of cancer and other forms of growth"; all in Box 5, Folder 7, Little Presidential Papers, Bentley Historical Library, University of Michigan (hereafter CCL-Mich); Little, "Genetic Investigations and the Cancer Problem," *Commonwealth Review* 8 (1926): 130–36. See also Jean Holstein, *The First Fifty Years at the Jackson Laboratory* (Bar Harbor, Maine, 1979), chap. 1.

68. See "The Institutes" section of Walter A. Donnelly et. al., eds., *The University of Michigan: An Encyclopedic Survey,* vol. 4 (Ann Arbor, Mich., 1958), pp. 1538–39.

69. CCL to Kellogg, October 11, 1927, Box 13, Folder 12, CCL-Mich; Little, "Opportunities for Research in Mammalian Genetics," *Scientific Monthly* 26 (1928): 521–34, quote from pp. 526–27. On Little's AES presidency, see "Little," *Who Was Who,* vol. 5 (1969–73). Little's other eugenic activities in the period from 1926 to 1929 include writing a series of articles for the *Birth Control Review* and chairing the Committee on Formal Education of the AES, which ultimately recommended that university-based eugenics classes be "intimately associated" with research work in genetics laboratories. See Rader, "Making Mice," pp. 151–52. On the state of the American eugenics movement in the interwar period, see Daniel J. Kevles, *In the Name of Eugenics* (Berkeley, Calif., 1985). Based on case studies of several prominent eugenics trials in the United States, Dvora Kamrat-Lang has argued that "By the early 1920s, medicine monopolized eugenic practice": see her "Healing Society: Medical Language in American Eugenics," *Science in Context* 8 (1995): 175–96.

70. "Expenditures and Proposed Budget, July 1927–June 1929," Box 742, CCL-UMO.

71. L. C. Strong, "Indication of Tissue Specificity in a Transplantable Sarcoma," *Journal of Experimental Medicine* 39 (1924): 447–56; C. C. Little and L. C. Strong, "Genetic Studies on the Transplantation of Two Adenocarcinomas," *Journal of Experimental Zoology* 41 (1924): 93–114; C. C. Little and B. W. Johnson, "The Inheritance of Susceptibility to Implants of Splenic Tissue in Mice," *Proceedings of the Society for Experimental Biology and Medicine* 19 (1922): 163–67; L. C. Strong, "On the Occurrence of Mutations

within Transplantable Neoplasms," *Genetics* 11 (1926): 294–303, and "Genetic Studies on the Nature of Cancer," *American Naturalist* 60 (1926): 201–26.

72. Jackson to CCL, January 23, 1929, and CCL to Jackson, January 25, 1929; both in Box 8-7, Jackson Laboratory Archive, Joan Staats Library, Jackson Laboratory, Bar Harbor (hereafter JLA-BH); C. C. Little, "Your Question #5": another one of the set of typescript answers to a reporter's questions, Box 737, CCL-UMO. Roberta Gallant Clark suggests that Little began discussing possible arrangements for a Bar Harbor facility with Jackson in 1927, but the earlier letters between Jackson and Little she cites as evidence were more likely in reference to negotiations with John D. Rockefeller over the existing Mount Desert Island summer biological institute project. See Clark, "Social Uses," p. 159 n. 20.

73. Little, "Evidence that Cancer Is Not Simple Mendelian Recessive," *Journal of Cancer Research* 12 (1928): 45. Little makes this same point more generally in "Agents Modifying the Germ Plasm," *Surgery, Gynecology, and Obstetrics* 46 (1928): 155–58.

74. C. C. Little, "Your Question #5."

75. Jackson to Little, February 13, 1929, Box 8-7, JLA-BH; "Statement of Income and Expenditures, 27 May 1929 to 30 June 1930," Box 739, CCL-UMO.

76. Roscoe Jackson's obituary is featured on the front page of the *Detroit News,* March 20, 1929. For many years, it appears, Little incorrectly recalled Jackson's cause of death as a heart attack. See "Your Question #5"; CCL to Herbert Neal, April 5, 1929, Box 8-7, JLA-BH.

77. Excerpt of letter from CCL to Dorr, March 28, 1929; CCL to Edsel Ford, April 22, 1929: both in Box 8-7, JLA-BH; CCL to George Dorr, April 2, 1929, and April 22, 1929; Box 739, CCL-UMO; "Will Be Memorial for R. B. Jackson," *Bar Harbor Times,* May 8, 1929, also Box 8-7, JLA-BH.

78. "To Begin Cancer Research," *New York Times.*

79. Also, Strong used transplantation of multiple tumors from an individual inbred mouse to examine the genetic variability of tumors themselves. Arthur Cloudman, another Michigan Ph.D., then extended this work in a study of eight individualistic transplantable tumors arising in the same inbred mouse line. In 1935, Bittner published an optimistic review article on tumor transplantation genetics that featured mostly work from the Jackson Laboratory: see L. C. Strong, "Transplantation Studies on Tumors Arising Spontaneously in Heterozygous Individuals," *Journal of Cancer Research* 13 (1929): 103–15; Bittner, "A Genetic Study of the Transplantation of Tumors Arising in Hybrid Mice," *American Journal of Cancer* 15 (1931):

2202–47, and "Linkage in Transplantable Tumors," *Journal of Genetics* 29 (1934): 17–27; Cloudman, "A Comparative Study of Transplantability of Mammary Tumor Arising in Inbred Mice," *American Journal of Cancer* 16 (1932): 568–630; cf. Bittner, "A Review of Genetic Studies in the Transplantation of Tumors," *Journal of Genetics* 31 (1935): 471–87.

80. Donald Shaughnessy, "The Story of the American Cancer Society" (Ph.D. diss., Columbia University, 1957), esp. chap. 8; "Clarence Cook Little," in Patterson, *Dread Disease,* pp. 121–26; "Clarence Cook Little," in *Who Was Who in America,* vol. 5 (1969–73), pp. 435–36.

81. C. C. Little, "The Role of Heredity in Determining the Incidence and Growth of Cancer," *American Journal of Cancer* 15 (1931): 2780–89.

82. Holstein, *The First Fifty Years,* p. 20.

83. Susie G. Barnes (secretary, NRC) to CCL, March 24, 1933, Box 735; CCL to Richard Webber, January 9, 1933, Box 740: all CCL-UMO.

84. The minutes of this second plenary session were published in Donald F. Jones, ed., *Proceedings of the Sixth International Congress of Genetics* (held in Ithaca, N.Y.), vol. 1 (Brooklyn, N.Y., 1932), pp. 17–19.

85. Ilana Löwy and Jean-Paul Gaudillière, "Disciplining Cancer: Mice and the Practice of Genetics Purity," in *The Invisible Industrialist: Manufacturers and the Production of Scientific Knowledge,* ed. Jean-Paul Gaudillière and Ilana Löwy (New York, 1998).

86. Cf. L. C. Strong, "The Production of the CBA Strain of Inbred Mice: Long Life Associated with Low Tumor Incidence," *British Journal of Experimental Pathology* 17 (1936): 60–63; "List of Stocks," in letter to Warren Weaver, December 4, 1937, RF Archive, RG 1.1, 200D, Box 143, 1774, Rockefeller Archive Center, Tarrytown, N.Y. (hereafter cited as RAC-NY); "Supply Department, Stock List," July 1, 1938, Box 7-2, JLA-BH; C. C. Little and P. A. Gorer, "The Genetics of Cancer," in *The Genetics of the Mouse,* ed. H. Gruneberg (Cambridge, Mass., 1943).

87. Interestingly, Little also solicited some small commercial mouse dealers to determine the "market price" for their non–genetically controlled mouse material, and he set the cost for his material below "industry" averages. CCL to Mary Russell and enclosed inbred mouse sales listing, March 3, 1933, Box 730; Irwin Wachtel to Mary Russell, March 18, 1933; C. D. Haedrich to CCL, n.d.; both Box 740, all CCL-UMO. See also Frederick Russell, March 20, 1933, and enclosed inbred mouse sales listing, "Animal Production Information" Box, JLA-BH; in May, Little would plead with the U.S. surgeon general to restore the USPHS's funding for mouse materials: see CCL to Hugh Cummings, May 25, 1933, Box 739, CCL-UMO.

88. For replies from medical researchers, see e.g. Halsey Bagg to CCL, April 4, 1933, Box 12, JLA-BH; Howard Andervont to William Murray,

March 29, 1933; C. W. Turner to CCL, May 8, 1933; CCL to C. W. Turner, June 12, 1933; all Box 739, CCL-UMO. There are no extant Jackson Lab annual or trustees' reports for the pre-1938 period, and I have found no other source which mentions the numbers of mice distributed during this initial period. One annual budget notes a figure of $2,519 in income from mouse sales during the first six months of 1933, but because of varying prices it is difficult to extrapolate from this. See also Holstein, *The First Fifty Years,* p. 79.

89. The Staff of the Jackson Laboratory, "The Existence of Non-Chromosomal Influence in the Incidence of Mammary Tumors in Mice," *Science* 18 (November 1933): 465–66.

90. Little and W. M. Murray, "The Genetics of Mammary Tumor Incidence in Mice," *Genetics* 20 (1935): 466–96; see also Little and Murray, "Further Data on the Existence of Extra-Chromosomal Influence in the Incidence of Mammary Tumors in Mice," *Science* 82 (1935): 228–30, and "Extrachromosomal Influence in Relation to the Incidence of Mammary and Non-Mammary Tumors in Mice," *American Journal of Cancer* 27 (1936): 516–18; J. J. Bittner, "Mammary Tumors in Mice in Relation to Nursing," ibid., 30 (1937): 530–38, and "Foster Nursing and Genetic Susceptibility for Tumors of the Breast in Mice," *Cancer Research* 1 (1941): 793–94. For a discussion of the mouse mammary tumor inciter work at JAX, see also Jean-Paul Gaudillière, "Circulating Mice and Viruses: The Jackson Memorial Laboratory, the National Cancer Institute, and the Genetics of Breast Cancer, 1930–1965," in *Practices of Human Genetics,* ed. Michael Forfujn and Everett Mendelsohn (Dordrecht, 1999), pp. 89–124.

91. Memo of Alan Gregg interview with CCL, April 7, 1933; and Gregg diary entry, April 7, 1933, both RG1.1, 200D, Box 143, 1773, RAC-NY; Hanson was more tuned in to the possible value of this production apparatus than Weaver or Gregg. See CCL to AG, May 24, 1933, including grant proposal, and CCL to Webber, June 14, 1933, both Box 740, CCL-UMO. Cf. Frank B. Hanson diary entries, October 5, 1933, and October 24, 1933; Warren Weaver to CCL, December 19, 1933; Natural Sciences Grant Resolution no. 33359, December 13, 1933; all RG 1.1, 200D, Box 143, RAC-NY; CCL to Webber, June 14, 1933, Box 740, CCL-UMO.

92. For my journal scans, I take as representative of experimental cancer studies the research work published in the official journal of the American Association of Cancer Research. This journal changed names three times over the period from 1900 to 1950: *Journal of Cancer Research* (1918–30); *American Journal of Cancer* (1931–39), and *Cancer Research* (1940 to the present). Beginning in 1941, I also include data from the newly created *Journal of the National Cancer Institute,* because the NCI by that time was sponsoring a significant proportion (exactly what amount remains to be

determined: see Strickland, *Politics, Science, and Dread Disease*) of the cancer research that went on in this country.

93. Robert Cook, "Roscoe B. Jackson Memorial Laboratory Celebrates Tenth Birthday," *Journal of Heredity* 30 (October 1939): 448–52; list of mouse mutants, arranged by gene, year observed, and researcher who published the "mutant report," in "Mouse Linkage Group Map Info" folder, JLA-BH.

94. Little and Strong subsequently warred over who would bear the expense of the wooden boxes needed to move his mice, and over whether or not he would leave representatives of his strains for JAX production: See "Protocol in re: transfer of L. C. Strong," n.d.; CCL to Dr. Kast, April 17, 1933: both Box 739, CCL-UMO. On Murray, see Holstein, *The First Fifty Years,* pp. 14–23.

95. On Snell's biography, see oral history interviews with George Snell, from both the Jackson Laboratory Oral History Collection, American Philosophical Society, and the oral history interviews at the Jackson Laboratory conducted by the author; see also Snell's curriculum vitae, provided to the author by Snell.

96. C. C. Little, "A New Deal For Mice: Why Mice Are Used in Research on Human Diseases," *Scientific American,* no. 152 (1935): 16–18. Little showed some awareness that perceptions of pain inflicted on the animals photographed might affect the public's reception of the argument: thus, the photo of the mouse undergoing an operation had a caption indicating that the animal was anesthetized.

97. On the various political maneuverings leading up to the passage of the 1937 NCI Act, see Strickland, *Science, Politics, and Dread Disease,* pp. 1–15; Patterson, *Dread Disease,* chap. 5. On similar struggles over the nature of government intervention for the case of polio research, see Creager, "The Life of a Virus."

98. House Committee on Interstate and Foreign Commerce, "Hearing on The National Cancer Act," *Congressional Record,* 75th Congress, 1st session, Report No. 1281, p. 4.

99. Little, "Genetics in Relation to Carcinoma," *Proceedings of the Staff Meeting of the Mayo Clinic* 11 (1936): 782–83, quote p. 782.

100. Little, "The Present Status of Our Knowledge of Heredity and Cancer," *Journal of the American Medical Association,* June 27, 1936, pp. 2234–35, esp. p. 2235.

101. C. C. Little, "Symposium on Cancer: the Biology of Cancer," *Proceedings of the Annual Congress on Medical Education and Licensure* (Chicago, 1941), February 15–16, 1937, pp. 12–14. On the use of this same language to promote the rat as an experimental organism, see Clause, "The Wistar Rat."

102. Quoted from a published excerpt of a Little speech in "Cancer Army," *Time,* March 22, 1937, pp. 40–41. In late 1936, he created the new office of publicity director in the ASCC, and appointed veteran media man Clifton Read to fill the position. Together, Little and Read engineered a virtual media blitz in many of the popular magazines of the day: see Patterson, *Dread Disease.*

103. "U.S. Science Wars Against Unknown Enemy: Cancer," *Life,* 1 March 1937, pp. 11–17.

104. "Cancer: The Great Darkness," *Fortune,* March 1937, pp. 112, 114, 162.

105. On the history of guaranteed pure reagents, first produced by the Merck company in Germany in the mid-nineteenth century, see William H. Brock, *Norton History of Chemistry* (New York, 1992), pp. 174–76.

106. Minutes of the Joint Hearings on Cancer Research, U.S. Senate, Committee on Interstate and Foreign Commerce, July 8, 1937, *Congressional Record.* Cf. Kenneth Adler's work on interchangeable machine parts and the rise of the metric system in France—e.g. his essay "A Revolution to Measure: The Political Economy of the Metric System in France," in *The Values of Precision,* ed. Norton Wise (Princeton, N.J., 1995).

107. Stephen P. Strickland, *Politics, Science and Dread Disease.*

108. National Advisory Cancer Council Minutes (Meeting Transcripts), vol. 1, November 9, 1937, esp. p. 55 (Conant quote) and discussion on pp. 59–82, Record Group 443, Box 6, National Archives (Archives II), College Park, Md.

109. National Advisory Cancer Council Minutes (Meeting Transcripts) (see note 108), November 11, 1937, pp. 28, 29, 131-B, 270/a-g (CCL's typed proposal to the NACC, addressed to Ludwig Hekoten, November 22, 1937), 278, 274–76, 272–73; and November 27, 1937, p. 291, 312.

110. Staff Conference Discussion Notes, November 7, 1935; Gregg to Howard Adams, April 22, 1936; RF Grant Dispersal Summary no. 35416, November 15, 1935: all RF Archives, RG 1.1, Series 200D, Box 143, Folder 1773; Warren Weaver's Summary of the JAX lab grants, December 20, 1937, same except Box 1774, RAC-NY.

111. JAX might have even been over its mouse housing capacity: Little told RF's Hanson that the building was meant to house sixty thousand mice but told a newspaper reporter that the figure was thirty thousand. See "Preparing Mount Desert House for 30,000 Mice to Be Used in Cancer Research Interests: Science to Play Role of Pied Piper for Humanity's Sake" (cf. Hanson diary entry, October 24, 1933) and *Portland Sunday Telegram,* n.d. but from content of article probably January or February 1930, JLA-BH, Box 1. The Rockefeller Institute's own W. H. Sawyer and J. H. Bauer,

for example, agreed in March 1937 to purchase 500–750 *C57Black* mice per week for their cancer research: CCL to W. A. Sawyer, March 19, 1937, and J. H. Bauer to CCL, March 24, 1937, both RF Archives, RG 1, Series 100, Box 10, Folder 81, RAC-NY; CCL to Warren Weaver, November 13, 1937, RF Archives, RG 1.1, 200D, Box 143, Folder 1774, RAC-NY.

112. Handwritten letter from Gregg to Weaver, c. November 11, 1937; Weaver diary excerpt, November 29, 1937; CCL to Weaver, December 4 and 14, 1937; Weaver memo to Raymond Fosdick, December 20, 1937; all RF Archives, RG 1.1, 200D, Box 143, Folder 1774, RAC-NY. RF did not actually disburse the money for the building until JAX Lab's earlier tapering grants fully expired in early 1940. Mammalian genetics was mentioned in passing in this resolution, but the appropriation was not officially listed under the NS's genetics program: see Natural Sciences Grant Resolution no. 38011, RF Archives, 1.1, 200D, Box 143, 1773, RAC-NY.

113. Hanson to CCL, December 3, 1938, and CCL to Hanson, December 13, 1938; CCL to Warren Weaver, July 7, 1939; George Gray to CCL, February 23, 1940; and CCL to GG, February 25, 1940; all RF Archives, 200D, Box 143, Folder 1774. "Eating Your Cake and Having It Too," excerpt from the RF trustees bulletin, March 1940, RF Archives, RG 1.1, 200D, Box 143, Folder 1775; all RAC-NY.

114. "Fundamental Cancer Research: Report of the NACC," *Public Health Reports* 53 (1938): 2121–30.

115. "H. B. Andervont," *Annual Obituary* (1981): 171–72. Cf. Andervont, "The Influence of Foster Nursing on Incidence of Spontaneous Mammary Tumor Cancer in Resistant and Susceptible Mice," *Journal of the National Cancer Institute* 1 (1949): 147–53; "The Influence of Hybridization on the Occurrence of Mammary Tumors in Mice," ibid., no. 3 (1943): 359–65; "The Milk Influence in the Genesis of Mammary Tumors," in *A Symposium on Mammary Tumors,* ed. F. R. Moulton, AAAS Publication no. 22 (Washington, D.C., 1945).

116. On Little's resistance to Bittner's interpretations of the mouse mammary tumor work, see Daniel J. Kevles, "The Strange Story of a Cancer Virus" (unpublished manuscript).

117. Jean-Paul Gaudillière, "NCI and the Spreading Genes: About the Production of Viruses, Mice, and Cancer"; cf. also Ilana Löwy and Jean-Paul Gaudillière, "Disciplining Cancer."

118. Holstein, *The First Fifty Years.*

119. "Cancer: National Research Center Explores Its Nature and Cause," *Life,* June 17, 1940, pp. 35–38, esp. 36–37.

120. Robert Cook, "Front Line Against Cancer," *Collier's Weekly,* February 8, 1941, pp. 21–23, esp. p. 21 and 22.

121. C. C. Little, *Civilization Against Cancer* (New York, 1939), esp. chaps. 4 ("The Hunt for Living Weapons") and 5 ("Making Cancer to Order").

122. These protests are chronicled in Patterson, *Dread Disease,* p. 129.

123. Critics quoted by the ASCC in "Answers to 'Questions on the Cancer Problem,'" March 1937, ASCC folder, James B. Murphy Papers, American Philosophical Society; cf. Rous to William Gye, June 14, 1937, Peyton Rous Papers, American Philosophical Society.

124. Peyton Rous, "A Sarcoma of the Fowl Transmissible by an Agent Separable from the Tumor Cells," *Journal of Experimental Medicine* 13 (1911): 397–403, 408–9, and Kevles, "Strange Story of a Cancer Virus."

125. H. Gideon Wells, Maud Slye, and Harriet Holmes, "The Occurrence and Pathology of Spontaneous Carcinoma of the Lung in Mice," *Cancer Research* 1 (April 1941): 259–61.

126. Practically, A-Z tests involved injecting a genetically homogeneous virgin female mouse with the female patient's urine: if a mouse's ovaries showed a change in size within forty-eight hours, the test was positive. See Little's form letter describing JAX Lab's Ascheim-Zondek testing program (preserved as CCL's scrap paper for the draft of a funding campaign letter!), c. 1940–44, Little Box no. 1 and Jackson Laboratory Annual Report, 1941, p. 4 (both JLA-BH); Jackson Lab Financial Statement, 1943–44, RF Archive, RG 1.2, 200A, Box 133, Folder 1189, RAC-NY. On the magnitude of JAX mice distribution to the government during World War II and on A-Z testing at JAX, see Holstein, *The First Fifty Years,* pp. 29–30; On JAX Lab's wartime sales, see Jackson Laboratory annual reports, fifth (1941) through ninth (1945), JLA-BH; and CCL to Weaver, May 3, 1946, RF Archives, RG 1.1, 200D, Box 144, Folder 1777, RAC-NY. On the building of the second RF "mouse house," see CCL to Weaver, January 18, 1946, March 22, 1946, and March 26, 1946; Weaver to CCL, March 21, 1946; Weaver to Raymond Fosdick, April 24, 1946; CCL to Weaver, May 23, 1946, March 24, 1947, and August 21, 1947, which notes that RF supplemented the original grant of ten thousand dollars for increased construction costs; all RF Archives, RG 1.1, 200D. Interestingly, when JAX got the go-ahead on the new building from RF, Little sent one of his scientists (George Snell) on site visits to other relatively large-scale mouse breeding facilities, looking for construction ideas; in this regard, he also encouraged Snell to look at an actual industrial lab—Bell Telephone Company: see Snell to Sidney Strickland, October 22, 1946, Box 7–6, JLA-BH.

127. Given the founding of the *Journal of the National Cancer Institute* in 1940, editorial bias was clearly at work in shaping this distribution. In the 1950 *JNCI,* for example, 100 percent of mouse articles deal with inbred

mice, and a significantly high proportion addresses milk influence and tumor transplantation issues.

128. Nineteenth annual report of the Jackson Laboratory, 1947–48, JLA-BH; see also Holstein, *The First Fifty Years,* chap. 2.

129. Telegrams contained in Box 735, CCL-UMO.

130. On responses to the JAX Lab fire, see Weaver to K. Compton, October 3, 1947; Weaver to Fosdick, November 7, 1947 (both RF Archives, RG 1.2, 200A, Box 134, Folder 1191, RAC-NY); "RBJ Lab," RF trustees bulletin, November 1947; CCL to Weaver, November 21, 1947, and November 28, 1947: all RF Archives, RG 1.1, 200D, Box 144, Folder 1777, RAC-NY. For a specific example of a "mouse return" see CCL to Muller, November 3, 1947; Muller to CCL, November 7, 1947; CCL to Muller, November 17, 1947: all in the H. J. Muller Mss., Manuscripts Department, Lilly Library, Indiana University–Bloomington; more generally, see 1948 JAX annual report (public version) and a film produced by JAX Lab on the subject: "Rx Mouse," c. 1950–52: both in JLA-BH. For Little on JAX Lab as a "spiritual force," see Elizabeth Russell, "The Roscoe B. Jackson Memorial Laboratory: Leaders in Cancer Research and the World's Largest Supplier of Inbred Mice," *Purina Laboratory Animal Bulletin,* spring 1955, in JLA-BH.

131. Richard W. Jackson (Roscoe's son) to Warren Weaver, October 7, 1949, plus enclosed endorsement, Rockefeller Foundation Archive, RG 1.1, Series 200D, Box 144, Folder 1778, RAC-NY.

132. Kenneth Endicott, "The Chemotherapy Program," *Journal of the National Cancer Institute* 19 (1957): 275–93. Cf. Bud, "Strategies," and Ilana Löwy, *Between Bench and Bedside: Science, Healing, and Interleukin-2 in a Cancer Ward* (Cambridge, Mass., 1996), chap. 1.

133. In a short and often-neglected article, historian Donald Swain makes a similar point about the general relationship between pre-and postwar biomedical policy at NIH: see Donald C. Swain, "The Rise of a Research Empire: NIH, 1930 to 1950," *Science,* December 14, 1962, pp. 1233–37.

134. Cf. Arnold Arluke, "'We Build a Better Beagle': Fantastic Creatures in Lab Animal Ads," *Qualitative Sociology* 17, no. 2 (1994): 143–57.

135. Donna Haraway, *Modest Witness@Second Millennium: FemaleMan Meets Oncomouse*™ (New York, 1997), p. 83.

13

A Rodent for Your Thoughts

The Social Construction of Animal Models

Kenneth J. Shapiro

Down the metal chute he slid. . . . A moment before, in a hygienically sealed room at Charles River Laboratories . . . a white-coated technician had weighed him, checked his sex . . . and tossed him down the open chute, onto a soft wood-chip bedding. . . . Now a packer picked him up and loaded him, with nine other rats, into a cardboard carton. . . . At Johns Hopkins, the rat would have a tiny, precisely demarcated region of his brain chemically destroyed, to simulate memory loss suffered by human victims of Alzheimer's disease. . . . {S}cientists would run him through a maze designed to test his short-term memory. Ultimately, he would be decapitated and his brain analyzed. In the laboratory records that documented his daily care, brain surgery, maze performance, and death, he would be identified as No. 1913.

—From "Specimen No. 1913," by Robert Kanigel (1987)

Some years ago I had a rat named Weebee. . . . He was never caged, except to travel. . . . He had his own bed next to my pillow, along with his food dishes. . . . I'd talk to him, usually nonsense talk. . . . The more I chattered, the more his teeth clicked. . . . The onset of his teeth-clicking was gradual, beginning with just his lips moving. . . . Then he'd flick his little tongue out a couple of times, and the teeth-clicks and eye-bulges would follow. Because it really looked like he was trying to talk, I called this "Weebee-talk-with-eye-pops." He would usually do this on command. He'd be sitting in my hand, looking me in the eye, and I'd say, "Weebee talk with eye-pops?" a couple of times, and sure enough, he'd oblige me. . . . He lived just about 18 months. . . . I've never gotten over losing him.

—From "Weebee Talk," by Elizabeth L. Fucci (1995)

In this paper I approach the issue of nonhuman animals (hereafter "animals") and medical models through an account of the social construction of the laboratory animal in psychology. It is framed by modern psychology's adoption of the strategy of developing animal models of human psychological disorders. The account of lab animals is historically based and keys on the use of rats in psychology labs. The discussion of animal models is empirically based, featuring the presentation of the results of a survey and citation analysis of selected animal models of eating disorders.[1]

In my use of a social constructionist perspective, I keep to the spirit of Berger and Luckmann's classical exposition in *The Social Construction of Reality*.[2] By doing so, I hope to skirt current arguments between proponents and critics of postmodernist thought. Some of these revolve around the extent to which a constructionist approach requires a radical ontological and/or epistemological position which either denies any reality beyond the construction or a veridical access to it. (See, for example the spirited exchanges in various intellectual fora occasioned by Sokal's hoax).[3]

Following Berger and Luckmann, then, social construction refers to the consensus, whether explicit or implicit, that a group of people arrive at as to the meaning to them of an object or class of objects. The term "construction" emphasizes the active role that people and institutions play in conferring that meaning and emphasizes the degree to which that meaning derives from and is made intelligible by the social contexts in which the actors and objects reside. In the case of the history of science, this perspective focuses explanation of scientific developments on the social world of which the scientific enterprise is part and parcel. In this limited usage, the account does not imply the idealist position that reality is a mental construct. Here I largely bracket ontological and epistemological concerns, while attending to relevant social processes—the work is historical and sociopsychological, not philosophical.

The "lab animal" is a construction achieved through the complex of philosophy, purpose, and practice that constitute the laboratory animal enterprise. It is a product of the variety of ways that scientists and other workers procure, breed, house, care for, experiment on, and dispose of laboratory animals. More subtly, that class of objects is also constructed through the philosophy of science underlying lab animal science; the language scientists use both formally in publications and research proposals, and informally, in the lab, to refer to lab animals; the ethics and policies they support and develop regulating their treatment of animals; and, finally, the architecture and layout of the labs they help design.

The White Rat

Both the black rat (*Rattus rattus*) and the brown rat (*Rattus norvegicus*, or the Norway rat) have a long history of involvement in human affairs before their selection as the standard laboratory animal in the field of psychology. In the fourteenth century, following the introduction of a bacterium hosted by black rats, the bubonic plague decimated the human population of Europe several times over.

Rats are among the most adaptive and successful animals. A social history dedicated to them documents their marvelous capacities for proliferation and survival.[4] Yet they have long been considered a scourge and pest—an apt symbol of our own villainous and scurrilous potential ("take that, you dirty rat"). There is considerable irony, then, in the selection of this nocturnal, and, from a human perspective, unruly and destructive species as the primary inhabitant of the highly controlled, rule-bound, broad-daylight laboratory of science.[5]

Only a few decades after the first recorded use of them in modern science, rats were used for the first time in psychological research, in 1898.[6] This was made possible by a process of selective breeding of albino forms of the Norway rat. Birke notes that the expertise developed in controlled livestock breeding in the eighteenth century and, later, in the selective breeding of "fancy" cats and dogs "[was] taken into the laboratory by the early twentieth century."[7] Quinn offers an intriguing account of how paintings of cattle provided an ideal image toward which breeders of so-called "fat cattle" could strive.[8] This process of domestication in the lab or, using the more specific term suggested by Lockyard, this process of "laboratorization" accelerated when Meyer and Donaldson selected albinos from a population of Norway rats, which became the now commonly used Wistar breed of laboratory rats.[9] In 1903, Watson, one of the founders of behaviorism, wrote a doctoral dissertation entitled "Animal Education: The Psychic Development of the White Rat" and Donaldson published *The Rat.* Through selective inbreeding for generations, strains such as the Wistar and Long-Evans hooded rat were established that have virtually identical genetic characteristics.

The differences between these constructed breeds and the Norway rat from which they were developed are striking. As Hediger states: "In the case of the white rat . . . , a domesticated animal may be fundamentally different from its ancestral wild stock. Since the wild rat was much more difficult to handle, it was replaced in 1895 by the far more docile standardized white rat in laboratory experiments in animal psychology."[10] As compared to a wild or even a first-generation captive Norway rat, an albino rat is less vicious, less fearful of people, and less excitable and intent on escape.[11] He or she also has lost the "threat posture," and fighting is reduced to playful behavior; mice killing is much reduced as is cautiousness in response to novelty and vocal communication. In general, the albino is more quiet and tractable.

By the mid-1920s a survey indicated that roughly 25 percent of psychology labs in the United States included animal facilities, and currently, the figure is closer to 50 percent.[12] The white rat became a common fixture of experimental psychology, both actually and symbolically. The animal laboratory was the place where psychologists did their work and the laboratory animal was the object of study. The construction of the laboratory rat and other species such as the mouse and guinea pig through selective breeding recently been supplemented by more sophisticated means of transformation. Through biogenetic engineering, transgenic animals have been developed that more efficiently and comfortably meet the needs of scientific investigators. On the level of cultural symbols, "oncomouse" (a cancer-prone variety) may soon vie in popularity with Mickey Mouse.

Before describing how and why an emerging scientific psychology constructed the laboratory animal, it is worth pointing out that another rather different psychology that emerged contemporaneously with it also utilized rats and other animals, if only on a symbolic level. In psychoanalysis, the most successful attempt at this interpretative approach to psychology, "animal" referred to that aspect of humanity which Freud offered as the motor force of individual behavior and personality, the unconscious. With its reservoir of instinctive yearnings, Freud's unconscious was the "animal" side of humans—instinct-dominated irrationality as distinguished from peculiarly human rationality and deliberative judgment. While scientific psychology featured the white rat, "psychoanalysts stalked the dark rat scurrying through the obscure shadows" of patients' dreams, as in Freud's seminal case study of the "Rat Man."[13] Ecopsychologists and feminist psychologists have developed a critique of the dark rat and the influence of this association on our views of nonhuman animals, nature, and women.[14]

EXPERIMENTAL PSYCHOLOGY

Viewed from the outside, most scientific fields appear unified, as if they consisted of one discrete subject matter and/or one general way of studying it. However, the view from within typically reveals a disparate array of subject matters and varied and often competing ways of investigating them. This is certainly true of the field of psychology both at its inception near the end of the nineteenth century and as it is today. In his monumental six-volume *Psychology: A Study of a Science,* one respected critic of the field, Sigmund Koch, denied that there is a single, unified field called psychology: "My argument has been that psychology is misconceived when seen as a coherent science or as any kind of coherent discipline devoted to the empirical study of human beings. Psychology, in my view, is not a single discipline but a collection of studies of varied cases, some few of which may qualify as science, whereas most do not."[15]

Merely naming a field only begins to answer the questions of what will be studied and how. Was this modern "psychology" to study behavior, what we do and why we do it; or was it to study what we experience, the meanings we find in and give to the world? Is behavior describable as the movement of a muscle, or as the execution of an individual's intention or of a family's communication style? Should we look inside the individual at the organization of his or her nervous system or outside, at the organization of the family and his or her role in it? Should we infer mental operations from behavior or look at the immediate environment that may control behavior, or perhaps at past environment? At the history of the individual or of the culture? These questions still are alive in psychology, so that the field is robust and vitally pluralistic according to some observers, and still immaturely struggling to define itself according to others.

By most accounts and certainly in the predominant view, the original impulse for this emerging field in the late nineteenth century was the desire to separate itself from the armchair speculations of philosophy by adopting an approach that would be closer to that of the natural sciences than to the humanities. However, from the beginning of this effort, there was another identity for psychology vying with this intention to achieve the legitimacy and status accorded a natural science. There were and are at least "two psychologies": one "based on the methods of the natural sciences, and a second, subjective psychology based on the human sciences."[16] This "second psychology" developed its own methods distinct from those of the experimental method of the natural sciences. Its interpretative enterprise involves introspection and reflection, interview and attitude survey, individual case study and observation in real life settings. Generally, these are methods that focus on the task of clarifying and understanding the meaning of experiences, on questions of what people experience and how an individual's purposes and intentions form experience and behavior.

By contrast, the construction of psychology as a natural science largely restricted itself to the study of directly observable events. Instead of a focus on the meaning of behavior, there was an emphasis on predicting its occurrence, frequency, and intensity—all of which lent themselves to quantitative description. With this more objective or external focus, the investigator applies his or her ingenuity to the development of apparatus that induce behavior in such a way that it can be directly recorded and expressed in mathematical or statistical form. The investigator then makes predictions about these quantitative results, using strict rules of logical inference. These "hard" methods, as distinguished from the "soft" approaches of interpretative or human science psychology, require that the investigator be impartial and detached from what he or she observes, like a physicist recording measurements on a gauge.

This natural scientific method was in accord with both the practice and the philosophy of science of the time. The positivist philosophy of science held that science is founded on impartial observations which can be quantified and from which inferences can be drawn governed by strict rules of logic.[17] Only by using such a method can one be positive or certain of one's findings. Any other approach reverts to the uncertainty and speculation of the earlier efforts at psychologizing in philosophy.

Another tenet of positivism is that it is partial to reductionist explanations.[18] Rather than describing complexity as it appears (the watchword of phenomenology, the philosophy underlying the interpretative approach to psychology), or employing terms borrowed from common sense,[19] or searching for principles of organization that only arise with increased complexity (emergentism), complex phenomena are understood by breaking them down into simpler components. As physics had its microparticles, chemistry its periodic table of basic elements, and biology its genes and later its DNA sequences, the new natural science of psychology would understand complex human experience and behavior in terms of building blocks of simple linked stimuli and responses.

Experimental psychology embodied this positivistic approach and offered the best chance to emulate the success of physics and biology. Despite the fact that several prominent early thinkers (James, Baldwin, Wundt, and Dewey) subscribed to some version of the "two psychologies" idea and held that at least part of modern psychology should be an interpretative science distinct from the natural sciences, experimentalism quickly became the established view of this modern psychology.

The experimental approach built on the positivistic idea of inquiry: the investigator restricts study to directly observable objects, observes that object from a detached or impersonal point of view, measures it in quantitative terms, and explains it by reducing it to simple component parts. To this ideal, experimentalism adds the notion that the observed, and particularly the conditions under which it occurs, must be varied systematically. To effect this experimental variation, it is necessary for the investigator to have control over the object of study.

How could these criteria of being observable, manipulable, measurable, and replicable be met? These investigations could not take place in the bedroom with its personal disclosures and intimacies; or the family dinner table and the street corner with their complex interactions and negotiations;

or the classroom, which at least in theory is the dedicated site of that important psychological phenomenon, learning; or even the clinic, where breakdowns in all of these are examined and treated.

Rather, early psychologists selected the laboratory as what Capshew calls the "locus classicus for scientific psychology"; indeed, "[T]he enduring motif in the story of modern psychology is neither a person nor an event but a place—the experimental laboratory."[20] Psychologists realized that the achievement of the primary features of experimentalism could be assured most fully in the laboratory. Physically, the laboratory allowed detachment of the investigator, unimpaired observation, and relative control of the object of study. It also provided a site in which technology to enhance and extend the limits of observation, and instrumentation to quantify and record results of observation could be developed. The lab was a place away from the buzzing confusion, the uncontrollable flow of events of the ordinary world. It was, as it were, outside of history and culture. Events observed and measured there could gain a claim to universal generality. In the simplified and controlled setting of the lab, psychology could become an exact or "hard" or "pure" science.

Beyond these physical fits with the parameters of an experimental science, the laboratory had strong symbolic value for the emerging field. In his review of the history of the laboratory and of accounts of it in psychological literature, Capshew points out that "in symbolic terms, the laboratory provided a focus for the disciplinary aspirations of psychologists, functioning as an icon of the transcendent power of scientific knowledge. In the process, it became as much a spiritual home as an experimental workplace." He also establishes that it had a "central role in the construction and maintenance of psychology's scientific identity," that it became a "metaphor for scientific psychology itself." Clearly, the establishment of laboratories was seized upon as the way toward the establishment of psychology as a science.[21]

Who would be the objects of investigation? Humans were too complex, unpredictable, uncontrollable, and resistant to manipulation. In any case, society would not abide inducing suffering in them or keeping them in the lab for a lengthy stay against their will.

Enter the rat. It is important to understand that at that time, in the late nineteenth century, "laboratory animals" were a new class of animals. There were "wild," "farm," and "companion" animals, but only recently laboratory animals. A new category had to be constructed to fit the laboratory experimental situation. To create this class of animals required more than simply catching some wild animals and putting them in cages in the lab. Three approaches were developed and applied: selective breeding, certain forms of socialization (or desocialization), and innovative architectural features of the lab (tiers of stacked cages, restraining devices, and Skinner boxes). As described, through the combined effect of these, psychologists transformed *Rattus rattus* and *Rattus norvegicus* into the docile, manipulable white rat of the laboratory.

THE ANIMAL MODEL STRATEGY

While laboratory animals were the direct object of observation in this scientific enterprise, human, not rat, psychology was the primary object of study. Early psychologists were aware of attempts in biomedical science to study human diseases, particularly infectious diseases, by creating animal models of them. Robert Koch, one of the developers of the germ theory of disease, had recently demonstrated that an animal model of an infectious disease, tuberculosis, could be created by isolating the tubercle bacillus, infecting an animal with it, and then attempting to affect the course of the disease through drug or vaccine intervention. Experimental psychologists readily grafted the animal model strategy onto the construction of the laboratory animal. By borrowing this strategy they hoped to understand the pathologies of the psyche or, to use more contemporary language, behavioral disorders.

While it is the adoption of this strategy that concerns us in this discussion of the construction of the lab animal, there are at least two other strategies involving animals in psychology. Comparative psychologists, who make up only a small proportion of the researchers using animals,[22] study animals as an end in themselves and to understand differences among species. The animals are not considered models of human behavior and any extrapolation to the human condition is considered incidental. Another group of psychology researchers studies animals to understand allegedly universal processes. For these researchers, in effect, any animal is a model for all other animals, including humans, since the working assumption is that the basic principles of learning, perception, and other psychological phenomena apply equally to all animals.

The decision to adopt the animal model strategy was one taken early by psychologists and pursued to the present. Sometimes behavioral disorders, or at least arguably similar conditions, occur naturally in nonhuman animals. These, then, can be induced in laboratory animals. However, more often, comparable conditions do not occur naturally in nonhuman animals and investigators must attempt to devise situations which induce what are at least analogous versions of the human condition under study. A recent publication by Overmier and Burke lists over twenty-five hundred abstracts of studies involving animal models.[23] To illustrate the scope of the contemporary animal model enterprise, table 1 lists selected examples of typical conditions for which animal models have been sought, taken from the subject index of this publication. Clearly, psychologists have attempted to develop an animal model for virtually any problem in the human condition that has a psychological cast.

The scope of the procedures devised to create the models as broad as the conditions to be modeled. In general, the procedures extend from (a) genetic manipulation—animals are bred for certain dispositions, such as susceptibility to seizures from auditory stimulation, or, more recently, biological engineering techniques manipulate certain susceptibilities or deficits; to (b) manipulations of the nervous system—different parts of the brain are implanted with electrodes, electrically stimulated, lesioned, or cut, surgically ablated or removed, or destroyed through toxic injection; to (c) invasive manipulation of other parts of the body—different systems and organs of the body are monitored through fistulas (surgically created holes), stimulated through substances introduced into cannulae (surgically implanted tubes), or destroyed; to (d) behavioral and environmental manipulations—food or drink are made conditional on the performance of certain tasks; avoidance of painful conditions such as shock, heat, or cold are made conditional on performance of certain tasks (or abruptly applied independent of any performance or made conditional on another animal's performance); all manner of basic, even life-sustaining social supports and environmental stimulation and enrichments are withdrawn or otherwise manipulated; and all manner of painful, distressing and/or harmful social and environmental stimuli are presented.

Models are analogies.[24] A model is analogous, not identical, to the original. Models have to do with similarity but they do not imply identity. The statement, "a rat is a model of a human being," makes the limited although undeniable claim that there are some ways in which the two are similar. However, it does not imply or, really, allow that they are identical. For, in the other direction, if a rat is merely a model of and not identical to a human being, it necessarily follows that there are some qualitative differences between the two. The relation model/modeled, then, consists of a set of similarities and differences.

Some philosophers of science have distinguished between heuristic and causal analogies, where the latter implies the stronger claim that the model actually duplicates discrete causal mechanisms in the phenomenon of interest. LaFollette and Shanks argue that identity is not theoretically possible, as systemic disanalogies necessarily come into play.[25] Following LaFollette and Shanks, I take the position that, however strong the claim, in the final

TABLE 1. Typical Conditions for Which Animal Models in Psychology Have Been Sought

affective disorder	Huntington's Chorea
aging	hyperactivity
alcoholism	insomnia
Alzheimer's disease	lead poisoning
amenorrhea	learning disorders
amnesia	liver disorders
anorexia nervosa	malnutrition
anxiety	maternal deprivation
asthma	memory disorders
attachment disorder	mental retardation
attention deficit disorder	motion sickness
autoimmune disorders	movement disorders
brain disorder	multiple sclerosis
bulimia nervosa	muscular disorders
cerebral vascular disorder	neonatal disorders
child abuse	neuromuscular disorders
cigarette smoking	neuroticism
cold exposure	noise effects
colon disorders	obesity
concussion	obsessive compulsive disorder
coronary disorders	pain
coronary-prone personality	panic disorder
crowding effects	Parkinson's disease
depression	phantom limbs
dermatitis	phobias
deviant behavior	posttraumatic stress disorder
diabetes	premature birth
drug addiction	prenatal exposure
drug side effects	reading disabilities
drug withdrawal	rheumatism
emotional trauma	schizophrenia
endocrine disorder	self-mutilation
environmental stress	sensory deprivation
epilepsy	separation anxiety
failure to thrive	sexual function disturbance
gender disorders	sleep deprivation
genetic disorders	social isolation
hallucination	stress disorders
headache	ulcers
hearing disorders	visual disorders
helplessness	vitamin deficiency disorders

Source: Overmier and Burke, *Animal Models of Human Pathology.* Reproduced in Kenneth J. Shapiro, *Animal Models of Human Psychology: A Critique of Science, Ethics, and Policy* (Seattle, 1998), p. 162. © 1998 Hogrefe & Huber Publishers, Seattle, Toronto, Göttingen, and Bern.

analysis any model is limited functionally to a heuristic and any insights so generated must be verified in the original.[26]

This is particularly true when the objects of study are behavioral disorders, which are, to a greater extent than biomedical disorders, shaped and conditioned by socially conferred meanings unlikely to be duplicated in the laboratory. In any case, constructive contribution is not necessarily a direct function of the veridicality of the model. High fidelity of the model to the original may or may not be critical. In fact, differences between the original and the model may be the occasion of new understanding and/or effective treatment intervention—for example, an animal model, by contrast with the modeled, may be immune from a disorder so that differences rather than similarities between the two provide the basis for gains in understanding and/or treatment. On the other hand, a high-fidelity model may not add any new understanding or treatment advance.

In the final analysis, the function of a model is limited to that of a heuristic device. Through the development of models, investigators generate hypotheses which then must be tested in the phenomenon of interest, here human psychology. The evaluation of the contribution of any animal model is an empirical and, more broadly, a historical problem. I provide an application of suggested requisite analytic methods illustrative of the empirical evaluative component below.

By contrast to this limited, albeit important, prospective role of animal models, there has been some inclination to adopt a more ambitious claim. Historically, the example of the use of models of infectious diseases occasioned this misleading concept of models in science. An animal model of a disease resulting from experimentally induced infection with the germ, say for tuberculosis, has a strong face validity not apparent for an animal model of a behavioral disorder, say depression or bulimia or an anxiety condition—or even for an animal model of a biomedical disease, say heart disease or a family of diseases like cancer, neither of which have a comparable discrete germ aetiology.

While even tuberculosis in a rat is clearly distinguishable from tuberculosis in a human, the disease process and course are very similar. There is an inclination to think of the laboratory animal as a host for a disease process that is identical to that found in the human setting. This leads to the notion of animal as stand-in for human. The idea of animal model, then, comes to refer to a *re-presentation,* a simple and direct transfer of the phenomenon of interest into another being. The animal model is merely a conduit or vehicle through which the disease, itself a discrete physiological process, is expressed. The concept of animal model here is inflated in the sense that it is seen as a full-blown instantiation of the original. The animal serving as model is at once inflated to a stand-in for human being and reduced to a conduit for certain relevant physiological processes. I would speculate that recent technologies that produce transgenic animals[27]—ones that carry aspects of the genetic code of other animals, particularly human animals—may trigger another period of a similarly inflated conception of the animal model.

With respect to the laboratory animal, this transcendence/reduction antinomy is largely resolved in the direction of reductions—as the following fuller description of the social construction of the lab animal will show. In the final section, I will provide evidence that, inflated conceptions notwithstanding, ironically, psychological research does not meet even the heuristic or generative function proper to the strategy of using models in science.

THE SOCIAL CONSTRUCTION OF THE LABORATORY ANIMAL

Psychologists constructed both laboratories and "lab animals" under the rubric of positivism and the ideal of experimentalism. What kind of being emerged from this project? Here I describe the peculiar kind of being that was the unwitting by-product of the experimentalist program with its attempt to gain experimental control, investigator detachment, objective observability, measurability, and the reduction to simple components.

Every animal is an individual and a member of a particular species. His or her identity consists of both those aspects uniquely individual and those shared within a species. Animals of a particular species vary in the degree of individual and species-specific behaviors that constitute their identity.

While we all probably nod to these generalizations, the predominant Western modernist philosophy, humanism, is built on their denial. It is a basic tenet of humanism that individuality requires rationality and that only human beings are rational.[28] In the humanist tradition, nonhuman animals are instinct- and need-dominated; they are not autonomous, self-determining beings; they are not individuals.

Modern science is also known as scientific humanism as it is a direct development of the humanistic philosophy beginning with the Enlightenment. It is no surprise, then, that the social construction of lab animals features the "loss" of their individuality. Some of the features identified that effect this *deindividuation* include: the large numbers of animals used; the brief time many of them live; the focus on parts of animals rather than on whole animals; the press for genetically identical animals; the need to kill the animals; and the production of animals bred solely for research.[29] Typically, in the experimental situation, it is assumed that there are no individual differences. Rather, observed differences are understood as results of the experimental manipulation itself, while any residual differences are attributed to errors in measurement or age or species.

Phillips describes a more subtle way in which "individual identities are consciously and systematically obliterated."[30] Both her own ethnographic study of twenty-three contemporary biomedical and behavioral research laboratories and Arluke's study of twenty labs show that "animals were almost never named."[31] Fouts observes the same reluctance to give animals names among research scientists in zoos, and also attributes it to "an attempt to de-individualize the animal."[32] In the few instances where they were named, the names "functioned more as jokes than as true proper names."[33] Names were selected that had personal relevance to the namer and they were not used to address the animal. Even this limited use of names was "generally confined to undergraduate and nonprofessional staff."[34] In their more proper use, "proper names are linked to the social emergence of personality, which engenders a matrix of ideas and behaviors peculiar to one individual."[35] Through naming, we construct an individual. Phillips goes on to suggest that without a name a narrative account of that life is precluded; that life is not a biography.

Phillips cites an account by Kanigel of a rat referred to as "No. 1913" (part of which appears as an epigraph to this essay).[36] While presenting the scientific context of the rat's life, Kanigel provides a deliberately ironical biography of "No. 1913" and allows us to get to know him as an individual, thereby demonstrating what is absent in the standard scientific account in which laboratory animals figure. Failure to name is a part of the process of deindividuation in the social construction of laboratory animals.

The rhetoric used in scientific reports further supports this deindividuation. Investigators describe what was done to the animals rather than what the animal was doing. Subsequent "action" by the animals is reduced to an effect of the investigator's action in manipulating the environment and the animal. The lab animal is not an agent or autonomous being but a product of the experimental manipulation. Even the description of the animal as passive respondent focuses more on the "technical procedures" than on the behavior of the animal: "[W]hat is actually happening to *the animal* is subordinated to the descriptions of what was done in *order to make measurements.*"[37] When the behavior of the animal is described, the description is couched in terms that make the animal as individual "disappear."[38] An act by a rat, say to avoid something (an aversive stimulus), becomes an "avoidance reaction." The rat does not act; rather, under certain stimulus conditions, the avoidance reaction is effected. The rat as individual, in the sense of autonomous agent, has disappeared.

Another view of laboratory animals which sustains this deindividuation is that any particular animal is replaceable. This is consistent with the emphasis, in the experimental approach, on control and reduction of variability. Clause describes the "ideal of a standardized mammal," whereby animals are produced for the laboratory that are as similar to each other as possible. Ideally, they are "interchangeable parts."[39] When animals die, there are no individuals lost, for lab animals are interchangeable and replaceable members of a "population," not unique individuals. Certainly, there is no loss in the more laden sense of mourned "loss," for there is no individual relation or bond between animal and investigator. What is lost is a replaceable member of a species. (Phillips notes in her study of research scientists that in one of the few instances in which an animal was named, "Sylvia" the monkey was still Sylvia even when the original Sylvia had been replaced by a second one.)[40] But is there even this more limited "loss"?

If there are no genuine individuals in the lab, surely there are different species and there are then differences among animals based on species-specific behavior. However, psychologists and other scientists using lab animals commonly refer to them by that name: they are not "the rat or pigeon or dog" but "the animal."[41] The irrelevance and denial of species differences to psychologists is revealed by Beach's alarm when he observed that, by 1950, psychologists were "dedicating over 50% of their research to an animal that represented only .001% of the types of living creatures that might be studied."[42]

In more formal settings such as scientific publications, another term that is used is "organism." This expresses a prevalent attitude that an animal under study is an organic process, a physiological and behavioral system. Rather than an individual or even a species-specific physiology or behavior, the lab animal is reduced to a general process. The animal is "prepared" so as to eliminate any individual or species identity and disclose for study a biological organism, a generic animal, a "preparation."[43] This process of preparation culminates in the use of apparatus and procedures such as stereotaxic fixation, anesthetization, the Pavlovian sling, the plastic restraining tube, and other forms of restraint (see figure 1).

However, the construction of lab animal as generic animal is a more extensive program that is not limited to the experimental procedure. It begins with selective breeding and, now, genetic engineering, and continues with the provision of certain housing conditions, both of which further effect the stripping away of species-specific behavior. I earlier described this process in the case of the selective breeding which led to the production of the white rat, the standard bearer of the new lab-based science of psychology.

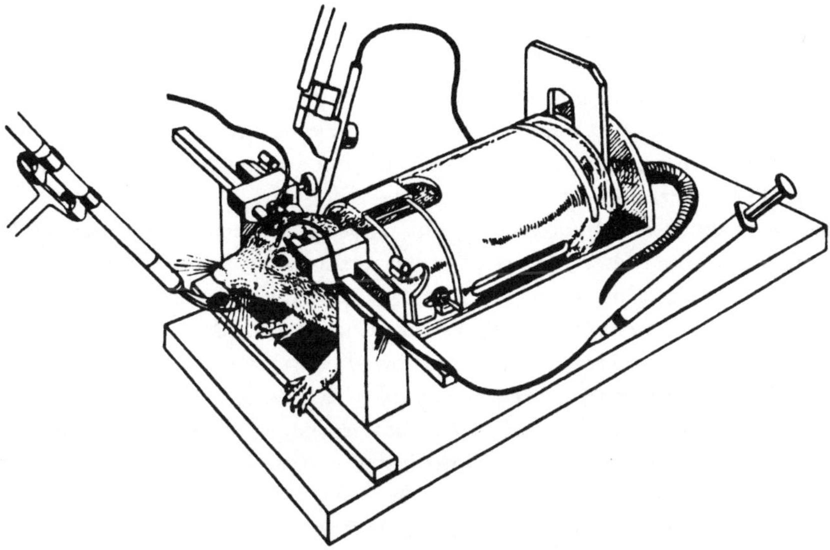

1. Apparatus for chronic recording of rat's "near-natural ingestion." R. Norgren, H. Nishijo, and S. P. Travers, "Taste Responses from the Entire Gustatory Apparatus," in *The Psychobiology of Human Eating Disorders: Preclinical and Clinical Perspectives,* ed. H. Schneider, S. J. Cooper, and K. A. Halmi (New York, 1989), pp. 246–63. Adapted from Kenneth J. Shapiro, *Animal Models of Human Psychology: A Critique of Science, Ethics, and Policy* (Seattle, 1998), p. 162. © 1998 Hogrefe & Huber Publishers, Seattle, Toronto, Göttingen, and Bern.

In terms of housing, the cage plays a major actual and symbolic role in the construction of generic animals. Caging an animal for much of his or her life and without the company of other animals has a profound effect on most animals. Wemelsfelder describes the subsequent condition as "boredom."[44] This boredom is more than an occasional passing mood which the animal can readily shake off. The behavior of a chronically caged animal is largely limited to bored behavior; at the extreme, it consists of repeated stereotypical movements and postures which, it is important to note, are similar for animals across species. A bored animal no longer demonstrates much of his or her species-specific behavior.

In an animal that has been caged for his or her entire life or even chronically caged for much of his or her development, this lack of species-specific behavior signals more than the immediate absence of an environment that would stimulate it. The animal has literally lost interest in his or her surroundings, and its species-specific identity is lost. A lion that has no lion country is no longer a lion.[45] The scientist has succeeded in constructing a generic animal, an organism bereft of species-specific identity.

A particular phase in the recent history of psychology provides a variant that illustrates this despeciation of the lab animal. According to behaviorism, which was the dominant approach to psychology in the 1940s and 1950s, animals (and humans) were completely understandable in terms of observed behavior. Mind, subjective experience, and consciousness did not count in this attempt at a scientific psychology.[46]

An investigator fully adopting this approach assumed the attitude toward animals that they consist of conglomerations of observable behaviors. Further, these behaviors are more or less exclusively a product of the animals' immediate environment. Behavior is "shaped" and maintained through aspects of the environment that "reinforce" particular behaviors. The identity of a given animal is predominantly a function of this shaping rather than of any membership in a particular species. All animals are organisms or generic animals in that their learning and their behavior is governed by the same rules of environmental contingency.

It is precisely this environment which the investigator can, at least in theory, completely control. The investigator can shape the animal to perform any behavior, limited only grossly by species-specific capabilities. In this behaviorist psychology, the social construction of lab animals as generic animals includes the "discovery" of the law governing organisms—they are eminently shapable by external forces, particularly the kind the investigator can readily control and manipulate.

A spin-off of the construction of animals as preparations is that the animals and their by-products are marketable. Of course, the laboratory produces a number of products—literary (published studies), technological, and methodological—and has been likened to a factory.[47] However, the animals and their by-products are also literally products or commodities of the laboratory. As early as the turn of the century, Pavlov "bottled the gastric juice drawn from esophagotomized dogs" and sold it to the international medical community.[48]

More generally, animal laboratories produce standardized animals or preparations that are sold as supplies to other laboratories. A major supplier of inbred mice, the Jackson Lab, was developed, in part, to provide standardized animals for biomedical research. Ironically, that lab was influenced from its outset by the "industrial strategy" of mass production, as its major benefactors were from the automotive industry.[49]

From individual to species to commodifiable preparation, there is yet one more turn of the screw in the social construction of the lab animal. The initial impulse of the effort to construct a scientific psychology involved a dedication to direct observation. Psychologists attempted to utilize technology and instrumentation to enhance and extend the range and depth of what they could observe. In a subtle way, this effort in practice eventually blurred the boundary between the instrument and the object of measurement.

Consider a rat that is chronically implanted with an electrode in his or her brain and is connected by a tether to machinery that sends stimuli and receives and records responses. The rat is more a part of the instrumentation than a discrete object of study. The animal is merely a conduit for certain energies, fluids, and electrical impulses. Certain relations between brain function, external stimuli, and movement are the object of study. The animal has lost his or her integrity in that only certain parts of the animal are the focus of interest.[50] The animal is not viewed or acted toward as a whole creature let alone as an individual.

The chronically implanted animal is not so much put "under the microscope" as made part of it. This exceeds fitting the technology to the animal. The integrity of the animal, of his or her individual and species-specific behavior, is dissolved; even the integrity of the generic organism presents no boundaries and is regularly violated. The rats are "laboratory animals" in the sense that they are part of the laboratory; they are part of this complex of sophisticated apparatus, instrumentation, and recording device that constitutes the site and object of study. They are "highly specialized scientific instruments," "a special kind of technology," "particular kinds of 'machines.'"[51] They are *deanimalized.*

The attempt to enhance observation through instrumentation, to record results in strictly quantitative terms, and to remove the person and bias of the investigator, all in the service of achieving objective and positive (certain) results, raises the question: What is the relation between model and modeled? When we selectively breed, genetically engineer, deindividuate, despecify, and deanimalize an animal, do we know what we have left? Ironically, the ideal of objectivity pursued in this manner may result in mystification rather than positive understanding. As indicated, in the final analysis, assessment of this relation requires empirical and historical evaluation.

Cracks and Contradictions in the Reductionist Construction

To this point, I have presented evidence that strongly supports the conclusion that the social construction of the laboratory animal is largely reductionist—that laboratory rats are in various ways diminished in this construction, by comparison to natural or wild animals of the same or similar species. (Of course, as the term is being used in this study, it is recognized that these latter are constructions as well.) Here I briefly review evidence from various quarters that suggests tensions or cracks in the present construction and I consider an alternative account.

One such tension issues from the scientific enterprise itself, including but not limited to the animal research laboratory. Despite the construction of laboratory animals as deindividuated, despecified, and deanimalized, not surprisingly there is substantial evidence that they are not in fact "degenerate forms." Although constructed as instruments through selective breeding and the design of the laboratory, they still retain "complex behavioral systems such as territoriality, sexuality and aggression." For example, it has been shown that white laboratory rats allowed to live in a large outdoor enclosure quickly dig burrows just like their wild forebears, and produce offspring who survive a cold winter.[52]

More generally, particularly with the displacement of the behaviorist by the cognitivist program in psychology, but also with the modern development of ethology and, most recently, a cognitive ethology, animals, including laboratory animals, have been "re-minded."[53] Most vestiges of the Cartesian view that all but human animals are mindless automatons have been displaced.[54] In both the wild and the lab, investigations have demonstrated the considerable capabilities of nonhuman animals—e.g., autonomy, agency, protolinguistic communication, symbolic representation, deceit, and sophisticated social organization.[55]

A second tension in this construction is that provided by contemporary ethical theories and their political counterpart, the animal rights movement. Whether arguing that animals have rights or interests, or that they must be counted in any utilitarian calculus, or that we must "care" for them, in the feminist sense of that term, the general consensus of moral philosophy is that nonhuman animals are necessarily objects of our moral consideration.[56]

Whatever the relative influence of these forces, changes in the laboratory are beginning to occur. For example, there is a considerable literature that demonstrates and tests efforts both to reduce the suffering and enhance the welfare of laboratory animals by developing alternative procedures and enriched housing arrangements. These include the provision of various objects for play and manipulation, species-appropriate social arrangements, and housing that allows a broader repertoire of behaviors in more naturalistic settings.[57] Providing an empirical basis for these changes in life in the laboratory, there is the emergence of an animal welfare science with supporting journals, faculty positions, and funding sources devoted to this subfield.

Finally, there is a growing realization among lab technicians and at least those research investigators who actually spend significant periods of time with laboratory animals outside the operating room that there is an "inevitable bond" between scientists and animals.[58] Human and nonhuman animals form complex, affectively based relations. The acceptance of this fact throws into question the ideal of an impartial, detached spectator observing an animal as preparation or instrument; for it suggests that we can only study animals in and through complex social structures.

In his ethnographic study of the subculture of the laboratory, Arluke found support for at least an ambivalence toward animals in this regard. While supporting the findings of Phillips's study, cited above, of the prevalence of the view of lab animals as objects, Arluke found that lab personnel perceive animals as subjects as well. Particularly lab technicians, those workers who deal with the animals on a daily basis (as distinguished from the principal investigator who designs and oversees the experiment), relate to animals as both objects and subjects. As subjects, lab technicians often form attachments to selected animals that are comparable to that seen in a typical human–companion animal relation.[59]

Arluke indicated that the lab animal becomes "pet" when the "process that transforms the animal into object is not fully effective." This suggests that the construction of the objectified animal has priority while that of the animal as pet is reactive and complementary. Arluke notes that the occasional practice of adopting, that is actually taking, a lab animal as a pet involves "subverting the protocol." Elsewhere he speaks of "objectifications that may . . . be defied."[60] It is clear that the animal as subject and, specifically, as pet is largely a development of a subordinate class (the lab tech) within the laboratory culture. It is what sociologists call an "underlife,"[61] a part of a subculture that does not fully accept the official culture. Apparently, lab technicians and student research assistants often balk at attempts to socialize them into the regnant construction in which the lab animal is object—data point or instrument. Still, taken together with the other developments just discussed, this subversion indicates cracks, contradictions, and changes in the predominant construction.

While some of the conditions discussed comprising the social construction of lab animals were inherent in the situation from the outset of the introduction of animals into the lab, most are of recent origin. However, from the outset, there was an alternative construction of lab animals—at least in the received or official view of the enterprise—which featured transcendence rather than reduction. In both accounts, animals in the lab are not themselves, not "natural" or "wild" animals. However, in this second account rats have more, not less, than their mundane nature.

Rats, for example, transcend their scurrilous reputation to become scientific objects, objects of scientific investigation. In fact, through this elevation, by becoming the repository of phenomena under rigorous investigation, rats make that investigatory enterprise possible. Rats are transcendent objects because the enterprise is dedicated to transcendent ends, those of enhancing understanding and human welfare.

In the narrower context of the strategy of developing animal models of human psychology, rats are further elevated. Beyond the status of scientific facts, of data points, they are models of some aspect of human being. Through modeling certain psychological and/or physiological processes, they are transformed into analogues of us. Rats are models in that they are stand-ins for human beings. Through this construction, in viewing rats we even forget our species' differences.

A primary metaphor in support of this account of laboratory animals is "sacrifice." Following Lynch, science is a set of transcending secular rituals through which a "naturalized" animal is lifted to a higher plain to become an "analytic animal," "a bearer of generalized knowledge."[62] When scientists induce experimental conditions in animals and study their results, they transform rats into "analytics," scientific objects existing within a space that is technological, abstract, and mathematized. Within this transcendent context, the suffering and death of animals is a sacrifice. The animals themselves are transcended as they become the means to further understanding.

A number of questions may be raised about this notion of laboratory rat as sacrificial lamb. On the mundane level, is the term still used? In the specific context of the killing of laboratory animals, Nace reported that the term "kill" is commonly used while the term "sacrifice" is not.[63]

Luke raises the provocative question whether the metaphor has deeper roots than those indicated by Lynch. Lynch keyed on the concept of transcendence as the limited basis of the parallel between religion and science. Killing in both is directed to a higher purpose—veneration of a god or increasing knowledge and human welfare. However, Luke suggests that science and religious rituals involving blood sacrifice have many other "striking structural correspondences." For examples, for both, sacrifice is "the rule-bound manipulation of an animal's body"; for both, "sacrifice is believed to be socially necessary"; and for both, "the rules must be followed exactly to achieve the purpose."[64]

Leaving this more speculative analysis aside, may it not be suggested that the transcendent account is complementary, not antithetical, to the reductionistic? Should we not say simply that the animals are reduced to preparations, instruments, or data points in order to establish the conditions necessary for their transcendence? Are not, then, the reductions just part of the ritual preparation of the object of sacrifice? If that were the case, we would see veneration or, at least, respect for the animals in these rituals. But our earlier description of the process of the three reductions clearly demonstrates that this is not the view. The animals are not viewed or treated as if they are approaching the altar, even the altar of science. I maintain that the construction features a reduction of the laboratory animals—a reduction that remains just that as it is not lent substantial support by a transcendent frame.

ASSESSING COSTS TO ANIMALS

In this and the final sections, I return to the strategy of animal model research. Following the argument that animal models, properly understood and utilized, should serve a heuristic function, here I provide a demonstration of how such research can be evaluated. In doing so, I bracket for the moment the debate among various contemporary moral philosophies, for many if not most of which the question of effectiveness has little purchase on the issue—certain acts involving nonhuman as well as human animals being intrinsically unethical.[65]

While the body of this paper has focussed on the complex of meanings socially conferred on laboratory animals, that social construction is suggestive of the nonhuman animals' actual experience. Deindividuation, despecification, deanimalization are ugly words that have ugly implications for the life-world of these beings. I have already indicated that while lab animals were socially constructed as less than whole individual beings, the literature in ethology and even lab-based animal research show that nonhuman animals in general, the species from which lab animals were constructed in particular, and lab animals themselves have considerable intellectual and social capabilities. While constructed as cooperative and malleable subjects to fit the requirements of experimental science, lab animals are not "degenerate."

Neo-Cartesian claims notwithstanding, there is a near consensus in the scientific community that individuals of most species of nonhuman animals, including lab animals, feel pain. There also is considerable agreement that they are capable of suffering as well—of experiencing distress, anxiety, fear, deprivation, and, as noted, even boredom.[66] While there is disagreement on the point, many argue that harm and injury, even when not accompanied by pain, also should be considered suffering. The broadening of the notion of costs to the animal from a narrow to a broad one and, in some quarters, to the inclusion of harm and death is a clear development in the animal rights debate over the past two decades.[67] Accepting the attribution of suffering to lab animals, the debate centers on whether their suffering is necessary.[68]

Turning to the psychology lab, Viney, King, and Berndt report that the following species were utilized in 1988 (in the order of frequency indicated by articles listed in *Psychological Abstracts,* beginning with the most frequent): rats, mice, pigeons, cats, rabbits, hamsters, dogs, chimpanzees, gerbils, baboons, snakes, bats and guinea pigs (tied), dolphins, gorillas, and lemurs and seals (tied). In terms of numbers of animals involved, while it is difficult to arrive at more than an estimate, elsewhere I have argued that a reasonable estimate is that from 1.25 to 2.5 million animals are used in psychological research in the United States annually.[69]

I have indicated earlier the considerable scope of the conditions and procedures utilized. Surely, it is reasonable to assume that an enterprise that claims to have developed animal models for virtually every conceivable harmful, distressing, and/or painful human condition involves very considerable costs to animals in terms of suffering.[70] Further, several studies have found evidence that scientific publications employ language that conceals the suffering involved in animal research. Lederer, as well as Birke and Smith, found a rhetorical laundering that sanitizes and minimizes animal suffering, that directs attention away from it to descriptions of apparatus and measures, and that obscures and minimizes description of death.[71] Field found that introductory psychology textbooks also minimize the presentation of animal suffering, though by providing appealing and cute photographs rather than by emphasizing technical features.[72]

Fortunately, rating scales have been developed that measure the degree of pain, distress, and harm associated with various experimental procedures.[73] One of these instruments, the Scale of Invasiveness, specifically measures the invasiveness of typical psychological experimental manipulations on a six-point scale, ranging from 0 (little or no invasiveness) to 5 (highest level).[74]

While the results are mixed, the several available studies of trends in invasiveness in psychological research, all employing this scale, indicate that the level is typically moderately high and has not changed in recent years. Examining research published in both physiological and comparative psychology journals, Field found an increase from 2.60 in 1947 to 3.12 in 1967 and then a decline to 2.34 in 1987.[75] However, both Lindquist and Schmorrow found no decline for a comparable recent period.[76] Lindquist rated the procedures reported in five journals, featuring behavioristic, physiological, and comparative psychological research, for invasiveness. The comparison of studies in 1975 and 1989 revealed no difference that reached statistical significance. The average score was around 3. There was a statistically insignificant trend to less invasiveness and a trend to fewer studies scored at 4 and 5, while more were scored at 3. Schmorrow rated studies in a behaviorist journal from 1958 to 1992 and found no change in level of invasiveness, which remained between 3 and 4 (except for two years in the 1960s when mean invasiveness levels were over 4).

Although formidable, the power of social construction is limited. Even when utilized and experienced unreflectively as instruments, clearly laboratory animals are sentient beings who are fully capable of suffering and, in fact, do suffer the considerable pain, distress, stress, fear, deprivation, and harm to which they are subjected.

ASSESSING BENEFITS TO HUMANS

To provide a focus to the evaluation on the benefits side, I have selected several current animal models from one area of research—animal models of the eating disorders bulimia and anorexia nervosa, which have reached near epidemic levels in the last two decades.

The essential features of bulimia, in particular, include lack of control over eating behavior leading to recurrent binge eating, induced vomiting, and overconcern with body shape and weight. Accordingly, scientists have attempted to reproduce these features in animals by the procedure of sham or mock feeding, whereby surgery is performed on an animal to produce a fistula. These tube-like passages from an interior cavity to the surface of the body are made in the stomach (gastric fistula), the esophagus (a canal above the stomach), or both. Sham feeding has been used on rats, the predominantly utilized species, as well as on dogs, monkeys, and rabbits.

In a typical sham feeding experiment, following anesthesia, a stainless steel tube (cannula) is sutured into the stomach of a rat. Coming out of the muscle wall of the stomach, the cannula emerges on the back of the animal, under the shoulder, where it is connected to a needle tube cemented to a fiberglass screen that provides anchoring under the skin. The cannula is fitted with a screw or cap so that it may be opened or closed. The needle tube is connected to a plastic tube, which extends up through a hole in the ceiling of the animal's testing cage and can be connected to a syringe to draw out ingested foodstuffs.

In terms of invasiveness, the sham feeding model of bulimia involves a surgical procedure from which an animal is allowed to recover for at least some period. A surgery that causes an extended period of pain/distress or moderate long-term harm is scored 4 on the six-point scale. An animal that has a fistula fitted with a cannula in his or her stomach has been significantly harmed. According to Webster, in addition to the distress during recovery from the surgery, a permanent fistula disrupts peristaltic movements and is the functional equivalent of chronic indigestion.[77] Further, many of the variables explored in conjunction with this model increase this base level of invasiveness, as a study is scored for the most invasive procedure involved. Some of these involve additional surgical procedures such as brain lesioning, fitting with an intracranial cannula, and implanting wire electrodes into muscles involved in ingestion; most of them are scored 3 or 4 even without the intragastric cannula requisite to sham feeding. Other procedures involve deprivation of food, physical immobilization, and other stress inductions such as tail pinching and cold-water immersion.

Sham feeding grossly mimics the bulimic behavior of bingeing and purging: in both food is ingested without being fully digested. It has been found that, even after the fistula has been closed, animals subjected to sham feeding eat more and with shorter intervals between meals. Such initial findings and the analogy that both bulimia and sham feeding involve "eating without calories" provide the basis for asserting that sham feeding is a model of bulimia.

Based on this rough analogy, the more or less inexhaustible set of variables have been explored that affect appetite and food intake. Using the sham feed model, subsequent research explored the considerable complexities of eating behavior—in the mouth, chewing and taste, orosensory experience; in the gut, absorption and satiation; in the brain, neurotransmitters, effects of pleasure, and stress.

What contribution have this and other animal models of bulimia (e.g., inducing stress through tail pinching) and anorexia nervosa (e.g., inducing starvation through provision of an activity wheel) made to the treatment of eating disorders? First, do effective treatments of these disorders exist? With respect to bulimia, the answer is yes and no. A review of nineteen controlled studies found that bingeing and purging are less frequent (although still present) and depression is abated immediately following treatment.[78] However, two or more years after treatment, 16–50 percent of subjects are still diagnosable as bulimic.[79]

In my reading of the literature, none of the animal models of bulimia have made any addition to the forms of psychotherapy currently utilized in the treatment of this disorder. While antidepressants are part of the current regimen, the several families of such drugs did not originate in animal models of eating disorders.

To provide a more systematic evaluation, I utilized a citation analysis and an attitude survey. In the citation analysis, I selected nine investigators from among those publishing studies related to one of three major animal models of eating disorders—sham feeding, tail pinch, and activity wheel. The number and annual frequency of citations of those studies published that involved animals in the laboratory and purported to be related to eating disorders were ascertained using the *Science Citation Index* (*SCI*). Taken together, the studies on eating disorders published by the nine investigators were cited .69 times per year in the years following their publication, during the nine-year period considered (1986–94). By comparison, the average annual frequency of all the references in the *SCI* is 1.87,[80] or more than two and a half times the rate of those examined in the present study. This indicates that the level of citation—and, by implication, the level of contribution to the field—of those investigators who develop animal models of eating disorders is low relative to published scientific studies generally.

Finally, following a suggestion in the literature on citation analysis,[81] citations were judged as to their significance. The investigator may cite an earlier study only in passing in a generalized introduction to the relevant area, or—a more significant use—he or she may use an earlier study to interpret additional animal or human data or to provide an innovative method. When only those citations judged significant are counted, the overall annual frequency rate drops to .31. This means that, on average, in a given year seven in ten of the studies examined received no significant citations in the scientific literature.

The more critical question in the present evaluation is, what influence did these studies have on clinicians who treat eating disorders? To answer this question, I developed and administered a questionnaire that assessed the degree of awareness and rudimentary understanding of animal models among clinicians specializing in treating these disorders.

Of the thirty such specialists (defined as a clinical practice 20 percent of which involved working with people whose diagnosis is an eating disorder) completing the survey, more than half (60 percent) indicated that they were not aware of the existence of animal models of eating disorders. When asked if they could name and describe such a model, 67 percent said they could not do so. Of the 33 percent who said they could, most gave inadequate or vague descriptions. When asked to identify and describe sham feeding, 87 percent were unable to do so.

When asked their view of whether animal models of eating disorders influenced their treatment approach to these disorders, 87 percent replied in the negative. Of those replying in the affirmative, one indicated that the research was useful for conceptualization but not for intervention, one indicated that it was "indirectly" an influence, and one stated, "only helpful in reinforcing my explanations to patients about how what's happening to them is a natural physiological response to food deprivation."

The clinicians also were asked to indicate the journals that they found most helpful in their work. When the list of journals citing the studies of the nine investigators involved in research related to animal models of eating disorders is compared to the list of journals identified by the clinicians as "most helpful for clinicians that specialize or work with eating disorders," there is no overlap. Not one of the journals in which the studies of the nine investigators are cited is mentioned as helpful by these practitioners specializing in the treatment of eating disorders.

Clearly, clinicians specializing in the treatment of eating disorders generally are unaware both of the existence and of the particular results of animal model research purporting to provide further understanding and effective treatment of these disorders.

A recent attitude survey of psychologists provides evidence that clinicians have a similar view of animal research more generally. Based on a survey of about four thousand randomly selected psychologists, Plous found that 92 percent of psychologists who are mental health workers indicate that they rarely, never, or only occasionally use findings from psychological research on animals. Nearly 95 percent of these psychologists indicated that a ban on such research would not seriously hamper their work.[82]

CONCLUSION

In its effort to become a modern field comparable to biological science, psychology adopted the strategy of developing animal models of human psychological disorders. Influenced by the then regnant philosophy of science, positivism, early psychologists articulated an experimentalist approach in which the results of systematic variation could be accurately recorded and then quantitatively expressed. To facilitate this form of inquiry, psychologists constructed laboratories and laboratory animals. While the lab was shaped architecturally to fit the animals, it is equally true that the animals were shaped to fit the lab. The resultant lab animal was a peculiar construction with only a reduced individual- or species-based identity, and even a limited identity as an animal as such.

The evaluation of the strategy of using animal models requires empirical analysis and historical interpretation. Pain or invasiveness scales are available which indicate that the costs to the animals involved are considerable in terms of suffering and harm. Citation analysis and survey instruments applied to a selected set of animal models of eating disorders indicate that the function of generating hypotheses to be tested in clinical settings is not being met.

If the animal models are not meeting this proper heuristic function, how are they being used? One use is as a degraded heuristic. There is a failure to validate the model in the primary context of interest, the clinic. Rather, models and research procedures become established as areas of research within animal behavior psychology. A literature gets built up in which a model, say for human depression, is established through many variations on the basic model. Instead of checking the model against the original behavior for which it is a model, the "validation" occurs within the paradigm of the model. Verification is limited to testing the degree of consistency of the new findings with other research within the literature on the same model. In this way, an entire research enterprise becomes ingrown and may be of little or no value in furthering understanding of the original object of study.

A second use is even more insidious. Sometimes, although no verification is obtained of it in the actual object of study, the model provides an image to describe the original. The model merely supplies a loose analogy for the object of study. For example, an animal in whom depression has been induced becomes a graphic image of depression. Our talk and imaginings about depression are influenced but it is not clear that any understanding has been gained for the model is never validated. The use of the model as a generative device degenerates to that of a rhetorical device.

NOTES

1. Kenneth J. Shapiro, *Animal Models of Human Psychology: A Critique of Science, Ethics, and Policy* (Seattle, 1998). Sources of epigraphs are Robert Kanigel, "Specimen No. 1913: A Rat's Brief Life in the Service of Science," *The Sciences* 27, no. 1 (1987): 30–37; and Elizabeth L. Fucci, "Weebee Talk," *The Rat Report* 38 (1995): 1–2.

2. Peter Berger and Thomas Luckmann, *The Social Construction of Reality* (Garden City, N.Y., 1966).

3. Alan Sokal, "Transgressing the Boundaries: Toward a Transformative Hermeneutics of Quantum Gravity," *Social Text* 14, nos. 1–2 (1996): 217–52.

4. Robert Hendrickson, *More Cunning than Man: A Social History of Rats and Men* (Briarcliff, N.Y., 1983).

5. Fred Wertz, "The Rat in Psychological Science," *The Humanistic Psychologist* 14, no. 3 (1986): 143–68.

6. Robert B. Lockyard, "The Albino Rat: A Defensible Choice or a Bad Habit?" *American Psychologist* 23 (1968): 734–42.

7. Lynda Birke, review of *The Sexual Politics of Meat, Meat: A Cultural Symbol,* and *Beyond Beef: The Rise and Fall of the Cattle Culture, Society and*

Animals 1, no. 2 (1993): 195.

8. Michael Quinn, "Corpulent Cattle and Milk Machines: Nature, Art and the Ideal Type," *Society and Animals* 1, no. 2 (1993): 145–59.

9. Wertz, "The Rat in Psychological Science," p. 151; B. T. Clause, "The Wistar Rat as the Right Choice: Establishing Mammalian Standards and the Ideal of a Standardized Mammal," *Journal of the History of Biology* 26, no. 2 (1993): 329–49.

10. Donald R. Griffin, *Animal Thinking* (Cambridge, Mass., 1984), p. 13.

11. Lockyard, "The Albino Rat," pp. 735–36.

12. James H. Capshew, "Psychologists on Site: A Reconnaissance of the History of the Laboratory," *American Psychologist* 47, no. 2 (1992): 136; G. Gallup and T. Eddy, "Animal Facilities Survey," ibid., vol. 45 (1990): 400–401.

13. Wertz, "The Rat in Psychological Science," p. 152.

14. Theodore Roszack, Mary Gomes, and Allen Kanner, eds., *Ecopsychology: Restoring the Earth, Healing the Mind* (San Francisco, 1995).

15. Sigmund Koch, "'Psychology' or 'the Psychological Studies'?" *American Psychologist* 48 (1993): 902–4.

16. Emily D. Cahan and Sheldon H. White, "Proposals for a Second Psychology," *American Psychologist* 47, no. 2 (1992): 224–35.

17. Kenneth J. Shapiro, "Verification: Validation or Understanding," *Journal of Phenomenological Psychology* 17, no. 2 (1986): 167–81.

18. Dan Bannister, "The Fallacy of Animal Experimentation in Psychology," in *Animals in Research,* ed. D. Sperlinger (New York, 1981), p. 308.

19. Bernard E. Rollin, *The Unheeded Cry: Animal Consciousness, Animal Pain, and Science* (New York, 1989), pp. 1–23.

20. Capshew, "Psychologists on Site," p. 132.

21. Ibid.

22. Donald A. Dewsbury, "Comparative Psychology and Ethology: A Reassessment," *American Psychologist* 47, no. 2 (1992): 208–15.

23. J. B. Overmier and P. D. Burke, eds., *Animal Models of Human Pathology: A Bibliography of a Quarter Century of Behavioral Research, 1967–1992* (Washington, D.C., 1992).

24. Rom Harré, *The Philosophies of Sceince* (New York, 1972), p. 172.

25. Hugh LaFollette and Niall Shanks, "Animal Studies in Biomedical Research: Some Epistemological Worries," *Public Affairs Quarterly* 7 (1993): 129–32.

26. Shapiro, *Animal Models of Human Psychology,* pp. 85–111.

27. Jon Gordon, "Impact of Transgenic Technology on Laboratory Use," Paper presented at the Second World Congress on Alternatives and Animal

Use in the Life Sciences, Utrecht, The Netherlands, 1996.

28. Kenneth J. Shapiro, "Animal Rights v. Humanism: The Charge of Speciesism," *Journal of Humanistic Psychology* 30, no. 2 (1990): 9–37.

29. Mary T. Phillips, "Proper Names and the Social Construction of Biography: The Negative Case of Laboratory Animals," *Qualitative Sociology* 17, no. 2 (1994): 135–36.

30. Ibid., p. 130.

31. Ibid., p. 126.

32. Roger Fouts, "Animals in Zoos: Arrogance of Knowledge vs. the Humility of Ignorance," in *Ethics on the Ark: Zoos, Animal Welfare, and Wildlife Conservation,* ed. B. Norton, M. Hutchins, E. Stevens, and T. Maple (Washington, D.C., 1995).

33. Phillips, "Proper Names," p. 127.

34. Ibid., p. 129.

35. Ibid., p. 123.

36. Kanigel, "Specimen No. 1913."

37. Lynda Birke and Jane Smith, "Animals in Experimental Reports: The Rhetoric of Science," *Society and Animals* 3 (1995): 31 (emphasis in original).

38. Marian Scholtmeijer, *Animal Victims in Modern Fiction: From Sanctity to Sacrifice* (Toronto, 1993): p. 72.

39. Clause, "The Wistar Rat," p. 341.

40. Phillips, "Proper Names," p. 129.

41. Ibid., p. 138.

42. F. Beach, "The Snark Is a Boojum," *American Psychologist* 5 (1950): 115.

43. George Devereux, *From Anxiety to Method in the Behavioral Sciences* (The Hague, 1967), p. 236.

44. Françoise Wemelsfelder, *Animal Boredom: Toward an Empirical Approach of Animal Subjectivity* (Utrecht, 1993).

45. Kenneth J. Shapiro, "The Death of an Animal: Ontological Vulnerability and Harm," *Between the Species* 5, no. 4 (1989): 183–95.

46. Rollin, *The Unheeded Cry,* pp. 206–25.

47. Daniel P. Todes, "Pavlov's Physiology Factory," *Isis* 88 (1997): 205–46.

48. Ibid., p. 243.

49. See Karen. Rader's essay in this volume, above, pp. 389–438.

50. Phillips, "Proper Names," p. 131.

51. Ibid., p. 136; Robert Kohler, *Lords of the Fly: Drosophila Genetics and the Experimental Life* (Chicago, 1995), p. 6; Todes, "Pavlov's Physiology Factory," p. 221.

52. Michael Fox, *Laboratory Animal Husbandry: Ethology, Welfare, and Experimental Variables* (Albany, N.Y., 1986), p. 101.

53. Kenneth J. Shapiro, "Reminding Animals: Developments in the Scientific Study of Nonhuman Animals" (Review of M. Bekoff and D. Jamieson, eds., *Interpretation and Explanation in the Study of Animal Behavior,* vol. 1 [Boulder, Colo. 1990]), *Between the Species* 8, no. 2 (1992): 74–80.

54. But see John S. Kennedy, *The New Anthropomorphism* (Cambridge, 1992), for a neo-Cartesian view.

55. See the essays in Bekoff and Jamieson, *Interpretation and Explanation in the Study of Animal Behavior.*

56. For the first view, see Tom Regan, *The Case for Animal Rights* (Berkeley, Calif., 1983); for the second, Peter Singer, *Animal Liberation: A New Ethic for Our Treatment of Animals* (New York, 1975); and for the third, Josephine Donovan and Carol Adams, eds., *Beyond Animal Rights: A Feminist Caring Ethic for the Treatment of Animals* (New York, 1996).

57. United States Department of Agriculture and International Fund for Animal Welfare, *Environmental Enrichment Information for Laboratory Animals 1965–1995* (Beltsville, Md., 1995).

58. Hank Davis and Dianne Balfour, eds., *The Inevitable Bond: Examining Scientist-Animal Interactions* (Cambridge, 1992).

59. Arnold Arluke, "Sacrificial Symbolism in Animal Experimentation: Object or Pet?" *Anthrozöos* 2, no. 2 (1988): 105–11.

60. Ibid., pp. 99, 108, 115.

61. Erving Goffman, *Asylums: Essays on the Social Situation of Mental Patients and Other Inmates* (Garden City, N.Y. 1961), p. 171.

62. Michael E. Lynch, "Sacrifice and the Transformation of the Animal Body into a Scientific Object: Laboratory Culture and Ritual Practice in the Neurosciences," *Social Studies of Science* 18 (1988): 266–69.

63. Patrick Nace, "The Use of Euphemisms in Research," *Science and Animal Care* 4, no. 2 (1993): 1–4.

64. Brian Luke, "Animal Experimentation as Blood Sacrifice," *ISAZ* 12 (1996): 4.

65. Shapiro, *Animal Models of Human Psychology,* pp. 213–46.

66. David DeGrazia and Andrew Rowan, "Pain, Suffering, and Anxiety in Animals and Humans," *Theoretical Medicine* 12 (1991): 193–211.

67. Shapiro, *Animal Models of Human Psychology,* pp. 247–85.

68. U.S. Congress, Office of Technology Assessment, *Alternatives to Animal Use in Research, Testing, and Education,* OTA-BA-273 (Washington, D.C., 1986).

69. Shapiro, *Animal Models of Human Psychology,* pp. 19–26.

70. Some psychologists have disputed this conclusion; see, for example, Gordon Gallup and Susan D. Suarez, "Antivivisection: Questions of Logic, Consistency, and Conceptualization," *Theoretical and Philosophical Psychology* 7 (1987): 81–94. For a reply, see Kenneth J. Shapiro, "Galloping Sophistry: A Rat in the Lab is Worth Two in the Dump," ibid., vol. 8, no. 2 (1987): 47–53.

71. Susan E. Lederer, "Political Animals: The Shaping of Biomedical Research Literature in Twentieth-Century America," *Isis* 83 (1992): 61–79; Birke and Smith, "Animals in Experimental Reports."

72. Peter Field, "Animal Images in College Psychology Textbooks," *Between the Species* 9 (1993): 193–203.

73. F. Barbara Orlans, "Research Protocol Review for Animal Welfare," *Investigative Radiology* 22 (1987): 253–58.

74. Kenneth J. Shapiro and Peter Field, "A New Scale of Invasiveness in Animal Experimentation," *PSYeta Bulletin* 7 (1987): 5–8.

75. Peter Field, "Surveying the Behavior of Researchers with a Scale of Invasiveness," paper presented at the American Psychological Association Annual Convention, Atlanta, Ga., August 1988.

76. Terry Lindquist, "The Effectiveness of the Animal Welfare Movement in Psychology: Evidence from APA Publications in Experimental Animal Behavior" (unpublished manuscript, 1990); Dylan Schmorrow, "The Use of Nonhuman Subjects in Behavior Analysis: A Review of *Journal of Experimental Analysis of Behavior* Studies" (Ph.D. diss., Western Michigan University, Kalamazoo, 1993).

77. John Webster, *Animal Welfare: A Cool Eye toward Eden* (Oxford, 1994), p. 239.

78. James Mitchell and Nancy Raymond, "Cognitive-Behavioral Therapy in Treatment of Bulimia Nervosa," in *Psychobiology and Treatment of Anorexia Nervosa and Bulimia Nervosa,* ed. K. Haini (Washington, D.C., 1992), pp. 307–28.

79. L. K. George Hsu, "Critique of Follow-up Studies," ibid., pp. 125–50.

80. Eugene Garfield, "Citation Analysis and the Anti-Vivisection Controversy, Part II: An Assessment of Lester H. Aronson's Citation Record," in *Essays of an Information Scientist,* vol. 3 (Philadelphia, 1980), p. 317.

81. Eugene Garfield, "How to Use Citation Analysis for Faculty Evaluations, and When Is It Relevant? Part 2," *Current Comments,* no. 45 (November 7, 1983): 5–14.

82. Scott Plous, "Attitudes toward the Use of Animals in Psychological Research and Education: Results from a National Survey of Psychologists," *American Psychologist* 51 (1996): 1167–80.

NOTES ON CONTRIBUTORS

RICHARD W. BURKHARDT JR. is Professor of History at the University of Illinois at Urbana-Champaign. His writings include *The Spirit of System: Lamarck and Evolutionary Biology* (1977, 1995) and numerous articles and book chapters on the history of evolutionary biology and the study of animal behavior.

JONATHAN BURT is a freelance writer working on animals in the early twentieth century and has published articles on animals and visual culture. He is co-director of the Animal Studies group in Britain and his book *Animals in Film,* is forthcoming.

KEN C. ERICKSON is Lecturer at California State University, Long Beach, and is associated with Ethnographic Research, Inc., in Kansas City. He is co-author (with Donald D. Stull) of *Doing Team Ethnography: Warnings and Advice* (Beverly Hills: Sage, 1997), and of articles and essays on the ethnography of immigrants in the workplace.

KATHERINE C. GRIER is Assistant Professor in the Department of History, University of South Carolina. She is the author of *Culture and Comfort: Parlor Making and Middle-Class Identity* (1997), She was a Davis Center Fellow in Spring 1997 and has also received fellowship support from the National Endowment for the Humanities; the Robert C. and Grace A. Tanner Humanities Center at the University of Utah; and the Henry Francis du Pont Winterthur Museum, Library, and Gardens for the project "At Home with the Animals: Middle-Class Americans and the Animals in Their Lives, 1820–1920."

MARY J. HENNINGER-VOSS received her Ph.D. from Johns Hopkins University. She specializes in the history of early modern science from its roots in literary and mechanical practices of the late fifteenth century to its institutionalization in the eighteenth century. She is completing work on the development of a mathematical, experimental physics in the context of military culture of sixteenth-century Italy. Her current research interests include seventeenth-century scientific and popular conceptualizations of human passions, and the relationship between European science and European global colonization in the early modern period.

RICHARD C. HOFFMANN is Professor of History at York University in Toronto, a continuing member of the Graduate Centre for Medieval Studies in the University of Toronto, and a chartered fisher of the City of Ceské Budějovice, Czech Republic. His book *Land, Liberties, and Lordship in a Late Medieval Countryside: Agrarian Structures and Change in the Duchy of Wrocław* (1989) received the Adams Prize of the American Historical Association, and his article "Economic Development and Aquatic Ecosystems in Medieval Europe" (*American Historical Review,* 1996), was awarded the Hamilton Prize by the American Society for Environmental History. His most recent major work is *Fisher's Craft and Lettered Art: Tracts on Fishing from the End of the Middle Ages* (1997). His essay in this volume comes from current research on medieval European fisheries in light of today's global fisheries crisis.

ANDREW C. ISENBERG, an Assistant Professor of History at Princeton University, received his Ph.D. from Northwestern University. He is the author of *The Destruction of the Bison: An Environmental History, 1750–1920* (2000). He is currently studying the environmental consequences of the California Gold Rush.

JACQUELINE MILLIET is a cultural anthropologist at the Centre National de Recherche Scientifique, Paris, specializing in the anthropology of human-animal relationships. She has taught at the University of Massachusetts, the University of Vermont, the Shelby Cullom Davis Center at Princeton University, and the Collège de France, among others, and has published widely in anthropological and ethnographic journals in France and Switzerland. She is currently working on a book dealing with the breastfeeding of animals by women.

JOHN SOLOMON OTTO, although he grew up on a cattle farm, earned an M.A. and Ph.D. from the University of Florida. He is the author of *The Southern Frontiers* (1989), *Southern Agriculture during the Civil War Era* (1994), and *The Final Frontiers* (1999). Dr. Otto currently works and resides in Washington, D.C.

KAREN A. RADER teaches Science, Technology, and Society at Sarah Lawrence College. Her interests include the history of genetics, "Big Science," gender and science, and the relations between American science and popular culture. Her book, *Making Mice: Standardizing Animals for American Biomedical Ressearch,* is forthcoming with Princeton University Press.

HARRIET RITVO is the Arthur J. Conner Professor of History at the Massachusetts Institute of Technology. She is the author of *The Animal Estate: The English and Other Creatures in the Victorian Age* (1987), and *The Platypus and the Mermaid, and Other Figments of the Classifying Imagination* (1997). She is currently working on a study of Victorian environmental history.

NIGEL ROTHFELS is the Director of the Edison Initiative at the University of Wisconsin–Milwaukee, where he also teaches history. He is the author of *Savages and Beasts: The Birth of the Modern Zoo* (2002), and the editor of *Representing Animals* (2002).

KENNETH J. SHAPIRO is Executive Director and co-founder of Psychologists for the Ethical Treatment of Animals. He received the B.A. in intellectual history from Harvard University and the Ph.D. in clinical and personality psychology from Duke University. He has published scholarly work in the areas of phenomenological psychology and animal welfare. He is founding editor of *Society and Animals,* and co-founding editor of the *Journal of Applied Animal Welfare Science.* His most recent book is *Animal Models of Human Society: Critique of Science, Ethics, and Policy.*

EDWARD I. STEINHART is currently Associate Professor of History at Texas Tech University. He received his doctorate in African History from Northwestern University in 1971 and has taught at the University of Chicago and the University of Texas at Austin in the United States, and at the University of Zambia in Africa. He is the author of *Conflict and Collaboration: The Kingdoms of Western Uganda, 1890–1907* (1977). He is currently working on a history of black and white hunters in colonial Kenya as well as a study of the history of the Bacwezi cult in the Uganda kingdoms region.

Index